AF323319

Numerical Analysis of Power System Transients and Dynamics

Other volumes in this series:

Numerical Analysis of Power System Transients and Dynamics

Edited by Akihiro Ametani

The Institution of Engineering and Technology

Published by The Institution of Engineering and Technology, London, United Kingdom

The Institution of Engineering and Technology is registered as a Charity in England & Wales (no. 211014) and Scotland (no. SC038698).

© The Institution of Engineering and Technology 2015

First published 2015

This publication is copyright under the Berne Convention and the Universal Copyright Convention. All rights reserved. Apart from any fair dealing for the purposes of research or private study, or criticism or review, as permitted under the Copyright, Designs and Patents Act 1988, this publication may be reproduced, stored or transmitted, in any form or by any means, only with the prior permission in writing of the publishers, or in the case of reprographic reproduction in accordance with the terms of licences issued by the Copyright Licensing Agency. Enquiries concerning reproduction outside those terms should be sent to the publisher at the undermentioned address:

The Institution of Engineering and Technology
Michael Faraday House
Six Hills Way, Stevenage
Herts, SG1 2AY, United Kingdom

www.theiet.org

While the authors and publisher believe that the information and guidance given in this work are correct, all parties must rely upon their own skill and judgement when making use of them. Neither the authors nor publisher assumes any liability to anyone for any loss or damage caused by any error or omission in the work, whether such an error or omission is the result of negligence or any other cause. Any and all such liability is disclaimed.

The moral rights of the authors to be identified as author of this work have been asserted by him in accordance with the Copyright, Designs and Patents Act 1988.

British Library Cataloguing in Publication Data
A catalogue record for this product is available from the British Library

ISBN 978-1-84919-849-3 (hardback)
ISBN 978-1-84919-850-9 (PDF)

Typeset in India by MPS Limited
Printed in the UK by CPI Group (UK) Ltd, Croydon

Contents

3 Simulation of electromagnetic transients with EMTP-RV   **103**

J. Mahseredjian, Ulas Karaagac, Sébastien Dennetière and Hani Saad

Preface

Numerical analysis has become quite common and is a standard approach to investigate various phenomena in power systems. In the field of power system transients, numerical simulation started in the 1960s when digital computers became available. In 1973 the CIGRE Working Group (WG) 13-05, of which A. Ametani was a member, was organized to investigate the accuracy and application limit of various computer software developed in universities and industries. After three years of WG activities, all of the members realized that the electromagnetic transients program (EMTP) originally developed by Prof. H.W. Dommel in the Bonneville Power Administration (BPA), US Department of the Interior (and later US Department of Energy), was superior to any other software in the world at that time. Since then, the BPA-EMTP became used by researchers, engineers and university students worldwide, and at the same time many experts contributed to a further development of the EMTP. In 1980, the EMTP was kind of a standard tool to analyze the power system transients, and it was also applied to steady-state phenomena such as power/load flow and to dynamic behavior of ac/dc converters, i.e., power electronic circuits in general.

In 1984, the Development Coordination Group/Electric Power Research Institute started to restructure the BPA-EMTP and, in 1986, the first version of the EMTP-RV (restructured version) was completed by Hydro-Québec. In the same time period, Manitoba HVDC Research Center was also developing a new type of EMTP called the EMTDC and PSCAD especially for an HVDC (high-voltage direct current) transmission system, because of the Nelson River HVDC system operated by Manitoba Hydro. Also, Dr W. Scott-Meyer, who had taken care of the BPA-EMTP since 1973, started to develop a EMTP-ATP (alternative transients program) with his personal time/expenses to keep the EMTP in the public domain.

Thus, there exist three EMTP-type simulation tools from the 1990s which have been widely used all over the world. The BPA-EMTP was developed for power system transients such as switching/fault surges and lightning surges together with steady-state solutions. Since FACTS and Smart Grid became common and were installed into power systems, EMTP-type software is required to deal with a much longer time period, i.e., millisecond to second, even a minute. For this, the above-mentioned software is modified and revised, and also new simulation tools are developed by many industries to match demand. A typical example is XTAP, developed for Japanese utilities. Also, so-called real-time simulators such as RTDS and ARENE have been developed.

All the above tools are principally based on a circuit-theory which assumes a TEM (transverse electromagnetic) mode of electromagnetic wave propagation. To simulate a transient associated with the TEM and non-TEM mode propagation, such as a transient electromagnetic field within a building (horizontal and vertical steel structures) and mutual coupling between power lines and nearby lightning etc., a numerical electromagnetic analysis (NEA) method is becoming a powerful approach. There are well-known softwares based on the NEA method, such as NEC and VSTL.

In the first part of this book, basic theories of circuit-theory based simulation tools and of numerical electromagnetic analysis methods are explained in Chapter 1. Then, various simulation tools are introduced and their features, strengths and weaknesses, if any, are described together with some application examples.

EMTP-ATP is explained in Chapter 2, EMTP-RV in Chapter 3, EMTDC/PSCAD in Chapter 4, and XTAP in Chapter 5. Numerical electromagnetic analysis is described using the FDTD (finite-difference time-domain) method in Chapter 6, and using the PEEC (partial element equivalent circuit) method in Chapter 7.

In the second part, various transient and dynamic phenomena in power systems are investigated and studied by applying the numerical analysis tools explained in the first.

Chapter 8 deals with transients in various components related to a renewable system, such as a wind turbine tower/generator/grounding, solar power system or an electric vehicle, by adopting an FDTD method.

Chapter 9 describes surges on windfarms and collection systems. Modeling of the system is explained for EMTP-ATP, EMTDC/PSCAD, ARENE, and an NEA method. Then, surge analysis is carried out especially for a back-flow surge.

Chapter 10 discusses a numerical analysis of protective devices, focusing on simulations of a fault locator and high-speed switchgear. The EMTP-ATP is used in this chapter.

Chapter 11 describes overvoltages in a power system and methods of protection. Also, the reduction of the overvoltages by surge arresters is studied, and the insulation coordination of the power system is explained.

Chapter 12 deals with dynamic phenomena in FACTS, especially STATCOM (static synchronous compensator). Also a numerical analysis of harmonic resonance phenomena of a voltage-sourced converter is described.

Chapter 13 explains the application of SVC to a cable system. The voltage control of the cable system by the SVC is discussed, based on effective value analysis, i.e., RMS (root means square) value simulation.

Chapter 14 is dedicated to grounding systems. Basic concepts and the transient response of grounding electrodes are explained. This response is simulated and analyzed using a numerical electromagnetic model. A case example is explored, consisting of sensitivity analysis of the influence of grounding electrodes on the lightning response of transmission lines.

Akihiro Ametani
Emeritus Professor, Doshisha University
Kyoto, Japan

Chapter 1

Introduction of circuit theory-based approach and numerical electromagnetic analysis

*A. Ametani**

1.1 Circuit theory-based approach: EMTP

The electromagnetic transients program (EMTP) has been the most well-known and widely used simulation tool as a circuit theory-based approach since its original development in the Bonneville Power Administration of the US Department of Energy from 1966 to 1984 [1–4]. Presently there exist three well-known EMTP-type tools, i.e., (1) EMTP-ATP, (2) EMTP-RV, and (3) EMTDC/PSCAD. The details of these tools are explained in Chapters 2–4.

The EMTP-type tools are based on an electric circuit theory which assumes transverse electromagnetic (TEM) mode of wave propagation. Thus, these cannot solve phenomena associated with non-TEM mode wave propagation. This is not only from the viewpoint of circuit theories, but also from viewpoint of the parameters used in a circuit analysis. For example, if the impedance and the admittance of an overhead line are derived under the assumption of the TEM mode propagation, these are not applicable to phenomena involving non-TEM mode propagation.

1.1.1 Summary of the original EMTP

The EMTP is based on an electric circuit theory studied in a junior class of an engineering department in a university.

Among various circuit theories, a nodal analysis method is adopted in an EMTP-type numerical simulation tool to obtain unknown voltages and currents in a given circuit [1–4]. In general, the nodal analysis results in taking the inverse of a nodal admittance matrix, which is obtained as a solution of simultaneous nodal equations. Because the nodal admittance matrix is composed of resistances, inductances, and capacitances, i.e., the matrix is complex, and also its size is very large when analyzing phenomena in a real power system, a numerical calculation of the inverse matrix requires a large computational resource. In the 1960s and 1970s when the original EMTP was developed in the Bonneville Power Administration of

*Doshisha University, Japan and Ecole Polytechnique Montreal, Canada

the US Department of Interior (later, the Department of Energy: DOE), it was a terrible job to calculate the inverse of a large complex matrix by an existing computer in that time.

Because of the above fact, Prof. H.W. Dommel, called the father of the EMTP, adopted the idea of an equivalent circuit composed only of a resistance and a current source to represent any circuit element in a power system so that a nodal admittance matrix becomes a conductance matrix, i.e., the matrix becomes real but no more complex [1]. Because the admittance matrix is quite sparse, matrix reduction by the sparse matrix approach, which was very common in the field of power/load flow and stability analyses [5, 6], was also adopted.

To deal with a distributed-parameter line such as an overhead transmission line and an underground cable, so-called Schnyder–Bergeron method [7–9], mathematically a method of characteristics to solve a partial differential equation, was introduced in the EMTP [1, 10]. Later, methods of handling the frequency-dependence of a distributed line due to a conductor (including earth) skin effect were implemented into the EMTP [11–14].

From the late 1970s to the beginning of the 1980s, various power system elements, such as a rotating machine and an arrester, were installed into the EMTP [3, 4, 15–32] as described in Table 1.1. Among these, one of very significant elements is a subroutine called transient analysis of control systems (TACS) [15], later revised and modified as "MODELS" [32] developed by late L. Dube. The TACS and MODELS are a kind of computer languages, and deal with control circuits including mathematical equations interactively running with the EMTP main routine that calculates transient or dynamic behavior of a given power system. It was very unfortunate that the theory behind the TACS originated by L. Dube could not be understood by any reviewer of the IEEE Transactions in the 1970s and no IEEE Transaction paper describing the TACS was published, and thus L. Dube was not awarded "Ph.D.," although his work related the TACS/MODELS is far more than a Ph.D. research.

1.1.2   Nodal analysis

In general, a nodal analysis can be defined by the following equation.

$$(I) = [Y] \cdot (V) \tag{1.1}$$

where I: current, V: voltage, Y: nodal admittance

    ( ) for column vector, [ ] for full matrix

The nodal analysis results in forming the nodal admittance matrix from obtained simultaneous nodal equations. For example, let's obtain the nodal admittance matrix of a circuit illustrated in Figure 1.1. By applying Kirchhoff's current law to nodes 1 to 3 in the circuit, the following simultaneous equations are obtained.

$$(Y_a + Y_c + Y_d)V_1 - Y_c V_2 = J_1$$

$$-Y_c V_1 + (Y_b + Y_c + Y_e)V_2 = J_2$$

Table 1.1   Power system elements and subroutines prepared in the original EMTP

(a) Circuit elements

Element	Model	Remark
Lumped R, L, C	Series, parallel	
Line/cable	Multiphase π circuit	Transposed, untransposed
	Distributed line with constant	Overhead, underground
	parameters frequency-dependent line	Semlyen, Marti, Noda
Transformer	Mutually coupled R-L element	Single-phase, three-phase
	N winding, single-phase	Saturation, hysteresis
	Three-phase shell-type	
	Three-phase • 3-leg • core-type	
Load, nonlinear	Starecase $R(t)$ (type-97)	Nonlinear resistor
	Piecewise time-varying R (type-91,94)	Nonlinear inductor
	Pseudo-nonlinear R (type-99)	Time-varying resistance
	Pseudo-nonlinear L (type-98)	
	Pseudo-nonlinear hysteretic L (type-96)	
Arrester	Exponential function $Z_n 0$	Gapped, gapless
	Flashover type multiphase R	
Source	Step-like (type-11)	Voltage source
	Piecewise linear (type-12, 13)	Current source
	Sinusoidal (type-14)	Surge functions
	Impulse (type-15)	
	TACS controlled source	
Rotating machine	Synchronous generator (type-59)	Synchronous,
	Universal machine	induction, dc
Switch	Time-controlled switch	Circuit breaker
	Flashover switch	Disconnector
	Statistic/systematic switch	Vacumn switch
	Measuring switch	
	TACS controlled switch (type-12, 13)	
	TACS controlled arc model	
Semi-conductor	TACS controlled switch (type-11)	Diode, thyristor
Control circuit	TACS (MODELS)	Transfer function,
		control dynamics
		Arithmetics, logics

(b) Supporting routines

Name	Function	Input data
LINE CONSTANTS	Overhead line parameters	Frequency, configuration, physical parameters
CABLE CONSTANTS	Overhead/underground cable parameters	Frequency, configuration, physical parameters
XFORMER	Transformer parameters	Configuration, rating, %Z
BCTRAN	Transformer parameters	Configuration, rating, %Z
SATURATION	Saturation characteristics	Configuration, rating, %Z
HYSTERESIS	Hysteresis characteristics (type-96)	Configuration, rating, %Z
NETEQV	Equivalent circuit	Circuit configuration, Z, Y, frequency
Marti/Semlyen Setup	Frequency-dependent line	Given by LINE CONSTANTS or CABLE CONSTANTS

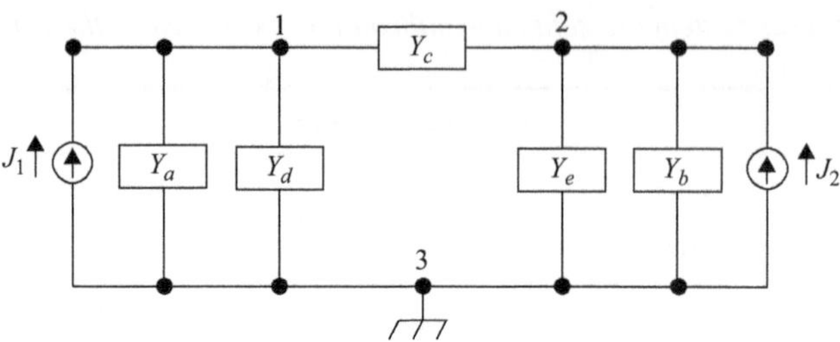

Figure 1.1   Nodal analysis

Rearranging the above equation and writing in a matrix form,

$$\begin{bmatrix} Y_{11} & Y_{12} \\ Y_{12} & Y_{22} \end{bmatrix} \begin{pmatrix} V_1 \\ V_2 \end{pmatrix} = \begin{pmatrix} J_1 \\ J_2 \end{pmatrix}$$

or,

$$[Y](V) = (J) \tag{1.2}$$

It should be clear from (1.2) that once the node admittance matrix is composed, the solution of the voltages is obtained by taking the inverse of the matrix, for current vector (J) is known. In the nodal analysis method, the composition of the nodal admittance is rather straightforward as is well known in a circuit theory. In general, the nodal analysis gives a complex admittance matrix, because the impedances of inductance L and capacitance C become complex, i.e., $j\omega L$ and $1/j\omega C$, respectively, where $j = \sqrt{-1}$.

However, in the EMTP, all the circuit elements being represented by a current source and a resistance as explained in the following section, the admittance matrix becomes a real matrix.

1.1.3   Equivalent resistive circuit

Inductance L, capacitance C, and a distributed-parameter line Z_0 are represented by a current source and a resistance as illustrated in Figure 1.2.

For example, the voltage and the current of the inductance are defined in the following equation.

$$v = L \cdot di/dt \tag{1.3}$$

Integrating the above equation from time $t = t - \Delta t$ to t,

$$\int_{t-\Delta t}^{t} v(t) \cdot dt = L \int_{t-\Delta t}^{t} \{di(t)/dt\} \cdot d_t = L[i(t)]_{t-\Delta t}^{t} = L\{i(t) - i(t - \Delta t)\}$$

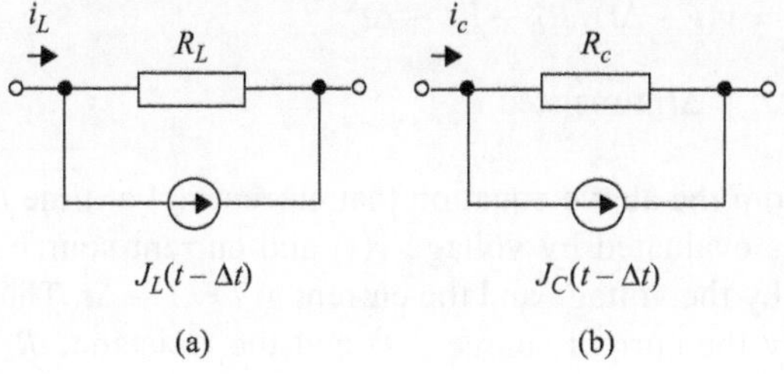

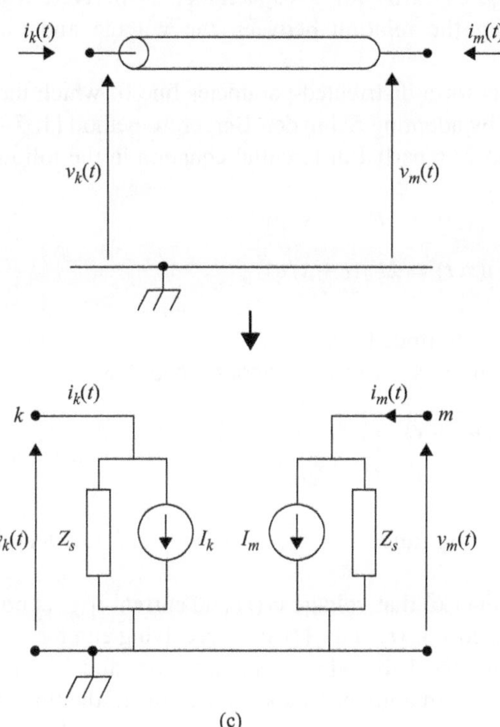

Figure 1.2   *Representation of circuit elements by a resistance and a current source: (a) inductance, (b) capacitance, (c) distributed line*

By applying Tropezoidal rule to the left-hand side of the equation,

$$\int v(t) \cdot dt = \{v(t) + v(t - \Delta t)\}\Delta t/2$$

From the above two equations,

$$i(t) = (\Delta t/2L)\{v(t) + v(t - \Delta t)\} - i(t - \Delta t) = v(t)/R_L + J(t - \Delta t) \tag{1.4}$$

where $J(t - \Delta t) = v(t - \Delta t)/R_L + i(t - \Delta t)$

$R_L = 2L/\Delta t, \quad \Delta t$: time step

It is clear from the above equation that current $i(t)$ at time t flowing through the inductance is evaluated by voltage $v(t)$ and current source $J(t - \Delta t)$, which was determined by the voltage and the current at $t = t - \Delta t$. Thus, the inductance is represented by the current source $J(t)$ and the resistance R_L as illustrated in Figure 1.2(a).

Similarly, Figure 1.2(b) for a capacitance is derived from a differential equation expressing the relation between the voltage and the current of the capacitance.

Figure 1.2(c) is for a distributed-parameter line of which the voltage and the current are related by adopting Schnyder–Bergeron method [1, 7–10] or method of characteristics to solve a partial differential equation in the following form.

$$v(x, t) + Z_0 \cdot i(x, t) = 2F_1(t - x/c)$$
$$v(x, t) - Z_0 \cdot i(x, t) = 2F_2(t + x/c)$$

$$(1.5)$$

where Z_0: characteristic impedance.

The above equation is rewritten at nodes 1 and 2 as

$$v_1(t - \tau) + Z_0 i(t - \tau) = v_2(t) - Z_0 i_2(t)$$
$$v_1(t) - Z_0 i(t) = v_2(t - z) + Z_0 i_2(t - \tau)$$

$$(1.6)$$

where $\tau = l/c$: traveling time from node 1 to node 2, l: line length, c: propagation velocity.

It is observed in (1.6) that voltage $v_1(t)$ and current $i_1(t)$ at node 1, the sending end of the line, influence $v_2(t)$ and $i_2(t)$ at the receiving end for $t \geq \tau$, where τ is the traveling time from node 1 to node 2. Similarly $v_2(t)$ and $i_2(t)$ influence $v_1(t)$ and $i_1(t)$ with time delay τ. In a lumped-parameter element, the time delay is Δt as can be seen in (1.4). In fact, Δt is not a time delay due to traveling wave propagation, but it is a time step for time descritization to solve numerically a differential equation describing the relation between the voltage and the current of the lumped element.

From the above equation, the following relation is obtained.

$$i_1(t) = v_1(t)/Z_0 + J_1(t - \tau), \quad i_2(t) = v_2(t)/Z_0 + J_2(t - \tau)$$

$$(1.7)$$

where $J_1(t - \tau) = -v_2(t - \tau)/Z_0 - i_2(t - \tau), J_2(t - \tau) = -v_1(t - \tau)/Z_0 - i_1(t - \tau)$.

The above results give the representation of a distributed-parameter line in Figure 1.2(c).

In (1.6), Z_0 is the chatacteristic impedance which is frequency-dependent. When the frequency dependence of a distributed-parameter line is to be considered, a frequency-dependent line such as Semlyen's and Marti's line models are prepared as a subroutine in the EMTP.

1.1.4   Sparse matrix

Sparsity is exploited intuitively in hand calculations. For example, to solve the following simultaneous equations:

$$x_1 - x_2 = 2$$

$$x_1 - x_3 = 6$$

$$2x_1 - 3x_2 + 4x_3 = 6$$

Anybody picks the first and second equations first, i.e., to express x_2 and x_3 as a function of x_1, and to insert these expressions into the third equation to find x_1. This is the basis of sparsity techniques. The sparsity techniques have been used in power system analysis since the early 1960s by W.F. Tinney and his coworkers [5, 6] in the B.P.A. There are a number of papers on the subject to improve the techniques.

Let us consider the node equations for the network in Figure 1.3. The node equations are given in the following form where an "x" is to indicate nonzero entries in the nodal admittance matrix in (1.2).

$$\begin{bmatrix} x & x & x & x & x \\ x & x & & & \\ x & & x & & \\ x & & & x & \\ x & & & & x \end{bmatrix} \begin{bmatrix} V_1 \\ V_2 \\ V_3 \\ V_4 \\ V_5 \end{bmatrix} = \begin{bmatrix} I_1 \\ I_2 \\ I_3 \\ I_4 \\ I_5 \end{bmatrix} \tag{1.8}$$

After triangularization, the equations have the following form:

$$\begin{bmatrix} x & x & x & x & x \\ & x & x & x & x \\ & & x & x & x \\ & & & x & x \\ & & & & x \end{bmatrix} \begin{bmatrix} V_1 \\ V_2 \\ V_3 \\ V_4 \\ V_5 \end{bmatrix} = \begin{bmatrix} I_1' \\ I_2' \\ I_3' \\ I_4' \\ I_5' \end{bmatrix} \tag{1.9}$$

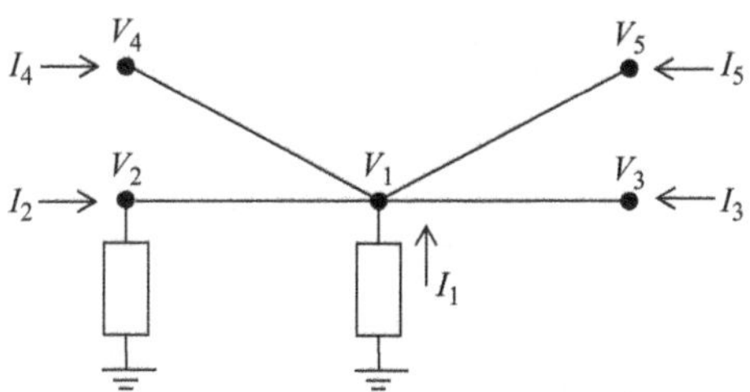

Figure 1.3   An example for sparsity technique

The triangular matrix is now full, in contrast to the original matrix which was sparse. The "full-in" is, of course, produced by the downward operations in the elimination process. This fill-in depends on the node numbering, i.e., on the order in which the nodes are eliminated.

The simplest "good" ordering scheme is: Number nodes with only one branch connected first, then number nodes with two branches connected, then nodes with three branches connected, etc. Better ordering schemes are discussed in [5, 6].

Exploitation of sparsity is extremely important in large power systems because it reduces storage requirements and solution time tremendously. The solution time for full matrices is proportional to N^3, where N is the order of the matrix. For sparse power systems, it increases about linearly. Typically, the number of series branches is about 1.6 × (number of nodes) and the number of matrix elements in the upper triangular matrix is about 2.5 to 3 times the number of nodes in steady-state equations. The node equations for the transient solutions are usually sparser because distributed-parameter lines do not contribute any off-diagonal elements.

1.1.5   Frequency-dependent line model

It is well known that the impedances of an overhead line and an underground cable show the frequency dependence due to the skin effect of any conductor as shown in Figure 1.4. It is not easy for a time domain transient analysis tool such as the EMTP to deal with the frequency-dependent effect.

When a step function voltage is applied to the sending end of a line, the receiving-end voltage is deformed due to the frequency-dependent effect of the line. The receiving-end voltage is obtained in the following form in a frequency domain.

$$S(\omega) = \exp(-\Gamma \cdot x) \tag{1.10}$$

where Γ: propagation constant in frequency domain.

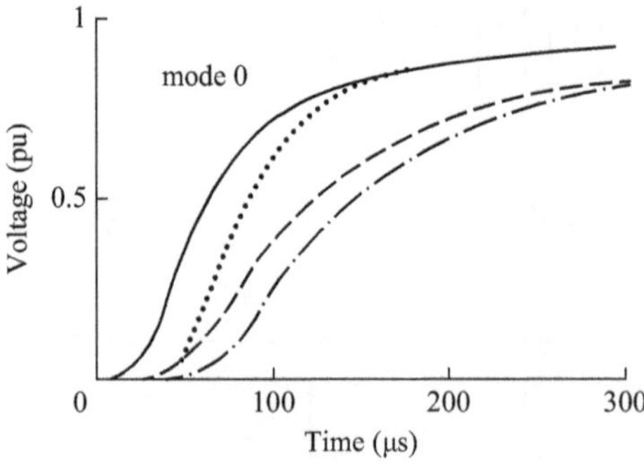

Figure 1.4   *An example of step responses s(t) of wave deformation due to earth skin-effect on a 500 kV horizontal line*

Assuming a traveling wave $E_0(\omega)$ propagates from the sending end to the receiving end, then at the receiving end, the traveling wave is given as

$$E(\omega) = S(\omega) \cdot E_0(\omega) \tag{1.11}$$

In a time domain, the above equation is expressed by a real time convolution.

$$e(t) = s(t) * e_0(t) = s(0) \cdot e_0(t) + \int_0^t s'(\tau) \cdot e_0(t - \tau) \cdot d\tau \tag{1.12}$$

where $e(t) = F^{-1}E(\omega)$, $e_0(t) = F^{-1}E_0(\omega)$, $s(t) = F^{-1}S(\omega)/j\omega$

F^{-1}: Fourier inverse transform

It should be noted that (1.11) is evaluated by single multiplication of two functions $S(\omega)$ and $E_0(\omega)$, while (1.12) in the time domain requires a number of multiplication, i.e., multiplication of two matrices. Thus, the inclusion of the frequency-dependent effect needs a large memory strotage and computation time. In fact, it required three full days to calculate switching surges on a three-phase line with the real time convolution to include the frequency-dependent effect of the propagation constant in 1968.

To avoid this large computer resource, a recursive convolution was developed by Semlyen using exponential functions in 1974 [12], and by Ametani using linear approximation of $s(t)$ in 1975 [13]. Later a more sophisiticated approach was developed by Marti [14].

1.1.6 Transformer

In the first model of a transformer in the EMTP, the transformer was presented by branch resistance and inductance matrices $[R]$ and $[L]$. The supporting routine XFORMER was written to produce these matrices from the test data of single-phase two- and three-winding transformers. Stray capacitances are ignored in these representations. However, it is easy to add the capacitances as branch data [22].

A star circuit representation for N-winding transformers was added later, which used matrices $[R]$ and $[L]^{-1}$ with the alternate equation in a transient analysis.

$$[L]^{-1}[v] = [L]^{-1}[R][i] + [di/dt] \tag{1.13}$$

This formulation also became useful when supporting routines BCTRAN and TRELEG were developed for inductance and inverse inductance matrix representations of three-phase units.

Saturation has been represented by adding extra nonlinear inductances and resistances to the above inductance (or inverse inductance) matrix representation. In the case of the star circuit, the nonlinear magnetizing inductance and iron-core resistance are added. A nonlinear inductance with hysteresis effects (pseudo-nonlinear) has been developed as well.

The simplest transformer representation in the form of an "ideal" transformer was the last model to be added to the EMTP in 1982, as part of a revision to allow for voltage sources between nodes.

1.1.7   Three-phase synchronous machine

The details with which synchronous machines must be modeled depend very much on the type of transient studies. The simplest representation of the synchronous machine is a voltage source E'' behind a subtransient reactance X_d''. This representation is commonly used in a short-circuit study with steady-state phasor solutions, and is also reasonably accurate for a transient study for the first few cycles of a transient disturbance. A typical example is a switching surge study. Another well-known representation is E' behind X_d' for a simplified stability study. Both of these representations can be derived from the same detailed model by making certain assumptions, such as neglecting flux linkage changes in the field structure circuits for E'' behind X_d'', in addition, assuming that the damper winding currents have died out for E' behind X_d'.

The need for the detailed model arose in connection with a fault of generator breakdown due to subsynchronous resonance in 1972 [23–25]. To analyze the subsynchronous resonance, the time span is too long to allow the use of the above explained simplified models. Furthermore, the torsional dynamics of the shaft with its generator rotor and turbine rotor masses have to be represented as well. The detailed model is now also used for other types of studies, for example, simulation of out-of-step synchronization. To cover all possible cases, the synchronous machine model represents the details of the electrical part of the generator as well as the mechanical part of the generator and turbine.

The synchronous machine model was developed for the usual design with three-phase ac armature windings on the stator and a dc field winding with one or more pole pairs on the rotor.

M.C. Hall, J. Alms (Southern California Edison Co.) and G. Gross (Pacific Gas & Electric Co.), with assistance of W.S. Meyer (Bonneville Power Administration), implemented the first model which became available to the general public. They opted for an iterative solution at each time step, with the rest of the system, as seen from the machine terminals, represented by a three-phase Thevenin equivalent circuit [24]. To keep this "compensation" approach efficient, machines had to be separated by distributed-parameter lines from each other. If that separation did not exist in reality, short artificial "stub lines" had to be introduced which sometimes caused problems. V. Barandwajn suggested another alternative in which the machine was basically presented as an internal voltage source behind some impedance [16]. The voltage source is recomputed for each time step, and the impedance becomes part of the nodal conductance matrix $[G]$ in (1.2). This approach depends on the prediction of some variables, which are not corrected at one and the same time step in order to keep the algorithm non-iterative. While the prediction can theoretically cause numerical instability, it has been refined to such an extent by now that the method has become quite stable and reliable. Numerical stability has been more of a problem with machine models partly because the typical time span of a few cycles in switching surge studies has grown to a few seconds in machine transient studies, with the step size Δt being only slightly larger, if at all, in the latter case.

The sign conventions used in the model are summarized.

(a)   The flux linkage λ of a winding, produced by current in the same winding, is considered to have the same sign as the current ($\lambda = Li$, with L being the self inductance of the winding).

(b)   The "generator convention" is used for all windings, i.e., each winding k is described by

$$v_k(t) = -R_k i_k(t) - \frac{d\lambda_k(t)}{dt} \qquad (1.14)$$

(with the "load convention," the signs would be positive on the right-hand side).

(c)   The newly recommended position of the quadrature axis lagging 90° behind the direct axis in the machine phasor diagram is adopted in the model [26]. In Park's original work, and in most papers and books, it is leading, and as a consequence terms in the transformation matrix $[T]^{-1}$ have negative signs there.

The machine parameters are influenced by the type of construction. For example, salient-pole machines are used in hydro plants, with two or more (up to 50) pole pairs. The magnetic properties of a salient-pole machine along the axis of symmetry of a field pole (direct axis) and along the axis of symmetry midway between two field poles (quadrature axis) are noticeably different because a large part of the path in the latter case is in air. Cylindrical-rotor machines used in thermal plants have long cylindrical rotors with slots in which distributed field windings are placed.

1.1.8   Universal machine

The universal machine was added to the EMTP by H.K. Lauw and W.S. Meyer [21], to be able to study various types of electric machines with the same model. It can be used to represent the following 12 major types of electric machines.

(1)   synchronous machine, three-phase armature;
(2)   synchronous machine, two-phase armature;
(3)   induction machine, three-phase armature;
(4)   induction machine, three-phase armature and three-phase rotor;
(5)   induction machine, two-phase armature;
(6)   single-phase ac machine (synchronous or induction), one-phase excitation;
(7)   same as (6), except two-phase excitation;
(8)   dc machine, separately excited;
(9)   dc machine, series compound (long shunt) field;

(10)   dc machine, series field;
(11)   dc machine, parallel compound (short shunt) field;
(12)   dc machine, parallel field.

The user can choose between two interfacing methods for the solution of the machine equations, with the rest of the network. One is based on compensation, where the rest of the network seen from the machine terminals is represented by a Thevenin equivalent circuit, and the other voltage source behind an equivalent impedance representation, similar to that of a synchronous machine, which requires prediction of certain variables.

The mechanical part of the universal machine is modeled quite differently from that of the synchronous machine. Instead of a built-in model of the mass-shaft system, the user must model the mechanical part as an equivalent electric network with lumped R, L, and C, which is then solved as if it were part of the complete electric network. The electromagnetic torque of the universal machine appears as a current source in this equivalent network.

Any electric machine has essentially two types of windings, one being stationary on the stator, and the other rotating on the rotor. Which type is stationary and which is rotating are irrelevant in the equations, because it is only the relative motion between two types that counts. The two types are

(1)   Armature windings (windings on "power side" in the EMTP Rule Book [3]). In induction and (normally) in synchronous machines, the armature windings are on the stator. In dc machines, they are on the rotor, where the commutator provides the rectification from ac to dc.
(2)   Windings on the field structure ("excitation side" in the EMTP Rule Book [3]). In synchronous machines, the field structure windings are normally on the rotor, while in dc machines they are on the stator. In induction machines they are on the rotor, either in the form of a short-circuited squirrel-cage rotor, or in the form of a wound rotor with slip-ring connections to the outside. The proper term is "rotor windings" in this case; the term "field structure winding" is only used here to keep the notation uniform for all types of machines.

These two types of windings are essentially the same as those of the synchronous machine. It is therefore not surprising that the system of equations for the synchronous machine describe the behavior of the universal machine along the direct and quadrature axes as well. The universal machine is allowed to have up to three armature windings, which are converted to hypothetical windings $d, q, 0a$ ("a" is for armature) in the same way as in Section 1.1.7. The field test structure is allowed to have any number of windings on the quadrature axis, which can be connected to external circuits defined by the user. In contrast to the synchronous machine, the field structure may also have a single zero sequence winding 0_f ("f" for the field structure) to allow the conversion of three-phase windings on the field structure (as in wound-rotor induction machines) into hypothetical $D, Q, 0$-windings.

With these minor differences to the synchronous machine, the voltage equations for the armature windings in d, q-quantities become

$$\begin{bmatrix} v_d \\ v_q \end{bmatrix} = - \begin{bmatrix} R_a & 0 \\ 0 & R_a \end{bmatrix} \begin{bmatrix} i_d \\ i_q \end{bmatrix} - \frac{d}{dt} \begin{bmatrix} \lambda_d \\ \lambda_q \end{bmatrix} + \begin{bmatrix} -\omega\lambda_q \\ +\omega\lambda_d \end{bmatrix} \tag{1.15}$$

with ω being the angular speed of the rotor referred to the electrical side, and in zero sequence,

$$v_{0a} = R_a i_{0a} - d\lambda_{0a}/dt \tag{1.16}$$

1.1.9   Switches

1.1.9.1   Introduction

Any switching operation in power systems can produce transients. For the simulation of such transients, it is necessary to model the various switching devices, such as circuit breakers, load breakers, dc circuit breakers, disconnectors, protective gaps, and thyristors.

So far, all these switching devices are presented as ideal switches in the EMTP, with zero current ($R = \infty$) in the open position and zero voltage ($R = 0$) in the closed position. If the switch between nodes k and m is open, then both nodes are represented in the system of nodal equations as in Figure 1.5(a). It is possible to add other branches to the ideal switch to more closely resemble the physical behavior, for example, to add a capacitance from k to m for the representation of the stray capacitance or the $R - C$ grading network of an acutual circuit breaker.

Switches are not needed for the connection of voltage and current sources if they are connected to the network at all times. The source parameters T_{START} and T_{STOP} can be used in place of switches to have current sources temporarily connected for $T_{START} < t < T_{STOP}$, as explained in Section 1.1.15. For voltage sources, this definition would mean that the voltage is zero for $t < T_{START}$ and for $t > T_{STOP}$, which implies a short-circuit rather than a disconeection. Therefore, switches are needed to disconnect voltage sources.

Switches are also used to create piecewise linear elements.

1.1.9.2   Basic switches

There are five basic switch types in the EMTP, which are all modeled as ideal switches. They differ only in the criteria being used to determine when they should open or close.

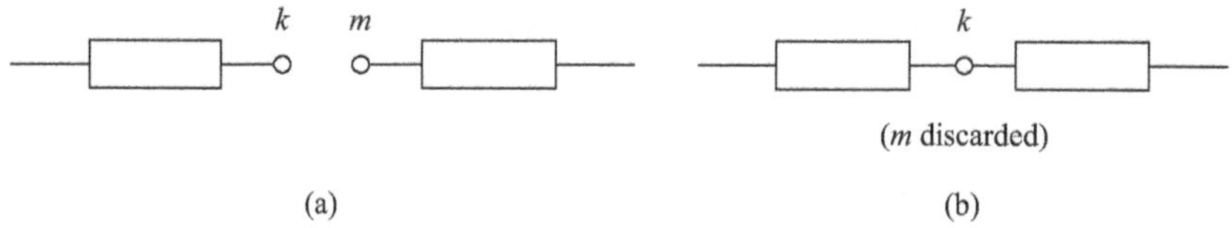

Figure 1.5   Representation of switches in the EMTP: (a) open, (b) closed

(a) Time-controlled switch

This type is intended for modeling circuit breakers, disconnectors, and similar switching devices, as well as short-circuits. The switch is originally open, and closes at T_{CLOSE}. It opens again after T_{OPEN} (if $< t_{max}$), either as soon as the absolute value of the switch current falls below a user-defined current margin, or as soon as the current goes through zero (detected by a sign change) as illustrated in Figure 1.6. For the simulation of circuit breakers, the latter criterion for opening should normally be used. The time between closing and opening can be delayed by a user-defined time delay.

The closing takes place at the time step where $t > T_{CLOSE}$. If the simulation starts from automatically calculated ac steady-state conditions, then the switch will be recognized as closed in the steady-state phasor solution.

The EMTP has an additional time-controlled switch type (TACS-controlled switch) in which the closing and opening action is controlled by a user specified TACS variable from the TACS part of the EMTP [3, 15]. With that feature, it is very easy to build more complicated opening and closing criteria in TACS.

(b) Gap switch

This switch is used to simulate protective gaps, gaps in surge arresters, flashovers across insulators, etc. It is always open in the ac steady-state solution. In the transient simulation, it is normally open, and closes as soon as the absolute value of the voltage across the switch exceeds a user-defined breakdown or flashover voltage. For this checking procedure, the voltage values are averaged over the last two steps, to filter out numerical oscillations. Opening occurs at the first current zero, provided a user-defined delay time has already elapsed. This close–open cycle repeats itself whenever the voltage exceeds the breakdown or flashover voltage again.

It is well known that the breakdown voltage of a gap or the flashover voltage of an insulator is not a simple constant, but depends on the steepness of the incoming wave. This dependence is usually shown in the form of a voltage-time characteristic as illustrated in Figure 1.7, which can be measured in the laboratory for standard impulse waveshapes.

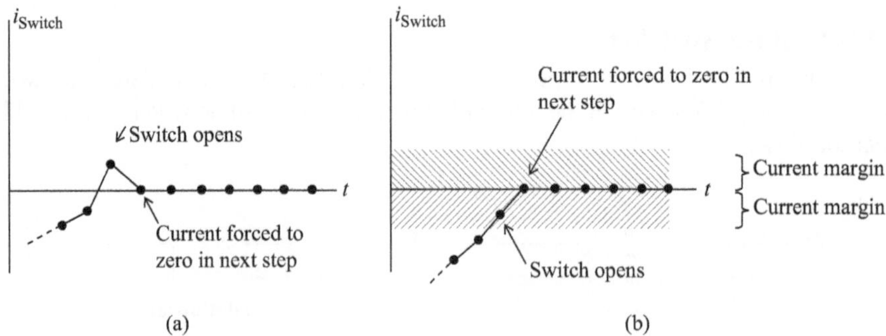

Figure 1.6   Opening of time-controlled switch: (a) current going to zero,
(b) current less than margin

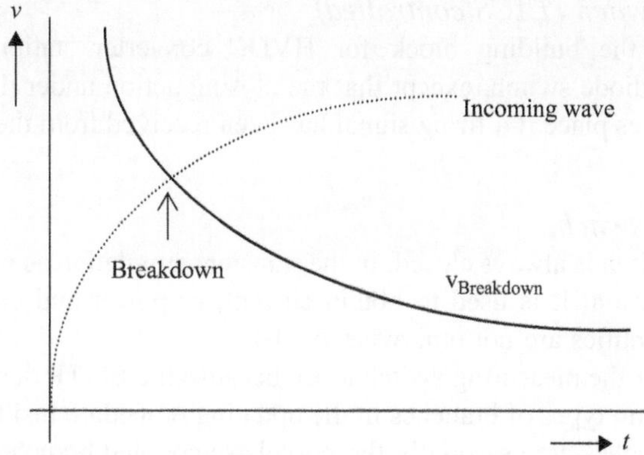

Figure 1.7   Voltage-time characteristic of a gap

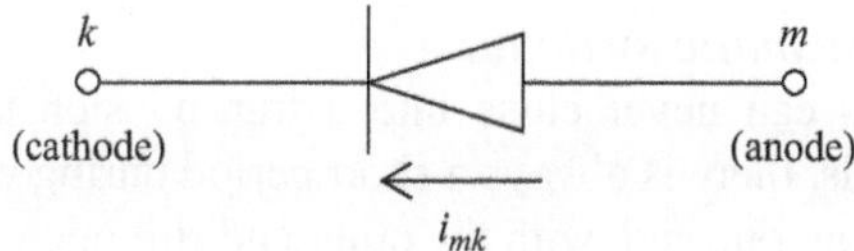

Figure 1.8   Diode switch

Unfortunately, the waveshapes of power system transients are usually very irregular, and therefore voltage-time characteristics can seldom be used. Analytical methods based on the integration of a function could easily be implemented.

$$F = \int_{t_1}^{t_2} (v(t) - v_0)^k \cdot dt \tag{1.17}$$

In (1.17), v_0 and k are constants, and breakdown occurs at instant t_2 where integral value F becomes equal to a user-defined value. For $k = 1$, this is the "equal-area criterion" of D. Kind [27].

The EMTP has an additional gap switch type (TACS-controlled switch), in which the breakdown or flashover is controlled by firing signal received from the TACS of the EMTP in Section 1.1.12. With that feature, voltage-time characteristics or criteria in the form of (1.17) can be simulated in TACS by skilled users.

(c) Diode switch

This switch is used to simulate diodes where current can flow in only one direction, from anode m to cathode k in Figure 1.8. The diode switch closes whenever $v_m > v_k$ (voltage values averaged over two successive time steps to filter out numerical oscillations), and opens after the elapse of a user-defined time delay as soon as the current i_{mk} becomes negative, or as soon as its magnitude becomes less than a user-defined margin.

In the steady-state solution, the diode switch can be specified as either open or closed.

(d) Thyristor switch (TACS controlled)

This switch is the building block for HVDC converter stations. It behaves similarly to the diode switch, except that the closing action under the condition of $v_m > v_k$ only takes place if a firing signal has been received from the TACS part of the EMTP.

(e) Measuring switch

A measuring switch is always closed, in the transient simulation as well as in the ac steady-state solution. It is used to obtain current, or power and energy, in place where these quantities are not otherwise available.

The need for the measuring switch arose because the EMTP does not calculate currents for certain types of branches in the updating procedure inside the time step loop. These branches are essentially the polyphase-coupled branches with lumped or distributed parameters. The updating procedures could be changed fairly easily to obtain the currents, as an alternative to the measuring switch.

(f) Statistics and systematic switches

Since circuit breakers can never close into a transmission line exactly simultaneously from both ends, there is always a short period during which the line is only closed, or reclosed, from one end, with the other end still open. Traveling waves are then reflected at the open end with the well-known doubling effect, and transient overvoltages of 2 p.u. at the receiving end are therefore to be expected. In reality, the overvoltages can be higher for various reasons; for example, the line is three-phase with three different mode propagation velocities.

The design of transmission line insulation is somehow based on the highest possible switching surge overvoltage. Furthermore, it is impossible or very difficult to know which combination of parameters would produce the highest possible overvoltage. Instead, 100 or more switching operations are usually simulated, with different closing times and possibly with variation of other parameters, to obtain a statistical distribution of switching surge overvoltages.

The EMTP has special switch types for running a large number of cases in which the opening or closing times are automatically varied. There are two types, one in which the closing times are varied statistically (statistics switch), and the other in which they are varied systematically (systematic switch). How well these variations represent the true behavior of the circuit breaker is difficult to say. Before the contacts have been completely closed, a discharge may occur across the gap and create "electrical" closing slightly ahead of mechanical closing ("prestrike").

1.1.10   Surge arrester and protective gap (archorn)

To protect generator and substation equipment and also transmission lines against levels of overvoltages which could break down their insulation, surge arresters are installed as close as possible to the protected device. Short connections are important to avoid the doubling effect of traveling waves on open-ended lines, even if they are short buses.

1.1.10.1 Protective gap (archorn)

Protective gaps are crude protection devices. They consist of air gaps between electrodes of various shapes. Examples are horns or rings on insulators and bushings, or rod gaps on or near transformers. They protect against overvoltages by collapsing the voltage to practically zero after sparkover, but they essentially produce a short-circuit which must then be interrupted by circuit breakers. Also, their voltage-time characteristic in Figure 1.7 rises steeply for fast fronts, which makes the protection against fast-rising impulses questionable.

The protective gap is simulated in the EMTP with the gap switch described in Section 1.1.9.

1.1.10.2 Surge arrester

There are two basic types of surge arresters, namely silicon-carbide surge arresters, and metal-oxide (Z_n0) surge arresters. Until about 30 years ago, only silicon-carbide arresters were used, but nowadays most arresters are the metal-oxide arrester.

Metal-oxide surge arresters are highly nonlinear resistors, with an almost infinite slope in the normal-voltage region, and an almost horizontal slope in the overvoltage protection region as illustrated in Figure 1.9. They were originally gapless, but some manufacturers have reintroduced gaps into the design. Its nonlinear resistance is represented by an exponential function of the form,

$$I = k(V/V_{ref})^n \tag{1.18}$$

where k, V_{ref}, and n are constans (typical values for $n = 20$ to 30). Since it is difficult to describe the entire region with one exponential function, the voltage region has been divided into segments in the EMTP, with each segment defined by its own exponential function. For voltage substantially below V_{ref}, the currents being extremely small, a linear representation is therefore used in this low voltage region. In the meaningful overvoltage protection region, two segments with exponential functions in (1.18) are usually sufficient.

The static characteristic of (1.18) can be extended to include dynamic characteristics similar to hysteresis effects, though the addition of a series inductance L, whose value can be estimated once the arrester current is approximately known [28].

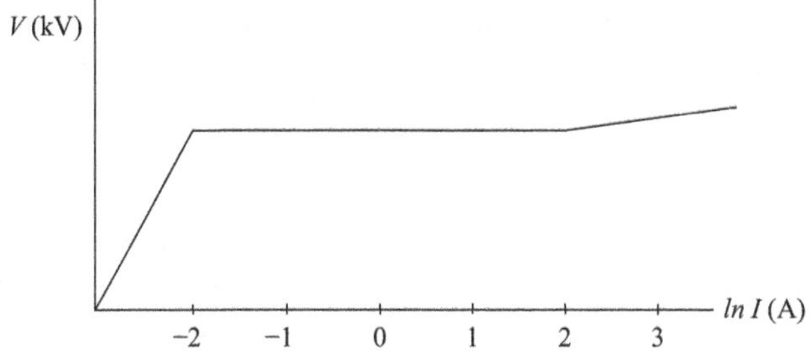

Figure 1.9   Voltage-current characteristic of a gapless Z_n0 arrester

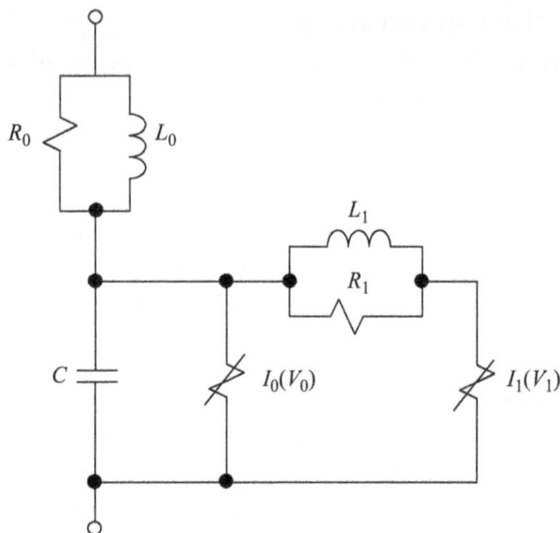

Figure 1.10   Two-section surge arrester model for fast front surges

A metal-oxide surge arrester model for fast front current surges with the time to crest in the range of 0.5 to 10 μs was proposed and compared against laboratory tests by Durbak [4, 29]. The basic idea is to divide the single nonlinear resistance into m parallel nonlinear resistances, which are separated by low pass filters, as illustrated in Figure 1.10 for two parallel nonlinearities, which is usually sufficient in practice. The R_1–L_1 circuit is the low pass filter which separates the two nonlinear resistances defined by $I_0(V_0)$ and $I_1(V_1)$. The inductance L_0 represents the small but finite inductance associated with the magnetic fields in the immediate vicinity of the surge arrester, while R_0 is used only to damp numerical oscillations. C is the stray capacitance of the surge arrester. The model of Figure 1.10 can easily be created from existing EMTP elements. If three models were connected to phases a, b, and c, then the six nonlinear resistances would have to be solved with the compensation method with a six-phase Thevenin equivalent circuit.

The energy absorbed in them is an important design factor, and should therefore be computed in whatever type of model is used. Since energy absorption may change as the system is expanded, it is important to check whether ratings which were appropriate initially may possibly be exceeded in the future years.

Metal-oxide surge arresters are generally solved with the compensation method in the EMTP, with iterations using Newton's method.

1.1.11   Inclusion of nonlinear elements

The most common types of nonlinear elements are nonlinear inductances for the representation of transformer and shunt reactor saturation, and nonlinear resistances for the representation of surge arresters. Nonlinear effects in synchronous and universal machines are handled in the machine equations directly, and are therefore not described here.

Usually, the network contains only a few nonlinear elements. It is therefore sensible to modify the well-proven linear methods more or less to accommodate nonlinear elements, rather than to use less efficient nonlinear solution methods for the entire network. This has been the philosophy which has been followed in the EMTP. Three modification schemes have been used over the years, namely

- current-source representations with time lag Δt (no longer used),
- compensation methods, and
- piecewise linear representations.

1.1.11.1   Current-source representation with time lag Δt

Assume that the network contains a nonlinear inductance with a given flux/current characteristic $\lambda(i)$, and that the network is just being solved at instant t. All quantities are therefore known at $t - \Delta t$, including flux $\lambda(t - \Delta t)$, which is found by integrating the voltage across the nonlinear inductance up to $t - \Delta t$. Provided Δt is sufficiently small, one could use $\lambda(t - \Delta t)$ to find a current $i(t - \Delta t)$ from the nonlinear characteristic, and inject this as a current source between the two nodes to which the nonlinearity is connected for the solution at instant t. In principle, any number of nonlinearities could be handled this way.

Since this method is very easy to implement, it may be useful in special cases, provided that the step size Δt is sufficiently small. It is not a built-in option in any of the available EMTP versions.

1.1.11.2   Compensation method

In earlier version of the EMTP, the compensation method worked only for a single nonlinearity in the network, or in case of more nonlinearities, if they were all separated from each other through distributed-parameter lines. It appears that the type-93 nonlinear inductance in the EMTP [3] still has this restriction imposed on it, but for most other types, more nonlinearities without travel time separation are allowed now.

The extension of the compensation method to more than one nonlinearity was first implemented for metal-oxide surge arresters, and later used for other nonlinear elements as well.

In comparison-based methods, the nonlinear elements are essentially simulated as current injections, which are super-imposed on the linear network after a solution without the nonlinear elements has first been found.

When the EMTP tries to calculate the Thevenin equivalent resistance for the nonlinear branch by injecting current into node m, a zero diagonal element will be encountered in the nodal conductance matrix if the nonlinear element is removed from the circuit during a simulation. Then, the EMTP will stop with the error message "diagonal element in node-m too small." The remedy in this case is: represent the nonlinear branch as a linear branch in parallel with a (modified) nonlinear branch. The EMTP manual also suggests the insertion of high-resistance path where needed, but warns that the resistance values cannot be arbitrarily large.

1.1.12   TACS

1.1.12.1   Introduction

The program called TACS was developed in 1977 by late L. Dube [15]. In 1983/84, Ma Ren-Ming did a thorough study of the code, and made major revisions in it, particularly with respect to the order in which the blocks of the control system are solved [30].

TACS was originally written for the simulation of HVDC converter controls, but it soon became evident that it had much wider applications. It has been used for the following simulations for example.

(a)   HVDC converter controls,
(b)   excitation systems of synchronous machines,
(c)   current limiting gaps in surge arresters,
(d)   arcs in circuit breakers.

TACS can be applied to the devices or phenomena which cannot be modeled directly with the existing network components in the EMTP.

Control systems are generally represented by block diagrams which show the interconnections among various control system elements, such as transfer function blocks, limiters. A block diagram representation is also used in TACS because it makes the data specification simple. All signals are assigned names which are defined by six alphanumeric characters (blank is included as one of the characters). By using the proper names for the input and output signals of blocks, any arbitrary connection of blocks can be achieved. There exists no uniform standard for describing the function of each block in an unambiguous way, except in the case of linear transfer functions. Users of the EMTP should be aware of this.

The control systems, devices and phenomena modeled in TACS, and the electric network are solved separately at this time. Output quantities from the network solution can be used as input quantities in TACS over the same time step, while output quantities from TACS can become input quantities to the network solution only over the next time step. TACS accepts as input network voltage and current sources, node voltages, switch currents, status of switches, and certain internal variables such as rotor angles of synchronous machines. The network solution accepts output signals from TACS as voltage or current sources (if the sources are declared as TACS controlled sources), and as commands to open or close switches (if the switch is a thyristor or a TACS controlled).

The present interface between the network solution and TACS is explained in the following section.

1.1.12.2   Interface between TACS and electric networks

To solve the models represented simultaneously with the network is more complicated than for models of power system components such as generators or transformers. Such components can essentially be represented as equivalent resistance matrices with parallel current sources, which fit directly into the nodal network (1.2). The equations of control system are quite different in that aspect. Their matrices are unsymmetrical and they cannot be represented as equivalent networks.

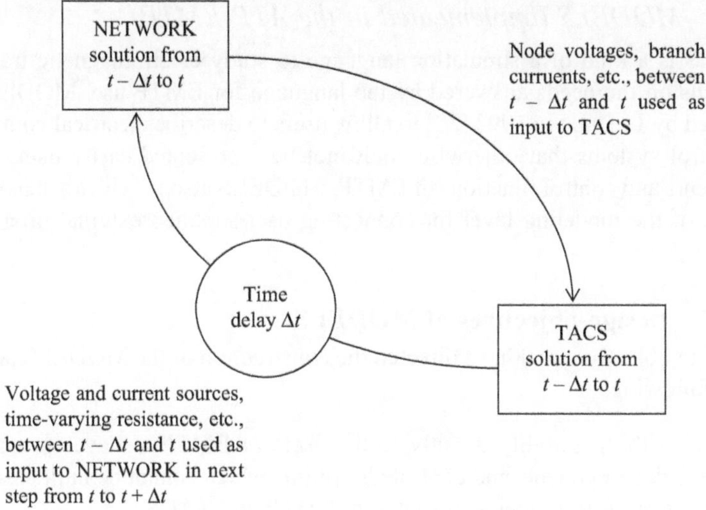

Figure 1.11   Interface between NETWORK and TACS solution

Because of these difficulties, L. Dube decided to solve the electric network (briefly called NETWORK from here on) and the TACS models (briefly called TACS from here on) separately. This imposes limitations which the user should be aware of. As illustrated in Figure 1.11, the NETWORK solution is first advanced from $(t - \Delta t)$ to t as if TACS would not exist directly. There is an indirect link from TACS and NETWORK with a time delay of Δt, inasmuch as NETWORK can contain voltage and current sources defined between $(t - \Delta t)$ and t which were computed as output signals in TACS in the preceding step between $(t - 2\Delta t)$ to $(t - \Delta t)$. NETWORK also receives commands for opening and closing switches at time t, which are determined in TACS in the solution from $(t - 2\Delta t)$ to $(t - \Delta t)$. In the latter case, the error in the network solution due to the time delay of Δt is usually negligible. First, Δt for this type of simulation is generally small, say, 50 μs. Second, the delay in closing a thyristor switch compensated by the converter control, which alternately advances and retards the firing of thyristor switches to keep the current constant in steady-state operation. With continuous voltage and current source functions coming from TACS, the time delay can become more critical, however, and the users are aware of its consequences. Cases have been documented where this time delay of Δt can cause numerical instability, i.e., in modeling the arc of circuit breakers with TACS [31].

Once NETWORK has been solved, the network voltages and currents specified as input to TACS are known between $(t - \Delta t)$ to t. No time delay occurs in this part of the interface, except that TACS itself has built-in delays which may not always be transparent to the user.

1.1.13   MODELS (implemented in the ATP-EMTP)

MODELS is a kind of a simulation language to study electromagnetic transients, with focus on the needs answered by the language for EMTP use. MODELS was developed by L. Dube in 1992 [32] to allow users to describe electrical components and control systems that otherwise could not be represented easily using regular components and control functions of EMTP. MODELS also provides a standardized interface at the modeling level for connecting user-supplied external programs to EMTP.

1.1.13.1   Design objectives of MODELS

The design objectives that have directed the construction of the MODELS language are the following:

(a)   to give the possibility directly to the users of EMTP to develop models of network components and control algorithms which cannot be built easily with the set of basic elements available in EMTP and TACS;
(b)   to provide to the user the flexibility of a full programming language without requiring the users to interact with EMTP at the programming level;
(c)   to let the user describe not only how a component operates, but also how the initial state (initial values and initial history) of the component is established, at a detail level selected by the user;
(d)   to provide standardized interface to EMTP defined at the modeling level in terms of voltages, currents, and control quantities (as opposed to an interface which would be defined at the programming level in terms of arrays of values and internal EMTP execution flow), giving directly to the user the possibility of connecting external programs to EMTP for modeling of components, access to measurements, or interaction with equipment, without requiring the user to have a programming knowledge if the internal operation of EMTP or to make any modification to the source code of EMTP.

In order to provide maximum power and flexibility when building models, it was considered necessary to maintain the advantages offered by regular programming languages. In particular, the language had to include the following possibilities:

(a)   separating the description of a component from its actual use in a simulation;
(b)   allowing the use of multiple instances of the same component, each possibly supplied with different parameters, different inputs, or different simulation directives;
(c)   providing a formal mechanism for the division of a large component description into smaller procedures, for reduced model complexity, explicit identification of interfaces, and functionality, easier testing, and use of prototyping during model development;
(d)   providing a separate naming scope for each procedure, to allow the user to assign meaningful names to the variables used in a description;

(e)   supporting the use of comments within a description, as an aid to producing models and procedures that are self-documenting:

(f)   supporting the use of conditions and loops (if, while, for, etc.) for building the structure of the execution flow of a procedure;

(g)   providing a regular syntax for the use of variables, arrays, expressions, and functions.

1.1.13.2   Implementation of the design objectives

In order to support the above design objectives, the simulation language MODELS has been constructed as a Pascal-like procedural language with additional applications-specific syntax and functions for representing the continuous-time operation of dynamic systems. Its grammar is keyword-directed and non-formatted.

In addition to the block diagram approach of TACS, MODELS accepts component descriptions in terms of procedures, functions, and algorithms. The user is not limited to predefined set of components, but can build libraries of components and submodels as required by each application. Similarly, the user is not limited to an input–output signal flow presentation, but can also combine this with numerical and logical manipulation of variables inside algorithmic procedures.

The free format of the description and the possibility of using arbitrarily long names facilitate self-documentation, valuable especially when large systems and team work are of concern. Comments and illustrations can be introduced throughout a model description for clearer documentation.

Numerical and Boolean values are supported, in both scalar and array forms. Values are described using algebraic and Boolean expressions, differential equations, Laplace transfer functions, integrals, and polynomials of derivatives. A large selection of numerical and Boolean functions can be used inside expressions. In addition, simulations-specific functions provide access to the time parameters of the simulation and to past and predicted values of the variables of a model. Additional functions can be defined by the user, in the form of parameterized expressions, point lists, and externally programmed functions.

Dynamic changes in the operation of a model are described by use of IF, WHILE, and FOR directives applied to groups of statements and to the use of submodels. Groups of simultaneous linear equations are solved in matrix form, using Gaussian elimination. Arbitrary iteration procedures can also be specified by the user for solving groups of simultaneous nonlinear equations. All other value assignments are calculated and applied sequentially in the order of their definitions in the model description.

The description of a model can be written directly as data in an EMTP data case, or accessed at simulation time from libraries of pre-written model files.

Self-diagnosis is especially important when large models have to be set up, debugged, and fine-tuned. Detection of the existence of possible misoperation can be built directly into the model by the user, forcing an erroneous simulation to be halted via the ERROR statement with clear indication of the problem that occurred. Warnings and other observations can also be recorded during the simulation for later analysis.

The language provides a mechanism for explicitly structuring a system into separate sub-components, supporting the definition and use of embedded sub-models. The same model can be used at multiple locations in a simulation, with parameters, inputs, outputs, and simulation directives specified individually for each use.

The simulation of each instance of a model is conducted in two phases. The initialization phase establishes initial values and initial history when the instance is first accessed. The model then repeatedly updates the values of the variables and of the accumulated history of the instance, in function of the changing values applied to its inputs. A detailed procedure can be defined to describe separately with full flexibility both the initialization and the operation of the model.

External programs can be directly defined and used as a model via a standardized interface, simply linked with the program without requiring any modification to be made to the EMTP source code. The interface is defined by the user in terms of voltages, currents, and control variables of the simulation. No knowledge of the EMTP at the programming level is required.

1.1.14   *Power system elements prepared in EMTP*

Table 1.1(a) summarizes various power system elements, and Table 1.1(b) shows supporting routines prepared in the original EMTP (Mode 31, April 1982 [3]). The supporting routines are a subroutine installed in the EMTP and calculate parameters required for the power system elements such as a transformer and a transmission line in Table 1.1(a).

1.1.15   *Basic input data*

Nowadays, any numerical simulation tool provides a user-friendly graphic interface of input (and output) data. Such an interface makes data input much easier from the users' viewpoint. At the same time, however, the users miss to catch physical meaning of each input data. It happens quite often that an input value is smaller or larger by 10^3 or 10^6.

In this section, basic input data are described for a simulation of a switching transient in a circuit of Figure 1.12 in the input data format specified in the original BPA EMTP [3] so that a use can catch the physical meaning of the input data.

Figure 1.13(a) shows the input data necessary for the simulation. The data set, however, is not easy to understand for the user except an experienced one.

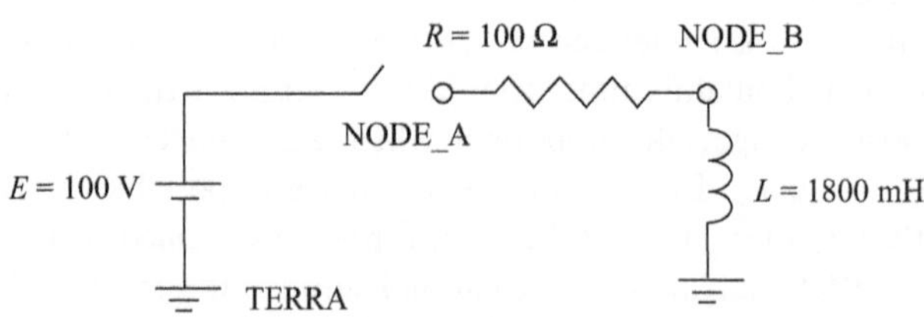

Figure 1.12   A transient on an R-L circuit

```
①BEGIN NEW DATA CASE
② 2.0E-3  50.E-3
③      1     1      0      0      1
④  NODE_ANODE_B              100.                              1
⑤  NODE_B                          1800.                      2
⑥BLANK ending BRANCH
⑦BLANK ending SWITCH
⑧11NODE_A        100.
⑨BLANK ending SOURCE
⑩  NODE_A
⑪BLANK ending OUTPUT Specification
⑫BLANK ending PLOT
⑬BEGIN NEW DATA CASE
⑭BLANK
```

(a)

```
①BEGIN NEW DATA CASE
  C *****************************
  C *          MISC CARDS       *
  C *****************************
  C [ DT][ TMAX ][ XOPT ][ COPT ][EPSILN][TOLMAT]
② 2.0E-3  50.E-3
  C  IOUT][ IPLOT][IDOUBL][KSSOUT][MAXOUT][ IPUN ][MEMSAV][ ICAT ][NENERG][IPRSUP]
③      1     1      0      0      1
  C
  C ***************************************************
  C  *        BRANCH CARD            *
  C ***************************************************
  C [BUS1][BUS2][BUS3][BUS4][  R ][  L ][  C ]                 !
④  NODE_ANODE_B              100.                              1
⑤  NODE_B                          1800.                      2
  C
⑥BLANK ending BRANCH
  C ***************************************************
  C *    SWITCH   CARDS             *
  C ***************************************************
  C [NODE][NODE][T  CLOSE][T   OPEN][      ]        [MEASURING]         !
  C
⑦BLANK ending SWITCH
  C
  C ***************************************************
  C *         SOUCE CARDS           *
  C ***************************************************
  C [NODE] I[AMPLITUD][FREQUENC][TINE-0 ][  AMPL  ][TIME-1 ][ TSTART ][ TSTOP  ]
⑧11NODE_A        100.
  C
⑨BLANK ending SOURCE
  C
  C ***************************************************
  C *    OUTPUT CARDS               *
  C ***************************************************
  C #### NODE VOLTAGE OUTPUT ####
  C [BUS1][BUS2][BUS3][BUS4][...
⑩  NODE_A
  C
⑪BLANK ending OUTPUT Specification
⑫BLANK ending PLOT
⑬BEGIN NEW DATA CASE
⑭BLANK
```

(b)

Figure 1.13   EMTP input data: (a) input data, (b) with in-line comments

Figure 1.13(b) shows a more user-friendly version of the data set with in-line comments explaining each data. Figure 1.14 gives the EMTP output to the input data in Figure 1.13. In Figures 1.13(b) and 1.14, "C" in the first column indicates that the line is for comments but not for the input data. ① to ⑮ in Figure 1.14 are identical to those in Figure 1.13(b).

```
  —    42   cards of disk file read into card cache cells  1    onward.
Alternative Transients Program (ATP).  GNU Linux or DOS.  All rights reserved by Can/Am user group of Portland, Oregon, USA.
 Date (dd-mth-yy) and time of day (hh.mm.ss) = 21-Apr-14  13:09:22   Name of disk plot file is  C:¥ATPWNT¥AAAAAAA.pl4
Consult the 860-page ATP Rule Book of the Can/Am EMTP User Group in Portland,  Oregon, USA.  Source code date is 23 October 2006.
Total size of LABCOM tables = 10913313 INTEGER words.  31 VARDIM  List Sizes follow:  6002  10K  192K 900  420K  1200  15K
   120K  2250  3800  720  2K  72800  510  90K  800  90  254  120K  100K  3K  15K  192K  120  45K  260K  600  210K  600  19  200
-------------------------------------------------+--------------------------------------------------------------------------
Descriptive interpretation of input data cards.  | Input data card images are shown below, all 80 columns, character by character
                                                  0         1         2         3         4         5         6         7        8
                                                  01234567890123456789012345678901234567890123456789012345678901234567890123456 7890
-------------------------------------------------+--------------------------------------------------------------------------
Comment card.    NUMDCD = 1.                      |C data:C:¥ATPWNT¥AAAAAAA.DAT
Marker card preceding new EMTP data case.        ①|BEGIN NEW DATA CASE
Comment card.    NUMDCD = 3.                       |C ****************************
Comment card.    NUMDCD = 4.                       |C *          MISC CARDS        *
Comment card.    NUMDCD = 5.                       |C ****************************
Comment card.    NUMDCD = 6.                      ②|C [  DT][ TMAX ][ XOPT ][ COPT ][EPSILN][TOLMAT]
Misc. data.      2.000E-03   5.000E-02   0.000E+00 |  2.0E-3  50.E-3
Comment card.    NUMDCD = 8.                       |C  IOUT][ IPLOT][IDOUBL][KSSOUT][MAXOUT][ IPUN ][MEMSAV][ ICAT ][NENERG][IPRSUP]
Misc. data.       1   1  0  0  1  0  0  0  0  0  ③|      1        1       0       0       1
Comment card.    NUMDCD = 10.                      |C
Comment card.    NUMDCD = 11.                      |C *****************************************************
Comment card.    NUMDCD = 12.                      |C  *       BRANCH CARD            *
Comment card.    NUMDCD = 13.                      |C *****************************************************
Comment card.    NUMDCD = 14.                      |C [BUS1][BUS2][BUS3][BUS4][ R ][ L ][ C ]                              !
Series R-L-C.    1.000E+02  0.000E+00  0.000E+00 ④| NODE_ANODE_B                100.                                      1
Series R-L-C.    0.000E+00  1.800E+00  0.000E+00 ⑤| NODE_B                                 1800.                          2
Comment card.    NUMDCD = 17.                      |C
Blank card ending branches.   IBR, NTOT = 2  3   ⑥|BLANK ending BRANCH
Comment card.    NUMDCD = 19.                      |C *****************************************************
Comment card.    NUMDCD = 20.                      |C *     SWITCH  CARDS              *
Comment card.    NUMDCD = 21.                      |C *****************************************************
Comment card.    NUMDCD = 22.                      |C [NODE][NODE][T  CLOSE][T   OPEN][      ]           [MEASURING]        !
Comment card.    NUMDCD = 23.                      |C
Blank card ending switches.   KSWTCH = 0.        ⑦|BLANK ending SWITCH
Comment card.    NUMDCD = 25.                      |C
Comment card.    NUMDCD = 26.                      |C **********************************************
Comment card.    NUMDCD = 27.                      |C *          SOUCE CARDS            *
Comment card.    NUMDCD = 28.                      |C **********************************************
Comment card.    NUMDCD = 29.                      |C [NODE] I[AMPLITUD][FREQUENC][TINE-0 ][ AMPL ][TIME-1 ][ TSTART ][ TSTOP ]
Source.    1.00E+02  0.00E+00  0.00E+00  0.00E+00⑧|11NODE_A            100.
Comment card.    NUMDCD = 31.                      |C
Blank card ends electric sources.   KCONST = 1   ⑨|BLANK ending SOURCE
Comment card.    NUMDCD = 33.                      |C
Comment card.    NUMDCD = 34.                      |C ***********************************************
Comment card.    NUMDCD = 35.                      |C *     OUTPUT CARDS              *
Comment card.    NUMDCD = 36.                      |C ***********************************************
Comment card.    NUMDCD = 37.                      |C ##### NODE VOLTAGE OUTPUT #####
Comment card.    NUMDCD = 38.                      |C [BUS1][BUS2][BUS3][BUS4][...
Card of names for time-step loop output.         ⑩|  NODE_A
Comment card.    NUMDCD = 40.                      |C

Blank card ending requests for output variables.⑪|BLANK ending OUTPUT Specification
```

(a)

Figure 1.14   (Continued)

The basic structure of the input data of the EMTP is

(1)   BIGIN NEW DATA CASE: shows a new data set or the end of the data set; ① and ⑮

(2)   MISC (miscellaneous) data: to specify I/O formats; ② and ③

(3)   BRANCH data: data for the branch specified by two nodes; ④ and ⑤

(4)   SWITCH data: data for a switch (no switch in this example)

(5)   SOURCE data: data for a source; ⑧

(6)   OUTPUT data: to specify the output; ⑩

(7)   PLOT data: to specify graphs of calculated results (not used in this example)

(8)   BLANK cards: to be inserted in between the data (2) to (7); ⑥, ⑦, ⑨, ⑪, ⑫, ⑭ in Figure 1.13.

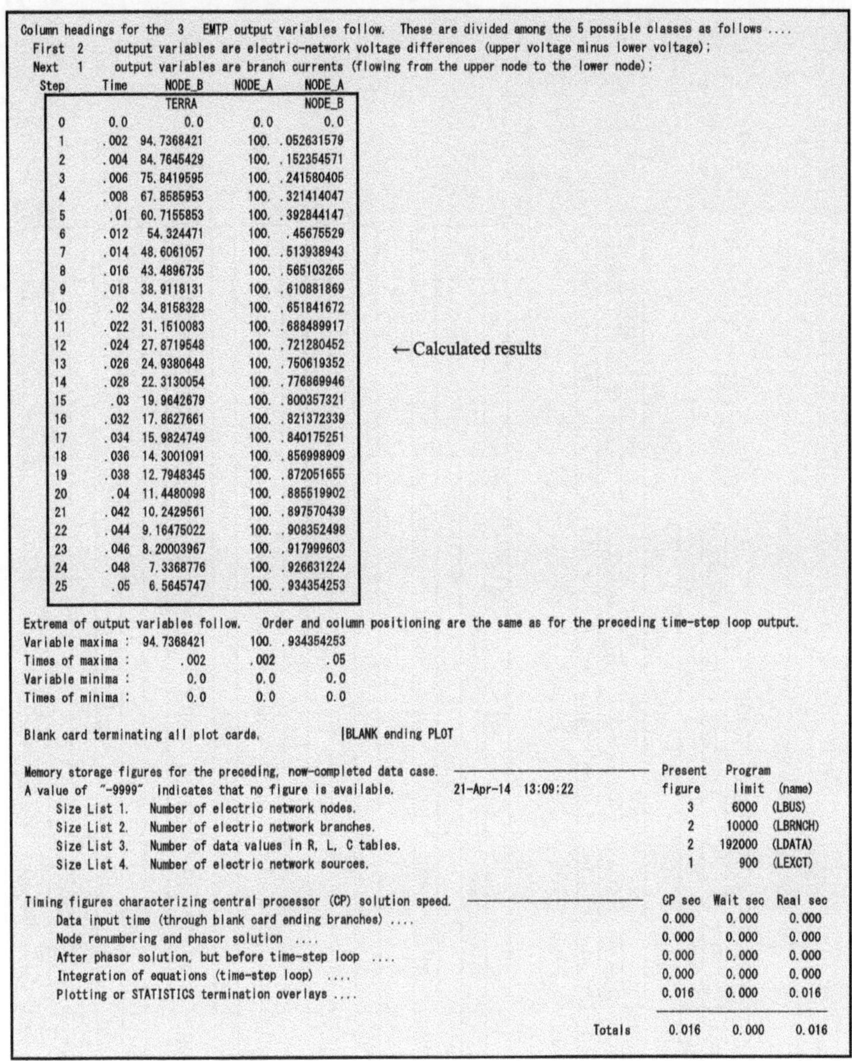

←Calculated results

(b)

Figure 1.14   EMTP output: (a) input data copied by EMTP, (b) output

Each data line is composed of 80 columns as shown in Figure 1.15. It should be noted that each column is specified to a specific data name or value. Thus, a user should be very careful of the name, or of the value being correctly positioned in the specified columns. This is a problem and a difficulty of producing a data set in the original EMTP, and a reason for a user-friendly graphic interface has been provided in the EMTP-ATP, EMTP-RV, and PSCAD.

(1) MISC data

②

1 8	9 16	17 24	25 32	33 40	41 48	49 56
DELTAT	TMAX	XOPT	COPT	EPSILN	TOLMAT	TSTART
E8.0	E8.0	E8.0	E8.0	E8.0	E8.0	E8.0

③

1 8	9 16	17 24	25 32	33 40	41 48	49 56	57 64	65 72	73 80
IOUT	IPOLT	IDOUBL	KSSOUT	MAXOUT	IPUN	MEMSAV	ICAT	NENERG	IPRSUP
I8	I8	I8	I8	I8	I8	I8	I8	I8	I8

(2) BRANCH data

④

12	3 8	9 14	15 26	27 32	33 38	39 44	45 79	80
ITY-	node name	node name	reference name		branch data			I-
PE	NODE_A	NODE_B			R [Ω]	L [mH]	C [μF]		OUT
I2	A6	A6	A6	A6	A6	E6.2	E6.2		I1

Ex. 1

	NODE_1	NODE_2					1.2		

Ex. 2-1

	NODE_A	TERRA			100.	1800.			

EX. 2-2

	NODE_A				100.	1800.			

Ex. 3

	BUS_1	BUS_2			10.		0.05		

Ex. 4

	IN	OUT			1.5	.1	.65		

Ex. 5

1	NODE1A	NODE1B			0.1	1.0	0.01		

Ex. 6

	IN_A	OUT_A			1.5	0.1	1.2		
	IN_B	OUT_B	IN_A	OUT_A					
	IN_C	OUT_C	IN_A	OUT_A					

(3) SWITCH data

1 2	3 8	9 14	15 24	25 34
ITY- PE	node name SWIT_1	node name SWIT_2	time to close T_{close} [s]	time to open T_{open} [s]
I2	A6	A6	E10.0	E10.0

Ex. 1

	IN_AS	IN_A	.0005	.002

(4) SOURCE data

1 2	3 8	9 10	11 20	21 30	31 40	41 50	51 60	61 70	71 80
(a) ITY- PE	(b) node name	(a) V/I	(c) AMPLI- TUDE	(d) FREQUEN- CY	(e) T_0 [sec] θ_0 [degree]	(f) only for TYPE 13, 14 A_1	(f) only for TYPE 13 T_1 [s]	(g) T_{start} [s] for TYPE$\neq$1,10	(h) T_{stop} [s]
I2	A6	I2	E10.6	E10.6	E10.6	E10.6	E10.6	E10.6	E10.6

Ex. 1-1

11	VOLT_A		1000.						

Ex. 1-2

11	VOLT_A	−1	1.5						

Ex. 2-1

11	VOLT_A		100.		.005				

Ex. 2-2

12	BUS_2	−1	−50.		.002				

Ex. 3-1

13	BUS_A		100.		4.E−6	50.	30.E−6		40.E−6

Ex. 3-2

13	BUS_A		1000.			2000.	50.E−6		

Ex. 4-1

14	GENE_A		100.	50.	45.				

Ex. 4-2

14	GENE_B		100.	50.	45.			−1.	

Ex. 4-3

14	GENE_C		100.	50.	.0025	1			

Figure 1.15   Basic structure of the EMTP input data

Now each data set, (2) to (6), is explained.

(1)   BIGINΔNEWΔDATAΔCASE (19 characters, Δ = space)
      ① To show data case and also a new data set in (2) to (7).
(2)   MISC data
      ② First MISC data: explanation given in the line right above the data
      –   DELTAT = time step Δt in sec. [s], to be in columns 1st to 8th (total
          8 characters).

          example: $2E - 3 = 2$ ms

      –   TMAX = maximum observation time in sec., 9th to 16th columns
          (8 characters).
      –   XOPT = input format of an inductance. When it is made "blank" ($= 0$),
          defaulted format which means that the inductance value should be in the
          unit of [mH]. In this chapter, XOPT, which defines the unit (mH or Ω) of
          the inductance, is kept to be "blank." XOPT is to be input in the columns
          17th to 24th.
      –   COPT = input format of a capacitance. For COPT is "blank" (or zero),
          the capacitance value should be given in the unit of [μF]. In this chapter,
          it is kept "blank." The data COPT in 25th to 32nd columns.
      –   EPSLN, TOLMAT = variables related to the inverse calculation of
          admittance matrix, and to judge the singularity. These are neglected in
          this example.
      –   TSTART = the time to start a transient simulation, neglected in this
          example.
          ③ Second MISC data: eight columns for every data
      –   IOUT = sampling span of calculated results, i.e., the results are printed
          out for every Δt when IOUT $= 1$. When IOUT $= 10$, printout of the
          results for every $10 \cdot \Delta t$. Thus, IOUT = TMAT/DELTAT. 9th to 16th
          columns.
      –   IPLOT = same as IOUT but for curve plotting. When IPLOT is defined
          negative, no curve plotting.
      –   IDUBLE $= 0$: no output of the circuit connection diagram.

                  $= 1$: output circuit connection diagram.

          In general, keep IDOUBLE $= 1$ to check the circuit connection is correct
          or not.
      –   KSSOUT $= 0$: no output of the steady-sate solution for $t \leq 0$.

                      $= 1$: output the steady-state solution. When TSTART < 0, the
                          steady state is calculated to start a transient calculation.

      –   MAXOUT $= 0$: no output of maximum voltages and currents.

                      $= 1$: The maximum voltages and currents are output after a
                          transient simulation.

     –   IPUN $= 0$: no output of node voltages and branch currents at $t = T_{max}$ which are used in the next transient in the same circuit.

       Also, there are the following MISC data which are required when a special calculation or a specific output is necessary. In most cases, just leave them "blank."

       MEMSAV $=$ memory save, ICAT $=$ plot file save, NENERG $=$ statistical surge calculation, IPRSUP $=$ output of intermediate results.

(3)   BANCH data

    ④, ⑤ BRANCH data

   (a)   ITYPE: 1st and 2nd columns

      ITYPE is the data to specify if the branch data are lumped parameters or a distributed line.

       ITYPE $= 0$ or blank: lumped $-$ parameter circuit

                $=$ negative value: distributed $-$ parameter line

                -1: for the first phase, -2: for the second phase, etc.

   (b)   node names: 3rd to 14th columns

      In the circuit of Figure 1.12, voltage source $e(t)$ and inductance L are grounded to the zero potential node (TERRA in the EMTP which is expressed by "blank"). In the BRANCH data, the six columns from the 3rd to the 8th column ("<BUS1>" in the comment line) are used for the node name of the sending end of a branch, and the next six columns (from the 9th to the 14th column, "<BUS2>" in the comment line) are used for the other end of the branch.

   (c)   Copied branch

      <BUS3> (from the 14th to the 19th column) and <BUS4> (from the 20th to 25th column) columns are used when the branch data for <BUS1> to <BUS2> branch are copied from <BUS3> to <BUS4> branch data.

   (d)   Branch data

      Because the branch from NODE_A to NODE_B is a resistance as illustrated in Figure 1.12, the value of the resistance, 100 Ω, is to be input in the columns from the 27th to the 32nd as shown in Figure 1.13. There exists an inductance between NODE_B to ground (TERRA $=$ blank), and the value of the inductance 1,800 mH is input in the columns from the 33rd to the 38th corresponding to the comment <L mH> in the comment line.

       Similarly if there is a capacitance in a circuit, its value is input in the columns from the 39th to the 44th corresponding to <C μF> in the comment line.

   (e)   Series circuit of lumped-parameter elements, R, L, and C can be input in one BRANCH data line. If there is no capacitance, then the columns corresponding to the capacitance should be made "blank."

       Some examples are given in Figure 1.15 (2) corresponding to the circuits illustrated in Figure 1.16.

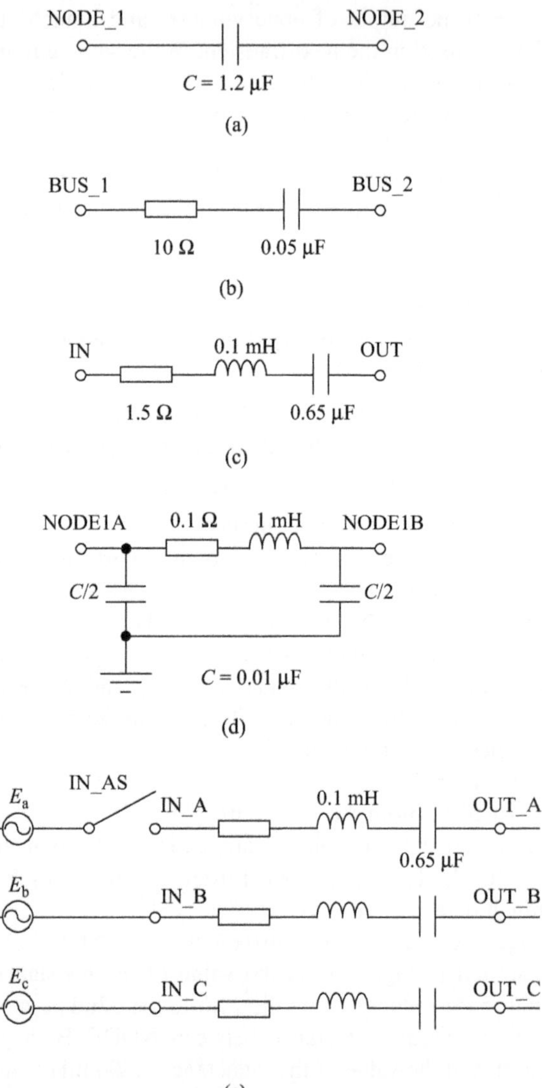

Figure 1.16   Examples of BRANCH: (a) Ex. 1 Capacitance C, (b) Ex. 3 R-C series circuit, (c) Ex. 4 R-L-C series circuit, (d) Ex. 5 π-equivalence, (e) Three-phase R-L-C circuit

(f)   Multiphase π circuit

By setting ITYPE to be positive sequential numbers similarly to a multiphase distributed line, π circuit equivalence of a multiphase distributed line is input, i.e.,

ITYPE = 1: first phase data, 2: the second phase, etc.

⑥ BLANK CARD: to show the end of the BRANCH data explained above.

(4) Switch data

"SWITCH" data are to specify the type of a switch, the closing time, the opening time, etc. The columns 1–2 of the SWITCH data are to define the type of switches as explained in Section 1.1.9. For example,

ITYPE $= 0$ (or blank): ordinary switches for which the voltage drop is zero when closed, and the current is zero when open. The ordinary switches involve the following options (called "Class").

Class 1 $=$ time-controlled switch (circuit breaker pole) which requires the input data of closing time "T_{close}" and opening time "T_{open}" as illustrated in Figure 1.15 (3).

The switch is open originally, close at $t \geq T_{close}$ and tries to open after $t \geq T_{open}$. There are two options; see Figure 1.6 in Section 1.1.9 (2).

Option A: the opening is successful as soon as the current i_{sw} has gone through zero detected by a sign change in i_{sw}.

Option B: the opening is successful as soon as $|i_{sw}| <$ current margin or as soon as i_{sw} has gone through zero.

Class 2 $=$ voltage-controlled switch (gap) which requires the input data "V_{flash}." When the voltage across the switch (open) terminals exceeds V_{flash}, then the switch is closed.

Class 3a $=$ statistic switch: time-controlled switch where the closing time is a random variable.

Class 3b $=$ systematic switch: time-controlled switch where the closing time is systematically varied.

Class 4 $=$ measuring switch: permanently closed, just for monitoring current, power, or/and energy.

Class 5 $=$ TACS controlled switch.

In the example illustrated in Figure 1.13, there is no switch although a switch is shown in Figure 1.12. This is because the switching operation is included in the voltage source E which is connected to NODE_A at $t = 0$.

⑦ BLANK CARD: to show the end of the switch data.

(5) SOURCE data

⑧ SOURCE data: composed of ITYPE, NODE name, V/I, AMPLITUDE, FREQUENCY, etc.

(a) ITYPE: 1st and 2nd columns to specify the type of sources.

ITYPE $= 11$: dc source shown as Ex. 1 in Figure 1.15 (4) and Figure 1.17(a-1) and (a-2). Remind that the voltage (or current) at $t = 0$ is zero unless the initial conition is given in the circuit. Thus, the voltage (or current) reaches the given amplitude at $t = \Delta t$ (time step) as shown in Figure 1.17(a).

ITYPE $= 12$: ramp function with linear rise from $t = 0$ to $t = T_0$, and constant apmlitude thereafter; see Ex. 2 in Figures 1.15 (4) and 1.17(b).

ITYPE $= 13$: ramp function with linear decay; see Ex. 3-1 in Figure 1.15 (4) and Figure 1.17(c-1). If A1>AMPLITUDE as in Ex. 3-2 in Figure 1.15(4), then the waveform in Figure 1.17(c-2) is given.

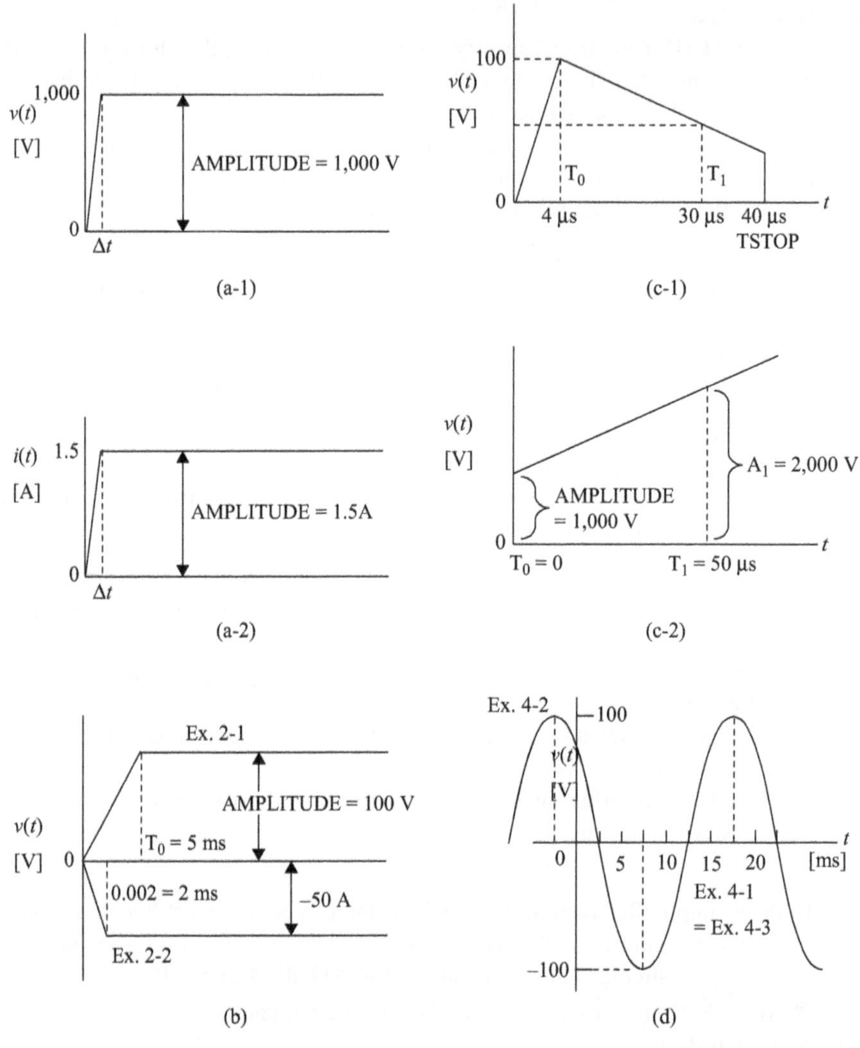

Figure 1.17   Various source waveforms: (a-1) Ex. 1-1 Voltage source, (a-2) Ex. 1-2 Current source, (b) Ex. 2, (c-1) Ex. 3-1, (c-2) Ex. 3-2, (d) Ex. 4

ITYPE $= 14$: sinusoidal function, see Ex. 4 in Figures 1.15 (4) and 1.17(d). In Ex. 4-2, $T_{start} < 0$ means that the circuit is in a steady state for $t < 0$ as shown in dotted line (Ex. 4-2) in Figure 1.17(d). Unless this is done, no printout of a steady-state solution.

In Ex. 4-3, $A_1 > 0$ means that the sinusoidal function is defined by

$$f(t) = \text{AMPLITUDE} * \cos[2\pi f(t + T_0)], T_0 \text{ in sec.}$$

When $A_1 = 0$: $f(t) = \text{AMPLITUDE} * \cos[2\pi f t + \theta_0], \theta_0$ in degrees.

(b)   Node name: columns 3–8 to specify the node to which a source is connected. The other node of the source is grounded (to TERRA).

(c) V/I: columns 9–10, $V/I = 0$ (blank or positive integer): voltage source

$$= \text{negative integer: current source, see Ex. 2-2}$$
$$\text{in Figure 1.15 (4).}$$

(d)   AMPLITUDE: columns 11–20 to give the amplitude of a voltage [V] or current [A].

In the case of a dc source (ITYPE $= 11$), the following data are not necessary.

(e)   FREQUENCY: columns 21–30 to specify the ac source frequency, only for ITYPE $= 14$.

(f) T_0[s] or θ_0[deg.]: columns 31–40

For ITYPE $= 14$ (ac source): see the above explanation in (a).

For ITYPE $= 12$: time T_0 to reach amplitude, see Ex. 2 in Figures 1.15 and 1.17(b).

For ITYPE $= 13$: time T_0 to the peak (amplitude), see Ex. 3-1 in Figures 1.15 (4) and 1.17(c-1). If T_0 is given as "blank," then, data for A_1 and T_1 are required.

(g) A_1: columns 41–50 only for ITYPE $= 13$ and 14, T_1: columns 51–60 only for ITYPE $= 13$.

For ITYPE $= 14$: see (a)

For ITYPE $= 13$: linear rise for positive A_1 at $t = T_1$ greater than AMPLITUDE, see Ex. 3-2 in Figures 1.15 (4) and 1.17(c-2).

(h) T_{start}: columns 71–80, blank in general. If a positive value of T_{stop} is given, then source $e(t) = 0$ (or $i(t) = 0$) at $t = T_{stop}$.

⑨ BLANK CARD: to end the SOURCE data.

(6)   OUTPUT data

⑩ OUTPUT data: to specify the node voltage output of an EMTP simulation.

In this example (see ⑩ in Figures 1.13 and 1.14), the voltage at NODE_A is printed out. For a branch current and a branch voltage, the output is requested at the 80th column of BRANCH data; see data ④ for a branch current output, and data ⑤ for a branch voltage (NODE_B to TERRA) in Figure 1.14(a).

The results output are shown in Figure 1.14(b).

⑪ BLANK CARD: to end the OUTPUT data

⑫ BLANK CARD: to end the PLOT data

⑬ BLANK CARD: to end this data case

⑭ BLANK CARD: to show no more input data

Explained in the above is only a very basic component in the EMTP. There are many other components and functions as described in the previous sections. For example, a user-defined source is prepared as a SOURCE data (ITYPE $= 1$ to 10). When a field test result is to be simulated, an applied voltage (or current) in the test is used as a SOURCE data by transferring the measured voltage waveform to a PC in an electronic format.

$ VINTAGE, 1

1 2	3 8	9 14	15 20	21 26	27 42	43 58	59 74	75 79	80
ITY PE	node name	node name	reference branch		branch parameters				IOUT
	NODE_1	NODE_2	node name	node name	$R\,[\Omega]$	$L\,[\text{mH}]$	$C\,[\mu\text{F}]$		
I2	A6	A6	A6	A6	E16.0	E16.0	E16.0		I1

$ VINTAGE, 0

Figure 1.18   A high precision format of a BRANCH data

The basic data format of a BRANCH data shown in Figures 1.13 and 1.14 is very tight. To avoid this, a high precession format of a BRANCH data is prepared as shown in Figure 1.18. To use this format, "$VINTAGE, 1" data is to be input before the BRANCH data, and "$VINTAGE, 0" comes after the data to declare the end of the BRABCH data.

More details of the input format and the output can be found in [3]. Also, a number of test cases are described in [33].

1.2   Numerical electromagnetic analysis

It is hard to handle a transient associated with non-TEM mode propagation by conventional circuit theory-based tools such as the EMTP, because the tools are based on TEM mode propagation. To overcome the problem, a numerical electromagnetic analysis (NEA) method looks most promised among existing transient analysis approaches for it solves Maxwell's equation directly without assumptions often made for the circuit theory-based approach.

This section describes the basic theory of two representative methods, i.e., method of moment (MoM) and finite-difference time-domain (FDTD) method, of the NEA.

1.2.1   Introduction

An NEA method is becoming a powerful tool to solve surge phenomena in power systems such as an induced lightning surge, a surge response of a grounding electrode and corona pulse propagation on a gas-insulated bus which are very hard to solve by an existing circuit theory-based tool such as the EMTP, because these surge phenomena involve transverse magnetic (TM) and transverse electric (TE) modes of wave propagation rather than TEM mode propagation which is the basis of the existing circuit theory-based tool. Also the circuit parameters, which are necessary to analyze a phenomenon by the circuit theory-based tool, are quite often not clear or not available. Because the NEA method solves Maxwell's equation directly based on the boundary conditions and the physical and geometrical parameters of a surge phenomenon to be solved, it requires no circuit parameter. Also, the method can solve the TE, TM, and TEM mode propagation without any difficulties.

The IEE Japan Working Group of Numerical Electromagnetic Analysis Method for Surge phenomena (WG here after, convenor: Ametani) started in April 2004 [34] and CIGRE WG C4.501 (convenor: Ametani) started in October 2009 [35] has surveyed the basic theory and assumption of various approaches of the NEA published in literatures, and has collected the application examples. Furthermore, bench-mark tests are carried out by the WG members and finally the WG has made clear the applicable limit, advantage, and disadvantage of each approach.

This section summarizes the works related to NEA including the WG activities.

1.2.2 Maxwell's equations

Before NEA methods are explained, Maxwell's equations, which are the basis of electromagnetics, are described. There are two types of the equations: one is in time domain and the other is in frequency domain.

In the time domain, Maxwell's equations are expressed as

$$\nabla \times \boldsymbol{E}(\boldsymbol{r},t) = -\frac{\partial \boldsymbol{B}(\boldsymbol{r},t)}{\partial t} \tag{1.19}$$

$$\nabla \times \boldsymbol{H}(\boldsymbol{r},t) = \frac{\partial \boldsymbol{D}(\boldsymbol{r},t)}{\partial t} + \boldsymbol{J}(\boldsymbol{r},t) \tag{1.20}$$

$$\nabla \cdot \boldsymbol{D}(\boldsymbol{r},t) = \rho(\boldsymbol{r},t) \tag{1.21}$$

$$\nabla \cdot \boldsymbol{B}(\boldsymbol{r},t) = 0 \tag{1.22}$$

where $\boldsymbol{E}(\boldsymbol{r},\,t)$: electric field, $\boldsymbol{H}(\boldsymbol{r},\,t)$: magnetic field, $\boldsymbol{D}(\boldsymbol{r},\,t)$: electric flux density, $\boldsymbol{B}(\boldsymbol{r},\,t)$: magnetic flux density, $\boldsymbol{J}(\boldsymbol{r},\,t)$: conduction current density, $\rho(\boldsymbol{r},\,t)$: volume charge density, each at the space point $\boldsymbol{r}$ and at the time t.

Equation (1.19) represents Faraday's law, and (1.20) represents Ampere's law. Equations (1.21) and (1.22) represent Gauss's law for electric and magnetic fields, respectively.

In the frequency domain, Maxwell's equations are expressed as

$$\nabla \times \boldsymbol{E}(\boldsymbol{r}) = -j\omega \boldsymbol{B}(\boldsymbol{r}) \tag{1.23}$$

$$\nabla \times \boldsymbol{H}(\boldsymbol{r}) = \boldsymbol{J}(\boldsymbol{r}) + j\omega \boldsymbol{D}(\boldsymbol{r}) \tag{1.24}$$

$$\nabla \cdot \boldsymbol{D}(\boldsymbol{r}) = \rho(\boldsymbol{r}) \tag{1.25}$$

$$\nabla \cdot \boldsymbol{B}(\boldsymbol{r}) = 0 \tag{1.26}$$

where j: imaginary unit, ω: angular frequency.

The relations of the electric and magnetic flux densities to the electric and magnetic fields are given as

$$D(r) = \varepsilon E(r) = \varepsilon_r \varepsilon_0 E(r) \tag{1.27}$$

$$B(r) = \mu H(r) = \mu_r \mu_0 H(r) \tag{1.28}$$

where ε_0: permittivity of vacuum $= 8.854 \times 10^{-12}$ F/m,

μ_0: permeability of vacuum $= 4\pi \times 10^{-7}$ H/m, ε_r: relative permittivity of medium,

μ_r: relative permeability of medium, ε: permittivity of medium,

μ: permeability of medium.

In a conductive medium, the following relation of the electric field to the conduction current density, Ohm's law, is fulfilled:

$$J(r) = \sigma E(r) \tag{1.29}$$

where σ is the conductivity of the medium.

1.2.3   NEA method

1.2.3.1   Various methods

Table 1.2 categorizes various methods of NEA which are presently used.

The method of moments (MoM) in the frequency and time domains [36–41], and the FDTD method [42, 43], both for solving Maxwell's equations numerically, have frequently been used in calculating surges on power systems. Applications of the finite element method (FEM) and the transmission matrix method (TLM) to surge calculations have been rare at present. The MoM and the FDTD methods are, therefore, two representative approaches in surge calculations.

1.2.4   Method of Moments in the time and frequency domains

1.2.4.1   MoM in the time domain

The MoM in the time domain [36, 37] is widely used in analyzing responses of thin-wire metallic structures to external time-varying electromagnetic fields. The entire conducting structure representing the lightning channel is modeled by a combination of cylindrical wire segments whose radii are much smaller than the wavelengths of interest. The so-called electric-field integral equation for a perfectly conducting thin wire in air as in Figure 1.19, assuming that current I and charge q are confined to the wire axis (thin-wire approximation) and that the boundary condition on the tangential electric field on the surface of the wire (this field must be equal to zero) is fulfilled, is given by

$$\hat{s} \cdot E_{inc}(r,t) = \frac{\mu_0}{4\pi} \int_C \left[\begin{array}{c} \dfrac{\hat{s} \cdot \hat{s}'}{R} \dfrac{\partial I(s',t')}{\partial t'} \\[2mm] + c\dfrac{\hat{s} \cdot R}{R^2} \dfrac{\partial I(s',t')}{\partial s'} - c^2 \dfrac{\hat{s} \cdot R}{R^3} q(s',t') \end{array} \right] ds' \tag{1.30}$$

where $q(s',t') = -\int_{-\infty}^{t'} \frac{\partial I(s',\tau)}{\partial s'} d\tau$.

C is an integration path along the wire axis, E_{inc} denotes the incident electric field that induces current I, $R = r - r'$, r, and t denote the observation location (a point on the wire surface) and time, respectively, r' and t' denote the source location (a point on the wire axis) and time, respectively, s and s' denote the distance along the wire surface at r and that along the wire axis at r', $\hat{s}$, and $\hat{s}'$ denote unit

Table 1.2   Various methods of NEA

Partition	Space					Boundary
Discretization/ domain	Finite difference		Boundary length		Finite element	
Time-domain	FDTD	TD-FI	3D circuit	TLM	TD-FEM	MoM (TWTDA)
Frequency	–	FI	–	–	FEM	MoM
Base equation	Maxwell differential	Maxwell integral	Maxwell characteristic	D'Alembert solution		Field integral
			Circuit theory extension			
Feature	Easy programming	Multimedia	Hard program	Easy program	Wide application	Small CPU, nonlinear in time domain

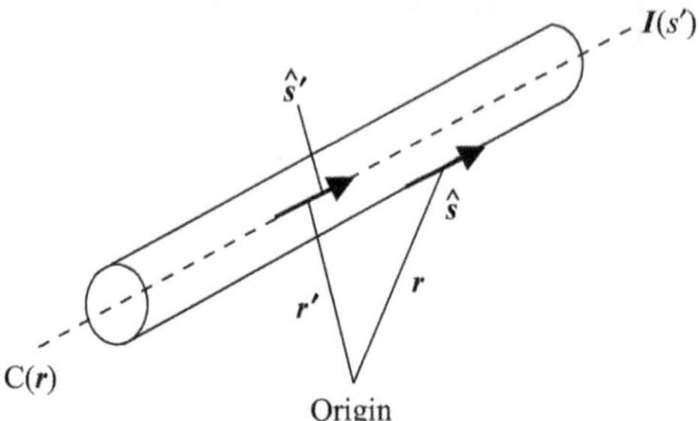

*Figure 1.19   Thin-wire segment for MoM-based calculations. Current is confined
to the wire axis, and the tangential electric field on the surface of the
wire is set to zero*

vectors tangent to path C in (1.19) at r and r', μ_0 is the permeability of vacuum, and
c is the speed of light. Through numerically solving (1.19), which is based on
Maxwell's equations, the time-dependent current distribution along the wire
structure (lightning channel), excited by a lumped source, is obtained.

The thin-wire time-domain (TWTD) code [36] (available from the Lawrence
Livermore National Laboratory) is based on the MoM in the time domain. One of
the advantages of the use of the time-domain MoM is that it can incorporate nonlinear effects such as the lightning attachment process [38], although it does not
allow lossy ground and wires buried in lossy ground to be incorporated.

1.2.4.2   MoM in the frequency domain

The MoM in the frequency domain [39] is widely used in analyzing the electromagnetic scattering by antennas and other metallic structures. In order to obtain the
time-varying responses, Fourier and inverse Fourier transforms are employed. The
electric-field integral equation derived for a perfectly conducting thin wire in air as
in Figure 1.19 in the frequency domain is given by

$$-\hat{s} \cdot E_{inc}(r) = \frac{j\eta}{4\pi k} \int_C I(s')\left(k^2\hat{s} \cdot \hat{s}' - \frac{\partial^2}{\partial s \partial s'}\right) g(r, r')ds' \tag{1.31}$$

where $g(r, r') = \exp\left(\frac{-jk|r-r'|}{|r-r'|}\right)$, $k = \omega\sqrt{\mu_0\varepsilon_0}$, $\eta = \sqrt{\frac{\mu_0}{\varepsilon_0}}$.

ω is the angular frequency, μ_0 is the permeability of vacuum, and ε_0 is
the permittivity of vacuum. Other quantities in (1.31) are the same as those in
(1.30). Current distribution along the lightning channel can be obtained numerically
solving (1.31).

This method allows lossy ground and wires in lossy ground (for example,
grounding of a tall strike object) to be incorporated into the model. The commer-
cially available numerical electromagnetic codes [40, 41] are based on the MoM in
the frequency domain.

1.2.5   Finite-difference time-domain method

The FDTD method [42] employs a simple way to discretize Maxwell's equations in differential form. In the Cartesian coordinate system, it requires discretization of the entire space of interest into small cubic or rectangular-parallelepiped cells. Cells for specifying or computing electric field (electric field cells) and magnetic field cells are placed relative to each other as shown in Figure 1.20. Electric and magnetic fields of the cells are calculated using the discretized Maxwell's equations given below.

$$E_z{}^n\left(i,j,k+\frac{1}{2}\right) = \frac{1 - \sigma(i,j,k+1/2)\Delta t/[2\varepsilon(i,j,k+1/2)]}{1 + \sigma(i,j,k+1/2)\Delta t/[2\varepsilon(i,j,k+1/2)]} \times E_z{}^{n-1}\left(i,j,k+\frac{1}{2}\right)$$

$$+ \frac{\Delta t/\varepsilon(i,j,k+1/2)}{1 + \sigma(i,j,k+1/2)\Delta t/[2\varepsilon(i,j,k+1/2)]}\frac{1}{\Delta x \Delta y}$$

$$\times \begin{bmatrix} H_y{}^{n-\frac{1}{2}}(i+1/2,j,k+1/2)\Delta y - H_y{}^{n-\frac{1}{2}}(i-1/2,j,k+1/2)\Delta y \\ -H_x{}^{n-\frac{1}{2}}(i,j+1/2,k+1/2)\Delta x + H_x{}^{n-\frac{1}{2}}(i,j-1/2,k+1/2)\Delta x \end{bmatrix}$$

$$(1.32)$$

$$H_x{}^{n+\frac{1}{2}}\left(i,j-\frac{1}{2},k+\frac{1}{2}\right) = H_x{}^{n-\frac{1}{2}}\left(i,j-\frac{1}{2},k+\frac{1}{2}\right) + \frac{\Delta t}{\mu(i,j-1/2,k+1/2)}\frac{1}{\Delta y \Delta z}$$

$$\times \begin{bmatrix} -E_z{}^n(i,j,k+1/2)\Delta z + E_z{}^n(i,j-1,k+1/2)\Delta z \\ +E_y{}^n(i,j-1/2,k+1)\Delta y - E_y{}^n(i,j-1/2,k)\Delta y \end{bmatrix}$$

$$(1.33)$$

Equation (1.32), which is based on Ampere's law, is an equation updating z component of electric field, $E_z(i, j, k+1/2)$, at point $x = i\Delta x$, $y = j\Delta y$, and $z = (k+1/2)\Delta z$, and at time $t = n\Delta t$. Equation (1.33), which is based on

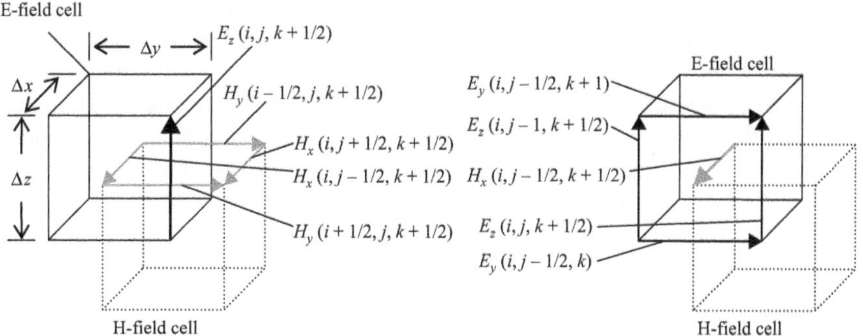

Figure 1.20   Placement of electric-field and magnetic-field cells for solving discretized Maxwell's equations using the FDTD method

Faraday's law, is an equation updating x component of magnetic field, $H_x(i, j - 1/2, k + 1/2)$, at point $x = i\Delta x$, $y = (j - 1/2)\Delta y$, and $z = (k + 1/2)\Delta z$, and at time $t = (n + 1/2)\Delta t$. Equations updating x and y components of electric field, and y and z components of magnetic field can be written in a similar manner. Note that $\sigma(i, j, k + 1/2)$ and $\varepsilon(i, j, k + 1/2)$ are the conductivity and permittivity at point $x = i\Delta x$, $y = j\Delta y$, and $z = (k + 1/2)\Delta z$, respectively, $\sigma\,(i, j - 1/2, k + 1/2)$ is the permeability at point $x = i\Delta x$, $y = (j - 1/2)\Delta y$, and $z = (k + 1/2)\Delta z$. By updating electric and magnetic fields at every point using (1.32) and (1.33), transient fields throughout the computational domain are obtained. Since the material constants of each cell can be specified individually, a complex inhomogeneous medium can be analyzed easily.

In order to analyze fields in unbounded space, an absorbing boundary condition has to be set on each plane which limits the space to be analyzed, so as to avoid reflections there. The FDTD method allows one to incorporate wires buried in lossy ground, such as strike-object grounding electrodes, and nonlinear effects [43, 44].

1.3 Conclusions

This chapter describes first a circuit theory-based approach, EMTP in particular. Since it is based on the assumption of TEM mode propagation, a phenomenon associated with non-TEM mode propagation cannot be dealt with. Thus, the necessity of an NEA has become clear. The chapter explains existing NEA methods and the basic theory. The methods are categorized by either in time domain or in frequency domain. The latter one necessitates frequency to time transform such as Fourier transform to obtain a transient response. Also, the methods are classified by a given medium being partitioned by space or by the boundary between media.

Phenomena involving scattering, radial wave propagation other than axial one, and non-TEM mode propagation cannot be solved by the circuit-theory approaches which are based on TEM mode propagation.

The NEA is very advantageous to solve a transient involving non-TEM mode propagation which is three-dimensional in general. It should be noted that the NEA is very useful to calculate the impedance and the admittance of a given circuit which are essential in a transient simulation by the circuit-theory approach, but often are not available. This suggests adopting the NEA as a subroutine for calculating input data of the circuit-theory approach, when the data are not available.

It should be pointed out that the NEA requires a large amount of computer resources, i.e., memory and CPU time. Also, existing codes are not general enough to deal with various types of transients especially in a large network.

References

[1] Dommel H. W. "Digital computer solution of electromagnetic transients in a single- and multiphase networks." *IEEE Trans. Power App. Syst.* 1969; **PAS-88** (4): 388–98.

[2] Dommel H. W., Scott-Meyer W. S. "Computation of electromagnetic transients." *Proc. IEEE*. Jul 1974; **62**: 983–93.

[3] Scott-Meyer W. *EMTP Rule Book*. Portland, Oregon: Bonneville Power Administration (BPA); Apr 1982.

[4] Dommel H. W. *EMTP Theory Book*. Portland, Oregon: Bonneville Power Administration (BPA); 1986.

[5] Tinney W. F., Walker J. W. "Direct solution of sparse network equations by optimally ordered triangular factorization." *Proc. IEEE*. Nov 1967; **55**: 1801–9.

[6] Tinney W. F., Meyer W. S. "Solution of large sparse systems by ordered triangular factorization." *IEEE Trans. Auto. Control*. Aug 1973; **AC-18**: 333–46.

[7] Schnyder O. "Druckstosse in Pumpensteigleitungen." *Schweiz Bauztg*. 1929; **94** (22): 271–86.

[8] Bergeron L. "Etude das variations de regime dans les conduits d'eau: Solution graphique generale." *Rev. Generale de L'hydraulique*. 1935; **1**: 12–69.

[9] Parmakian J. *Waterhammer Analysis:* New York: Dover; 1963.

[10] Frey W., Althammer P. "The calculation of electromagnetic transients on lines by means of a digital computer." *Brown Boveri Rev*. 1961; **48** (5/6): 344–55.

[11] Meyer W. S., Dommel H. W. "Numerical modeling of frequency-dependent transmission parameters in an electromagnetic transients program." *IEEE Trans. Power App. Syst*. 1974; **PAS-93**: 1401–9.

[12] Semlyen A., Dabuleanu A. "Fast and accurate switching transient calculations on transmission lines with ground return using recursive convolution." *IEEE Trans. Power App. Syst*. 1975; **PAS-94**: 561–71.

[13] Ametani A. "A highly efficient method for calculating transmission line transients." *IEEE Trans. Power App. Syst*. 1976; **PAS-95**; 1545–51.

[14] Marti J. R. "Accurate modeling of frequency-dependent transmission lines in electromagnetic transient simulations." *IEEE Trans*. 1982; **PAS-101** (1): 147–55.

[15] Dube L., Dommel H. W. "Simulation of control systems in an electromagnetic transients program with TACS." *IEEE PES PICA Conference Record*. 1977; **10**: 266–71.

[16] Brandwajn V. *Synchronous Generator Models for the Simulation of Electromagnetic Transients*. Ph.D. Thesis, University of British Columbia, Canada, Apr 1977.

[17] Carroll D. P., Flugan R. W., Kalb J. W., Peterson H. A. "A dynamic surge arrester model for use in power system transient studies." *IEEE Trans. Power App. Syst*. 1972; **PAS-91**: 1057–67.

[18] Phadke A. G. (ed.). *Digital Simulation of Electrical Transient Phenomena*. IEEE Tutorial Course, Course Text 81, EH0173-5-PWR. Piscataway, NJ: IEEE Service Center; 1980.

[19] Avila-Rosales J., Alvarado F. L. "Nonlinear frequency-dependent transformer model for electromagnetic transient studies in power systems." *IEEE Trans. Power App. Syst*. 1982; **PAS-101**: 4281–8.

[20]   Frame J. G., Mohan N., Liu. T. "Hysterisis modeling in an electromagnetic transients program." *IEEE Trans. Power App. Syst.* 1982; **PAS-101**: 3403–12.

[21]   Lauw H. K., Meyer W. S. "Universal machine modeling for the presentation of rotating electric machinery in an electromagnetic transients program." *IEEE Trans. Power App. Syst.* 1982; **PAS-101**: 1342–51.

[22]   Funabashi T., Sugimoto T., Ueda T. "A study on frequency characteristics of a generator model for transformer transfer voltage simulation." *Int. Workshop of High-Voltage Engineering (IWHV 99)*, paper HV-99-46, 1999.

[23]   Shulz R. P., Jones W. D., Ewart D. N. "Dynamic models of turbine generators derived from solid rotor equipment circuits." *IEEE Trans. Power App. Syst.* 1973; **PAS-92**: 926–33.

[24]   Cross G., Hall M. C. "Synchronous machine and torsional dynamics simulation in the computation of electromagnetic transients." *IEEE Trans. Power App. Syst.* 1978; **PAS-97**: 1074–86.

[25]   IEEE Subsynchronous Resonance Task Force. "First benchmark model for computer simulation of subsynchronous resonance." *IEEE Trans. Power App. Syst.* 1999; **PAS-96** (5): 1565–72.

[26]   IEEE Working Group Report. "Recommended phasor-diagram for synchronous machines." *IEEE Trans. Power App. Syst.* 1969; **PAS-88**: 1593–610.

[27]   Kind D. "The equal area criterion for impulse voltage stress of practical electrode configurations in air." *ETZ-A*. 1958; **79**: 65–9.

[28]   Tominaga S., Azumi K., Shibuya Y., Imataki M., Fujiwara Y., Nishikawa S. "Protective performance of metal-oxide surge arrester based on the dynamic V-I characteristics." *IEEE Trans. Power App. Syst.* 1979; **PAS-98**: 1860–71.

[29]   Durbak D. W. "Zinc-oxide arrester model for first surges." *EMTP Newsletter*. 1985; **5** (1): 1–9.

[30]   Ren-Ming M. "The challenge of better EMTP-TACS variable ordering." *EMTP Newsletter*. 1984; **4** (4): 1–6.

[31]   Lima J. A. "Numerical instability due to EMTP-TACS inter-relation." *EMTP Newsletter*. 1985; **5** (1): 21–3.

[32]   Dube L., Bonfanti I. "MODELS: a new simulation tool in the EMTP." *ETEP (European Trans. on Electrical Power Engineering)*. 1992; **2** (1): 45–50.

[33]   Ametani A, Nagaoka N. "EMTP Test Cases." Japanese EMTP Committee, Jun 1987.

[34]   Working Group of IEE Japan (chair A. Ametani). "Numerical electromagnetic analysis method for surge phenomenon." IEE Japan; Tokyo, Apr 2004 to Sep 2007.

[35]   CIGRE WG C4.501 (convenor A. Ametani). "Guideline for Numerical Electromagnetic Analysis Method and its Application to Surge Phenomena." CIGRE Technical Brochure 543, Jun 2013.

[36]   Van Baricum M., Miller E. K. *TWTD—A Computer Program for Time-Domain Analysis of Thin-Wire Structures*. UCRL-51-277, Lawrence Livermore Laboratory, California, 1972.

[37] Miller E. K., Poggio A. J., Burke G. J. "An integro-differential equation technique for the time-domain analysis of thin wire structures." *J. Computational Phys.* 1973; **12**: 24–48.

[38] Podgorski A. S., Landt J. A. "Three dimensional time domain modelling of lightning." *IEEE Trans. Power Delivery.* Jul 1987; **PWRD-2** (3): 931–8.

[39] Harrington R. F. *Field Computation by Moment Methods.* New York: Macmillan Co.; 1968.

[40] Burke G. J., Poggio A. J. "Numerical Electromagnetic Code (NEC)—Method of Moments." *Technical Document 116.* 1980; San Diego: Naval Ocean Systems Center.

[41] Burke G. J. *Numerical Electromagnetic Code (NEC-4)—Method of Moments.* UCRL-MA-109338, Lawrence Livermore National Laboratory, California, 1992.

[42] Yee K. S. "Numerical solution of initial boundary value problems involving Maxwell's equations in isotropic media." *IEEE Trans. Antennas Propagat.* Mar 1966; **AP-14** (3): 302–7.

[43] Tanabe K. "Novel method for analyzing dynamic behavior of grounding systems based on the finite-difference time-domain method." *IEEE Power Engineering Review.* 2001; **21** (9): 55–7.

[44] Noda T., Tetematsu A., Yokoyama S. "Improvements of an FDTD-based surge simulation code and its application to the lightning overvoltage calculation of a transmission tower." IPST05-138-24c, Montreal, Canada, Jun 2005.

Chapter 2
EMTP-ATP

M. Kizilcay[*] *and H.K. Hoidalen*[†]

2.1 Introduction

EMTP-ATP (ElectroMagnetic Transients Program, version Alternative Transients Program) is a non-commercial digital simulation program for electromagnetic transients particularly in electrical power systems based on the royalty-free development of EMTP by Bonneville Power Administration (BPA) in Portland, Oregon, USA [1–3]. BPA was an agency of the U.S. Department of Energy. The roots of EMTP-ATP can be traced to early 1984, when it became apparent to BPA's EMTP developers that Development Coordination Group (DCG) was not working as it was supposed to, and formed a threat to free EMTP. DCG for EMTP was founded by a coordination agreement in 1982, originally with the intention to keep the EMTP proper in public domain. At that point 12 years of 'EMTP Memoranda' were ended by Dr. Scott W. Meyer, and every available hour of his free time was switched from BPA's EMTP to the creation of a viable alternative that would be denied to those having commercial ambitions. ATP was the result during the fall of 1984. This followed Dr. Meyer's return from Europe, where the first European EMTP short course, in Leuven, Belgium, took place, and purchase of his first home computer, the new IBM PC AT, during August of 1984. The first annual meeting of Leuven EMTP Center (LEC) was held in Leuven on November 4, 1985, and Dr. Meyer agreed to bring the EMTP-ATP.

LEC as an institution of Katholieke University (K.U.) Leuven operated through 1993 without any formal connection to any power organization of North America. However, LEC did cooperate with BPA on an informal basis for several years and it remained willing to consider the establishment of a formal tie provided non-commercial EMTP activity at K.U. Leuven could thereby be assisted. In 1994 the European EMTP User Group was reformed as a non-profit scientific association 'European EMTP-ATP Users Group (EEUG)' according to German law in Germany by Prof. Mustafa Kizilcay. EEUG has been operating successfully for 20 years in

[*]University of Siegen, Germany
[†]Norwegian University of Science and Technology, Norway

close cooperation with the program developers and Can/Am EMTP User Group in North America.

In order to spread EMTP-ATP efficiently all over the world and help new users, national and regional EMTP user groups have been established in all continents. EEUG Association has been organizing since 1994 annual conferences and courses in Europe, and giving services besides its members to all ATP users by operating a mailing list and a web server for the distribution and maintenance of programs and documents. Any information about EMTP-ATP can be accessed or a contact can be established via websites http://www.emtp.org and http://www.eeug. org. EMTP-ATP licensing conditions are explained in detail at the official website http://www.emtp.org. In general, EMTP-ATP software is royalty-free, but it is not freeware.

2.2   Capabilities

2.2.1   Overview

EMTP-ATP consists, besides the actual time-domain simulation module, of several program modules, the so-called supporting programs that have been developed in the past four decades. ATP is capable to perform not only time-domain computations but also computations in the frequency-domain based on its steady-state phasor solution method. A schematic overview of the simulation modules and supporting programs is shown in Figure 2.1.

The supporting modules shown in Figure 2.1 have been embedded in EMTP-ATP program to generate model data of various power system components for electromagnetic computations in the simulation part. Increasingly, newer supporting modules have been embedded in ATPDraw – graphical pre-processor to EMTP-ATP (refer to section 2.5) – like the transformer model 'hybrid model'. Besides the simulation core for the computation of electromagnetic transients in an electrical network, there exist simulation modules TACS and MODELS, both developed by the same author [4–6]. In TACS (acronym for Transient Analysis of Control Systems), any control system can be built as a block diagram by connecting various devices. MODELS is a general-purpose description and simulation language for the representation of user-defined components and time-variant systems. Both TACS and MODELS can be executed as stand-alone modules.

2.2.2   Built-in electrical components

Due to continuing development of EMTP-ATP for more than five decades the extent of built-in components is huge. Starting with simple passive linear and non-linear branches, switches, static sources up to sophisticated models of components like transmission lines, rotating machines, and power transformers are available to model power grids, and also special electrical circuits to study certain transient phenomenon. Table 2.1 gives an overview of built-in components in EMTP-ATP.

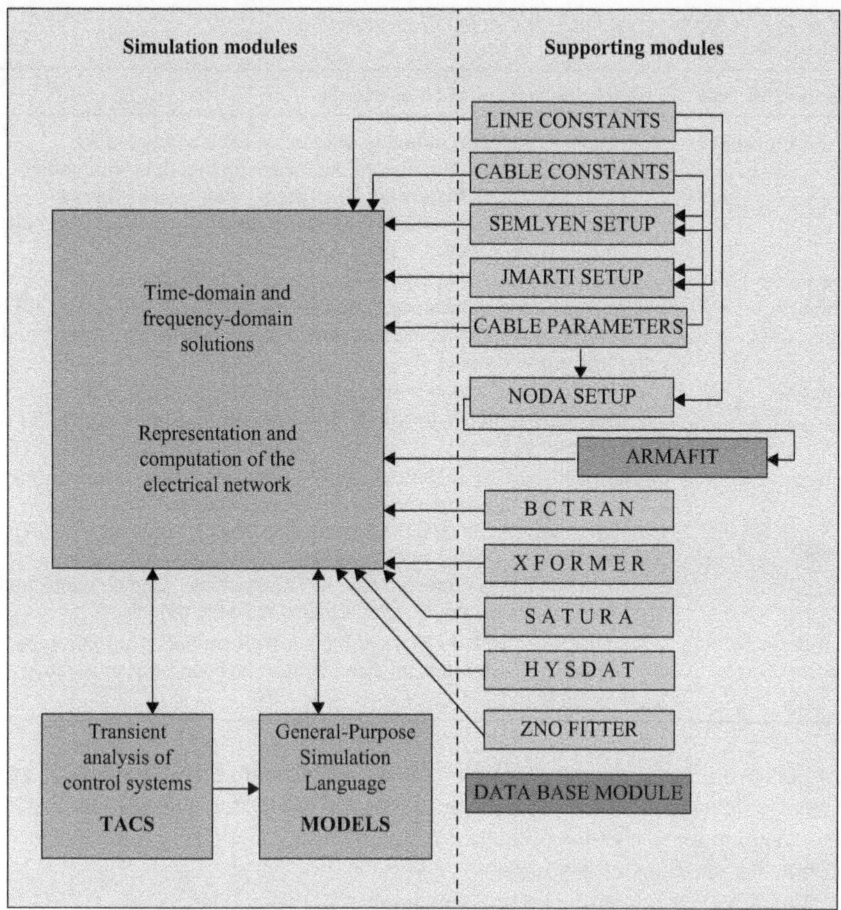

Figure 2.1   Simulation and supporting modules of EMTP-ATP

2.2.3   Embedded simulation modules TACS and MODELS

MODELS embedded in EMTP-ATP is a general-purpose description language supported by a large set of simulation tools for the representation and study of time-variant systems. The main features of MODELS can be summarized as follows:

- The description of each model is enabled using free-format, keyword-driven syntax of local context, which is largely self-documenting.
- MODELS allows the description of arbitrary user-defined control components and of user's own electrical component, providing a simple interface for connecting other programs/models to ATP.
- As a general-purpose programmable tool, MODELS can be used for signal processing either in time or frequency domain.

Table 2.1   Overview of built-in electrical components

Component type	Short description of components
Linear branches	Series RLC element, mutually coupled π-circuit, coupled RL elements, various distributed parameter line models with/without consideration of frequency dependency, various transformer models, high-order rational admittance function, cascaded π-circuits, and $[Y]$ matrix input for phasor solution
Non-linear branches	Pseudo-non-linear resistance and inductance, true non-linear resistance and inductance, time-varying resistance, TACS/MODELS controlled resistance, user-supplied Fortran non-linear element, and surge arrester model
Switches	Time-controlled switch, voltage-controlled switch, TACS/MODELS controlled switches for diode, and thyristor and transistor (IGBT) applications
Sources	Empirical sources, analytical sources (voltage/current) with various waveforms like step, ramp, cosine, and exponential type surge functions, TACS/MODELS controlled sources, rotating machine models with voltage and speed control like universal machine model enabling representation of DC machines, asynchronous and synchronous machines, synchronous machine models
User-definable components	MODELS-controlled electrical branch with variants Thevenin type, iterated type, non-transmission Norton type, and transmission Norton type

TACS is a simulation module for time-domain analysis of control systems. It was originally developed for the simulation of HVDC converter controls. In TACS, block diagram representation of control systems is used. It contains various devices and transfer functions of any order.

TACS/MODLES can be used, for example, for the simulation of

- HVDC converter controls;
- excitation systems of synchronous machines;
- power electronics and drives;
- electric arcs (circuit breaker and fault arc).

Interface between electrical network and TACS/MODELS is established by exchange of signals like node voltage, switch current, switch status, time-varying resistance, voltage and current sources. A special interface between MODELS and electrical circuit allows modelling of user's own electrical component as explained in section 2.4.3.

2.2.4   Supporting modules

LINE CONSTANTS is a supporting routine to compute electrical parameters of overhead lines in frequency-domain like per length impedance and capacitance matrices, π-equivalent, and model data for constant-parameter distributed line (CPDL) branch. LINE CONSTANTS in ATP is internally called to generate frequency data for the line models SEMLYEN SETUP, JMARTI SETUP and NODA SETUP.

CABLE CONSTANTS and CABLE PARAMETERS are supporting routines to compute electrical parameters of power cables. CABLE PARAMETERS is newer than CABLE CONSTANTS and has additional features like handling of conductors of arbitrary shape, snaking of a cable system and distributed shunt admittance model. CABLE PARAMETERS is linked to the supporting routines SEMLYEN SETUP, JMARTI SETUP and NODA SETUP to generate frequency-dependent electrical parameters of cables.

SEMLYEN SETUP is a supporting routine to generate frequency-dependent model data for overhead lines and cables. Modal theory is used to represent unbalanced lines in time-domain. Modal propagation step response and surge admittance are approximated by second-order rational functions with real poles and zeros.

JMARTI SETUP generates high-order frequency-dependent model for overhead lines and cables. The fitting of modal propagation function and surge impedance is performed by asymptotic approximation of the magnitude by means of a rational function with real poles.

BCTRAN is an integrated supporting program in the EMTP-ATP that can be used to derive a linear $[R]$, $[\omega L]$ or $[A]$, $[R]$ matrix representation for a single- or three-phase transformer using data of the excitation test and short-circuit test at rated frequency. For three-phase transformers, both the shell-type (low homopolar reluctance) and the core-type (high homopolar reluctance) transformers can be handled by the routine.

XFORMER is used to derive a linear representation for single-phase, two- and three-winding transformers by means of RL coupled branches. BCTRAN should be preferred to XFORMER.

SATURA is a conversion routine to derive flux–current saturation curve from either rms voltage–current characteristic or current–incremental inductance characteristic. Flux-current saturation curve is used to model a non-linear inductance, e.g., for transformer modelling. ATPDraw has this feature integrated in the model 'saturable three-phase transformer'.

ZNO FITTER can be used to derive a true non-linear representation for a zinc-oxide surge arrester, starting from manufacturer's data. ZNO FITTER approximates manufacturer's data (voltage–current characteristic) by a series of exponential functions of type $i = p \cdot \left(\frac{v}{V_{ref}}\right)^{q}$.

DATA BASE MODULE allows the user to modularize network sections. Any module may contain several circuit elements. Some data, such as node names or numerical data, may have fixed values inside the module, whereas other data can be treated as parameters that are passed to the database module when the module is connected to the data case.

2.2.5   *Frequency-domain analysis*

EMTP-ATP performs optionally a steady-state phasor solution, provided sinusoidal sources are present in the electrical network with a negative TSTART parameter, in order to start a time-domain computation from steady state. Frequency-domain solution methods implemented in EMTP-ATP are based on the steady-state solution of the linear network. Nodal equations are written using complex phasor quantities for currents and node voltages.

The FREQUENCY SCAN (FS) feature of the EMTP-ATP performs repetition of steady-state phasor solutions, as the frequency of sinusoidal sources is automatically incremented between a beginning and an ending frequency. Rather than conventional time-response output, a frequency-response output of desired quantities like node voltage, branch current or driving-point impedance/admittance is obtained. When plotted, the time axis of conventional ATP simulations becomes the frequency axis, with the result being a frequency curve. Polar coordinates (magnitude and angle) and/or rectangular coordinates (real and imaginary parts) of the phasor solution variables are used for output purposes.

Typical applications of the FREQUENCY SCAN are

- analysis and identification of resonant frequencies of power networks and individual system components;
- computation of frequency response of driving-point network impedances or admittances seen from a busbar, e.g., positive-sequence or zero-sequence impedance;
- analysis of harmonics propagation in a power system using extended feature Harmonic Frequency Scan (HFS).

HARMONIC FREQUENCY SCAN is a companion to FS. Both FS and HFS perform a series of phasor solutions. FS solves the network for the specified sources, incrementing in each subsequent step the frequency of the sources but not their amplitudes. HFS, on the other hand, performs harmonic analysis by executing a string of phasor solutions determined by a list of sinusoidal sources entered by the user [7]. This procedure is the same as the procedure for harmonic analysis used by all commercial harmonic analysis software. The main advantage of this approach compared with the time-domain harmonic analysis is a reduction in runtime of ten times or more, and avoidance of accuracy problems with Fourier analysis.

Models developed for HFS analysis using ATP are

- frequency-dependent R-L-C elements;
- frequency-dependent load based on the CIGRE type C model [8];
- harmonic current/voltage sources with frequency-dependent amplitude and phase.

2.2.6   *Power flow option – FIX SOURCE*

A three-phase power flow option was added to the EMTP in 1983. It adjusts the magnitudes and angles of sinusoidal sources iteratively in a sequence of steady-state solution, until specified active and reactive power, or specified active power and voltage magnitude, or some other specified criteria, are achieved. This will create the initial conditions for the subsequent transient simulation. The performance of the ATP load flow method is sensitive to initial conditions and requires generally high number of iterations.

2.2.7   Typical power system studies

Electrical transients in power systems are manifold. Over the years through the experience in the operation of power systems, especially due to malfunction of the system, different kinds of transient phenomena have been identified and investigated. The following is the list of transient and steady-state phenomena that can be efficiently computed by EMTP-ATP:

- Lightning overvoltages and protection
- Switching transients and overvoltages
    - Energization and re-energization of lines and cables
    - Energization of transformers
    - Capacitor and reactor switching
    - Circuit breaker duty, current chopping, restriking phenomena
    - Load rejection
    - Ferroresonance
    - Motor startup
- Power electronics applications
- HVDC studies
- Harmonic distortion
- Faults
    - Fault clearing
    - Short-line faults
    - Shaft torsional oscillations
    - Transient stability
- Very fast transients in GIS (gas-insulated switchgear)

2.3   Solution methods

ATP uses a nodal description and the trapezoidal rule of integration [1, 3] for solving the differential circuit equations. Sparsity of the system matrix is utilized with node ordering and the solution is generally very fast. Voltage sources are basically handled as grounded but are available as ungrounded via ideal transformer usage. The solution method has some inherited weaknesses in case of abrupt changes in the circuit, typically when switching inductive currents. This is a fundamental problem that all users of ATP will come across and need to manage.

2.3.1   Switches

ATP models switches as ideal (zero or infinite impedance) and re-establishes the system description for each switching operation. Augmentation techniques or compensation

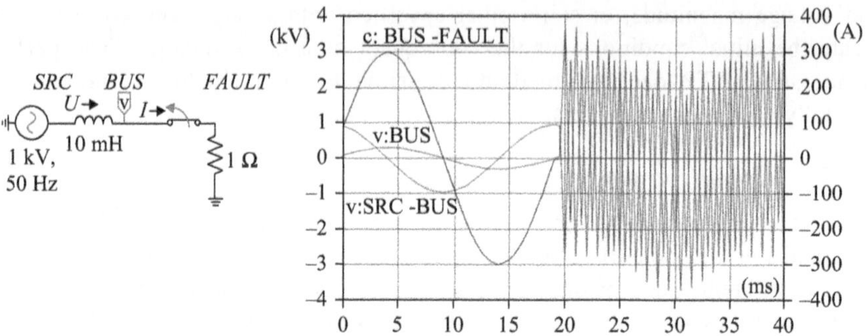

*Figure 2.2   Interruption of inductive current resulting in numerical oscillations;
time step h = 0.5 ms*

methods are not used. Switches are not allowed to be connected in parallel or even as delta as the current then becomes undefined. Switches have three different problems: trapezoidal rule oscillations, off-time-step switching and dependent switching.

The trapezoidal rule of integration for an inductance $(v = L \cdot \frac{\partial i}{\partial t})$ gives $v_k = \frac{2L}{h} \cdot (i_k - i_{k-1}) - v_{k-1}$ with h equal to the time step. With a sudden change in the inductor currents the voltage across the inductor will start to oscillate instead of going to zero. Figure 2.2 shows the simulation result of a simple interruption of an inductive fault current case.

Several ways to overcome this problem are as follows, but none of them are perfect (see Figure 2.3):

- The user can add a damping resistor across the inductor of a value typically $R_d = k \cdot 2L/h$ with $k = 5..10$ which will damp the transient but not change the circuit behaviour too much. ATPDraw has an option to add this automatically.
- The user can smooth out the plot by adding an 'AVERAGE OUTPUT' request card. This is a post-process solution with two averaged consecutive solution points. This will make the plot look better but will not fix the first time-step oscillation and the situation when the voltage is used in a more advanced circuit.
- The user can add an 'INTERPOLATE SWITCH ZERO CROSSING' request card. This will completely remove the oscillations, but the time samples after the interruption will be off the initial time-step grid.
- The user can add snubber circuits across switches to ensure a smooth transition.
- The user can of course also realize that the numerical oscillations really come from the unphysical and ideal switching and go ahead to model the arc voltage and stray capacitances.

Numerical oscillations can also be caused by a switching operation not occurring exactly at the time-step grid. Reducing the time step could in such cases help. The solutions outlined above will also address this situation.

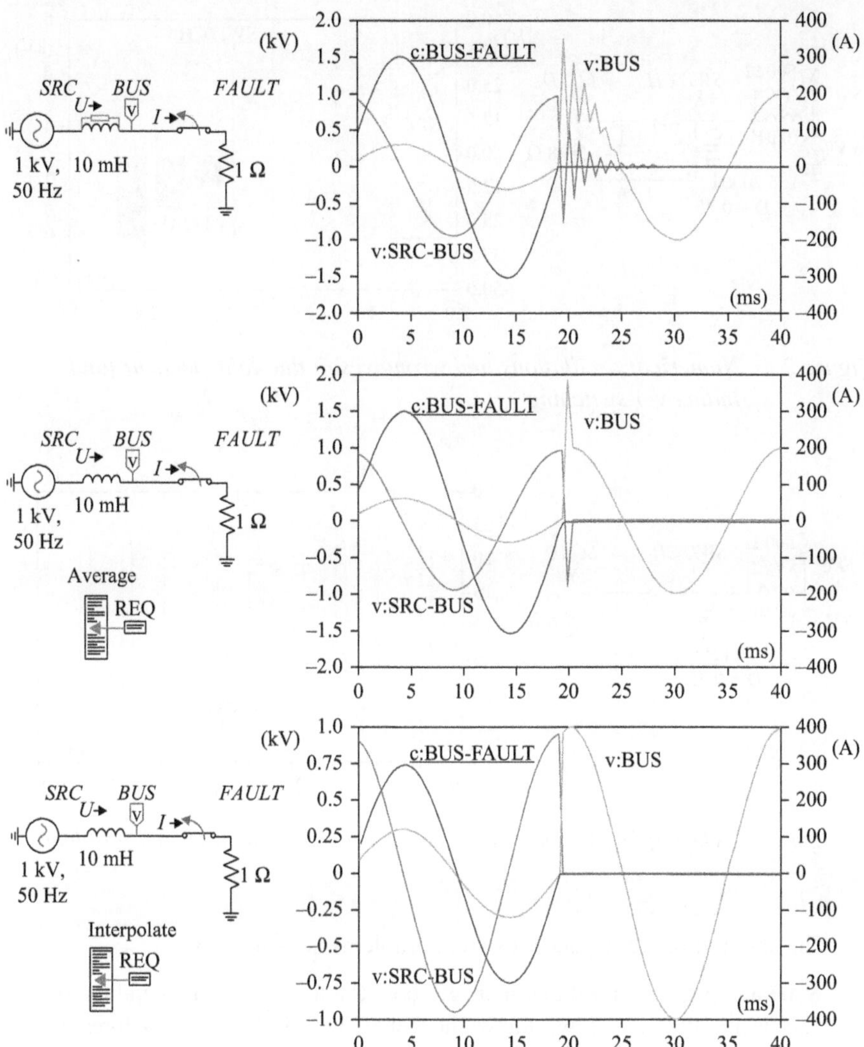

Figure 2.3   Solutions to remove inductive-switching numerical oscillations: damping resistor (top); output averaging (middle); interpolation (bottom)

Switches can in some circuits be dependent. This applies in particular to diodes which will change status simultaneously with other switching operations. However, the ATP program does not automatically know about this dependency and will figure out one time step later (when it is too late) that for instance a diode should have operated. The simple boost DC–DC converter in Figure 2.4 illustrates this. In this case the simulation is not just slightly troubled by some oscillations that can be damped or averaged. In fact none of the methods in Figure 2.3 works.

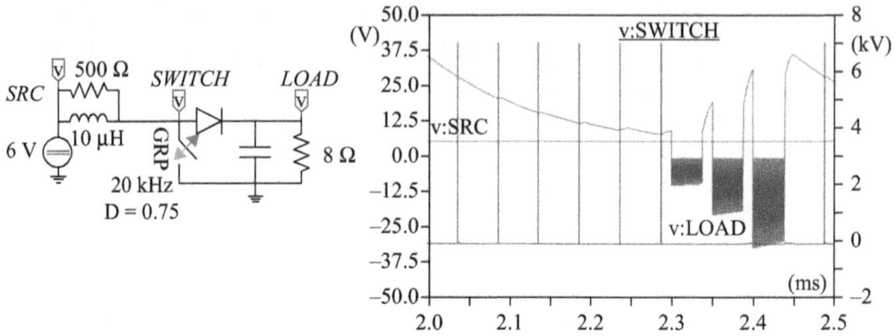

Figure 2.4   Numerical oscillations and wrong result due to dependent (and inductive) switching

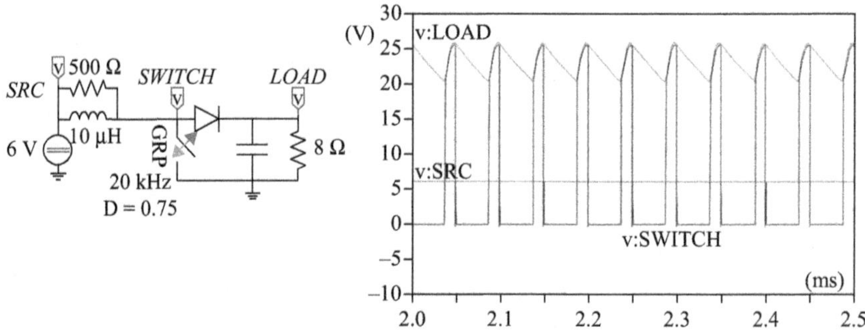

Figure 2.5   Using GIFU option to overcome the dependent switch problems

The solution to the dependent switch problems is as follows:

- Write a common control that manages both the involved switch and the diode. This is possible since we know the behaviour of the diode beforehand; it should conduct when the switching transistor opens and vice versa. This is the solution used in [9].
- ATP supports a feature called 'GIFU'. A TACS-controlled switch (thyristor, triac and igbt) can be specified as 'GIFU', and the diodes immediately following the switch are checked for change in status simultaneously with the 'GIFU' switch (actually the simulation goes one time step back to achieve this). The solution is illustrated in Figure 2.5.

The problem of dependent switches can also occur in a simple diode rectifier as shown in Figure 2.6. In addition to the inductive current interruption, opening of a diode should cause a diode in the other leg to start conducting. In this case there is no 'GIFU' option to help the user. The simulation result without any precautions becomes completely wrong, as shown in Figure 2.6.

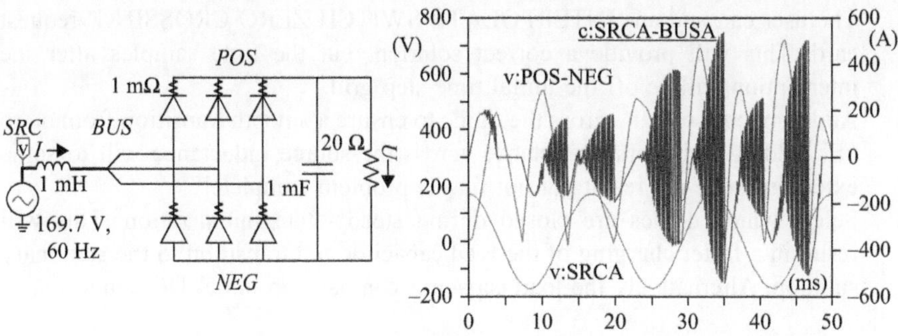

Figure 2.6   Diode rectifier modelled without any precautions, except forward resistor; time step h = 50 μs

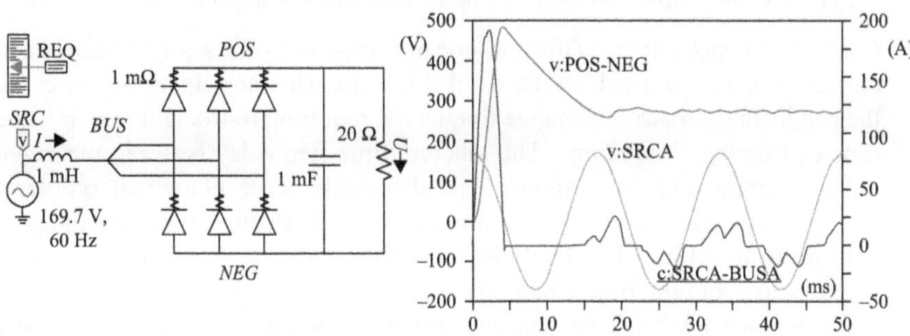

Figure 2.7   Diode rectifier with INTERPOLATE option

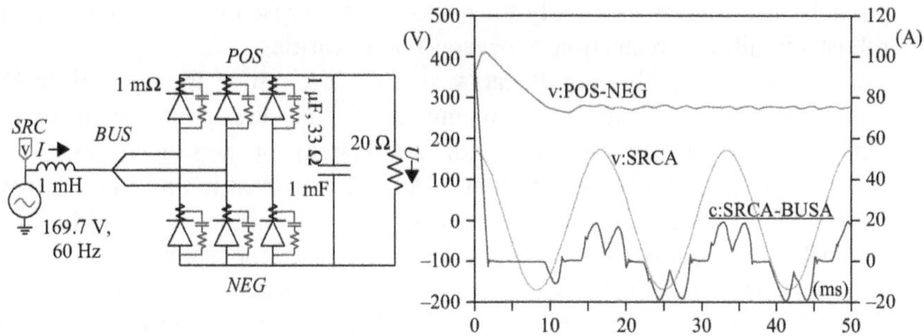

Figure 2.8   Diode rectifier with snubber circuits. Top-left and bottom-right diodes are closed in steady state

Several solution methods are needed, as illustrated in Figures 2.7 and 2.8:

- Add a forward resistance to the diode to avoid the problem of switches in parallel. Without this, ATP reports an error message 'Logic trouble within SUBROUTINE SWITCH of overlay 16'.

- The user can add an 'INTERPOLATE SWITCH ZERO CROSSING' request card. This will provide a correct solution, but the time samples after the interruption will be off the initial time-step grid.
- Add a snubber circuit across the diode to ensure a smooth transition from on to off. Adding a damping resistance across the source inductance will to some extent improve the result but not fix the problem completely.
- Select which diodes are closed during steady-state initialization. This will result in a faster charging of the load capacitor and transition to the stationary solution. Alternatively the load capacitor can have an initial DC condition.

2.3.2   Non-linearities

ATP supports several types of non-linearities both for resistances and inductances. All non-linearities are assumed to be linear during the steady-state initialization. Three fundamentally different types of non-linearities are as follows:

- *Controlled non-linearity*: This non-linearity involves values (of resistance and current source) specified by the control system. This non-linearity can cause the conductance matrix to change frequently, requiring re-triangularization and time-consuming simulations. The inherent time-step delay between measured voltage/current and the response can also cause some numerical problems and inaccuracies. Otherwise this approach is very flexible when it comes to non-linear modelling. The TYPE94 component combines the control and non-linearity to avoid the time delay.
- *Pseudo-non-linearity*: This non-linearity is supposed to operate on a linear segment modelled by a constant conductance and a history source until the segment boarder is violated. Then the conductance and current source are changed. This will result in a time-step delay between each segment shifts and possible inaccuracies especially for resistors. Otherwise this approach is quite robust and allows connection of several non-linearities.
- *True-non-linearity*: This non-linearity assumes the rest of the network to be linear and reduces this to Thevenin equivalent. This equivalent is then solved directly or iteratively together with the system of non-linearities. This approach is also called compensation in ATP. The current in this non-linearity is then used as a source in the final calculation step. This approach avoids time-step delays but requires the system to be solved twice (open circuit and with a current source). The final model as a current source also puts some topological constraints on the non-linearities; this model cannot be connected in series or delta, and even parallel connections can be a problem.

2.3.3   Transmission lines

Transmission line models in EMTP-ATP can be treated in two groups: lumped-element line models and distributed-parameter line models. Lumped-element line models – called *mutually coupled RLC elements* – are mainly nominal multiphase π-circuits as shown in Figure 2.9 for a three-phase line. Usually, the line parameters of an untransposed line, series impedance matrix and shunt admittance matrix per unit length are computed with either one of the two supporting modules LINE

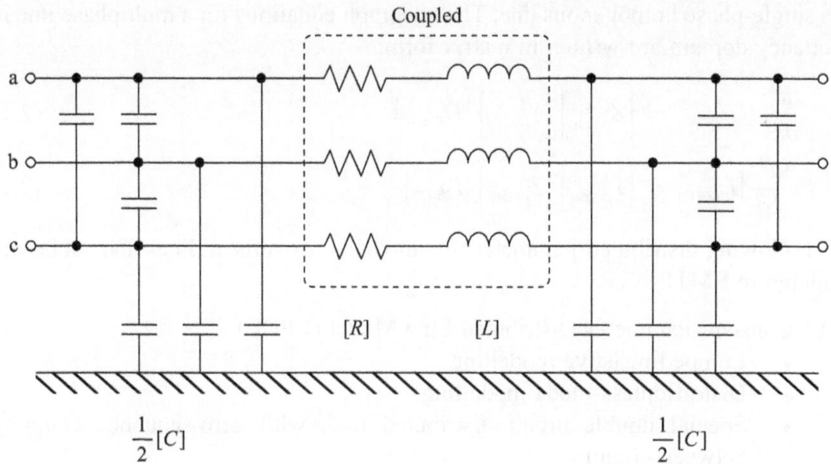

Figure 2.9   Three-phase nominal π-circuit representation of a line

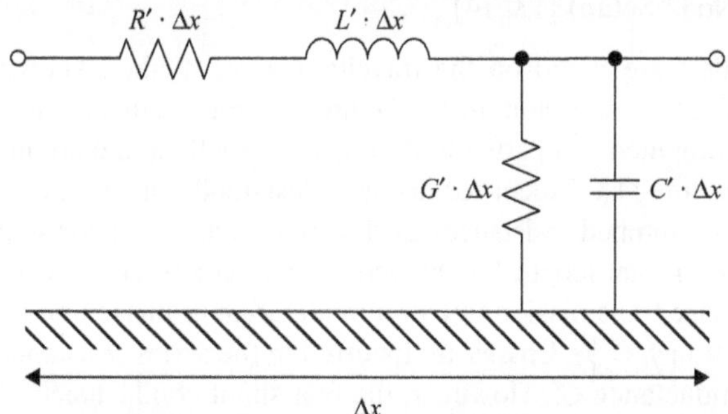

Figure 2.10   Infinitesimal section of a single-phase homogenous line

CONSTANTS and CABLE PARAMETERS (see section 2.2.4). In the case of assumption of perfectly balanced line, either overhead line or cable, the user can enter directly positive-sequence and zero-sequence values for the series impedance and shunt capacitance per unit length and the line length. Internally, the balanced line is represented by symmetrical $[R]$, $[L]$ and $[C]$ matrices.

Nominal π-circuits are generally not the best choice for transient solutions, specially travelling waves on lines cannot be reproduced accurately. Cascade connections of nominal π-circuits of short line sections may be useful for the computation of low-frequency transients or steady-state problems. For steady-state phasor solution, exact equivalent π-circuit model is also available in EMTP-ATP that considers well-known equations for long lines with the hyperbolic functions.

The distributed-parameter line models in EMTP-ATP are based on the differential equations, called *telegraph* equations, of a homogenous transmission line with regard to the length and time. Figure 2.10 shows an infinitesimal section of a single-phase homogenous line. The telegraph equations for a multiphase line in frequency domain are written in matrix form:

$$\frac{d^2}{dx^2}\left[U_{phase}\right] = \left[Z'_{phase}\right]\left[Y'_{phase}\right]\left[U_{phase}\right] \tag{2.1}$$

$$\frac{d^2}{dx^2}\left[I_{phase}\right] = \left[Y'_{phase}\right]\left[Z'_{phase}\right]\left[I_{phase}\right] \tag{2.2}$$

Following distributed-parameter line models for overhead lines and cables are available in EMTP-ATP:

(i) Constant-Parameter Distributed Line Model (CPDL)
  - Lumped-resistive modelling
  - Distortionless-mode modelling
  - Special double-circuit distributed line, with zero-sequence coupling between circuits
(ii) Second-Order, Recursive-Convolution Line Model (Semlyen Setup) [10]
(iii) Rigorous, Frequency-Dependent Line Model (JMarti Setup) [11, 12]
(iv) Frequency-Dependent ARMA (Auto-Regressive Moving-Average) Line Model (Noda Setup) [13, 14]

CPDL models are based on the travelling wave method known as Bergeron method for lossless transmission lines. The line parameters that are in fact frequency dependent are computed at a particular frequency, usually at the dominant resonance frequency of the line [1]. Except for distortionless-mode line model, the line losses are represented by lumped resistances as shown in Figure 2.11 for single-phase line. The lossless line of total length l is divided in two parts of equal length.

A single-phase transmission line or a mode of an m-phase line is distortionless under the condition $\frac{R'}{L'} = \frac{G'}{C'}$. Losses are incurred in the series resistance R' as well as in the shunt conductance G'. However, the real shunt conductance of an overhead line is very small (close to zero). If its value must be artificially increased to make the line distortionless, with a resulting increase in shunt losses, then it is best to compensate for that by reducing the series resistance losses. The EMTP does this automatically by regarding the input value R'_{input} as an indicator for the total losses and uses only half of it for R':

$$\frac{R'}{L'} = \frac{G'}{C'} = \frac{1}{2}\frac{R'_{input}}{L'}$$

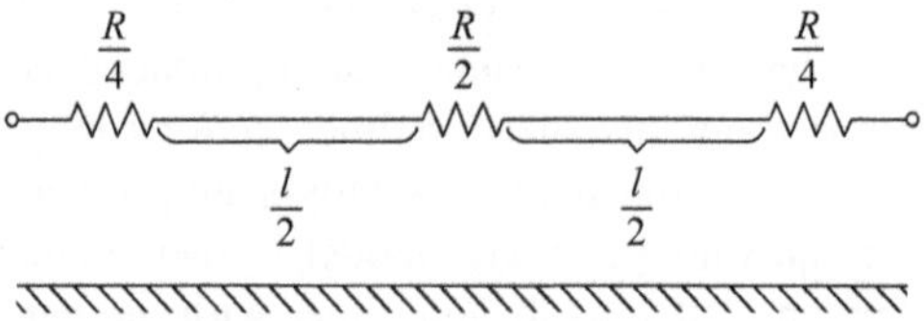

Figure 2.11   *Representation of line losses by lumped resistances*

With this formula the AC steady-state results are practically identical for the line being modelled as distortionless or with R lumped in three places (Figure 2.11); the transient response differs mainly in the initial rate of rise [1].

For an unbalanced multiphase transmission line the line equations are first transformed into modal domain, by means of which an m-phase transmission line can be treated like decoupled m single-phase lines. In general the transformation matrices for current and voltage are dependent on frequency, and the matrix elements are complex. Usually, the modal transformation matrices and modal parameters of the transmission line are computed at the dominant resonance frequency of the line for transient studies. The real part of the modal transmission matrix is to be used for time-domain computations. By means of eigenvector rotation the imaginary part of the elements of the transformation matrix is minimized.

The transmission line models (ii), (iii) and (iv) consider the frequency depen-dence of line parameters due to skin effect in the conductors and in the earth as return medium. The first two models (ii) and (iii) make use of modal components for the time-domain solution of line equations. The modal propagation function and modal characteristic impedance/admittance are computed in a sufficiently wide frequency range. In the frequency-dependent line model (ii), these functions are approximated by a second-order rational function with real poles, whereas line model (iii) uses higher-order rational function fitting with real poles of the modal propagation functions and the characteristic impedances by asymptotic approx-imation of amplitudes [12]. The time-domain computation of the transients is performed by numerical recursive convolutions. The drawback of the line models that utilize modal components for time-domain computations is that modal trans-formation matrix is assumed to be constant and real. This assumption may intro-duce inaccurate results and in some cases may lead to instability of the line model because of violation of the passivity conditions.

The line model (iv) attempted a new approach, a direct phase domain model that is based on a nodal-conductance representation [13, 14]. The time-domain model uses Interpolated Auto-Regressive Moving Average (IARMA) convolution method. The modelling of a transmission line (overhead lines and cables) using the NODA SETUP requires following two steps:

1. Calculation of the frequency-dependent line constants of the transmission line, hereafter referred to frequency data, using CABLE PARAMETERS or LINE CONSTANTS supporting routine in ATP
2. Fitting the frequency data with IARMA models for the time-domain realization of the frequency dependence using an independent program ARMAFIT

Even if all the IARMA of the line model are stable, it is still possible that the line model behaves like an active circuit, and thus as a numerically unstable model. To ensure the complete numerical stability, the real part of the admittance matrix of the line model has to be positive-definite. This criterion has not been incorporated in the optimization procedure.

2.3.4   *Electrical machines*

EMTP-ATP supports several machine models [3, 4]. Two main solution method categories are constant inductance through park transformation that requires estimation of next time–step states, and variable inductances in the phase domain that requires the machine to be solved as a non-linear component. All machine models in ATP support saturation.

The first model implemented was the synchronous machine (SM) model (type 59) [15, 16]. This is based on park transformation and usage of the compensation [15] and later prediction [16] method. Input to the model can either be manufacturer's data or direct electrical data. The SM is always automatically initialized based on the armature phase voltage and angle provided by the user. The control of this machine is carried out by specifying the per unit field voltage or mechanical power from the control system. Internal machine variables can be brought into the control system.

Later the universal machine (UM) module was implemented that generalized the machine equations also with park transformation [17]. The UM comes in 12 different variants: synchronous (1: three-phase armature, 2: two-phase armature); induction (3: three-phase armature, cage rotor, 4: three-phase armature, three-phase field, 5: two-phase armature, cage rotor); single-phase AC (6: one-phase field, 7: two-phase field); and DC (8: separate excitation, 9: series compound field, 10: series field, 11: parallel compound field, 12: self-excitation). Unlike the SM module which has an embedded mechanical system, the mechanical system of the UM module is external and connected as shown in Figure 2.12.

The UMs can be initialized either automatically like the SMs by providing terminal phase voltage and angle for SMs or slip for induction machines, or manually by providing internal winding currents. For automatic initialization the required mechanical torque (and field voltage) is calculated by ATP, and the torque (and field) source has to be of AC-type with a very low frequency, a negative Tstart, and connected to the machine. The solution method to handle the need for estimated values in the park transformation is user-selectable (compensation or prediction). Control of the UMs is by setting the absolute values of the mechanical torque and the field voltage. Internal machine variables can be brought into the control system.

Two more machine models support modelling directly in the phase domain and thus avoid the estimation required by the park transformation approach. These are the type-58 SM and the type-56 induction machine. The type-58 machine is transparent to the user, while the type-56 machine has no direct park transformed

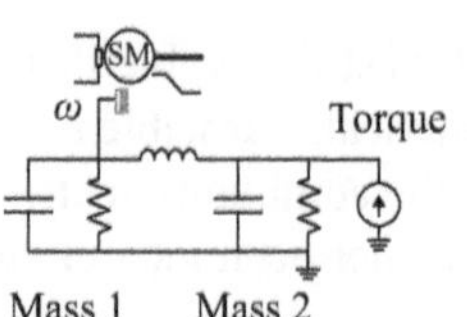

• Moment of inertia (km/m^2) ↔ Capacitance (F)
• Shaft section spring constant (Nm/rad) ↔ Inv. inductance (1/H)
• Shaft friction, viscous damp (Nm/rad/s) ↔ Conductance (1/ohm)
• Angular speed (rad/s) ↔ Voltage (Volt)
• Torque (Nm) ↔ Current (A)
• Angle (rad) ↔ Charge (Coulomb)

Figure 2.12   Universal machine and the external mechanical system

equivalent implemented. These machines are supposedly more stable for higher time steps. Both machines were developed by TEPCO.

2.4 Control systems

EMTP-ATP supports two different control systems TACS [6] and MODELS [5]. TACS consists of predefined blocks categorized into sources, transfer functions and devices as well as a Fortran statement. MODELS is a symbolic language with special features to handle time-domain signals, and this is what makes ATP very flexible and adaptable to various technologies. The circuit and the controls are solved separately resulting in a one time-step delay between the solutions. A special class of MODELS components called type-94 addresses the time delay problem.

2.4.1 TACS

Extra time delays due to non-linearities in the controls are reduced by optimal sorting of the control blocks. The categorization of control variables as internal, output or input is no longer required.

The sources in TACS can come from the circuit or from the other control system, MODELS. Circuit variables can be node voltage, switch current, internal machine variables and switch status. Special, internal sources in TACS can also be defined are supported. In addition Fortran statements can be used to define sources and relations between inputs and outputs. Table 2.2 shows the various sources.

Transfer function with maximum order 9 and optional with static and dynamic limiters are supported. ATPDraw also supports some simplified variants for convenience. Table 2.3 shows the various transfer functions.

Fortran statements are used to define variables with a Fortran-like syntax. Arguments in operations and functions can be other TACS functions. The sequence of the calculations is by the order of the blocks in the input file to ATP.

Algebraic operators:	$+, -, *, /, **$
Logical operators:	.OR., .AND., .NOT.
Relational operators:	.EQ., .NE., .LT., .LE., .GE., .GT.

Table 2.2   Sources in TACS

Source	Icon	Source	Icon
Circuit variable *Node voltage, Switch current,* *Machine variable, Switch status*	T	**MODELS variable** *MODELS output*	M
DC step – 11	T	**AC cosine – 14**	T
Pulse train – 23	T	**Ramp sawtooth – 24**	T

Table 2.3   Transfer functions in TACS

Transfer function	Icon	Transfer function	Icon
General *Order 0–9, Input 1–5,* *Fixed and named Limits*	Σ G(s)	**Order 1** *Order 0–1, Input 1, Dynamic icon* *Fixed and named Limits*	$5\dfrac{S}{1+10s}$ $\dfrac{1+s}{1+10s}$
Integral	$\dfrac{K}{s}$	**Derivative**	$K{\cdot}s$
Low pass	$\dfrac{K}{1+T{\cdot}s}$	**High pass**	$\dfrac{K}{1+T{\cdot}s}$

Table 2.4   Devices in TACS

Device	Icon	Device	Icon
Frequency sensor – 50	f 50	**Simple derivative – 59**	$G\dfrac{du}{dt}$ 59
Relay switch – 51	51	**Input IF – 60**	60
Level triggered switch – 52	52	**Signal selector – 61**	61
Transport delay – 53	53	**Sample & track – 62**	Samp Track 62
Pulse delay – 54	54	**Inst min/max – 63**	MIN MAX 63
Digitizer – 55	55	**Min/max track – 64**	MIN MAX 64
User def nonlin – 56	56	**Accumulator and counter – 65**	ACC 65
Multiple open/close switch – 57	57	**Rms meter – 66**	RMS 66
Controlled integrator – 58	$\int$ 58		

Fortran functions:	sin, cos, tan, cotan, sinh, cosh, tanh, asin, acos, atan (arguments in radians)	
Special functions:	tranc, minus, invers, rad, deg, sign, not, seq6, ran	

The devices are numbered from 50 to 66 and have, with a few exceptions, five inputs that are first summed with a user-defined polarity. Table 2.4 shows the devices.

2.4.2 MODELS

MODELS has a Fortran-like syntax and all variables (INPUT, OUTPUT, VAR) are assumed to be of double precision signals or array of signals with history values available via the DELAY function. Special functions like INTEGRAL, DERIV, LAPLACE etc. are available for converting the signals. MODELS also supports logic (IF-THEN-ELSE) and control (FOR-WHILE) statements and iteration (COMBINE). MODELS components are not present during steady-state initialization and must be initialized manually. As any other computer language, the syntax has to be mastered to take full advantage of the functionality as parser feedback is limited. The MODELS language is not compiled but interpreted and this slows down the simulation speed. The structure of a MODELS component looks as follows:

```
MODEL <name>
INPUT i1, i2[1..3]
OUTPUT o1[1..3], o2
VAR o1[1..3], o2, v1,v2,v3[1..4]
DELAY CELLS
  --specify required number of past values for variables. 100 default.
HISTORY
  --assign variables for t<0. Required for laplace, delay etc.
INIT
  --assign variables for t=0
ENDINIT
EXEC
  --execute every local time step, time T available
ENDEXEC
ENDMODEL
```

A MODELS component has also an interface to the circuit in ATP so that the INPUTs and OUTPUTs are assigned. This is done in two steps; first there is main interface for all MODELS components and then each model is interfaced in what is called a USE statement. The input to the MODELS can be of the following types: current, voltage, switch status, machine variable, tacs, Im(V), Im(I), Model, ATP.

2.4.3 User-definable component (type 94)

The type-94 component is a MODELS component that connects directly into the circuit description in a way similar to the non-linearities. The usual time-step delay is thus somewhat avoided. The component comes in several flavours:

- *Thevenin*: Similar to a true non-linear component, with the same restrictions. The input to the model from the circuit is its Thevenin equivalent (resistance matrix and voltage vector). The model must then solve this reduced system and return the branch current vector.
- *Norton*: Similar to a pseudo-non-linear component. The model receives the node voltage and returns a Norton equivalent (conductance matrix and current

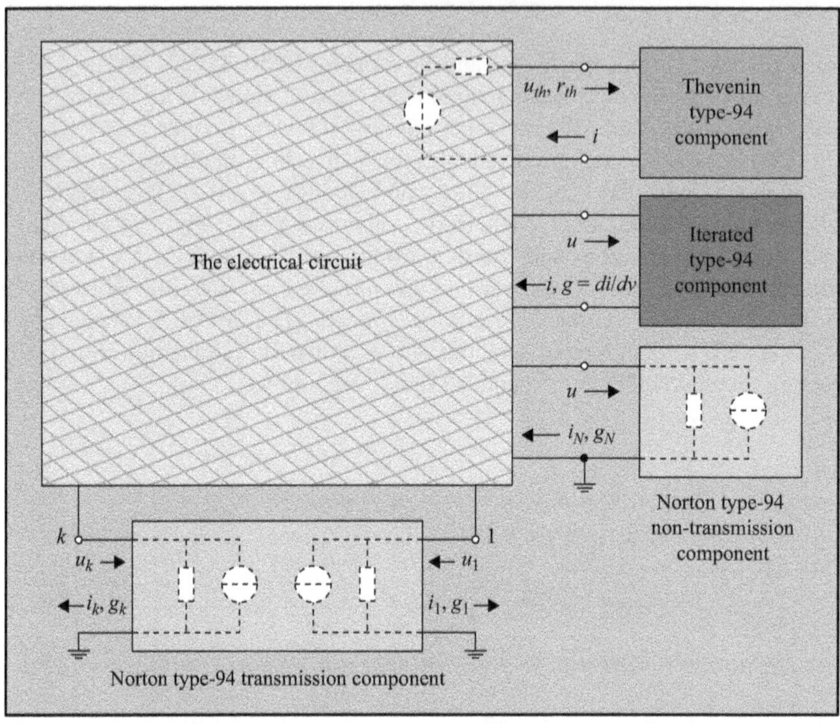

Figure 2.13   Various interfaces between type-94 component and the electrical circuit

source) to be solved together with the circuit. The model is here connected phase to ground.

- *Norton transmission*: The model receives the left and right node voltages and returns a Norton equivalent (conductance matrix and current source) for both the left and right sides to be solved together with the circuit. The model here has two terminals.
- *Iterated*: The model gets the node voltage and returns the current and the *di/dv* conductance matrix to be solved iteratively together with the circuit.

The interface between type-94 component and the electrical circuit is shown for these four variants in Figure 2.13.

2.5   Graphical preprocessor ATPDraw

ATPDraw is a graphical preprocessor to EMTP-ATP on the MS-Windows platform. The program is written in CodeGear Delphi 2007 (object-oriented Pascal) and runs under Windows 9x/NT/2000/XP/Vista/7/8. In ATPDraw the user can construct an electrical circuit using the mouse and selecting components from menus. ATPDraw generates the ATP input file in the appropriate format based on 'what you see is what you get'. ATP itself and plotting programs can be integrated with ATPDraw.

ATPDraw supports multiple documents that make it possible to work on several circuits simultaneously and copy information between them. All kinds of standard circuit editing facilities (copy/paste, grouping, rotate, export/import, undo/redo) are available. In addition, ATPDraw supports the Windows clipboard and metafile export. The circuit is stored on disk in a single project file, which includes all the simulation objects and options needed to run the case. The project file is in zip-compressed format that makes the file sharing with others very simple. ATPDraw supports hierarchical modelling by replacing selected group of objects with a single icon in an almost unlimited numbers of layers. Components have an individual icon in either bitmap or vector graphic style and an optional graphic background. ATPDraw supports circuit sizes up to 10,000 components each with maximum 64 data and 32 nodes.

Most of the standard components of ATP as well as TACS are supported, and in addition the user can create new objects based on MODELS or $Include (Data Base Module). Line/Cable modelling (KCLee, PI equivalent, Semlyen, JMarti and Noda) is also included in ATPDraw where the user specifies the geometry and material data and has the option to view the cross section graphically and verify the model in the frequency domain. Special components support the user in machine and transformer modelling based on the powerful UM and BCTRAN components in EMTP-ATP. In addition the advanced hybrid transformer model XFMR and Windsyn support are included.

ATPDraw is developed by Dr. Høidalen continuously since 1991. The initial motivation was for educational purposes. Initially various DOS versions were developed until the first Windows 3.1 version became available in 1997. Initially Bonneville Power Administration sponsored much of the development. Today ATPDraw covers most of the possibilities in ATP and is a huge program with more than 100,000 code lines. The latest year's development has focused on transformer modelling, electrical machine support and power frequency components, in addition to higher level optimization and variable solutions. ATPDraw has a copy installation and does not rely on external libraries. Even if ATPDraw supports most of the functionality of ATP, there are exceptions and users should not forget to study the ATP RuleBook [4]. Further information can be obtained from [18–21] and at the website http://www.atpdraw.net.

2.5.1   *Main functionality*

The user assembles the circuit by selecting components from the *Selection menu* (right-click in open space) or by copying them in from the 'All.acp' file or any other circuit. Components and a group of components can be selected and rotated in steps of 90 degrees. A *GridSnap* feature helps to fix components at 10 px intervals. Nodes are automatically connected if they overlap or if a *Connection* is drawn between nodes, and a Connection picks up intermediate nodes. Typically, the user stores the circuit in a project file .acp and continues to build and extend this in a process. Only the circuit itself with data is stored in the project and not any simulation results. Based on the graphical drawing and the given data, ATPDraw creates names of all nodes and writes the circuit specification to disk in a format readable by ATP.

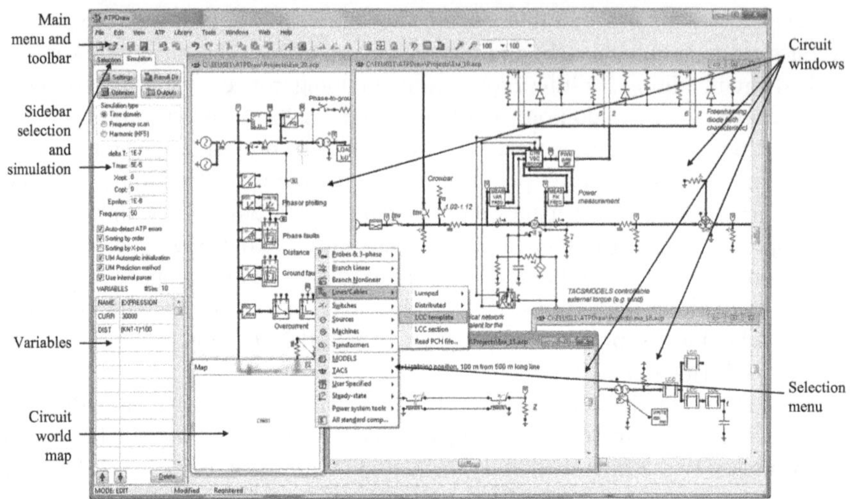

Figure 2.14   Main window of ATPDraw

ATPDraw will give warnings if a node has duplicate names or if the same name is given to non-connected nodes. Figure 2.14 shows the main windows of ATPDraw.

The tool bar at the top of the main window is customizable and by default contains menu buttons for the most frequently used menu options (Table 2.5).

ATPDraw supports multiple (up to 100) undo-redo steps in strict sequence. Auto-saving and backup of the circuit can be performed at a user-selectable interval. ATPDraw offers a large help file system in the compiled html help format (ATPDraw.chm). All default values of standard data of all components can be set by the user under the *Library* menu (ATPDraw.scl). This applies also to the icon. The icon can be in vector or bitmap graphic and have an accompanied bitmap (bmp, png) attached.

2.5.2   Input dialogues

Most of the components have the input dialogue shown in Figure 2.15 with data values given in the top-left grid. Node names are listed in the top-right grid but are typically given by clicking individual nodes of special interest.

2.5.3   Line and cable modelling – LCC module

ATPDraw supports direct input of electrical parameters of lumped RL and RLC (π-equivalent) lines and fixed frequency and distributed transposed and un-transposed lines (1, 2, 3, 6 phases).

In addition ATPDraw supports (LCC module) LINE CONSTANTS, CABLE CONSTANTS and CABLE PARAMETERS of ATP with the restriction of maximum 42 overhead line conductors and 16 cables. The maximum number of ungrounded conductors is 28. In this module geometrical and material data are

Table 2.5 Toolbar in ATPDraw with explanation

Item[a]	Menu	Shortcut	Description
New	File|New	–	Open an empty circuit file.
Open	File|Open	CTRL+O	Load a circuit file into a new window.
Save	File|Save	CTRL+S	Save the active circuit window to the current project file.
Save As	File|Save As	–	Save the active circuit window to a new project file.
Import	File|Import	–	Insert a stored circuit into the current circuit.
Export	File|Export	–	Export the selected circuit to an external project file.
Undo	Edit|Undo	CTRL+Z	Undo the previous operation.
Redo	Edit|Redo	CTRL+Y	Redo the previous undo operation.
Cut	Edit|Cut	CTRL+X	Copy the current selected circuit to the clipboard and then delete it.
Copy	Edit|Copy	CTRL+C	Copy the current selected circuit to the clipboard.
Paste	Edit|Paste	CTRL+V	Paste the ATPDraw-content from the clipboard into the circuit.
Duplicate	Edit|Duplicate	CTRL+D	Copy+Paste.
Edit text	Edit|Edit text	CTRL+T	Enter Edit text mode for adding and selecting text.
Select all	Edit|Select|All	CTRL+A	Select the entire circuit.
Rotate R	Edit|Rotate-R	CTRL+R	Rotate 90 deg. clock-wise.
Rotate L	Edit|Rotate-L	CTRL+L	Rotate 90 deg. counter clock-wise.
Flip	Edit|Flip	CTRL+F	Flip left-to-right. The nodes change postion. Vector text is not flipped.
Refresh	View|Refresh	CTRL+Q	Redraw circuit.
Centre	View|Centre circuit	–	Draw circuit centred in window.
Lock	View|Lock circuit	–	Turn on 'child safety'; only input, no edit.
run ATP	ATP|run ATP	F2	Make node names + write the ATP file+ run ATP.
run Plot	ATP|run Plot	F8	Run the Plot Command (Tools| Options/Preferences), send pl4 file.
Result	ATP|Sub-process...	–	View and select the result folder (atp/ pl4 etc).
Zoom in	View|Zoom in	NUM +	Zoom in 20%.
Zoom out	View|Zoom out	NUM –	Zoom out 20%.
Zoom	–	–	Zoom in %.
Node	–	–	Node size in %.

[a]Tools in the column "Item" are listed in order as appearing in the toolbar above the table, from left to right.

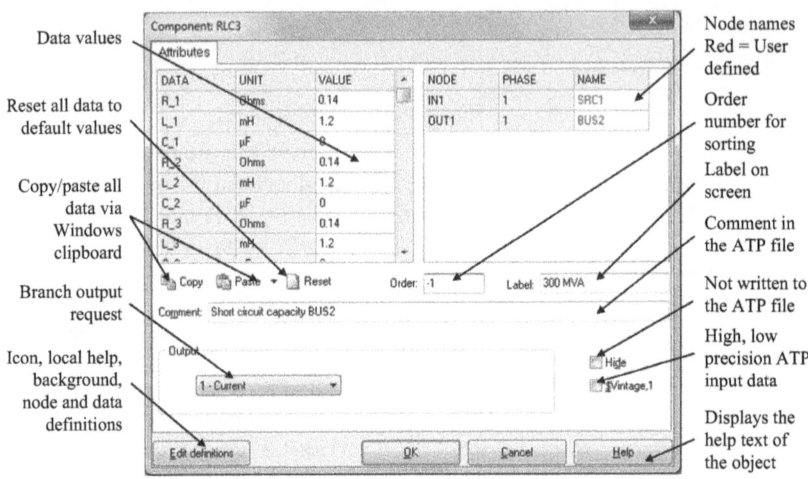

Figure 2.15   The Component dialogue box

entered and the electrical parameters (model) are calculated by ATP. ATPDraw supports PI equivalents, fixed frequency-distributed parameter lines (Bergeron) and frequency-dependent-distributed parameter lines (Semylen, JMarti and Noda). The input dialogue of the LCC module, Figure 2.16, offers to plot the cross section (*View*) and model verification options (*Verify*).

The *Verify* feature calculates the open- and short-circuit input impedance either at power frequency or as function of frequency (LINE MODEL FREQUENCY SCAN feature of ATP) and lets the user compare the model to the exact PI model for positive-sequence or zero-sequence excitations. For long lines this method gives inaccurate results, so a *LineCheck* feature is added externally that compensates for the receiving-end conditions and enables testing of multiple sections of transmission lines with, for instance, manual transpositions.

2.5.4   *Transformer modelling – XFMR module*

ATPDraw supports the saturable transformer component (one- and three-phase) with all couplings (Wye, Delta, Auto and Zigzag) and vector groups where the user has to specify the electrical quantities directly. The conversion from open circuit tests (Urms/Irms-inductive) to the flux-linkage/current peak characteristic is supported internally (SATURA routine of ATP).

ATPDraw also supports two modules with test report input: BCTRAN [22] and XFMR [23, 24]. The BCTRAN module supports some couplings (Wye, Delta and Auto) with all phase shifts for two to three windings. The XFMR module requires input similar to BCTRAN but also supports up to four windings and zigzag coupling. In addition the structure of the core (triplex, three-legged, five-legged, shell-form A/B) can be taken into account if relative core dimensions are provided. The magnetic B/H relationship is assumed to follow an expanded Frolich equation and is the same for both legs and yokes. This can be rewritten for

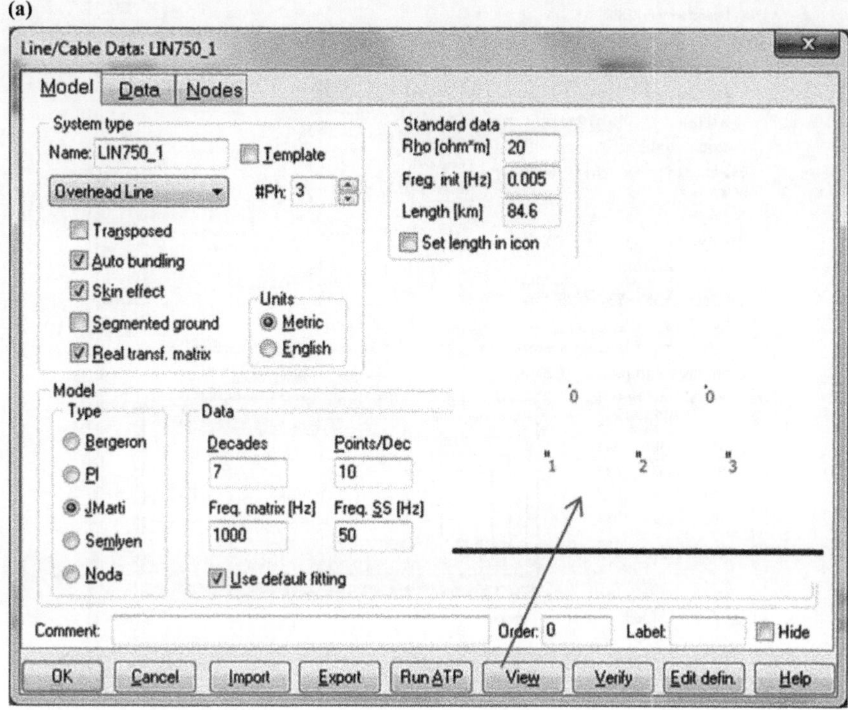

Figure 2.16 *(a) LCC dialogue – overhead line (JMarti model); (b) LCC dialogue – data for 745 kV overhead line*

the electrical quantities flux-linkage per relative area (λ/A_r) and current per relative length (i/l_r).

$$\lambda(i)/A_r = \frac{i/l_r}{a + b \cdot |i|/l_r + c \cdot \sqrt{|i|/l_r}} + L_a \cdot i/l_r \qquad (2.3)$$

The free variables *a, b, c* are then fitted to the open circuit test report data taking the core structure into account. The final slope L_a, which is a critical

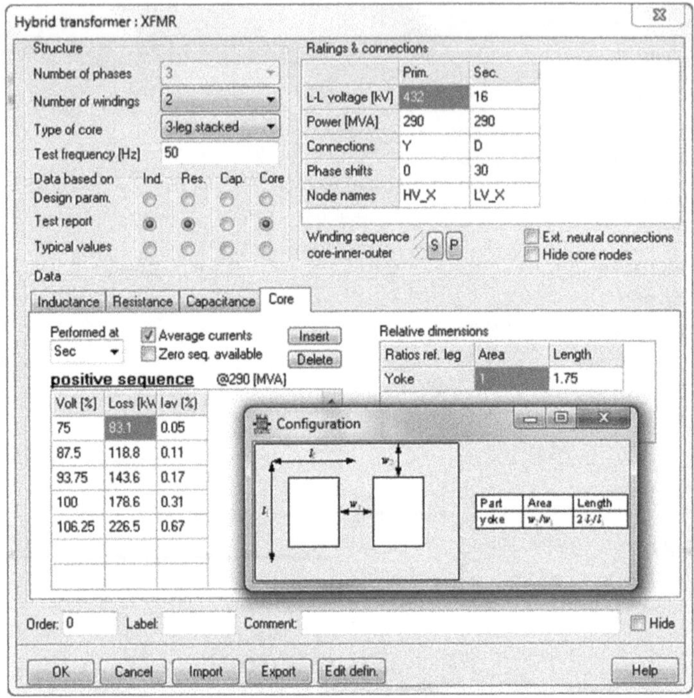

Figure 2.17   XFMR input dialogue

parameter for inrush studies, must be provided by the user either from design information or from estimation; $L_a = \mu_0 \cdot N^2 \cdot A_L/l_L = \mu_0 \cdot a_m/a$, where N is the number of turns of the innermost winding, A_L and l_L are the absolute area and length of the main leg, respectively, a_m is the initial inverse slope of the magnetic relation $a_m \approx \lim_{H \to 0} H/B$ and a is the same for the electric relation and known from the fitting process.

Figure 2.17 shows the XFMR input dialogue where data based on test report is chosen. The XFMR module also supports modelling based on typical values and design information.

2.5.5   *Machine modelling – Windsyn module*

ATPDraw supports ATP's SMs type 58 (phase) and type 59 (Park) with up to four masses and with manufacturer's data input. An initial support of the induction machine type 56 (phase) with electrical parameter input is also added. The UMs type 1 (synchronous), type 3 (induction), type 4 (doubly fed induction), type 6 (single phase) and type 8 (DC machine) with electrical parameter input are also supported. Control of all these machines is totally up to the user and considerable experience is required to accomplish this.

To facilitate the use of the UMs, the WIndSyn utility program was created and offered by Gabor Furst. The program provided calculation of the electrical parameters from manufacturer's data, added options regarding rotor type and

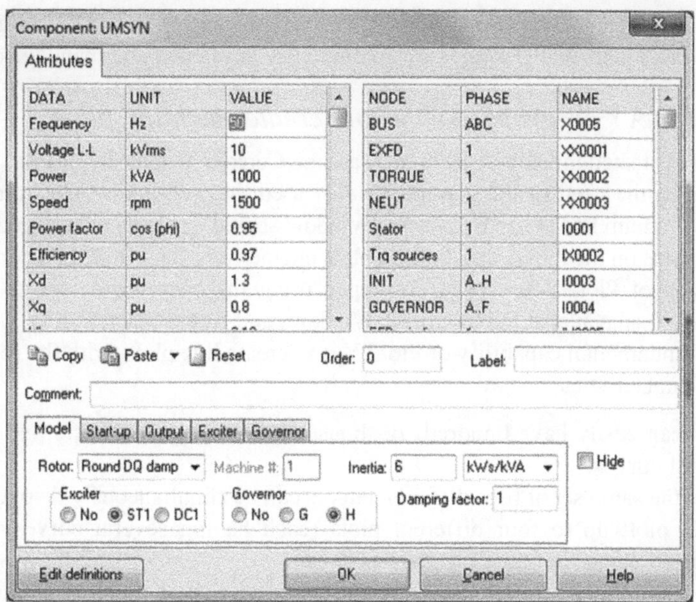

Figure 2.18   Embedded WIndSyn module dialogue for the universal synchronous machine

dampers and offered start-up and control functionalities. The concepts of this program are now directly integrated in ATPDraw [25] with input dialogue as shown in Figure 2.18. This module also offers exciter and governor controls and plotting of torque–slip characteristics for induction machines.

2.5.6   MODELS module

The MODELS language is one of the things that make ATP powerful and flexible. ATPDraw facilitates the integration of MODELS into ATP by automatic identification of INPUT, OUTPUT and DATA variables in the declaration section and conversion of this interface into a component with proper nodes and data. This makes it easy to customize and change a model. The icon is constructed so that inputs are placed to the left and outputs to the right. Array interface [1..n] is supported up to the order n = 26 (corresponds to the A..Z phase extensions). Array data is also supported to some extent (maximum total number of data is 64). The user must click on the input nodes to define the type of input (current, voltage, switch status, machine variable, tacs, Im(V), Im(I), Model, ATP). In order to use a model output as input to another model, the first one must also come first in the ATP file. This can be accomplished by the *Order* property of the MODELS component and *Sort by Order*.

2.6   Other post- and pre-processors

In the past three decades several interactive user interfacing tools have been developed running mainly at the Microsoft Windows platform. Those are PCPlot, TPPLOT,

GTPPLOT, PlotXY, ATP Analyzer, ATP Control Center and ATPDesigner. At present PlotXY and ATPDesigner are continuously developed and maintained.

2.6.1   *PlotXY program to view and create scientific plots*

PlotXY was created initially in 1998 by Massimo Ceraolo, University of Pisa, Italy, as an answer to the need of the community that used the well-known electromagnetic transients program EMTP-ATP to have a Windows-based fast and practical program to view diagrams on the monitor and create plots interactively [26]. In 2014 a completely new version of PlotXY has been developed using the open source compiler 'Qt', version 5.2. It is available for Microsoft Windows and Apple Macintosh.

The fundamental capability of PlotXY is to create line plots, with the following special characteristics:

- Plots can easily have hundreds of thousands of data points each, and contain several curves.
- From the same set of files different curves can be simultaneously displayed in a single plot; up to four different plots (each having several curves) can be managed using the same program instance.
- Besides plotting raw data from files it can also create new plots mixing those data using sums, products, integrals etc.
- It can perform Fourier analysis of a plot and create bar charts for the Fourier harmonic components.
- The produced plots can be exported as images in the system's clipboard or as a SVG (Scalable Vector Graphics) file. They can also be printed on an actual printer or as a pdf file.
- Underlying numerical data can be peeked out directly from the plots.
- Selected plot variables from a plot file can be exported as a subset into a new plot file in different formats.

A sample screen shot of PlotXY's user interface with two plot windows is shown in Figure 2.19. Formats of plot files recognized by PlotXY are as follows:

- ATP binary files (from Alternative Transients Program)
- MAT files (fully compatible with MATLAB® 4.0 format, and partially compatible with 6.0 format)
- COMTRADE files (both binary and textual)
- LVM files (output from National Instruments' Lab-view compatible software)
- ADF files (a very simple text file format explicitly defined for PlotXY use)

2.6.2   *ATPDesigner – design and simulation of electrical*
power networks

ATPDesigner is a control centre for Interactive Design and Simulation of Electrical Power Networks developed by Michael Igel, University for Technology and Economy of the Saarland, Germany. The first version was released in 2007. ATP-Designer provides the user-analysing faults and disturbances in electrical power networks. In addition a lot of highly sophisticated features are implemented to make EMTP-ATP-based network simulation easier. All these features are based on the

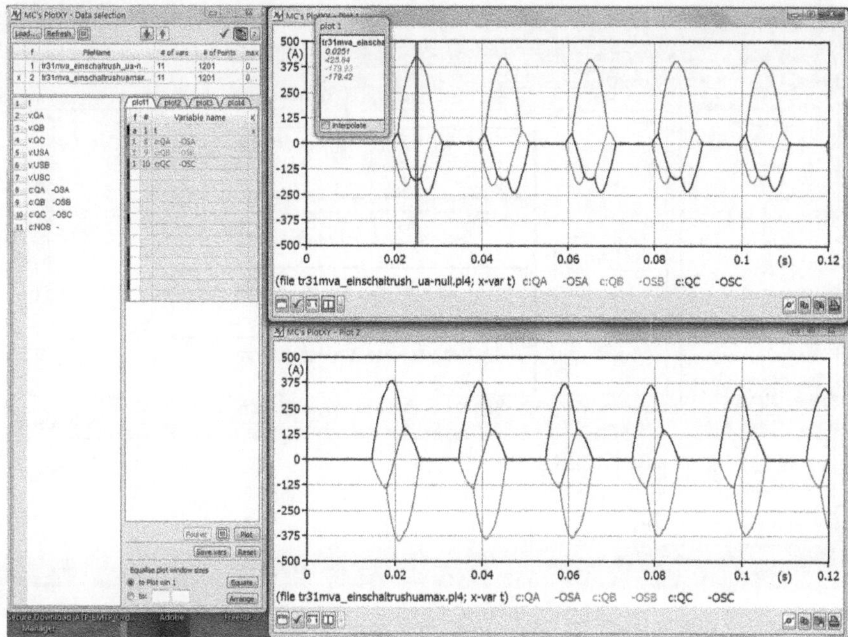

Figure 2.19   PlotXY user interface – comparison of two simulation cases with transformer inrush currents

processing of the ATP-specific .PL4 plot files or COMTRADE files. ATPDesigner is able to read and write both the formats COMTRADE 1999 and 2001 according to the international standard IEC 60255-24. ATPDesigner provides the conversion of the ATP plot files into the COMTRADE format. ATPDesigner has been approved for the operating systems Windows NT, Windows 9x, Windows 2000, Windows XP Home Edition and Professional Edition, Windows Vista, Windows 7 (32 and 64 bit), Windows 8 (64 bit) and Windows 8.1 (64 bit). The new user interface is shown in Figure 2.20.

Main features of the program are as follows:

- Diagram viewer
    - Creating diagrams based on the .PL4 files of the ATP

- Processor for COMTRADE files
    - Converting .PL4 file to COMTRADE file
    - Creating COMTRADE-based diagrams
    - Analysing COMTRADE-based signals
- Integrated signal analysis methods
    - Based on the discrete Fourier transformation
    - Typical for protection relays
    - DFT windowing
    - Frequency spectrum
    - Harmonic analysis presented as a bar diagram

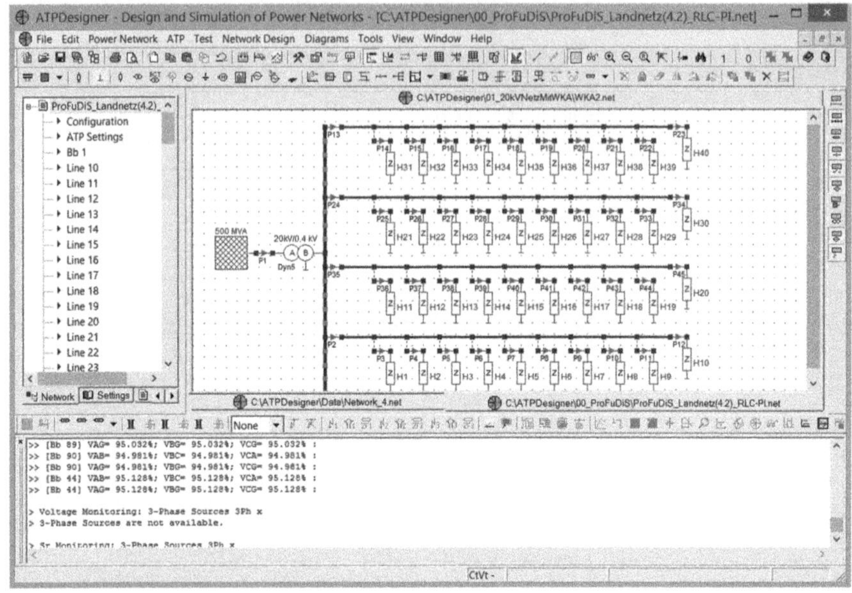

Figure 2.20   New user interface of ATPDesigner

- – Calculation of the frequency-dependent power network impedance Z(f)
- – Calculation of true rms values
- Short circuit analysis
  - – Overcurrent element I>
  - – Undervoltage element V<
  - – Overvoltage element V>
  - – Short-circuit data displayed in tooltips in the power network graphics
  - – Voltages and currents displayed in p.u. related to nominal values
  - – Phase-to-phase and phase-to-ground impedances as usual for distance protection relays
  - – Calculation of power losses, e.g., for lines and transformers
- Protective relaying
  - – Directional overcurrent protection using probes and short-circuit results
  - – Distance protection: phase-to-phase and phase-to-ground impedances calculated by short-circuit results and presented in tooltips
  - – Displaying Trip commands directly in the network graphics
  - – Testing protection schemes
  - – Testing the dynamic behaviour of protection algorithms using MODELS
  - – Protection relaying algorithms implemented in MODELS, e.g., PTOC: Protection Time Overcurrent
  - – Distance protection with two zones can be combined with a circuit-breaker to simulated real-time tripping.

Further information on ATPDesigner can be obtained at the website http://www.atpdesigner.de.

2.6.3   ATP Analyzer

The 32-bit ATP Analyzer at present available in version 4.15 has been developed by Bonneville Power Administration, Portland, USA. ATP Analyzer is a Windows-based program intended for observing and analysing analog signals and discrete channel data-associated power generation, transmission and distribution systems. The program is capable of reading and displaying analog signals produced by ATP as type PL4 output file data, industry standard COMTRADE file analog and digital data produced from protective relays and fault recording equipment, analog signals from table ASCII text data, and audio wave files. User interface and a sample overlay chart are shown in Figure 2.21.

Signals can be displayed in time domain on a single overlay chart, on multiple overlay charts showing the same time range and as single signals on multiple charts showing the same time range. One or more signals can also be displayed as a function of another on an X versus Y chart as a trace over time duration, or as instantaneous points or vectors. Relay characteristics with circular, polygon or threshold line characteristics may also be displayed with signals on the X versus Y chart.

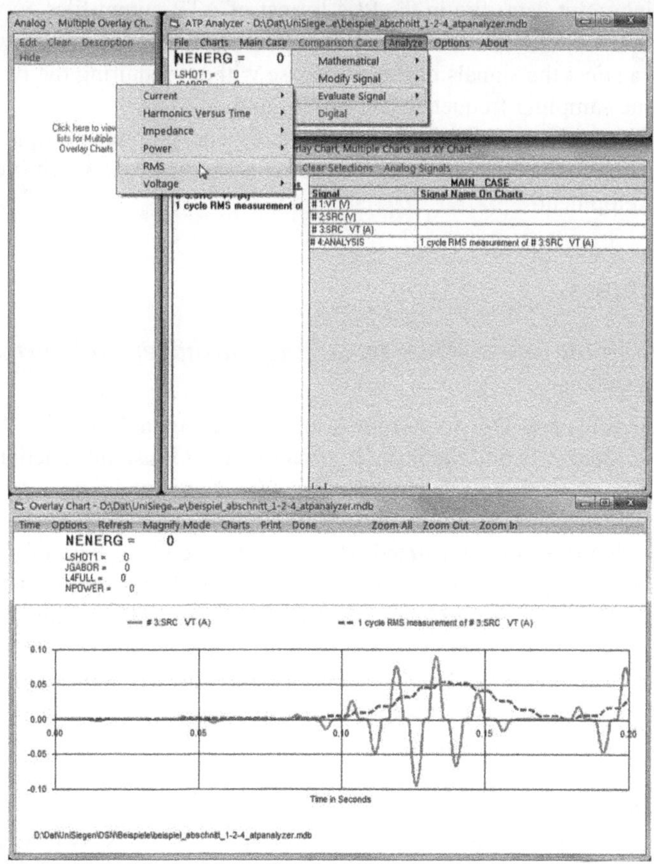

Figure 2.21   User interface and a sample overlay chart of ATP Analyzer

ATP Analyzer has extensive signal analysis functions. They are grouped as follows:

- *mathematical operations* (add, subtract, multiply, divide, log etc.)
- *signal modification* (clip, differentiate, filter, integrate, negate, offset, rectify etc.)
- *signal evaluation* (symmetrical components, park, Clarke transformations, measure phasors, sliding FFT window for harmonic analysis, impedance measurement using single phase signals, impedance measurement in symmetrical and d-q-0 components, power and rms computation etc.)

Charts may be printed on Windows compatible printers and can be made available to other Windows applications using the system clipboard. Printer output documents may be saved as Windows metafiles and bitmap images.

The data file can be resampled at any sampling frequency using either linear or cubic spline interpolation. Selected analog and digital signals can be exported in ANSI standard COMTRADE C37.111-1999 file format with either binary or ASCII data, and in C37.111-1991 format with ASCII data. Analog signals can also be exported in table form to an ASCII text file, or as an ASCII file or binary (IEEE 32 bit floating point data) file in the PL4 format of ATP output files.

A second PL4, COMTRADE, or table type ASCII data file can be viewed for comparison against the signals of the main case without requiring the two cases to have the same sampling frequency or starting time.

The ATP Analyzer program is royalty free. The proprietary rights of ATP Analyzer belong to the Bonneville Power Administration USA, which financed the program development.

2.7   Examples

2.7.1   *Lightning study – line modelling, flashover and current variations*

This example illustrates how to perform a lightning study in ATP using ATPDraw. Values are approximate and the primary objective is to illustrate functionality and guidelines. The example handles transmission line, tower and surge arrester modelling, as well as an insulator string flashover model. How to perform a current amplitude variation is also illustrated. However, Monte Carlo type of simulations with statistical parameter and lightning strike location variations are not covered here. For the transmission line modelling the *Template-Section* approach is used where the cross section is maintained in a single Template and section lengths are set in the Section. Figure 2.22 shows the base case with lightning hitting the sky-wire modelled as the fourth conductor of the transmission line. The cross section is added to the circuit as a png bitmap.

2.7.1.1   Line model

The overhead line is a tentative 420 kV triplex line with two sky-wires. Values are approximate as the purpose is mainly to demonstrate functionality. Figure 2.23

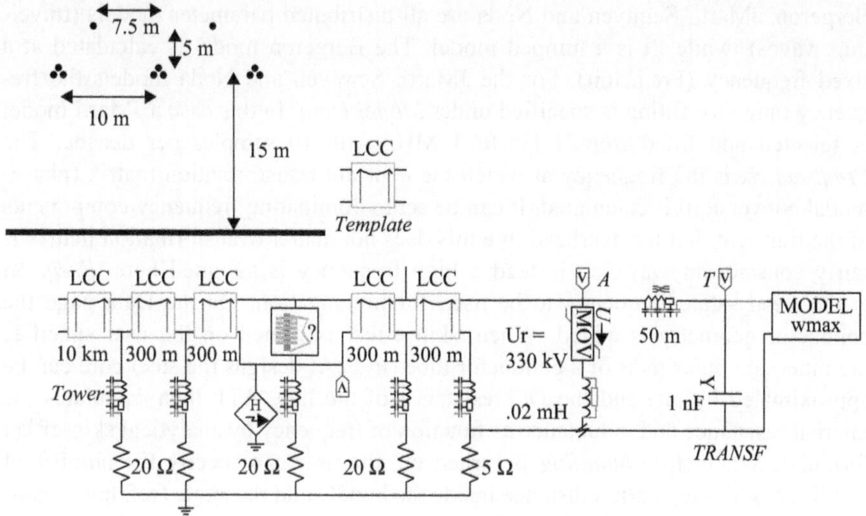

Figure 2.22   Lightning study base case

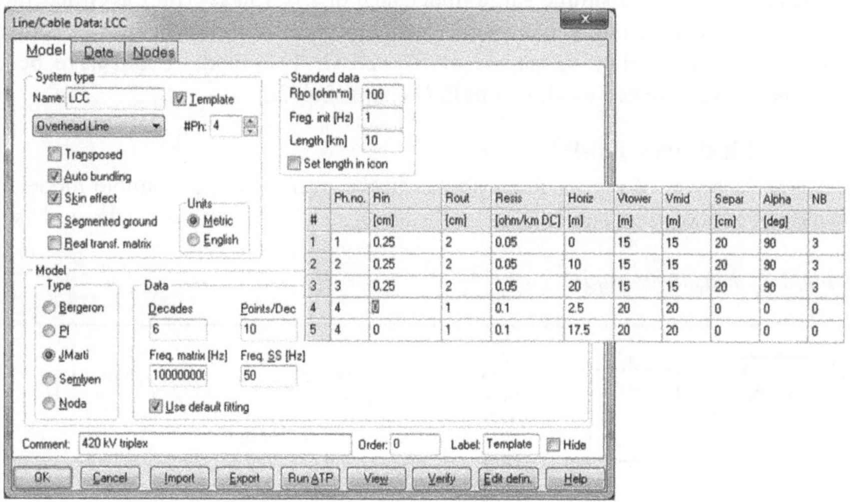

#	Ph.no.	Rin [cm]	Rout [cm]	Resis [ohm/km DC]	Horiz [m]	Vtower [m]	Vmid [m]	Separ [cm]	Alpha [deg]	NB
1	1	0.25	2	0.05	0	15	15	20	90	3
2	2	0.25	2	0.05	10	15	15	20	90	3
3	3	0.25	2	0.05	20	15	15	20	90	3
4	4	0	1	0.1	2.5	20	20	0	0	0
5	4	0	1	0.1	17.5	20	20	0	0	0

Figure 2.23   Line model; Model and Data pages; specification of a JMarti
overhead line

shows the *Model* and *Data* input for the LCC-Template component. Within *System type* the user selects the name of the object, the type of configuration (overhead line, single core cables, enclosing pipe), the number of phases (1–28) and some details regarding the overhead line modelling. Within Standard data the user specifies the ground resistivity (Rho), initial (or only) frequency and the length (does not matter for a Template). Under *Model Type* the user can choose the electrical model.

Bergeron, JMarti, Semlyen and Noda are all distributed parameter models (travelling waves), while PI is a lumped model. The Bergeron model is calculated at a fixed frequency (Freq. init). For the JMarti, Semlyen and Noda models the frequency range for fitting is specified under *Model Data*. In this case a JMarti model is selected and fitted from 1 Hz to 1 MHz with 10 samples per decade. The *Freq. matrix* is the frequency at which the constant transformation matrix (phase-modal conversion) is calculated. It can be set to dominating frequency components of the transient, but for overhead line this does not matter (transformation matrix is fairly constant anyway), so instead a high frequency is assumed here. *Freq. SS* is the steady-state frequency to be used for initialization. On the Data page the conductor geometry is added. When Skin effect is turned on the user specifies the inner and outer radii of a conductor tube (in FeAl designs the steel core can be approximated by air) and the DC resistance of the line. ATP then calculates the internal resistance and inductance as function of frequency by analytical skin-effect formulas. When *Auto bundling* is turned on, the user can specify the number of bundles *NB*, the separation distance inside the bundle and the angle (ref. horizontal) of the first conductor in the bundle.

2.7.1.2   Tower model

The tower consists of a simple surge impedance of $Z = 130\ \Omega$ with travelling time $\tau = 66.67$ ns (20 m height, wave velocity equal to speed of light). The tower is then grounded over a constant resistance of $R_t = 20\ \Omega$. This model could have been made more sophisticated as shown in [27] chaps. 2.4–2.5.

2.7.1.3   Flashover model

To study flashover of the insulator strings in the overhead line a simple model is developed as shown in Table 2.6. The flashover model consists of a (three-phase)

Table 2.6   Flashover model

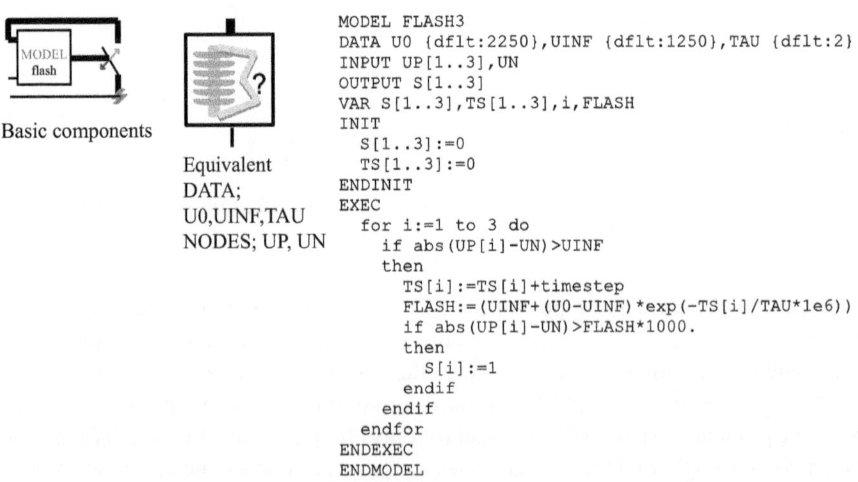


```
MODEL FLASH3
DATA U0 {dflt:2250},UINF {dflt:1250},TAU {dflt:2}
INPUT UP[1..3],UN
OUTPUT S[1..3]
VAR S[1..3],TS[1..3],i,FLASH
INIT
   S[1..3]:=0
   TS[1..3]:=0
ENDINIT
EXEC
   for i:=1 to 3 do
      if abs(UP[i]-UN)>UINF
     then
        TS[i]:=TS[i]+timestep
        FLASH:=(UINF+(U0-UINF)*exp(-TS[i]/TAU*1e6))
        if abs(UP[i]-UN)>FLASH*1000.
        then
           S[i]:=1
        endif
      endif
   endfor
ENDEXEC
ENDMODEL
```

TACS switch controlled by a MODELS component with a time–voltage characteristic. The switch is three-phase but connected to a common neutral point, and we accomplish this by checking the property *Short circuit* in the neutral node dialogue box. The three-phase positive terminal and the single-phase neutral terminal are then connected to a model FLASH3 that controls the closing, OUTPUT S [1..3]. The MODELS code is given below and it calculates a time–voltage characteristic (FLASH) as exponential decaying starting from U0 (2,250 kV) and ending at UINF (1,250 kV) with a time constant of TAU (2 µs). The flashover process is initiated when the voltage goes above UINF. The values in this characteristic can certainly be discussed, but the value of UINF should be well above the switching insulation level of typically 1,050 kV.

2.7.1.4   Surge arrester model

The surge arrester data are taken from the catalogue of ABB's Exlim-T 20 kA, line-discharge class 5. The catalogue provides data for switching and lightning protection levels 1–40 kA, and data for lower currents are not accessible. The data for the surge arrester with rated voltage 330 kV were chosen as input to the MOVN component as shown in Figure 2.24. These data are fitted by ATPDraw with exponential functions on the form $i = p(U/Vref)^q$ with so many segments as required to meet the *ErrLim* requirements if possible. *Ilim* is the value of current where the characteristic switch is from linear to exponential. On the Attributes page the *Vref* data must be chosen within the range of the characteristic to avoid large exponents in the fitted data.

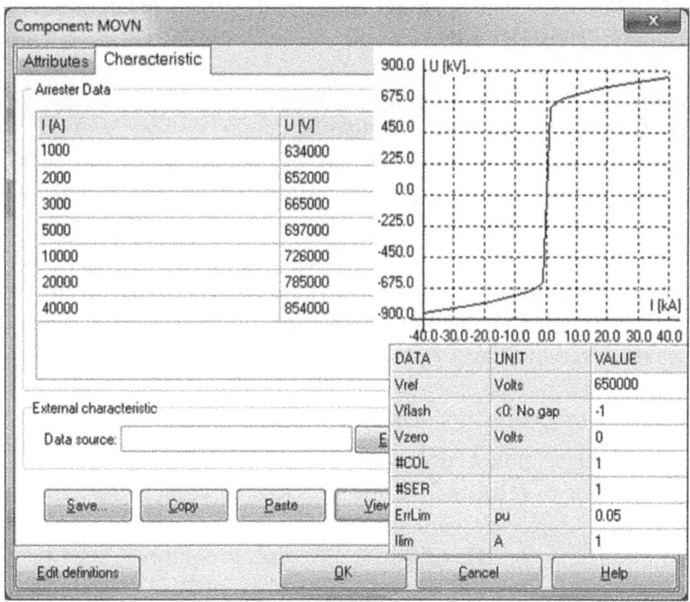

Figure 2.24   Surge arrester input dialogue

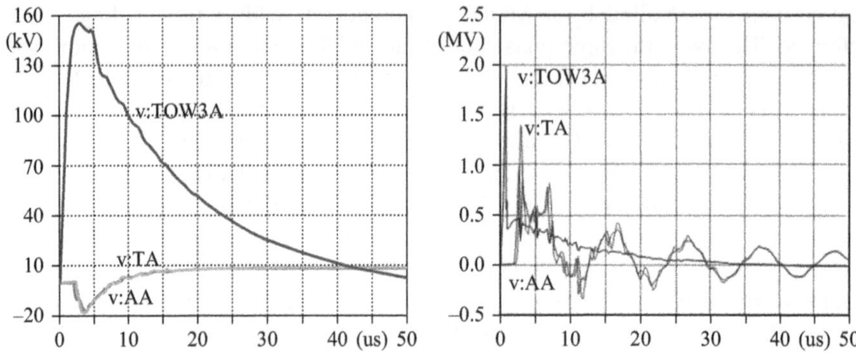

*Figure 2.25   Voltage at nodes A, T and TOW3 as function of time; strike to
            sky-wire (left) and phase conductor (right); 600 m strike distance
            to surge arrester, 50 m from arrester to transformer*

For convenience the MODEL-SWITCH components is compressed into a
group with a readable vector icon and with the data U0, UINF, TAU and nodes;
UPA-B-C and UN surfaced.

2.7.1.5   Voltage at substation – base case

Figure 2.25 shows the voltage at the substation (A and T nodes) for lightning strikes
of 30 kA to two towers away (600 m). The BIL for such 420 kV system is typically
1,425 kV and as we see there are no overvoltage problems in the base case. We see
from Figure 2.25 (right) that lightning strike to the phase conductor will cause a
back-flashover at TOW3 and that the initial 2 MV spike is limited by the surge
arrester.

2.7.1.6   Influence of current amplitude

We can study the maximum voltage at the observation point T (transformer) as
function of the lightning current directly in ATPDraw. We declare the current
amplitude as a variable CURR and assign this a value as function of the simulation
number, for instance CURR = 1000*KNT and KNT = 1..100. The study is performed with the lightning strike at 600 m (third tower) from the substation. The
voltages in this simulation did not exceed the basic insulation level (BIL) of
1,450 kV as observed in Figure 2.26.

2.7.2   Neutral coil tuning – optimization

This example illustrates how to use the Optimization module in ATPDraw to tune a
neutral point reactor to resonance. Dependent on the complexity of the network,
analytical solutions might be hard to identify and a simulation is required. One
thing we can utilize is that the voltage in the neutral point reaches a maximum at
resonance. Given the circuit in Figure 2.27 and regardless of complexity of the
system the solution for coil tuning is always the same: give a variable name to the

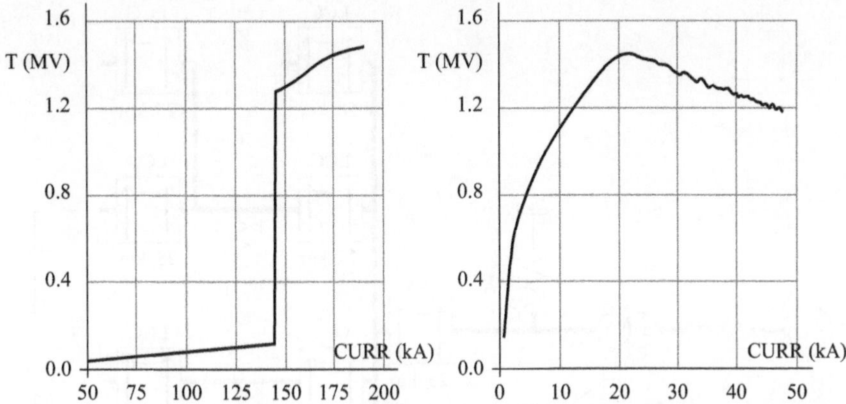

Figure 2.26   Voltage at node T as function of lightning current magnitude; strike to sky-wire (left) and phase conductor (right); 600 m strike distance to surge arrester, 50 m from arrester to transformer

inductance, for instance REACT. Then add a MODELS|WriteMaxMin component to the neutral point and measure the voltage there. Select to write out the MAX value. In the Variables section we can work directly with the REACT variable, but it is more common to use the equivalent current in the neutral reactance, $X = U_N/\sqrt{3}/I_L$. We can then let the current IL = CURR vary over the desired range, for instance from 1 A to 50 A in steps of 1 A. We do this simply by assigning the simulation number KNT to CURR and specify 50 simulations.

The WriteMaxMin module will write out the maximum value of simulation to the diagnosis file (LIS) and read back in the results after the ATP simulation, as shown in Figure 2.28. The module handles multiple runs and can plot the result as function of a variable (CURR in this case) or simply the simulation number KNT in *AsFuncOf* is zero. From Figure 2.28 we can identify that the inductor current resulting in resonance is approximately equal to 25 A. We could of course fine-tune the CURR range, but let us instead invoke the Optimization module (ATP|Optimization) as shown in Figure 2.29. In the optimization dialogue the user has to specify the object function as one of the WriteMaxMin components in the circuit and then specify the variable space with minimum and maximum values. The user can select any of the predefined variables here, but in our case we only have to look at the variable CURR and we specify the range 1–50 A as shown in Figure 2.29. Then we must select an optimization algorithm. The Gradient method is best for a convex and simple problem like this and the result converges quickly to the optimum, 24.6155 A. For more complex cases the Genetic Algorithm or the Simplex Annealing methods can be tested first to reach the global optimum [28].

After the optimization process the optimal value is written back to the variable (CURR in this case). To set the neutral reactor to 10% overcompensation we can now take advantage of the way we specified the reactance by simply dividing REACT by 1.1 in Figure 2.27b.

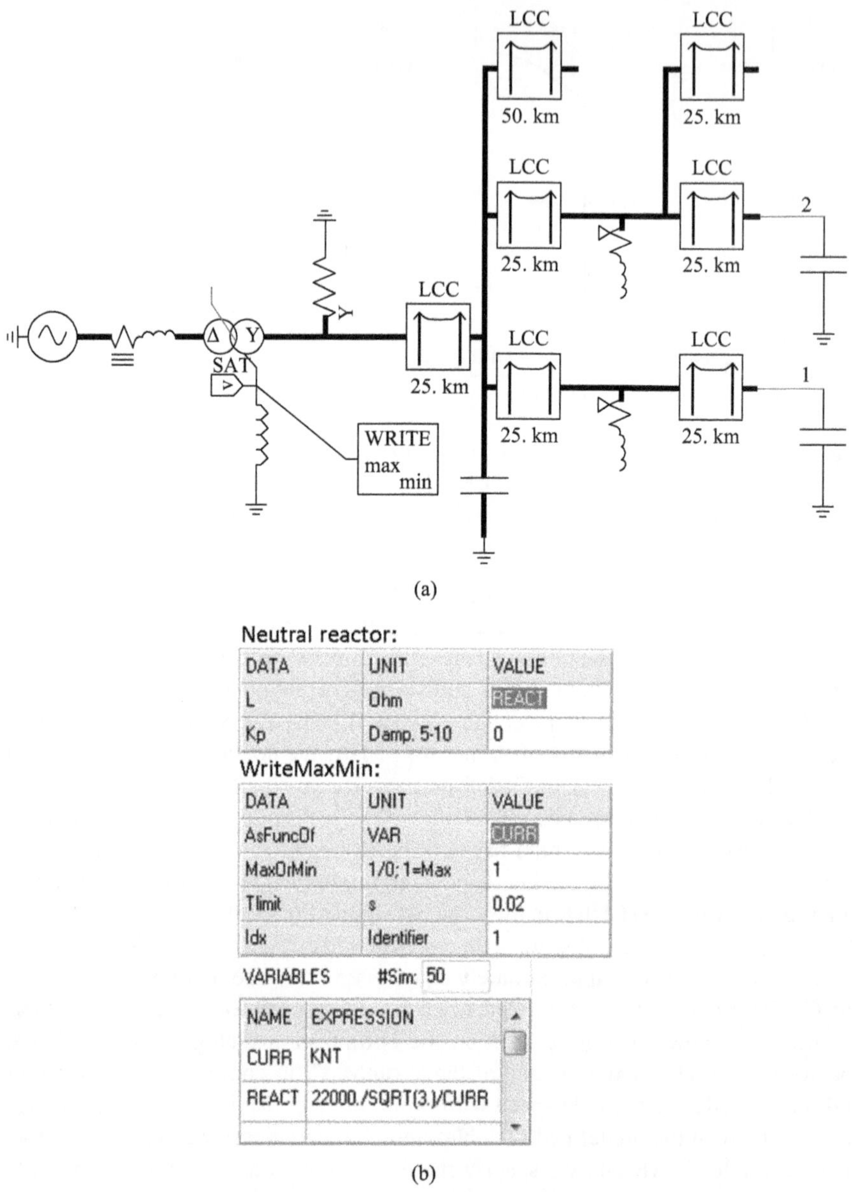

DATA	UNIT	VALUE
L	Ohm	REACT
Kp	Damp. 5-10	0

DATA	UNIT	VALUE
AsFuncOf	VAR	CURR
MaxOrMin	1/0; 1=Max	1
Tlimit	s	0.02
Idx	Identifier	1

NAME	EXPRESSION
CURR	KNT
REACT	22000./SQRT(3.)/CURR

Figure 2.27   (a) Circuit for calculation of neutral reactor (22 kV); (b) input dialogues

2.7.3   Arc modelling

The circuit-breaker (CB) arc model that is described here was represented first in 1986 using type-91 TACS-controlled time-varying resistor. The ATP input data is available as ATP benchmark DC-43.DAT. The CB arc is modelled by the type-94 component (see section 2.4.3). The fundamental difference and advantage of the

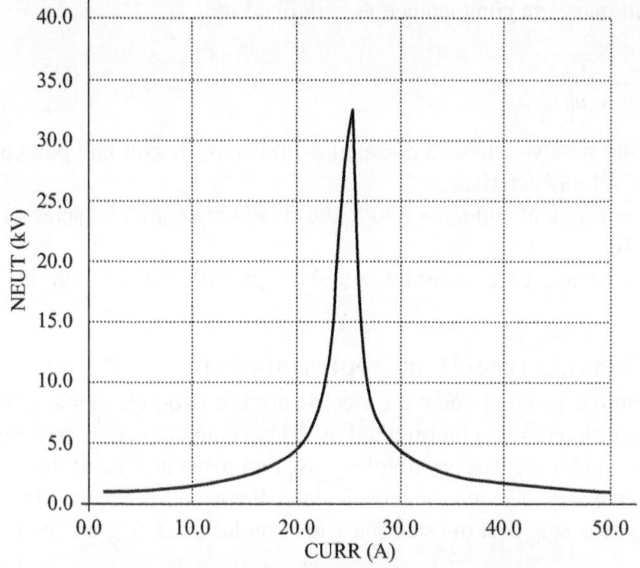

Figure 2.28   Result from the WriteMaxMin component (View)

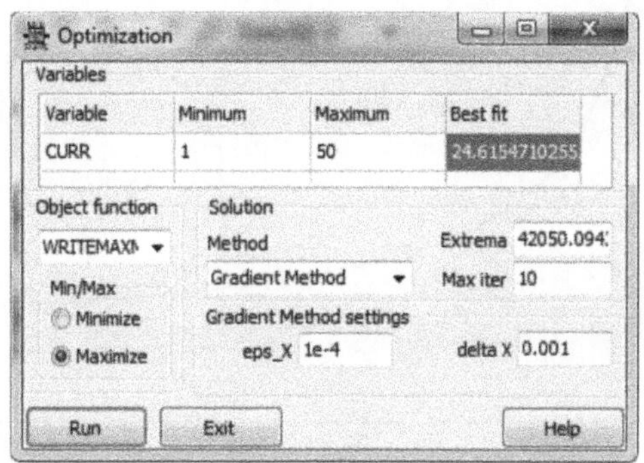

Figure 2.29   Optimization dialogue with final result

type-94 component compared to the type-91 time-varying resistor is that the arc equations and the equations of the electrical system are solved simultaneously.

2.7.3.1 Arc equations

The following form of the extended Mayr equation proposed by A. Hochrainer and A. Grütz is used for this application [29]. This model describes the arc by a differential equation of the arc conductance g:

$$\frac{dg}{dt} = \frac{1}{\tau}(G - g) \tag{2.4}$$

where τ is the arc time constant, g is the instantaneous arc conductance and G is the stationary arc conductance.

The stationary arc conductance G is defined as

$$G = \frac{i_{arc}^2}{p_0 + u_0 \cdot |i_{arc}|} \tag{2.5}$$

where p_0 is the steady-state heat dissipation and u_0 is the constant percentage of the steady-state V-I characteristic.

For $p_0 = 170$ kW and $u_0 = 890$ V the steady-state arc characteristic is shown in Figure 2.30.

A constant arc time constant $\tau = 0.29$ μs will be used in the following simulations.

2.7.3.2 Thevenin type-94 arc representation

In the Thevenin-type component the electric arc is a two-pole element, whereas the rest of the electric circuit is represented by a Thevenin equivalent (see Figure 2.31). Inputs to arc model are Thevenin voltage v_{th} and resistance r_{th} at the terminals for the current time step. The arc model calculates the value of the resulting arc current i_{arc}. At each time step first the stationary arc conductance G is updated using (2.5).

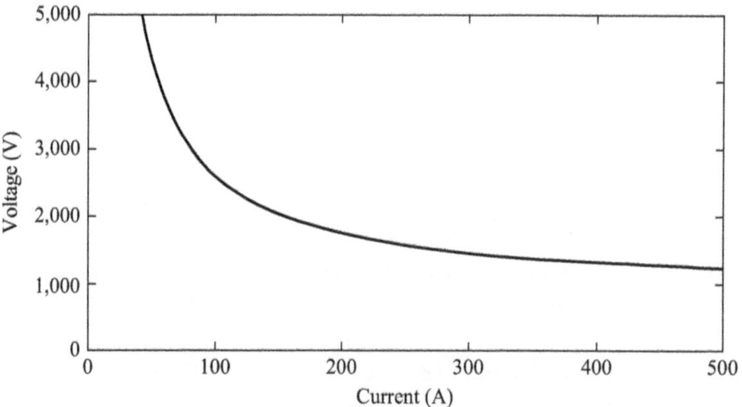

Figure 2.30   *Steady-state arc characteristic*

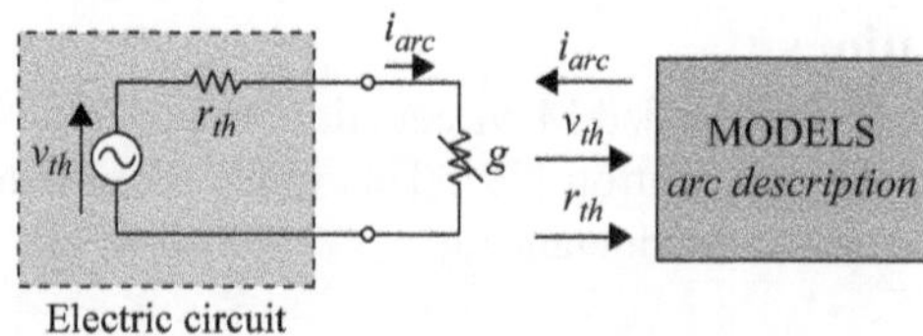

Figure 2.31   *Interaction between the electric circuit and the arc represented by Thevenin type-94 component*

The arc current i_{arc} can be expressed referring to Figure 2.31 as follows:

$$i_{arc} = \frac{g \cdot v_{th}}{1 + g \cdot r_{th}} \tag{2.6}$$

The arc equation (2.4) is solved using MODELS' LAPLACE function to obtain g:

$$g(s) = \frac{1}{1 + \tau \cdot s} \cdot G(s) \tag{2.7}$$

Equations (2.5)–(2.7) are solved simultaneously using an iterative method available in MODELS as 'COMBINE ITERATE … ENDCOMBINE'. The execution part of MODELS description of the arc is as follows:

```
EXEC
    IF NOT(extinguished) THEN
        COMBINE ITERATE AS group1
            i := g*vth / (1+g*rth)          -- arc current with Thevenin interface
            Gst := i*i / (p0+u0 * abs(i))  -- stationary arc conductance
            LAPLACE(g/Gst) := 1.0| / (1.0|+tau|S )  -- solve for g(t)
        ENDCOMBINE
        g_old := prevval(g);
        r := RECIP(g)
        IF (g <= gmin) AND (g_old <= gmin) THEN
            i := 0                      ----- after extinction the current is set to 0
            extinguished := true
        ENDIF
    ENDIF
ENDEXEC
```

If the condition, $g \leq g_{min}$, is satisfied, then the arc will extinguish. After arc extinction, the arc current will be set to 0.

The electric circuit as shown in Figure 2.32 consists of an AC voltage source (100 kV; $f = 50$ Hz; phase $= 170°$), an inductor ($L = 31.8$ mH) and the CB arc. Parallel to the arc there is a time-controlled switch that opens immediately after the first time step. In order to ensure switch opening, the current margin is set to a very

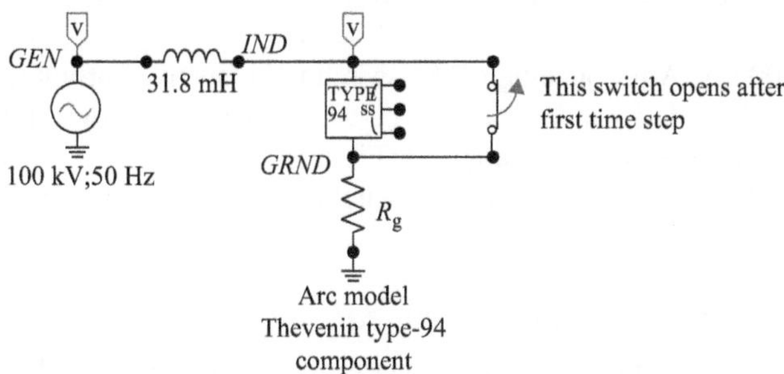

Figure 2.32   Inductive current switching using Thevenin-type arc model

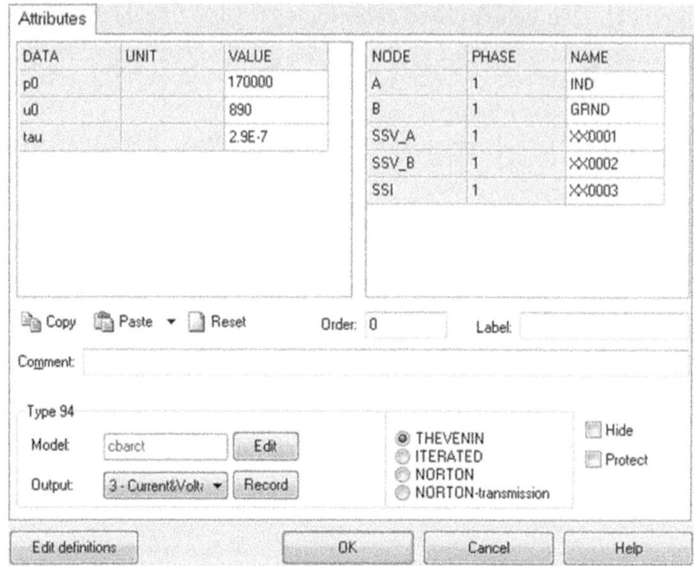

Figure 2.33   *Data window of the arc model*

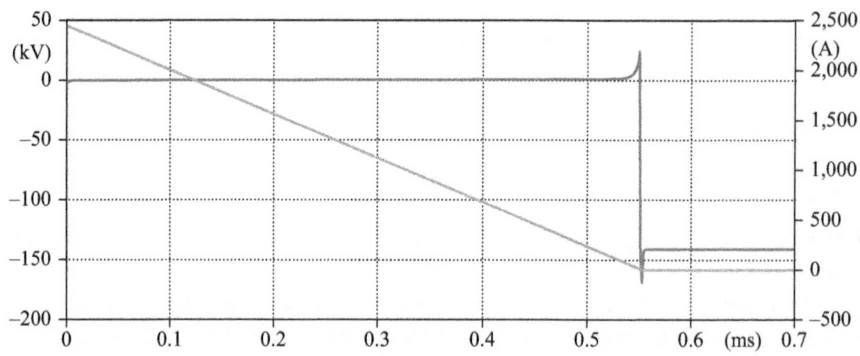

Figure 2.34   *Arc voltage (curve starting from 0 kV) and current (curve starting from 2,500 A) near current zero*

high value, 100 kA. After opening of the switch the inductive current is diverted to the electric arc. The data window of the type-94 component is shown in Figure 2.33.

The simulation results are shown in Figures 2.34 and 2.35. Contrary to the numerical oscillations expected due to switching of an inductance, with the arc model no numerical oscillations occur.

2.7.4   Transformer inrush current calculations

ATP supports natively the BCTRAN transformer component [3, 22] where manufacturer's (test report) data is used to calculate an inductance matrix transformer model. This model does not contain support for the non-linear magnetization

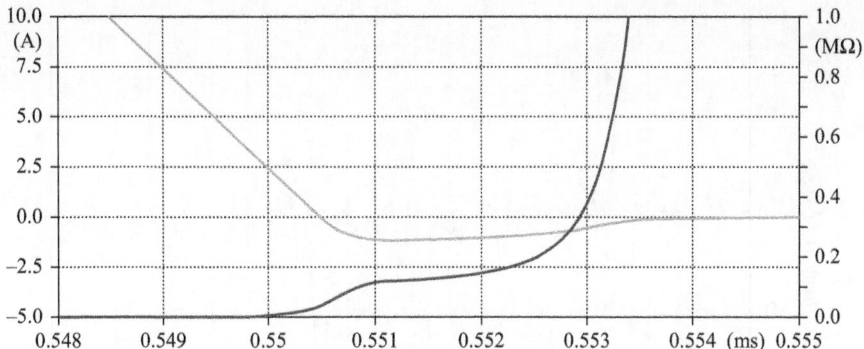

Figure 2.35   Arc current (thin line) and arc resistance (dark line) during arc extinction instant

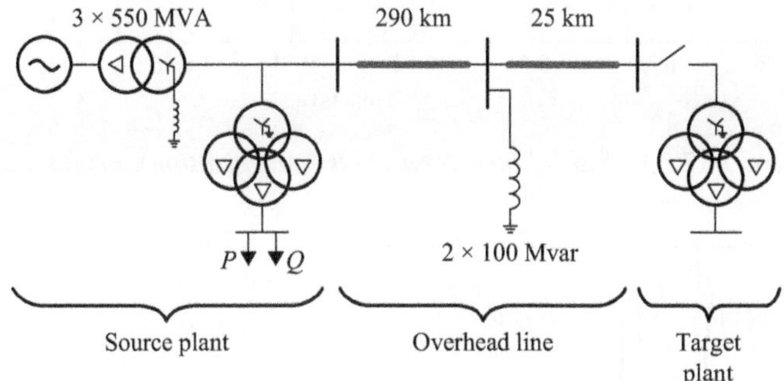

Figure 2.36   Circuit for inrush current calculations with ATP

characteristic, but ATPDraw offers to add this externally on one of the terminals. For inrush studies the final slope of this characteristic is of crucial importance and adding artificial points beyond the test report data is required to get correct results. Otherwise the inrush current will be substantially underestimated. Adding artificial point(s) is not straightforward and is most easily done in the final flux-linkage-current characteristic as shown in [30]. Another critical thing in inrush calculations is the residual flux which to a large extent can influence the inrush current. Initialization of the magnetization characteristic is often difficult to achieve in practice.

ATPDraw also supports the hybrid transformer model called XFMR [23, 24]. This model is based on the same type of input data as BCTRAN, but allows the core design to be taken into account and more direct access to adding a final slope of the magnetization characteristic. Here we bring an example from [30] that illustrates the differences between various transformer models. The circuit diagram is shown in Figure 2.36.

Actual measurements were performed for the circuit shown in Figure 2.36. The measured inrush current of the target transformer is shown in Figure 2.37 and the measured induced voltage when switching off the transformer on the 6.8 kV side is shown in Figure 2.38.

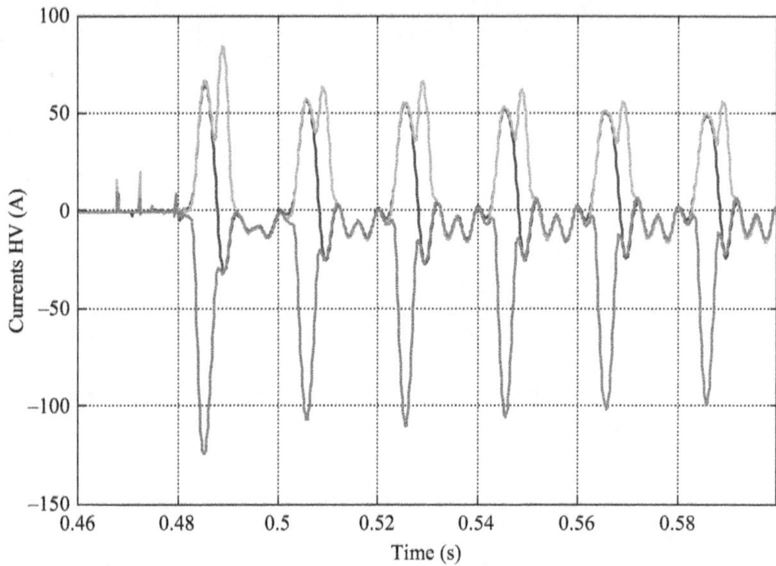

Figure 2.37   *Measured inrush currents (energization), primary side*

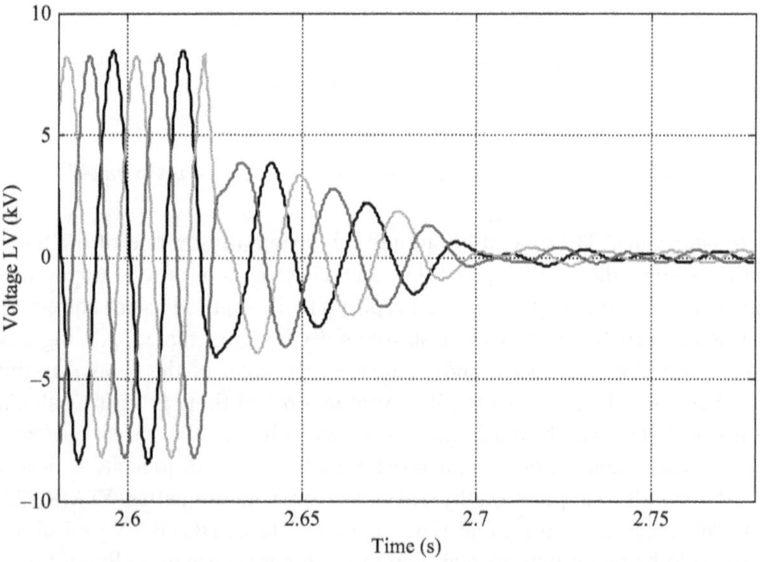

Figure 2.38   *Measured de-energization (prior to energization) transients, induced voltages on secondary side*

In order to evaluate the residual flux, the induced voltage on the secondary side for the de-energization transient prior to the inrush situation should be measured and integrated. Capacitances in the circuit (naturally in the winding and in connected bushings and cables) will create a ring-down transient that will reduce the

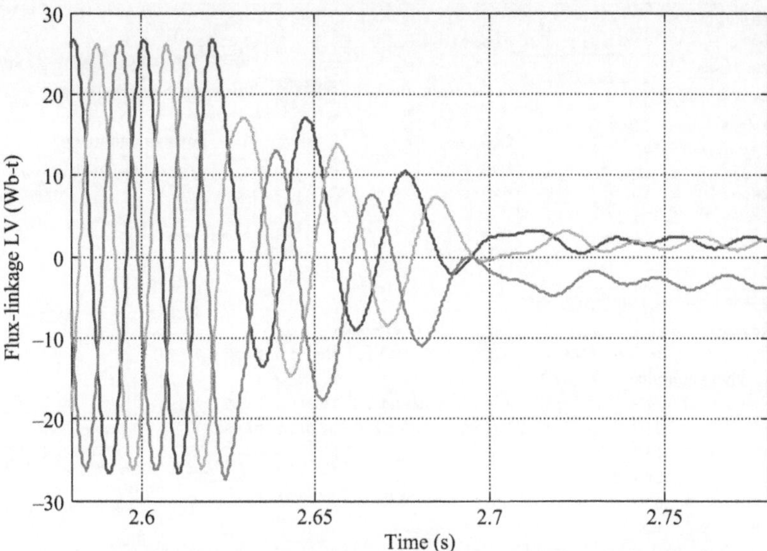

Figure 2.39   Calculated flux-linkage by integration of the induced voltage in Figure 2.38

residual flux. The integration of the induced voltage in Figure 2.38, shown in Figure 2.39, reveals that the residual flux here is below 10% and was ignored in the study [30]. Adding residual flux to the preferred inductance (pseudo-non-linear, type 98 used here) is not trivial as this device does not have a residual flux setting and that artificially application of voltage impulse is difficult in the complex network of inductances as seen in Figure 2.41. Jiles–Atherton non-linearities written in MODELS were instead used in [31].

2.7.4.1 *Transformer model*

The transformer used in the study had an extended test report with open circuit tests up to 115% excitation and in addition information about the relative core dimensions and the air-core inductance were provided by the manufacturer. The hybrid transformer model XFMR was used in this study with an input dialogue shown in Figure 2.40, corresponding to the data provided in [30].

The air-core inductance calculated from the HV side consists of contribution from both leakage and the main flux. Ignoring the winding thickness this can be written

$$L_{ac} = L_a + L_c + L_{LH} = L_a + (1 + k) \cdot L_{LH} \tag{2.8}$$

where L_a is the final slope of the magnetization inductance, L_c comes from the leakage channel between the inner winding and the core and L_{LH} is the main leakage channel. In the XFMR model L_c is estimated as $L_{LH}/2$ resulting in $k = 0.5$.

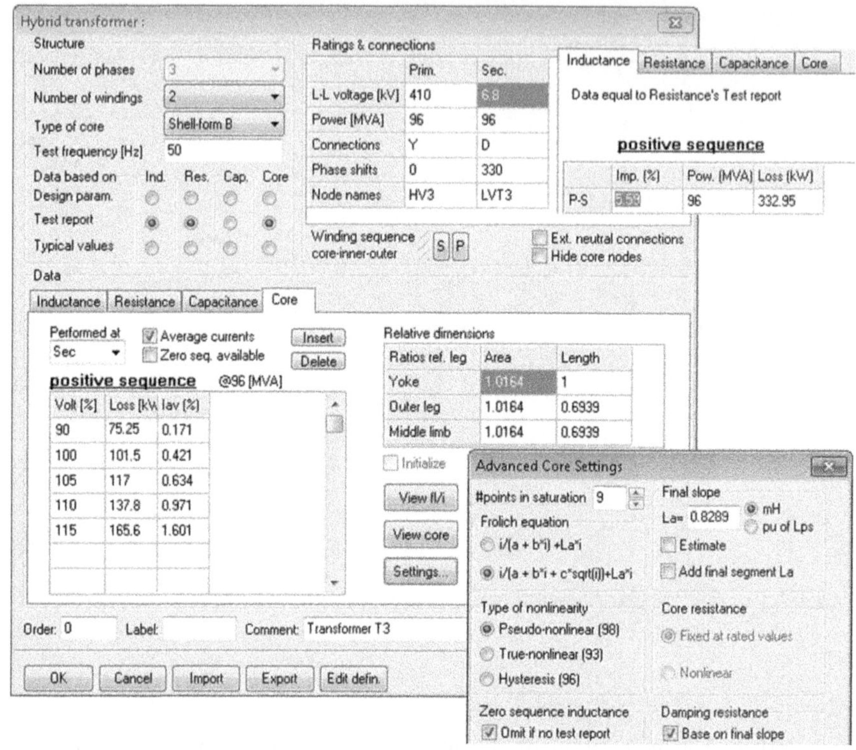

Figure 2.40   *XFMR transformer input dialogue*

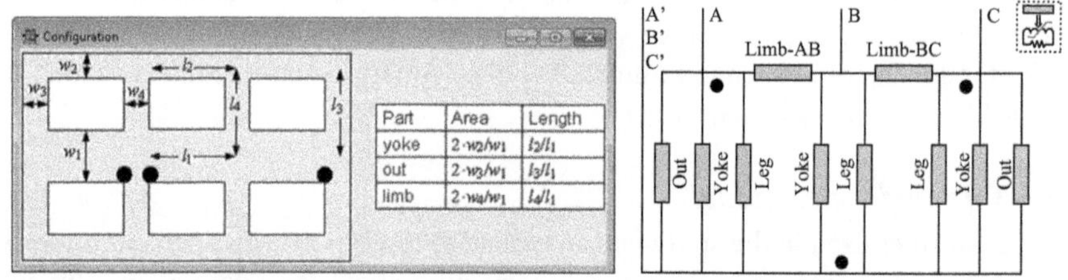

Figure 2.41   *Transformer core configuration for shell-form type-B design*

This gives the final slope inductance

$$L_a = \left[L_{ac,HV} \cdot \left(\frac{V_L}{V_H}\right)^2 - \frac{(1+k)}{2\pi f} \cdot \sqrt{\left(\frac{Z[\%]}{100}\right)^2 - \left(\frac{P[\text{MW}]}{S[\text{MVA}]}\right)^2} \cdot \frac{V_L^2}{S} \right] \cdot 3$$

$$= 0.8289 \text{ mH} \tag{2.9}$$

The relative core dimensions are given according to the manufacturer's data for the shell-form design and the core model given in Figure 2.41. Normally the relative areas are 10–15% above 1.

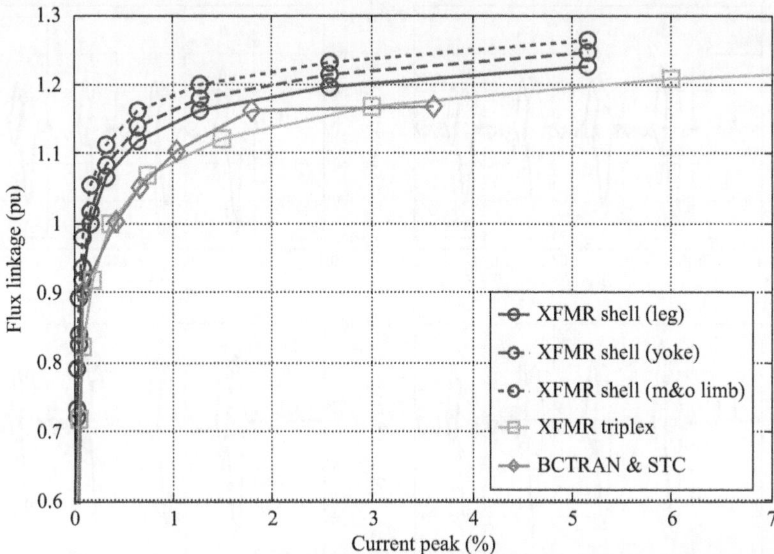

Figure 2.42   Magnetization characteristics of the transformer models, based on test report and extra final slope

The final slope calculation in (2.8) can also be calculated for other transformer models where the magnetization characteristic is connected to the artificial star point of the transformer as in the Saturable Transformer Component (STC) or at the terminal as in BCTRAN according to [30]. This will result in characteristics as shown in Figure 2.42. As the XFMR model is based on a Frolich equation (2.3) fitting to the test report, the final slope is added at a higher flux value compared to the BCTRAN and STC models that have to use the final test report value as its second last point.

2.7.4.2   Results

This section compares the measured inrush currents with the calculated ones. All models give fairly good agreement, but the XFMR with the detailed core information (shell-form) manages to reproduce most of the details better. The reason for the fairly well agreement in all models is that the test report goes up to 115% excitation with additional air-core information (Figure 2.43).

2.7.5   *Power system toolbox: relaying*

ATPDraw offers a power system toolbox with calculation of steady-state qualities and relay protection devices. The work horse of this toolbox is the DFT algorithm that is used for the phasor calculations. The components in the toolbox have an open source so that users can modify and add sophistication.

2.7.5.1   Phasor calculations

The work horse of the toolbox is a DFT algorithm in MODELS that calculates the fundamental (and harmonic) component of the time-domain signal by a

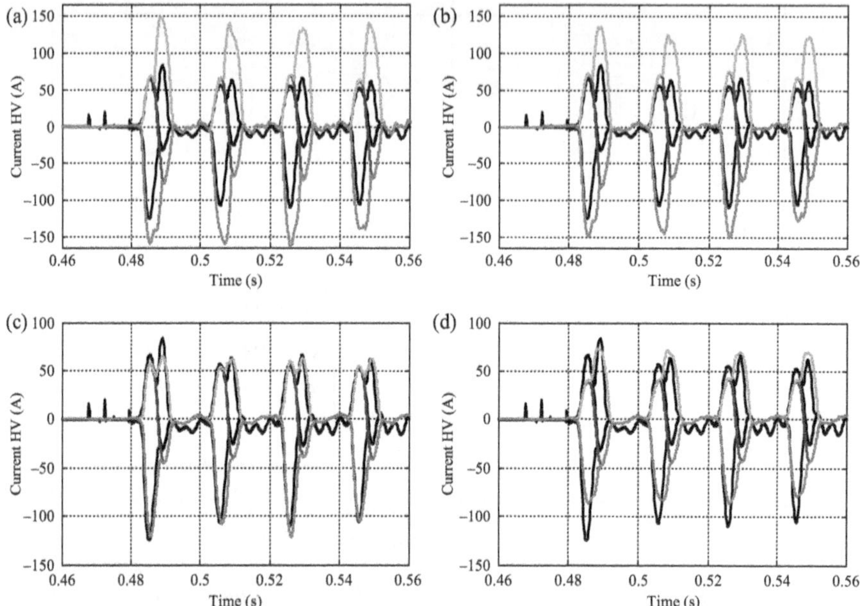

Figure 2.43   (a) STC; (b) BCTRAN; (c) XFMR (shell B); (d) XFMR (triplex)

moving window. The algorithm takes advantage of several of the signal processing specialities of the MODELS language as shown in the programs below.

Sample frequency control is added with the declaration `timestep min:1/ SampleFreq`, which simply puts an upper constraint of the local time step in the model and where `SampleFreq` is given in Samples/second.

Low-pass filtering is added with the line `claplace(Y[i]/X[i]):= (1|s0)/(1|s0+tau|s1)`, where X is the input signal and Y is the filtered signal that is used further in the DFT calculations and where tau is defined as `tau:=recip(2*PI*FilterFreq)`. `recip` is a 1/x function with zero division protection. The `claplace` function uses the |s0 and |s1 operators to define the transfer function.

Calculation of phasors is done with the syntax

```
D:=1/PI*((f4-f2)*sin(OMEGA*T)-(f3-f1)*sin(OMEGA*(T-timestep))
+(f4-f3-f2+f1)/(timestep*OMEGA)*(cos(OMEGA*T)-cos(OMEGA*(T-time-
step)))),
```

where D is the incremental contribution to the Fourier coefficients in which a few delay functions are used to throw out the sample one period ago and add the most recent sample.

The algorithm is based on assuming a linear behaviour of the signal $y(t)$ between two sample points so that it can be written $y(t) = \alpha_i \cdot t + \beta_i \wedge t$

$\in [i \cdot \Delta t, (i+1) \cdot \Delta t]$ which is the same assumption as used in the trapezoidal rule of integration. The Fourier series coefficients can then be written

$$a_h(t) = \frac{\omega}{\pi} \int_{t-T}^{t} y(\tau) \cdot \cos(h\omega \cdot \tau) \cdot d\tau \approx \frac{\omega}{\pi} \sum_{i=0}^{m} \int_{t-T+i\cdot\Delta t}^{t-T+(i+1)\Delta t} (\alpha_i \cdot \tau + \beta_i) \cdot \cos(h\omega \cdot \tau) \cdot d\tau$$

$$= \frac{1}{h \cdot \pi} \sum_{i=0}^{m} \left(\frac{\alpha_i}{h\omega} (\cos(h\omega(t+(i+1)\Delta t)) - \cos(h\omega(t+i\Delta t))) \right.$$

$$\left. + (\alpha_i\Delta t + \beta_i) \cdot (\sin(h\omega(t+(i+1)\Delta t)) - \sin(h\omega(t+i\Delta t))) \right)$$

$$(2.10)$$

Similarly $b_h(t) = \frac{\omega}{\pi} \int_{t-T}^{t} y(\tau) \cdot \sin(h\omega \cdot \tau) \cdot d\tau$, where T is the period, h is the harmonic number, ω is the angular frequency and $m = T/\Delta t + 1$.

By assuming a moving window of one period, the increment of the Fourier coefficients can be written $\Delta a_h(t) = a_h(t, i=m) - a_h(t, i=-1)$ and $\Delta b_h(t) = b_h(t, i=m) - b_h(t, i=-1)$ which is exactly what is calculated by the MODELS code below:

```
MODEL ABC2PHR2
INPUT X[1..3]              --input signal to be transformed
DATA FREQ {dflt:50}        --power frequency
     SCALE {dflt:1}        --multiply output by this number
     FilterFreq {dflt:400}--low-pass corner frequency
     SampleFreq {dflt:800}--actual sample frequency of algorithm
OUTPUT reF[1..3], imF[1..3]            -DFT signals
VAR reF[1..3], imF[1..3], NSAMPL, OMEGA, Y[1..3]
    D,F1,F2,F3,F4,i, h, tau, timestep min:1/SampleFreq
HISTORY Y[1..3] {dflt:0}
DELAY CELLS (Y[1..3]) : 1/(FREQ*timestep)+1
INIT
   OMEGA:= 2*PI*FREQ
   NSAMPL:=recip(FREQ*timestep)
   tau:=recip(2*PI*FilterFreq)
   reF[1..3] :=0
   imF[1..3] :=0
   h:=1 --harmonic number
ENDINIT
EXEC
   FOR i:=1 TO 3 DO
      claplace(Y[i]/X[i]):=(1|s0)/(1|s0+tau|s1)
      f1:=delay(Y[i],(NSAMPL+1)*timestep,1)
      f2:=delay(Y[i],NSAMPL*timestep,1)
      f3:=delay(Y[i],timestep,1)
      f4:=Y[i]
      D:=1/(h*PI)*((f4-f2)*sin(h*OMEGA*T)-(f3-f1)*sin(h*OMEGA*(T-timestep)))
```

```
    +(f4-f3-f2+f1)/(timestep*h*OMEGA)*
    (cos(h*OMEGA*T)-cos(h*OMEGA*(T-timestep))))
reFh[i]:=reFh[i]+D*scale
D:=1/(h*PI)*((f4-f2)*cos(h*OMEGA*T)-(f3-f1)*cos(h*OMEGA*(T-timestep))
    -(f4-f3-f2+f1)/(timestep*h*OMEGA)*
    (sin(h*OMEGA*T)-sin(h*OMEGA*(T-timestep))))
imFh[i]:=imFh[i]+D*scale
    ENDFOR
ENDEXEC
ENDMODEL
```

2.7.5.2   Power and impedance calculations

The algorithm shown in the program above is used also for PQ and RX calculations
and in that case the phasors of two three-phase input current I [1..3] and voltage
V [1..3] are calculated. This gives

$$R + jX = \frac{V_{re} + jV_{im}}{I_{re} + jI_{im}} \quad \text{and} \quad P + jQ = (V_{re} + jV_{im}) \cdot (I_{re} - jI_{im}) \qquad (2.11)$$

For distance relays the positive-sequence impedance is needed and this is found by
calculating line quantities as shown below for the (real part of the) voltage:

```
VL[n]:=(V[n]-V[(n mod 3) +1])
claplace(YV[n]/VL[n]):=(1|s0)/(1|s0+tau|s1)
```

2.7.5.3   Relay functionality

All the relays connect to a signal conditioning model (rms value, RX calculation,
phasor calculation). They have further a block-signal input, a diagnosis output and
a trip signal. All relays have also a zone and trajectory plot functionality. This is
accomplished by reading in data from the LIS file.

- The overcurrent relay has either a constant or inverse time characteristic. In
  both the cases, the rms value of the current is used as input.
- The distance relay uses an RX input and contains settings for three zones with
  quadrilateral characteristic.
- The differential relay contains also harmonic blocking functionality. The
  h-harmonic of the signal can easily be obtained from the above programs by
  setting *h* to typically 2 or 5 and use a criterion for how much harmonic is
  allowed before trip should be blocked.

2.7.5.4   Example

The example illustrates a two-phase fault in a 420 kV system and the response of a
distance relay. The circuit is shown in Figure 2.44 and consists of three 420 kV
transmission lines in a meshed network, one source and two loads at the secondary
side of 420:132 kV transformers. A distance relay is located as BUS1 looking into
the line to BUS2. A phase-to-phase fault is applied at BUS2.

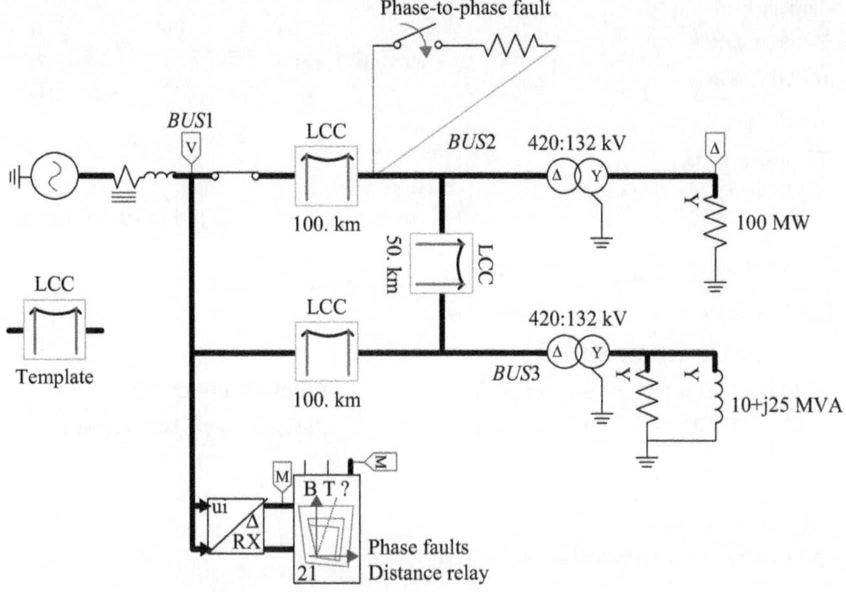

Figure 2.44   Distance relay example circuit

The transmission line modelling is equivalent to what is described in section 2.7.1, Figure 2.22, but the sky-wires are now assumed at ground potential by setting their *Ph. No = 0* in Figure 2.23. The line is still modelled as a JMarti line even if a simple PI equivalent could have been sufficient.

The transformers are modelled with the BCTRAN component with simple and typical data input. The data are similar to those in Figure 2.40, but simplified somewhat as saturation is not important here as shown in Figure 2.45.

The distance relay at BUS1 gets its RX signals from a DFT algorithm in a predefined MODELS component UI2RXL (line quantities) with a low-pass filter and sample frequency restrictions added as described above. The model has the input data as shown in Figure 2.46.

The UI2RXL model must be written to the ATP file prior to the relay model and to enforce this, the *Order* option can be used. The relay model has the input data as shown in Figure 2.47. This consists of the time delay for each of the three zones, four *RX* points for each of the three zones, an *RL-XL* point describing the line (just for plotting), a *DownSample* setting to reduce printout, a *t_init* setting that describes the minimum time for the relay to respond (typically one period required to get reliable DFT signals) and an *Idx* value to identify the relay if several values are used in the data case.

2.7.5.5   Result

The trajectory of all the relay can be plotted related to their zones of protection as seen in Figure 2.48. This shows initially the load position as seen from the relay

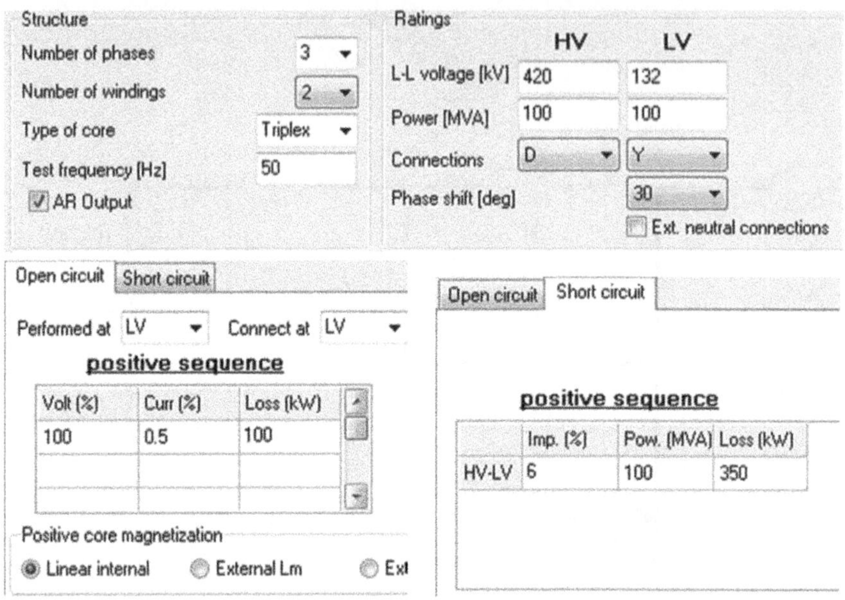

Figure 2.45   *Transformer model as BCTRAN component*

DATA	UNIT	VALUE
FREQ	Hz	50
ScaleV	pu	1
ScaleI	pu	1
FilterFreq	Hz	400
SampleFreq	S/s	800

Figure 2.46   *Input data to RX-calculation module* `UI2RXL`

DATA	UNIT	VALUE
t_zone1	sec	0.1
t_zone2	sec	0.2
t_zone3	sec	0.4
R1_z1	ohm	0
X1_z1	ohm	0
R2_z1	ohm	10
X2_z1	ohm	-2
R3_z1	ohm	15
X3_z1	ohm	20
R4_z1	ohm	-5
X4_z1	ohm	20

DATA	UNIT	VALUE
R4_z3	ohm	-15
X4_z3	ohm	40
RL	ohm	10
XL	ohm	100
DownSampl	time	1
t_init	sec	0.02
Idx	1..n	1

Figure 2.47   *Input data to distance relay module* `W1RELAY21P`, *three zones, four points per zone*

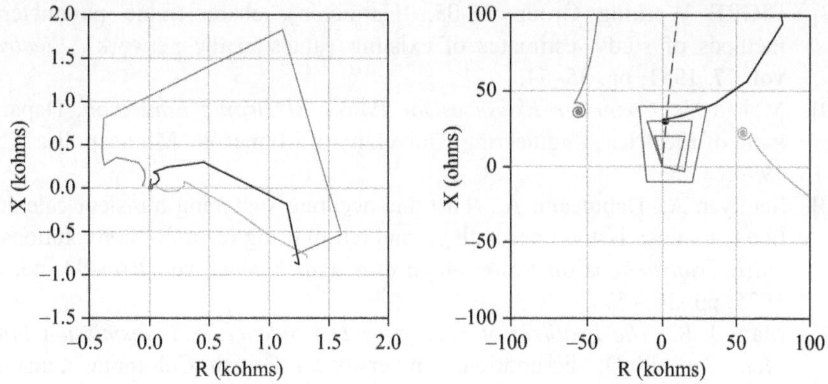

Figure 2.48   Distance relay example with trajectory plots

towards the steady-state fault location. In this case the phase A trajectory (that corresponds to phase A-B faults) will enter the trip zones around the boundary of zone 2. Zone 2 should thus have been extended in X-direction to better cover this fault.

References

[1]   Dommel H. W., 'Digital computer solution of electromagnetic transients in single and multiphase networks,' *IEEE Transactions on Power Apparatus and Systems*, **vol. PAS-88**, April 1969, pp. 388–399.

[2]   Meyer W. S., Dommel H. W., 'Numerical modelling of frequency-dependent transmission-line parameters in an electromagnetic transients program', *IEEE Transactions on Power Apparatus and Systems*, **vol. PAS-93**, no. 5, 1974, pp. 1401–1409.

[3]   Bonneville Power Administration. *EMTP Theory Book*, Branch of System Engineering, Oregon, USA, 1987 Available from the secure EEUG website http://www.eeug.org.

[4]   Canadian/American EMTP User Group. *ATP Rule Book*, Portland, Oregon, USA, revised and distributed by the EEUG Association, 2014. Available from the secure EEUG website http://www.eeug.org.

[5]   Dubé L., Bonfanti, I., 'MODELS: A new simulation tool in the EMTP', *European Transactions on Electrical Power*, **vol. 2**, no. 1, January/February 1992, pp. 45–50.

[6]   Dubé L., Dommel H. W., 'Simulation of control systems in an electro-magnetic transients program with TACS', *Proceedings of the IEEE PICA Conference*, May 1972, pp. 266–271.

[7]   Furst G., Kizilcay M., 'Frequency-domain harmonic analysis using the ATP-EMTP program', *International Conference on Electrical Engineering 2000 (ICEE)*, IEE Japan, Kitakyushu, Japan, July 24–28, 2000.

[8]   CIGRE Working Group 36-05. 'Harmonics, characteristic parameters, methods of study, estimates of existing values in the network', *Electra*, **vol. 77**, 1981, pp. 35–54.

[9]   Mohan N., *Computer Exercises for Power Electronic Education*, Department of Electrical Engineering, University of Minnesota, Minneapolis, MN, 1990.

[10]   Semlyen A., Dabuleanu A., 'Fast and accurate switching transient calculations on transmission lines with ground return using recursive convolutions', *IEEE Transactions on Power Apparatus and Systems*, **vol. PAS-94**, no. 2, 1975, pp. 561–571.

[11]   Marti J. R., *The Problem of Frequency Dependence in Transmission Line Modelling*, Ph.D. dissertation, University of British Columbia, Canada, 1981.

[12]   Marti J. R., 'Accurate modelling of frequency-dependent transmission lines in electromagnetic transient simulations', *IEEE Transactions on Power Apparatus and Systems*, **vol. PAS-101**, no. 1, 1982, pp. 147–155.

[13]   Noda T., Nagaoka N., Ametani A., 'Phase domain modeling of frequency-dependent transmission lines by means of an ARMA model', *IEEE Transactions on Power Delivery*, **vol. 11**, no. 1, January 1996, pp. 401–411.

[14]   Noda T., Nagaoka N., Ametani A., 'Further improvements to a phase-domain ARMA line model in terms of convolution, steady-state initialization, and stability', *IEEE Transactions on Power Delivery*, **vol. 12**, no. 3, July 1997, pp. 1327–1334.

[15]   Gross G., Hall M. C., 'Synchronous machine and torsional dynamics simulation in the computation of electromagnetic transients', *IEEE Transactions on Power Apparatus and Systems*, **vol. PAS-97**, no. 4, 1978, pp. 1074–1086.

[16]   Brandwajn V., Dommel H. W., 'A new method for interfacing generator models with an electromagnetic transients program,' *IEEE PES PICA Conference Record*, May 24–27, 1977, pp. 260–265.

[17]   Lauw H. K., Meyer W. S., 'Universal machine modeling for the representation of rotating electric machinery in an electromagnetic transients program', *IEEE Transactions on Power Apparatus and Systems*, **vol. 101**, no. 6, 1982, pp. 1342–1351.

[18]   Prikler L., Høidalen H. K., *ATPDraw for Windows 3.1x/95/NT version 1.0 – User's Manual*, 1998. Available from http://www.atpdraw.net.

[19]   Prikler L., Høidalen H. K., *ATPDRAW version 3.5 for Windows 9x/NT/2000/ XP*, 2002. Available from http://www.atpdraw.net.

[20]   Prikler L., Høidalen H. K., *ATPDRAW version 5.6 for Windows 9x/NT/2000/ XP/Vista*, 2009. Available from http://www.atpdraw.net.

[21]   Høidalen H. K., Mork B. A., Prikler L., Hall J. L., 'Implementation of new features in ATPDraw version 3', *Proceedings of the International Conference on Power System Transients (IPST)*, New Orleans, LA, 29 Sep–2 Oct 2003.

[22]   Brandwajn V., Dommel H. W., Dommel I. I., 'Matrix representation of three-phase N-winding transformers for steady-state and transient studies', *IEEE*

Transactions on Power Apparatus and Systems, **vol. PAS-101**, no. 6, 1982, pp. 1369–1378.

[23] Høidalen H. K., Mork B. A., Gonzalez F., Ishchenko D., Chiesa N., 'Implementation and verification of the hybrid transformer model in ATP-Draw', *Proceedings of the International Conference on Power System Transients (IPST)*, Lyon, France, 3–7 June 2007.

[24] Høidalen H. K., Chiesa N., Avendaño A., Mork B. A., 'Developments in the hybrid transformer model – Core modeling and optimization', *Proceedings of the International Conference on Power System Transients (IPST)*, Delft, the Netherlands, 14–17 June 2011.

[25] Høidalen H. K., 'Universal machine modeling in ATPDraw – Embedded Windsyn', *Proceedings of the EMTP-ATP Users Group meeting*, Ohrid, Makedonia, 25–28 September 2011.

[26] Ceraolo M., 'PlotXY – a new plotting program for ATP', *EEUG News*, February–May 1998. Available from http://www.eeug.org.

[27] Martinez-Velasco J. A. (Ed.), *Power System Transients – Parameter Determination*, CRC Press, New York, 2010.

[28] Høidalen H. K., Radu D., Penkov D., 'Optimization of cost functions in ATPDraw', *Proceedings of the EMTP-ATP Users Group meeting*, Delft, the Netherlands, 26–27 October 2009.

[29] Grütz A., Hochrainer A., 'Rechnerische Untersuchung von Leistungsschaltern mit Hilfe einer verallgemeinerten Lichtbogentheorie', *etz-a*, **vol. 92**, no. 4, 1971, pp. 185–191 (in German).

[30] Chiesa N., Høidalen H. K., Lambert M., Duro M. M., 'Calculation of inrush currents – benchmarking of transformer models', *Proceedings of the International Conference on Power System Transients (IPST)*, Delft, the Netherlands, 14–17 June 2011.

[31] Chiesa N., Mork B. A., Høidalen H. K., 'Transformer model for inrush calculations: simulation, measurements and sensitivity analysis', *IEEE Transactions on Power Delivery*, **vol. 25**, no. 4, 2010, pp. 2599–2608.

Chapter 3

Simulation of electromagnetic transients with EMTP-RV

J. Mahseredjian, Ulas Karaagac*, Şébastien Dennetière[†] and Hani Saad[†]*

3.1 Introduction

EMTP-RV is the restructured version of the Electromagnetic Transients Program (EMTP). It is a simulation tool for electromagnetic transients (EMTs) in power systems. The development of EMTP-RV was initiated and completed in the Development Coordination Group (DCG) of EMTP. It was funded by several international organizations. The objective of the development of EMTP-RV, RV standing for restructured version, was to rewrite and modernize the EMTP code using the latest programming techniques, and the most recent research on models and solution methods in the field of power system transients. The overall programming effort started in 1998 and was completed in 2003. Due to technical reasons, it was decided to abandon completely the old DCG-EMTP code and program EMTP-RV from scratch; that is why EMTP-RV does not contain any computer code lines from the old DCG-EMTP and innovates in several aspects as described below.

EMTP-RV is a commercial software package [1] used by many universities and industrial organizations worldwide.

Since the letters RV are only used to denote the new generation of EMTP, they are dropped in the following sections by using only the software acronym EMTP.

3.2 The main modules of EMTP

The main modules of EMTP, presented in Figure 3.1, are:

1. Graphical user interface (GUI)
2. Data input file
3. Multiphase load-flow solution
4. Steady-state solution
5. Initialization: automatic or manual initial conditions

*Ecole Polytechnique de Montreal, Canada
[†]RTE, France

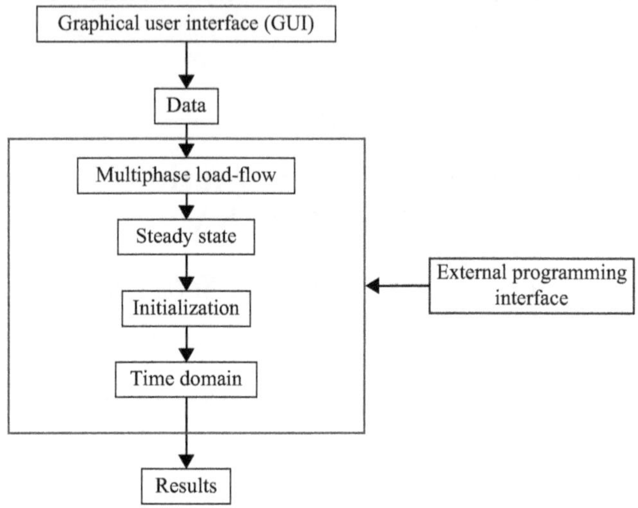

Figure 3.1   EMTP computation modules

6. Time-domain solution
7. Results: waveforms and outputs
8. External programming interface

The main modules are described in the following sections.

3.3   Graphical user interface

The graphical user interface (GUI) is the first entry level to the simulation process of EMTP. It is the simulated network schematic input. An example of EMTP schematic input is presented in Figure 3.2. The EMTP GUI is called EMTPWorks.

The EMTP GUI is based on a hierarchical design approach with subnetworks and masking. Subnetworks allow simplifying the drawing and hiding details while masking provides data encapsulation. The design of Figure 3.2 uses several sub-networks. The 230 kV network is interconnected with a 500 kV network evacuated with all its details into the subnetwork shown in Figure 3.2. In a hierarchical design, subnetworks can also contain other subnetworks. Subnetworks can be also used to develop models. The three-phase transformers shown in Figure 3.2 are based on the interconnection of single-phase units. The synchronous machine subnetworks contain load-flow constraints, machine data and subnetworks with voltage regulator and governor controls, as shown in Figure 3.2.

The EMTP GUI is programmed using three layers (Figure 3.3). The first layer is the hard code of the software (written in C/C++) containing all its operations, menus, drawing tools and manipulations of data. The second layer delivers interfacing methods for full access to schematic drawing functions and data scripting. This layer is based on JavaScript with extensions for EMTPWorks and constitutes

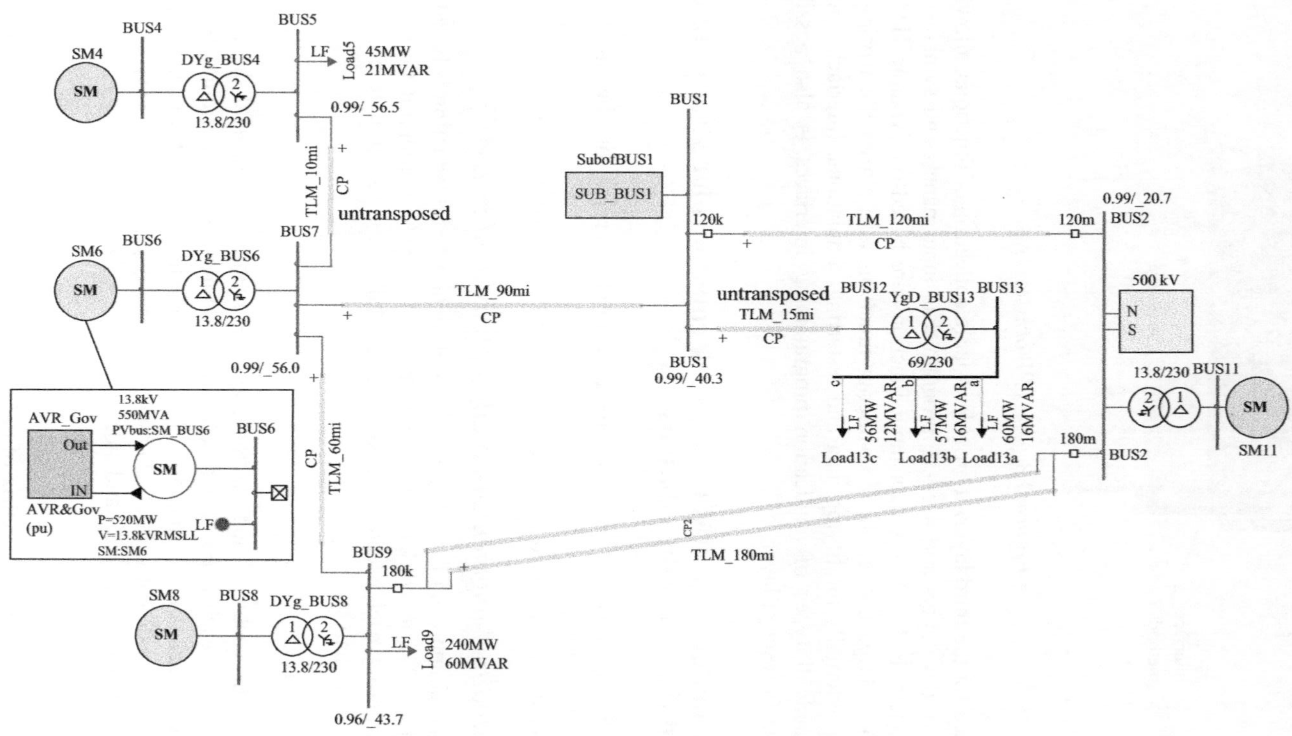

Figure 3.2 Sample 230 kV network simulation presented in a GUI

Figure 3.3   EMTPWorks layers

the interface for the third layer. It provides a large collection of functions, allowing the software developers and users to program data input panels and symbols for EMTP models. EMTPWorks data input functions are based on dynamic HTML (DHTML). A large collection of JavaScript functions also allows the writing of scripts for externally modifying data and restarting simulations. Interfacing with ActiveX and all types of application programming interfaces is also possible through the JavaScript layer.

3.4   Formulation of EMTP network equations for steady-state and time-domain solutions

The objective of this section is to present a brief overview on the numerical methods used in EMTP.

Bold characters are used hereafter for the representation of vectors and matrices.

3.4.1   Modified-augmented-nodal-analysis used in EMTP

The traditional approach for the formulation of main network equation is based on nodal analysis. The network admittance matrix $\mathbf{Y_n}$ is used for computing the sum of currents entering each electrical node and the following equation results from classical nodal analysis.

$$\mathbf{Y_n}\,\mathbf{v_n} = \mathbf{i_n} \tag{3.1}$$

where $\mathbf{v_n}$ is the vector of node voltages and the members of $\mathbf{i_n}$ hold the sum of currents entering each node. It is assumed that the network has a ground node at zero voltage which is not included in (3.1). Since the network may contain voltage sources (known node voltages), (3.1) must be normally partitioned to keep only the unknown voltages on the left-hand side:

$$\mathbf{Y_n'}\,\mathbf{v_n'} = \mathbf{i_n'} - \mathbf{Y_s}\,\mathbf{v_s} \tag{3.2}$$

where $\mathbf{Y_n'}$ is the coefficient matrix of unknown node voltages $\mathbf{v_n'}$, $\mathbf{i_n'}$ holds the sum of currents entering nodes with unknown voltages, $\mathbf{Y_s} \in \mathbf{Y_n}$ and relates to known voltages $\mathbf{v_s}$. It is noticed that $\mathbf{v_n} = \begin{bmatrix} \mathbf{v_n'} & \mathbf{v_s} \end{bmatrix}^T$.

Equation (3.1) has several limitations. It does not allow, for example, the modeling of branch relations instead of nodal relations and it basically assumes that every network model has an admittance matrix representation, which is not possible in many cases. That is why EMTP uses the modified-augmented-nodal analysis (MANA) [2–4] formulation. Equation (3.1) is augmented to include generic device equations and the complete system of network equations can be rewritten in the more generic form:

$$\mathbf{A_N}\, \mathbf{x_N} = \mathbf{b_N} \tag{3.3}$$

In this equation $\mathbf{Y_n} \in \mathbf{A_N}$, $\mathbf{x_N}$ contains both unknown voltage and current quantities and $\mathbf{b_N}$ contains known current and voltage quantities. The matrix $\mathbf{A_N}$ is not necessarily symmetric in EMTP and it is possible to directly accommodate non-symmetric model equations. Equation (3.3) can be written explicitly as

$$\begin{bmatrix} \mathbf{Y_n} & \mathbf{A_c} \\ \mathbf{A_r} & \mathbf{A_d} \end{bmatrix} \begin{bmatrix} \mathbf{v_n} \\ \mathbf{i_x} \end{bmatrix} = \begin{bmatrix} \mathbf{i_n} \\ \mathbf{v_x} \end{bmatrix} \tag{3.4}$$

where the matrices $\mathbf{A_r}$, $\mathbf{A_c}$, and $\mathbf{A_d}$ (augmented portion, row, column, and diagonal coefficients) are used to enter network model equations which are not or cannot be included in $\mathbf{Y_n}$; $\mathbf{i_x}$ is the vector of unknown currents in device models, $\mathbf{v_x}$ is the vector of known voltages, $\mathbf{x_N} = [\mathbf{v_n}\ \ \mathbf{i_x}]^T$ and $\mathbf{b_N} = [\mathbf{i_n}\ \ \mathbf{v_x}]^T$. In reality the network component equations can be entered in any order in (3.3), but the partitioning presented in (3.4) allows to simplify explanations. Further details can be found in [3] and [4].

Equation (3.3) (or (3.4)) can be used for both steady-state and time-domain solutions. A generic coupled admittance model is written as

$$\begin{bmatrix} \mathbf{i_k} \\ \mathbf{i_m} \end{bmatrix} = \begin{bmatrix} \mathbf{Y_{kk}} & \mathbf{Y_{km}} \\ \mathbf{Y_{mk}} & \mathbf{Y_{mm}} \end{bmatrix} \begin{bmatrix} \mathbf{v_k} \\ \mathbf{v_m} \end{bmatrix} \tag{3.5}$$

where $\mathbf{i_k}$ and $\mathbf{i_m}$ represent the current vectors for the sum of currents entering the k side and m side nodes, respectively, the vectors $\mathbf{v_k}$ and $\mathbf{v_m}$ represent the voltages on the k side and m side nodes, respectively. In most cases $\mathbf{Y_{km}} = \mathbf{Y_{mk}}$ and the resulting matrix $\mathbf{Y_n}$ is symmetric, but EMTP also allows to use $\mathbf{Y_{km}} \neq \mathbf{Y_{mk}}$.

For the case of a generic model equation such as

$$k_1 v_k + k_2 v_m + k_3 i_x + k_4 i_y + \cdots = b_z \tag{3.6}$$

where the unknown node voltages (v_k, v_m) and branch currents (i_k, i_m) are members of $\mathbf{x_N}$, while b_z is entered into $\mathbf{b_N}$. Any number of unknown node voltages and branch currents can be used and (3.6) can also include coupling by rewriting it with vectors and matrices. A three-phase voltage source, for example, connected between two arbitrary nodes k and m is expressed as:

$$\mathbf{v_k} - \mathbf{v_m} - \mathbf{Z_s}\mathbf{i_{km}} = \mathbf{v_{b\,km}} \tag{3.7}$$

where $\mathbf{Z_s}$ is the source impedance, $\mathbf{i_{km}} \in \mathbf{i_x}$ representing source currents (entering plus node k) and $\mathbf{v_{b\,km}} \in \mathbf{v_x}$ is the known source voltage. This equation contributes its own row into $\mathbf{A_r}$ with coefficients 1 and -1 on the columns k and m, respectively. The impedance $\mathbf{Z_s}$ (can be set to zero) is entered into a diagonal block of $\mathbf{A_d}$. The row contributed into $\mathbf{A_r}$ is transposed and placed into the corresponding column of $\mathbf{A_c}$. Equation (3.7) can be rewritten in its single-phase form using scalars.

For a single-phase ideal switch model there are two states. In the closed state:

$$v_k - v_m = 0 \tag{3.8}$$

and in the open state:

$$i_{km} = 0 \tag{3.9}$$

where i_{km} is the switch current entering node k and leaving node m. Equation (3.8) is entered into (3.4) by setting:

$$\begin{aligned}
A_{r_{ik}} &= 1, \quad A_{r_{im}} = -1 \\
A_{c_{ki}} &= 1, \quad A_{c_{mi}} = -1 \\
A_{d_{ii}} &= 0
\end{aligned} \tag{3.10}$$

where $A_r \in \mathbf{A_r}$, $A_c \in \mathbf{A_c}$, $A_d \in \mathbf{A_d}$ and the switch equations are entered into the ith row. In the case of (3.9), the formulation becomes:

$$\begin{aligned}
A_{r_{ik}} &= 0, \quad A_{r_{im}} = 0 \\
A_{c_{ki}} &= 1, \quad A_{c_{mi}} = -1 \\
A_{d_{ii}} &= 1
\end{aligned} \tag{3.11}$$

Equation (3.11) makes $\mathbf{A_N}$ non-symmetric and it is also noticed that any switch status change requires updating $\mathbf{A_N}$. It is also possible to model a non-ideal switch using:

$$v_k - v_m - R_S\, i_{km} = 0 \tag{3.12}$$

with R_S being the switch resistance. In this case, the contributions into $\mathbf{A_N}$ become:

$$\begin{aligned}
A_{r_{ik}} &= 1, \quad A_{r_{im}} = -1 \\
A_{c_{ki}} &= 1, \quad A_{c_{mi}} = -1 \\
A_{d_{ii}} &= -R_S
\end{aligned} \tag{3.13}$$

The resistance R_S is replaced by a very high value for the switch open state and by a very low value for the switch closed state. In the closed state, it is allowed to make $R_S = 0$ to obtain the ideal condition.

Single-phase and three-phase transformers can be built in EMTP using the ideal transformer unit shown in Figure 3.4. It consists of dependent voltage and current sources. The secondary branch equation is given by

$$v_{k2} - v_{m2} - g\, v_{k1} + g\, v_{m1} = 0 \tag{3.14}$$

where g is the transformation ratio. This equation contributes its own row into the matrix $\mathbf{A_r}$, whereas the matrix $\mathbf{A_c}$ contains the transposed version of that row. It is possible to extend to multiple secondary windings using parallel connected current

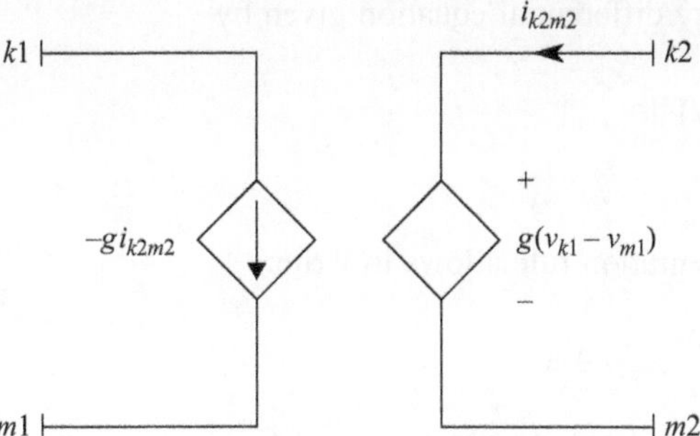

Figure 3.4   Ideal transformer model unit

sources on the primary side and series connected voltage sources on the secondary side. Leakage losses and the magnetization branch are added externally to this ideal transformer nodes.

Three-legged core-form transformer models or any other types can be included using coupled leakage matrices and magnetization branches.

The MANA formulation (3.3) of EMTP is completely generic and can easily accommodate the juxtaposition of arbitrary component models in arbitrary network topologies with any number of wires and nodes. It is not limited to the usage of the unknown variables presented in (3.4) and can be augmented to include different types of unknown and known variables. The MANA formulation is conceptually simple to realize and program.

3.4.1.1   Steady-state solution

The steady-state version of (3.3) is based on complex numbers. Equation (3.3) is rewritten using capital letters for the specific case of complex variables (phasors).

$$\mathbf{A_N X_N = B_N} \tag{3.15}$$

For a network containing sources with different frequencies, (3.15) can be independently solved at each frequency and, assuming that the network is linear (linear models are used for all devices), the solutions of $\mathbf{X_N}$ at each frequency are combined to derive the harmonic steady-state solution in the form of a Fourier series. EMTP is able to produce a steady-state linear harmonic solution.

In EMTP the steady-state solution is calculated for the frequencies of the studied network, but can be also calculated for a user-specified frequency range for performing frequency scans. The frequency scan option can be also used to calculate Thevenin impedances at arbitrary locations in the network. Sequence impedances can be calculated from three-phase buses.

3.4.1.2   Time-domain solution

EMTP uses the trapezoidal integration method with Backward Euler for the treatment of discontinuities. This approach is similar to [5, 6].

For an ordinary differential equation given by

$$\frac{dx}{dt} = f(x, t)$$

$$x(0) = x_0 \tag{3.16}$$

the trapezoidal integration rule allows to write:

$$x_t = \frac{\Delta t}{2}f_t + \frac{\Delta t}{2}f_{t-\Delta t} + x_{t-\Delta t} \tag{3.17}$$

The time-domain module of EMTP starts from 0-state (all devices are initially deenergized) or from given automatic or manual initial conditions and computes all variables as a function of time using a time-step Δt. The terms found at $t - \Delta t$ constitute history terms and all quantities at time-point t are also related through network equations. The computation process finds a solution at discrete time-points as illustrated in Figure 3.5.

During a discontinuity, (3.17) may cause numerical oscillations. Such oscillations can be eliminated by moving into two halved time-step integrations using the Backward Euler method. In this case, (3.17) becomes:

$$x_t = \frac{\Delta t}{2}f_t + x_{t-\frac{\Delta t}{2}} \tag{3.18}$$

where $t - \Delta t/2$ is the discontinuity detection time-point. After the completion of halved time-step solutions with (3.18), and if there are no other discontinuities, the solution method moves back to (3.17). Discontinuities can be automatically detected, for example, from ideal switches changing states, sources starting and stopping and nonlinear functions changing slopes. They become more complicated to detect with black-box type device models and generic nonlinear functions. Although the concept of this approach is relatively simple, its actual implementation causes several complexities and specific equations must be derived for each device model in EMTP.

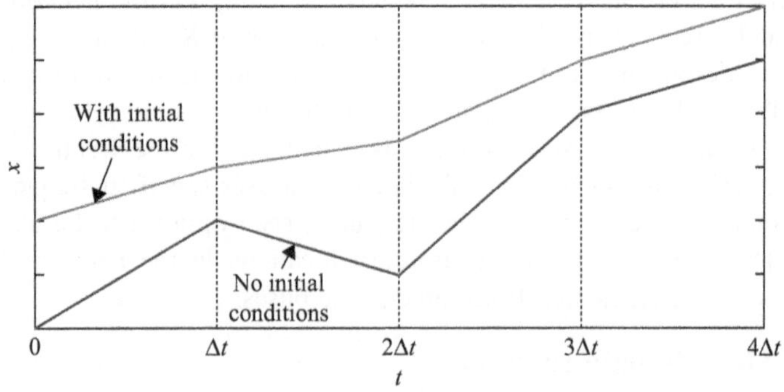

Figure 3.5   Discrete solution time-points

In the time-domain solution, EMTP uses the real version of (3.3) for solving the algebraic-differential equations of the network. At each solution time-point:

$$\mathbf{A}_{\mathbf{N}_t}\, \mathbf{x}_{\mathbf{N}_t} = \mathbf{b}_{\mathbf{N}_t} \tag{3.19}$$

This is achieved simply by converting or discretizing the device equations using a numerical integration technique such as (3.17). For a pure inductor branch connected between two nodes k and m:

$$v_{km} = L\frac{di_{km}}{dt} \tag{3.20}$$

and the discretized version (equivalent circuit model or companion branch) [7–9] is given by

$$i_t = \frac{\Delta t}{2L}v_t + \frac{\Delta t}{2L}v_{t-\Delta t} + i_{t-\Delta t} \tag{3.21}$$

where the km subscript is dropped to simplify notation. Since the last two terms of this equation represent computations available from a previous time-point solution, it can be written as

$$i_t = \frac{\Delta t}{2L}v_t + i_{h_t} \tag{3.22}$$

where i_{h_t} is a history term for the solution at the time-point t. The history term contributes into $\mathbf{b}_{\mathbf{N}_t}$, the coefficient of voltage contributes into $\mathbf{A}_{\mathbf{N}_t}$ and $v_t \in \mathbf{x}_{\mathbf{N}_t}$. More specifically, (3.22) is rewritten as follows at the time-point t:

$$\begin{bmatrix} i_k \\ i_m \end{bmatrix} = \begin{bmatrix} \hat{y} & -\hat{y} \\ -\hat{y} & \hat{y} \end{bmatrix} \begin{bmatrix} v_k \\ v_m \end{bmatrix} + \begin{bmatrix} i_h \\ -i_h \end{bmatrix} \tag{3.23}$$

where $\hat{y} = \frac{\Delta t}{2L}$.

An equation similar to (3.22) is written for capacitor branches. The inductor, capacitor, and resistor are primitive elements for building other models.

The companion branch approach is generalized in EMTP to include coupled branch models, rotating machines, and other components with differential equations. The generic nodal admittance formulation is similar to (3.5), but since it is done in time-domain, it includes history terms as in (3.23).

3.4.1.3 Nonlinear models

Nonlinearities occur due to nonlinear functions used in EMTP network device models. In most cases a nonlinear function can be modeled using piecewise linear segments. The positive part of a sample nonlinear function with 3 segments is shown in Figure 3.6 for a voltage $v_x(t)$ and a current $i_x(t)$. In EMTP such functions are monotonically increasing since it is assumed that there is a unique solution for a given voltage. Each segment j is represented by a liner equation:

$$i_{x_j} = K_j\, v_{x_j} + i_{q_j} \tag{3.24}$$

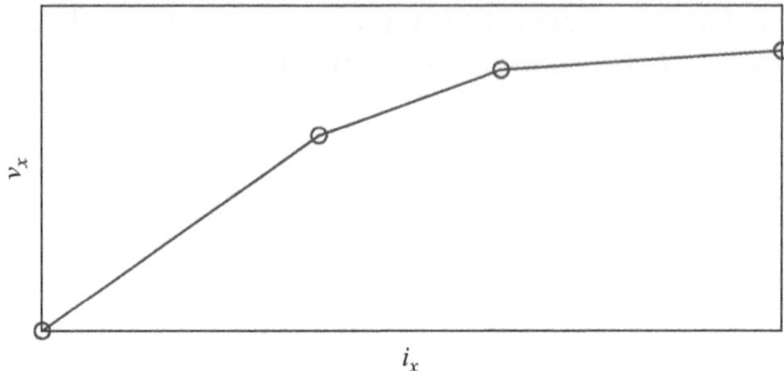

Figure 3.6   Generic nonlinear function in EMTP

which is in fact a Norton equivalent with admittance K_j and Norton current source i_{q_j}. This relation can be directly included into (3.19). It constitutes a linearization of the nonlinear function at the operation point for the voltage solution at the time-point t.

In EMTP, (3.24) is written in its vector-matrix form for coupled nonlinearities.

EMTP can also handle generic nonlinear functions directly, without initial transformation into segments. In some cases, such as for arrester models, fitting with exponential segments must be used. For generic nonlinear functions, the linearization equation (3.24) must be recalculated at each voltage solution since it is not available beforehand. Typical examples are the breaker arc model, the surge arrester models, and generic black-box devices. For generic black-box devices, EMTP can use numerical perturbation to calculate function derivatives.

There are two main categories of methods for solving nonlinear functions: with solution delays and without solution delays. EMTP uses a fully iterative solver and allows the elimination of all solution delays by solving all nonlinear functions of a network simultaneously.

The simultaneous solution means that if at a given solution time-point the node voltage of a device modifies its current (or equivalent model) then it is necessary to update and resolve the network equations until all voltages stop changing within a tolerance. The convergence of all voltages must occur before moving to the next solution time-point.

EMTP uses function linearization with full matrix updating (LFMA). At each time-point the linearized model equations (such as (3.24)) are updated through an iterative process until convergence. The updating of each device equation requires updating the main system of equations. In this approach the matrix $\mathbf{A_{N_t}}$ is actually the Jacobian matrix. Despite the fact that the iterations are performed with the full system matrix, this approach is programmed very efficiently and provides fast convergence properties due to the Newton method.

3.4.2   State-space analysis

In EMTP it is possible to use generic state-space equations. This option is for accommodating models (see [10], for example) based on this formulation. State-space equations are given by

$$\dot{\mathbf{x}} = \mathbf{A}\mathbf{x} + \mathbf{B}\mathbf{u} \tag{3.25}$$

$$\mathbf{y} = \mathbf{C}\mathbf{x} + \mathbf{D}\mathbf{u} \tag{3.26}$$

where $\mathbf{x}$ is the vector of state variables, $\mathbf{u}$ is the vector of inputs and $\mathbf{y}$ is the vector of outputs. The matrices $\mathbf{A}$, $\mathbf{B}$, $\mathbf{C}$, and $\mathbf{D}$ are called the state matrices. These matrices can be calculated for given ideal switch positions and piecewise linear device segments. Each topological change requires updating the state matrices. EMTP allows to use the more generic version:

$$\mathbf{y} = \mathbf{C}\,\mathbf{x} + \mathbf{D}\,\mathbf{u} + \mathbf{D}_1\,\dot{\mathbf{u}} \tag{3.27}$$

for output equations.

The state-space equations (3.25) and (3.26) can be solved in both steady-state and time-domain. Other theoretical details can be found in EMTP documentation and [11].

3.4.2.1 Steady-state solution

In steady-state conditions the differential of $\mathbf{x}$ is transformed into $s\tilde{\mathbf{X}}$ with the Laplace operator $s = j\omega$ (ω is the steady-state frequency in rad/s and j is the complex operator):

$$\tilde{\mathbf{X}} = (s\,\mathbf{I} - \mathbf{A})^{-1}(\mathbf{B}\,\tilde{\mathbf{U}}) \tag{3.28}$$

where tilde-upper-case vectors are used to denote phasors and $\mathbf{I}$ is the identity matrix. Equation (3.27) can be written in steady-state as

$$\tilde{\mathbf{Y}} = \left[\mathbf{C}(s\,\mathbf{I} - \mathbf{A})^{-1}\,\mathbf{B} + \mathbf{D} + s\,\mathbf{D}_1 \right] \tilde{\mathbf{U}} \tag{3.29}$$

In this equation $\tilde{\mathbf{U}}$ represents network voltages and $\tilde{\mathbf{Y}}$ corresponds to currents. This means that (3.29) has the same shape as (3.5) and can be directly entered into (3.15).

3.4.2.2 Time-domain solution

In time-domain the state-space equations are discretized using trapezoidal integration and Backward Euler for discontinuities. It can be shown that (3.25) results into

$$\mathbf{x}_t = \left(\mathbf{I} - \frac{\Delta t}{2}\mathbf{A}\right)^{-1}\left(\mathbf{I} + \frac{\Delta t}{2}\mathbf{A}\right)\mathbf{x}_{t-\Delta t} + \left(\mathbf{I} - \frac{\Delta t}{2}\mathbf{A}\right)^{-1}\frac{\Delta t}{2}\mathbf{B}(\mathbf{u}_t + \mathbf{u}_{t-\Delta t}) \tag{3.30}$$

when discretized through the trapezoidal integration method of (3.17). Equation (3.30) can be written as

$$\mathbf{x}_t = \hat{\mathbf{A}}\,\mathbf{x}_{t-\Delta t} + \hat{\mathbf{B}}\,\mathbf{u}_{t-\Delta t} + \hat{\mathbf{B}}\,\mathbf{u}_t \tag{3.31}$$

where hatted matrices result from the discretization process.

Once (3.25) and (3.27) (or (3.26)) are defined for a given model (or group), they can be inserted into the main network equation (3.19) and simultaneously interfaced with other model equations based on state-space or nodal formulations. It is first noticed that (3.27) can be rewritten at a solution time-point t in terms of current injections $\mathbf{i}_t$ and nodal voltages $\mathbf{v}_t$:

$$\mathbf{i}_t = \mathbf{C}_k\,\mathbf{x}_t + \mathbf{D}_k\,\mathbf{v}_t + \mathbf{i}_t^c \tag{3.32}$$

where trapezoidal integration for the capacitive current $(\mathbf{i}_t^c)$ results into

$$\mathbf{i}_t^c = \frac{2}{\Delta t}\mathbf{D}_{1_k}\,\mathbf{v}_t - \frac{2}{\Delta t}\mathbf{D}_{1_k}\,\mathbf{v}_{t-\Delta t} - \mathbf{i}_{t-\Delta t}^c \qquad (3.33)$$

The state equations are now written with input voltage:

$$\mathbf{x}_t = \hat{\mathbf{A}}_k\,\mathbf{x}_{t-\Delta t} + \hat{\mathbf{B}}_k\,\mathbf{v}_{t-\Delta t} + \hat{\mathbf{B}}_k\,\mathbf{v}_t \qquad (3.34)$$

Equations (3.33) and (3.34) can be replaced into (3.32) for creating a nodal formulation that can be directly integrated into (3.19).

3.5 Control systems

The simulation of control system dynamics is fundamental for studying power system transients. In EMTP, control elements can be transfer functions, limiters, gains, summers, integrators, and many other mathematical functions. In many applications the block-diagram approach is used in EMTP to build and interface user-defined models with the built-in power system and control system components.

A typical control diagram taken from the AVR_Gov block shown in Figure 3.2 is shown in Figure 3.7. Such diagrams are drawn in the EMTP GUI and solved directly. The GUI allows drawing control systems directly.

A complicated problem in oriented-graph systems is the capability to solve the complete system simultaneously without inserting artificial (one-time step) delays in feedback loops with nonlinear functions. A solution to this problem is available in EMTP [12]. A matrix formulation is used in EMTP to solve the complete control system equations. The system of equations is given by

$$\mathbf{A}_C\,\mathbf{x}_C = \mathbf{b}_C \qquad (3.35)$$

where the sparse matrix $\mathbf{A}_C$ is the coefficient matrix of control signals $\mathbf{x}_C$ and $\mathbf{b}_C$ is the vector of known variables. All linear functions (blocks) are entered directly

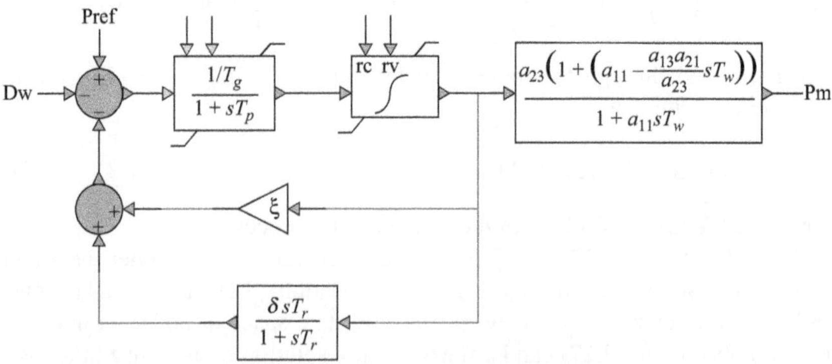

Figure 3.7 Sample control system diagram

into (3.35). The inclusion of nonlinear blocks is more complicated. Equation (3.35) is generically nonlinear and its solution is found through the minimization function

$$\mathbf{f}(\mathbf{x}) = \mathbf{A_C}\,\mathbf{x_C} - \mathbf{b_C} = \mathbf{0} \tag{3.36}$$

This equation can be solved through an iterative Newton method:

$$\mathbf{f}(\mathbf{x}^{(k)}) + \mathbf{J}^{(k)}\left(\mathbf{x}_\mathbf{C}^{(k+1)} - \mathbf{x}_\mathbf{C}^{(k)}\right) = \mathbf{0} \tag{3.37}$$

where k is the iteration count and $\mathbf{J}^{(k)}$ is the Jacobian matrix. The combination of (3.35) to (3.37) results into

$$\mathbf{J}^{(k)}\,\mathbf{x}^{(k+1)} = \mathbf{J}^{(k)}\,\mathbf{x}^{(k)} - \mathbf{A_C}\,\mathbf{x}^{(k)} + \mathbf{b}_\mathbf{C}^{(k)} \tag{3.38}$$

which is now equivalent to writing:

$$\hat{\mathbf{A}}_\mathbf{C}^{(k)}\,\mathbf{x}_\mathbf{C}^{(k+1)} = \mathbf{B}_\mathbf{C}^{(k)} \tag{3.39}$$

The matrix $\hat{\mathbf{A}}_\mathbf{C}^{(k)}$ cells are the same as in $\mathbf{A_C}$ for linear blocks and different from $\mathbf{A_C}$ for nonlinear blocks. This is best illustrated using the simple example of Figure 3.8. In this diagram, the input variable u is a known variable and the output signal is x_3. The function f_4 is nonlinear. The Jacobian matrix of this system is given by

$$\mathbf{J}^{(k)} = \begin{bmatrix} 1 & 0 & 0 & 0 \\ 1 & -1 & 0 & -1 \\ 0 & c & -1 & 0 \\ 0 & 0 & -2x_3^{(k)} & 1 \end{bmatrix} \tag{3.40}$$

Equation (3.40) is solved using sparse matrix techniques where decoupled block diagrams are treated independently. The matrix $\hat{\mathbf{A}}_\mathbf{C}^{(k)}$ must be continuously updated, refactorized, and resolved at each iteration and for each solution time-point. This updating is not necessary when the quiescent points of all nonlinear functions are not changing. The size of $\mathbf{J}^{(k)}$ is the total count of control blocks. In addition the derivative of each nonlinear block must be reevaluated either from its known formula or through perturbation for complex and black-box functions.

A relaxed non-iterative solution variant is also available in EMTP. It is based on updating the Jacobian matrix only once at each solution time-point. This approach

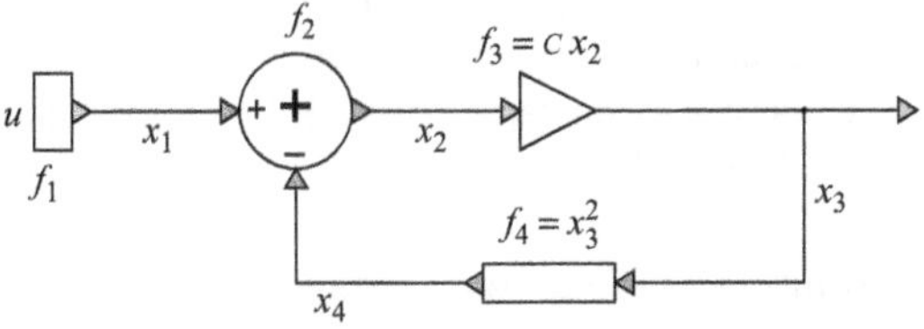

Figure 3.8   Block diagram example in EMTP

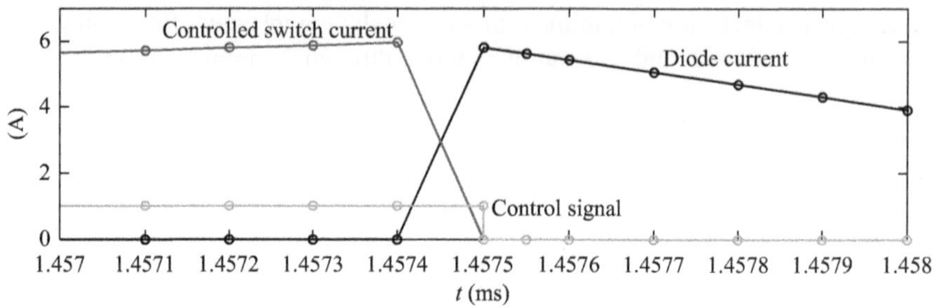

Figure 3.9   Simultaneous solution for a buck-boost converter circuit

is non-iterative and remains acceptable for a large class of problems, but may lose accuracy when the numerical integration time-step increases.

The automatic initialization of control system equations is an important research topic on numerical methods. Currently EMTP uses semi-automatic methods for propagating the required output values backwards in control system diagrams for the establishment of initial conditions in all blocks. The required output values can be manually imposed or automatically calculated from some device models.

In most applications the control system diagram equations are solved separately from network equations. Although this is not a significant source of errors in most cases, it can become an important drawback for user-defined network models and in the simulation of power electronics systems. The combination of both systems into a unique system of iteratively solved equations is complex. A fixed-point approach is programmed in EMTP, where both systems are solved sequentially is more efficient and acceptable in many cases [12].

As indicated above, the iterative solution of network equations with control system equations also allows to solve simultaneous switching problems encountered in power electronics circuits. The example of Figure 3.9 demonstrates the simultaneous transition of currents in a buck-boost converter. When the controlled switch is turned off by its control signal, a diode must simultaneously (within the same time-step) turn on since a completely wrong solution will result otherwise. Cascading effects can be accounted for when using this approach.

3.6   Multiphase load-flow solution and initialization

The load-flow module in EMTP is used to compute the operating conditions of the power system. EMTP uses a multiphase solution. In fact, the EMTP load-flow solution is circuit based and does not make any assumptions on the simulated network. The objective is to use the same network topology/data and initialize the time-domain network solution. The methodology presented below is the one used in [3, 4] (see other details in [13]).

The load-flow solution is based on constraints. The sources (synchronous machines or other types of generation) are replaced by PQ, PV, or slack bus constraints. The loads are replaced by PQ constraints. These are the main load-flow devices.

All network components must provide a load-flow solution model. For the majority of devices, this model is the same as the one used in the steady-state representation.

The system of (3.3) is used for representing the passive network equations, but it must be augmented with load-flow constraint equations. This is another major advantage of the augmented formulation approach. The modified-augmented load-flow equations are given by

$$\begin{bmatrix} \mathbf{A}_N^{LF} & \mathbf{A}_I \\ \mathbf{L}_{LA} & \mathbf{L}_d \end{bmatrix} \begin{bmatrix} \Delta\mathbf{x} \\ \Delta\mathbf{x}_{LF} \end{bmatrix} = -\mathbf{f}_{LF} \tag{3.41}$$

Equation (3.41) is real, since real and imaginary parts must be represented separately for load-flow constraints. $\mathbf{A}_N^{LF}$ is constructed from the original complex version of $\mathbf{A}_N$ by separating real and imaginary parts of each element, $\mathbf{A}_I$ is a connectivity matrix for accounting for load-flow devices, and $\mathbf{L}_{LA}$ and $\mathbf{L}_d$ provide load-flow device constraint equations. The vector function $\mathbf{f}_{LF}$ must be minimized (zero) for finding the load-flow solution. It is recalled that the solution of $\mathbf{f}_{LF}(\mathbf{x}) = \mathbf{0}$ can be found using the Newton method:

$$\mathbf{J}^{(j)}\Delta\mathbf{x}^{(j)} = -\mathbf{f}_{LF}(\mathbf{x}^{(j)}) \tag{3.42}$$

where (j) is the iteration count, $\mathbf{J}$ is the Jacobian matrix, and $\Delta\mathbf{x}$ is used in

$$\mathbf{x}^{(j+1)} = \mathbf{x}^{(j)} + \Delta\mathbf{x}^{(j)} \tag{3.43}$$

Equation (3.41) can be expanded and rewritten as

$$\begin{bmatrix} \mathbf{Y}_n & \mathbf{A}_c & \mathbf{A}_{IL} & \mathbf{A}_{IG} & 0 \\ \mathbf{A}_r & \mathbf{A}_d & 0 & 0 & 0 \\ \mathbf{J}_L & 0 & \mathbf{J}_{LI} & 0 & 0 \\ \mathbf{Y}_G & 0 & 0 & \mathbf{B}_{GI} & \mathbf{B}_{GE} \\ \mathbf{J}_{GPQ} & 0 & 0 & \mathbf{J}_{GPQI} & 0 \\ \mathbf{J}_{GPV} & 0 & 0 & 0 & 0 \\ \mathbf{J}_{GSL} & 0 & 0 & 0 & 0 \end{bmatrix}^{(j)} \begin{bmatrix} \Delta\mathbf{V}_n \\ \Delta\mathbf{I}_x \\ \Delta\mathbf{I}_L \\ \Delta\mathbf{I}_G \\ \Delta\mathbf{E}_G \end{bmatrix} = - \begin{bmatrix} \mathbf{f}_n \\ \mathbf{f}_x \\ \mathbf{f}_L \\ \mathbf{f}_{GI} \\ \mathbf{f}_{GPQ} \\ \mathbf{f}_{GPV} \\ \mathbf{f}_{GSL} \end{bmatrix}^{(j)} \tag{3.44}$$

where the submatrices $\mathbf{A}_{IL}$ and $\mathbf{A}_{IG}$ are adjacency matrices for interfacing the load ($\mathbf{I}_L$) and generator ($\mathbf{I}_G$) currents with the corresponding passive network nodes. The following submatrices [13] are used to include load-flow constraints for any number of phases: $\mathbf{J}_L$ and $\mathbf{J}_{LI}$ for PQ-loads, $\mathbf{Y}_G$, $\mathbf{B}_{GI}$, and $\mathbf{B}_{GE}$ for generator current equations, $\mathbf{J}_{GPQ}$ and $\mathbf{J}_{GPQI}$ for PQ-type generators, $\mathbf{J}_{GPV}$ for PV-type generators, and $\mathbf{J}_{GSL}$ for slack-type generators (buses). The unknown supplementary vectors are defined as follows: $\mathbf{I}_L$ is for PQ-load currents, $\mathbf{I}_G$ is for generator currents, and $\mathbf{E}_G$ represents generator internal voltages.

Equation (3.44) is the expanded version of (3.42) and the left-hand side matrix is the Jacobian matrix of the solved nonlinear function $\mathbf{f}_{LF}$.

The linear network coefficient submatrices are copied from the real version (3.4) and remain unchanged, except that now the right-hand side of (3.4) is eliminated due to the differentiation process for the derivation of the Jacobian matrix. There is no need to derive separate equations for linear network devices. For the ideal transformer case, for example, the constraint equation is taken directly from (3.14):

$$V_{k2} - V_{m2} - gV_{k1} + gV_{m1} = f_x \tag{3.45}$$

and results into

$$\Delta V_{k2} - \Delta V_{m2} - g\,\Delta V_{k1} + g\,\Delta V_{m1} = -f_x^{(j)} \tag{3.46}$$

This means that the existing submatrices $\mathbf{A_r}$ and $\mathbf{A_c}$ are not affected and already constitute the Jacobian terms allowing to account for all single-phase and three-phase transformer configurations. The tap positions will affect the ratio g.

3.6.1   Load-flow constraints

EMTP uses constraint equations of load-flow devices which are presented in this section to demonstrate how such arbitrary equations are entered into (3.44).

In the case of a single-phase load, the PQ constraint can be expressed in terms of load voltage and current. If a load is numbered as p in the list of load devices and connected between the nodes k and m, its voltage V_{L_p} and current I_{L_p} are given by

$$
\begin{aligned}
V_{L_p} &= V_k^R + jV_k^I - V_m^R - jV_m^I = V_{km}^R + jV_{km}^I \\
I_{L_p} &= I_{p_{km}}^R + jI_{p_{km}}^I
\end{aligned}
\tag{3.47}
$$

where the superscripts R and I stand for real and imaginary parts, respectively, and the subscript km means from k to m. The constraint equations become:

$$
\begin{aligned}
P_p - V_{km}^R\, I_{p_{km}}^R - V_{km}^I I_{p_{km}}^I &= f_{L_p}^R = 0 \\
Q_p - V_{km}^I\, I_{p_{km}}^R + V_{km}^R\, I_{p_{km}}^I &= f_{L_p}^I = 0
\end{aligned}
\tag{3.48}
$$

and the Jacobian terms can now be written as

$$
\begin{bmatrix} -I_{p_{km}}^{R(j)} & -I_{p_{km}}^{I(j)} \\ I_{p_{km}}^{I(j)} & -I_{p_{km}}^{R(j)} \end{bmatrix}
\begin{bmatrix} \Delta V_{km}^R \\ \Delta V_{km}^I \end{bmatrix}
+
\begin{bmatrix} -V_{km}^{R(j)} & -V_{km}^{I(j)} \\ -V_{km}^{I(j)} & V_{km}^{R(j)} \end{bmatrix}
\begin{bmatrix} \Delta I_{p_{km}}^R \\ \Delta I_{p_{km}}^I \end{bmatrix}
=
\begin{bmatrix} -f_{L_p}^{R(j)} \\ -f_{L_p}^{I(j)} \end{bmatrix}
\tag{3.49}
$$

The voltage and current coefficient matrices populate the matrices $\mathbf{J_L}$ and $\mathbf{J_{LI}}$ (of (3.44)), respectively. The right-hand side is entered into the vector $\mathbf{f_L}$ of (3.44). In EMTP the three-phase load devices are created from three single-phase loads. The load can be connected in arbitrary configurations.

In EMTP the current constraint of a three-phase generator numbered as p in the list of generators and connected between buses $\mathbf{k}$ and $\mathbf{m}$ takes the following form:

$$\mathbf{Y_{G_p}}(\mathbf{V_k} - \mathbf{V_m} - \tilde{\mathbf{E}}_{\mathbf{G}_p}) - \mathbf{I_{G_p}} = \mathbf{f_{GI}}p = 0 \tag{3.50}$$

This is a three-phase and complex system of equations. The generator currents in the vector $\mathbf{I}_{Gp}$ are entering bus $\mathbf{k}$, the matrix $\mathbf{Y}_{G_p}$ is the internal admittance matrix, and the vector $\tilde{\mathbf{E}}_{G_p}$ holds the internal positive sequence voltages and can be written as $\mathbf{E}_{Gp} = \begin{bmatrix} 1 & a^2 & a \end{bmatrix}^T E_{G_p}$ with $a = 1\angle 120°$ and the scalar E_{G_p} being the voltage of phase A. Equation (3.50) is rewritten as

$$\mathbf{Y}_{G_p}\Delta\mathbf{V_k} - \mathbf{Y}_{G_p}\Delta\mathbf{V_m} - \mathbf{B}_{GE_p}\Delta E_{G_p} - \Delta\mathbf{I}_{G_p} = -\mathbf{f}_{\mathbf{GI}_p}^{(j)} \tag{3.51}$$

In this case $\mathbf{Y}_{G_p}$ populates $\mathbf{Y_G}$, $\mathbf{B}_{GE_p}$ is entered into $\mathbf{B_{GE}}$, and the identity coefficient matrix of $\Delta\mathbf{I}_{G_p}$ contributes to $\mathbf{B_{GI}}$. In addition, $\Delta E_{G_p} \in \Delta\mathbf{E_G}$, $\Delta\mathbf{I}_{G_p} \in \Delta\mathbf{I_G}$ and $\mathbf{f}_{\mathbf{GI}_p} \in \mathbf{f_{GI}}$.

In addition to (3.50), each generator must provide its equations for controlled quantities. In the case of PQ control, the Jacobian terms are similar to (3.49), except that now the left-hand side voltage and current vectors are three-phase which results into two 6 by 6 coefficient matrices contributing into $\mathbf{J_{GPQ}}$ and $\mathbf{J_{GPQI}}$. The right-hand side is now related to $\mathbf{f_{GPQ}}$.

For the case of PV control, the Q row of PQ control is replaced by the voltage constraint equation. The control is on the voltage magnitude V_p (pth generator) of one phase and gives:

$$V_p - \sqrt{\left(V_p^R\right)^2 + \left(V_p^I\right)^2} = f_{GPV_p} = 0 \tag{3.52}$$

The contribution to (3.44) is found from:

$$-\left[\left(V_p^R\right)^2 + \left(V_p^I\right)^2\right]^{-\frac{1}{2}}\left(V_p^{R(j)}\,\Delta V_p^R + V_p^{I(j)}\,V_p^I\right) = -f_{GPV_p}^{(k)} \tag{3.53}$$

and affects $\mathbf{J_{GPV}}$ and $\mathbf{f_{GPV}}$. It is not complicated to extend (3.52) to control the positive sequence voltage magnitude by calculating it from the Fortescue transformation for positive sequence.

For the slack bus case it is necessary to write two equations (real and imaginary parts) for controlling the positive sequence voltage magnitude and phase.

3.6.2   Initialization of load-flow equations

To initialize its load-flow equations, EMTP uses a linearized version of (3.44). All loads are replaced by equivalent impedances calculated at nominal voltages and all generators are replaced by ideal sources behind their impedances. This approach yields sufficiently good initial conditions for the fast convergence of (3.44).

3.6.3   Initialization from a steady-state solution

Upon convergence of (3.44), all steady-state phasors become available in EMTP. The synchronous machine phasors are used to calculate internal state variables. The asynchronous machine requires the calculation of slip for a given mechanical power or torque.

The steady-state module starts with the load-flow solution and replaces all devices by lumped equivalents to proceed with a phasor solution. This is achieved with (3.15). The resulting solution is the same as with the load-flow module, except that now all device models have access to internal phasors for proceeding with initialization. The steady-state solution phasors are used for initializing all state-variables at the time-point $t = 0$. The solution at $t = 0$ is only from the steady-state and all history terms for all devices are initialized for the first solution time-point.

3.7 Implementation

The matrix $\mathbf{A_N}$ of an arbitrary network is normally sparse. A block diagonal structure results when the network contains transmission lines or cables. EMTP uses a sparse matrix solver for its load-flow (see (3.44)), steady-state (see (3.15)), and time-domain (see (3.19)) equations.

The direct approach is used with LU factorization. For the given sparse matrix $\mathbf{A}$:

$$\mathbf{Ax} = \mathbf{L}(\mathbf{Ux}) = \mathbf{b} \tag{3.54}$$

where $\mathbf{L}$ and $\mathbf{U}$ are lower and upper triangular matrices, respectively. Special ordering techniques can be used to generate row and column permutations and reduce fill-in during factorization. Minimum degree ordering is currently used in EMTP.

Recently the KLU solver has been established as one of the most effective solvers for circuit simulation problems [14–16]. The KLU solver permutes the matrix $\mathbf{A}$ to obtain the Block Triangular Form (BTF) before ordering. The BTF of $\mathbf{A}$ results into the new matrix $\mathbf{A'}$:

$$\mathbf{PAQ} = \begin{bmatrix} \mathbf{A'_{11}} & \mathbf{A'_{12}} & \cdots & \mathbf{A'_{14}} \\ \mathbf{0} & \mathbf{A'_{22}} & & \vdots \\ \mathbf{0} & \mathbf{0} & \ddots & \vdots \\ \mathbf{0} & \mathbf{0} & \mathbf{0} & \mathbf{A'_{nn}} \end{bmatrix} \tag{3.55}$$

where $\mathbf{P}$ and $\mathbf{Q}$ are column and row permutation matrices, respectively. The BTF provides important advantages. The diagonal blocks can become independent and hence only these blocks require factorization. The off-diagonal non-zeroes do not contribute to any fill-in. When a network contains delay-based line/cable models, the BTF allows us to automatically formulate the block-diagonal version of $\mathbf{A}$. The independent blocks can be solved in parallel. This method is currently available in EMTP for testing and generalization.

3.8 EMTP models

The basic modeling theory used in EMTP originates from [17], but several improvements have been made through ongoing research and within the new programming environment and computational engine of EMTP. A collection of built-in

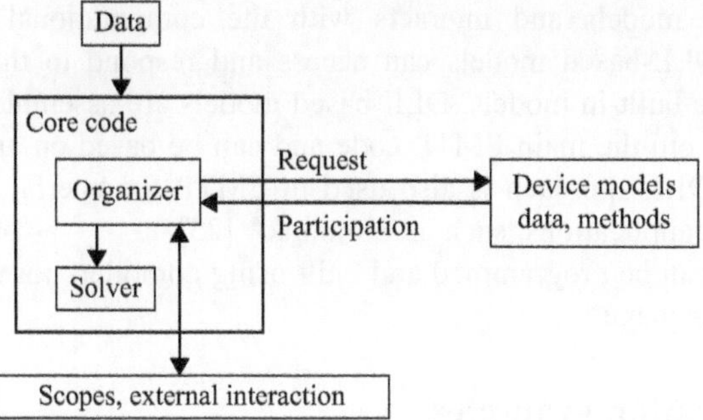

Figure 3.10   EMTP architecture for interfacing its core code with device models

models, referred to as primitive models, is used to construct assembled models. The primitive control block diagram models are used, for example, to construct governor and exciter models, metering devices, arc models, and various other models.

The models are programmed using an object-oriented approach. Each model is given its set of data fields and methods. The interaction with the core code is presented in Figure 3.10. The models receive requests (method calls) and may participate by sending back the required data. This approach decouples completely the models from the code and provides both flexibility and encapsulation. The model development is completely independent from the core code. The models can be implemented directly or through standard libraries and DLLs (see below).

The rotating machine models are based on Park's transformation (dq0). The machine models [18, 19] are interfaced using Norton equivalents. In addition to internal iteration loops [18, 19], it is optionally possible to iterate the machine equations with the main network equations to achieve highest accuracy levels [20]. The importance of such an interface in practical applications is demonstrated in [20]. The machine models participate into all solution modules and it is possible to initialize them under unbalanced network conditions.

In addition to constant parameter transmission line and cable models, EMTP uses frequency-dependent models based on [21–24].

The EMTP model libraries include numerous components, such as power electronics switching devices and converters; HVDC converters; breakers; ideal switches; transformers; arresters; rotating machines; wind generators; line/cable date calculation tools; lines and cables; measuring devices; control system components; and various source functions.

EMTP does not require any compiler to perform simulations with its built-in models.

3.9   External programming interface

EMTP allows the development of user-defined models and interacts with its computational core, using various interfacing techniques. The dynamic link library (DLL) setup offers an elaborate programming interface that allows the building of user-defined models and interacts with the computational engine options and matrices. DLL-based models can access and respond to the same requests (methods) as the built-in models. DLL-based models are assembled and compiled independently from the main EMTP code and can be based on any programming language. The DLL approach is also used in EMTP for interfacing with models built in external applications, such as Simulink® [25].

The DLLs can be programmed and built using compilers for the user selected programming language.

3.10   Application examples

EMTP can be used to simulate generic and very large-scale power systems. It does not have memory constraints and can automatically allocate memory according to simulated network requirements. EMTP applications include:

- switchgear, TRV, shunt compensation, current chopping, delayed-current zero conditions
- insulation coordination
- saturation and surge arrester influences
- harmonic propagation, power quality
- interaction between compensation and control
- wind generation, distributed generation
- precise determination of short-circuit currents
- detailed behavior of synchronous machines and related controls, auto-excitation, subsynchronous resonance, power oscillations
- protection systems
- distribution system analysis: unbalanced load-flow, protection, renewable energies
- HVDC systems, power electronics, FACTS and Custom Power controllers
  These applications are in a wideband range of frequencies.

The following illustrative examples are presented below to provide a short list of EMTP simulation capabilities. The objective of EMTP is to allow users to simulate various power system events for a wideband frequency range, using the same data set and with no user intervention for supporting the solution methods. It is not necessary, for example, to insert artificial delays between network portions and control system equations or to insert artificial devices for helping convergence or solving accuracy/robustness issues.

The graphical user interface of EMTP also allows us to work with various levels of complexity by scripting study scenarios and data.

3.10.1   Switching transient studies

A typical switching transient simulation case is presented in Figure 3.11. This transient is due to the energization of a transformer in the 500 kV network connected to BUS2 in Figure 3.2. A harmonic resonance is observed and the transient

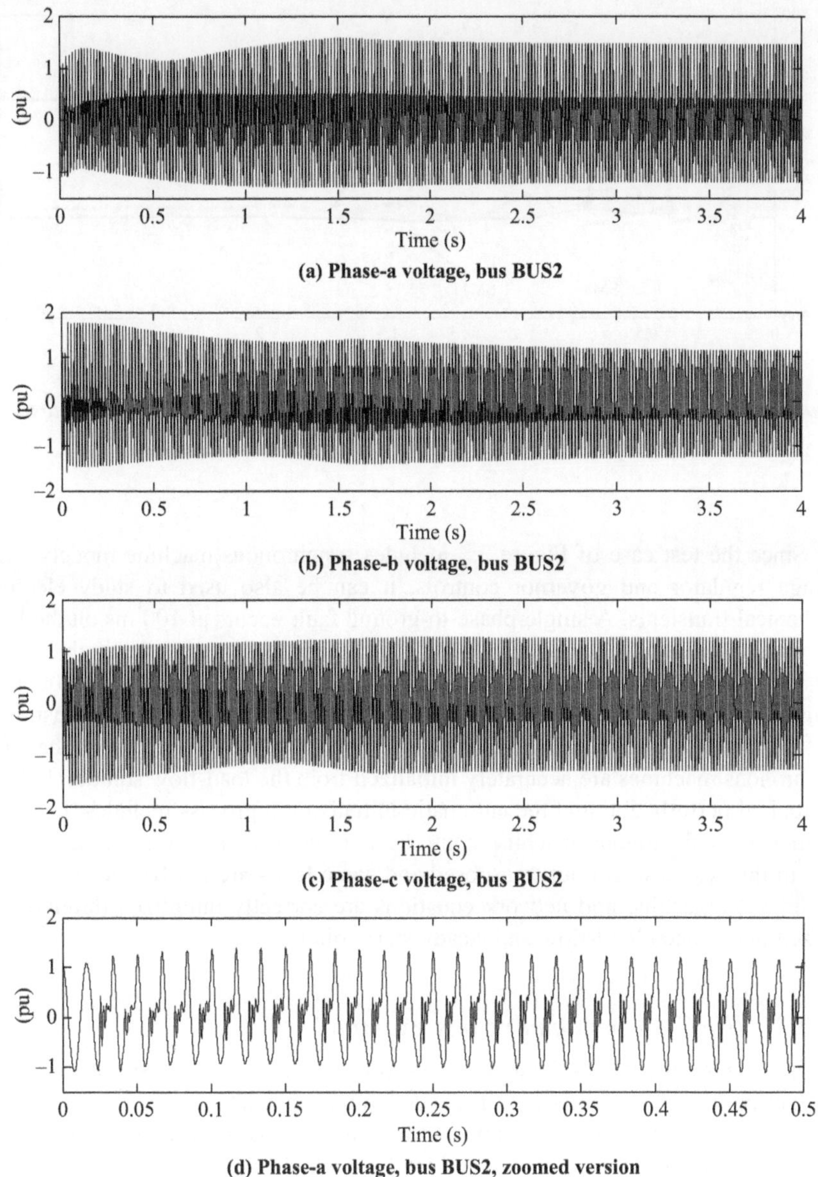

(a) Phase-a voltage, bus BUS2

(b) Phase-b voltage, bus BUS2

(c) Phase-c voltage, bus BUS2

(d) Phase-a voltage, bus BUS2, zoomed version

Figure 3.11   Transients resulting from transformer energization, test case of Figure 3.2

results into a temporary overvoltage condition. It is apparent from the zoomed voltage of phase-a in Figure 3.11(d) that the network is in steady-state (automatically initialized from its load-flow solution) before transformer energization at 20 ms.

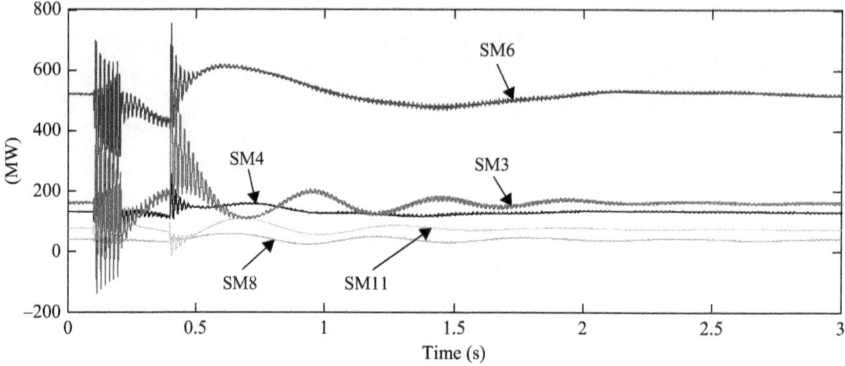

Figure 3.12   Electromechanical transients, synchronous machine powers, test case of Figure 3.2

Since the test case of Figure 3.2 includes synchronous machine models with voltage regulator and governor controls, it can be also used to study electro-mechanical transients. A single-phase-to-ground fault occurs at 100 ms on the line side of the breaker 120 k of the line TLM_120mi. The line is opened at both ends at 200 ms, the fault is eliminated and the line breakers are reclosed at 400 ms. Figure 3.12 presents the simulation results for generator powers and demonstrates that the system is able to regain stability for this event. It is apparent that the synchronous machines are accurately initialized from the load-flow solution before the applied perturbation and the automatic initialization process includes the initi-alization of synchronous machine controls. It is noticed that some transmission lines in this test case are not transposed and some loads are not balanced, but the synchronous machine and network equations are correctly initialized through the utilized unbalanced load-flow and steady-state solutions.

3.10.2   IEEE-39 benchmark bus example

In this test case, the IEEE-39 bus system shown in Figure 3.13 is assembled using constant-parameter transmission lines, synchronous machines with magnetization data, governor, and exciter controls for synchronous machines, transformer models with magnetization branch and load models. In this case, the load models are dynamic voltage and frequency-dependent models [26]. This type of system setup can be used for both electromagnetic and electromechanical transient simulations.

The simulation results presented in Figures 3.14 and 3.15 are for a single-phase-to-ground fault occurring near bus B3 on the line side of the breaker (see Figure 3.13). The permanent fault occurs at 200 ms and the line breakers are tripped at both ends at 300 ms. Once again, the network solution is in complete steady-state at the startup of the time-domain solution.

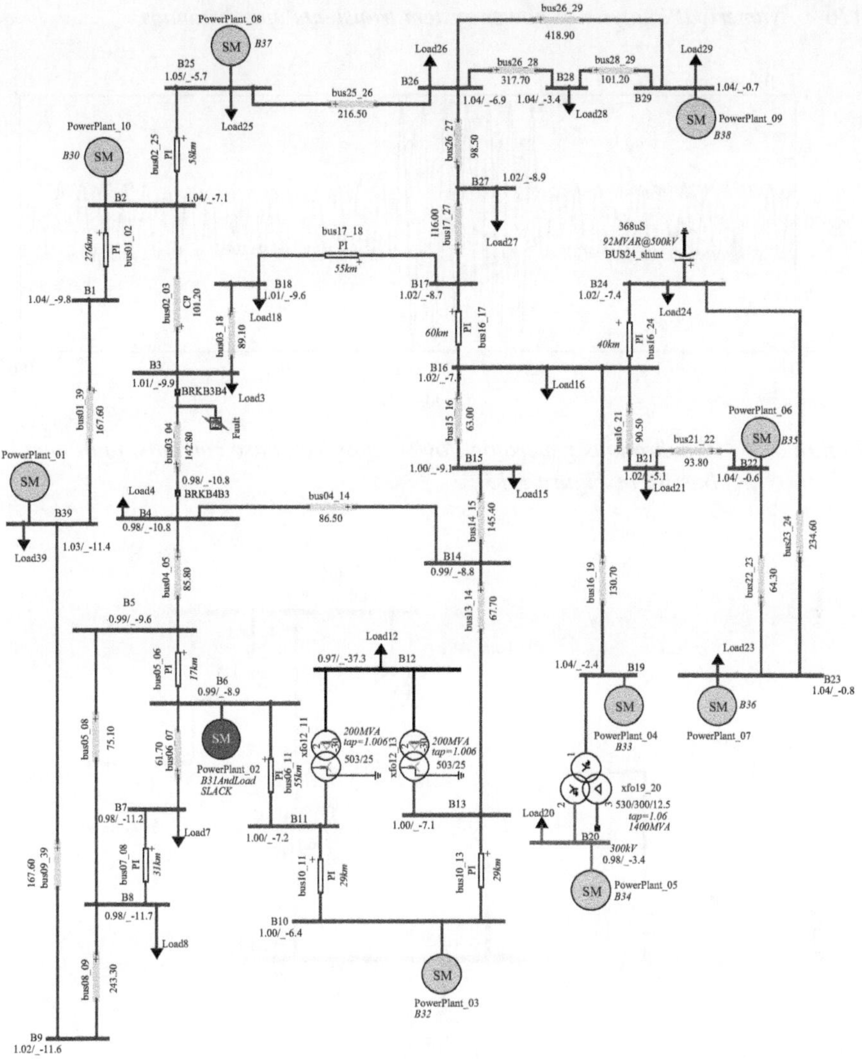

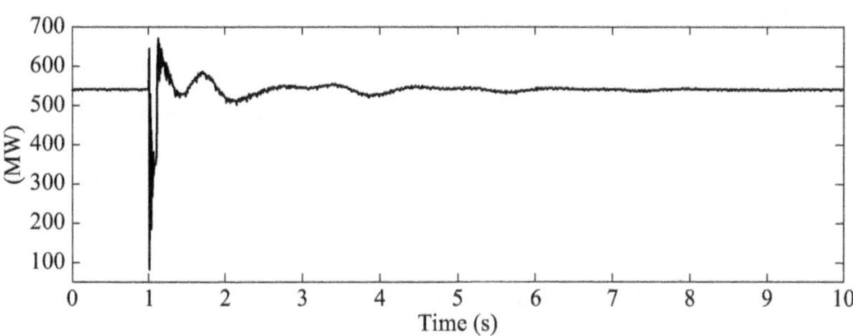

Figure 3.13   IEEE-39 benchmark version simulated in EMTP

Figure 3.14   Power from synchronous generator PowerPlant_08, fault near bus
B3 in Figure 3.13

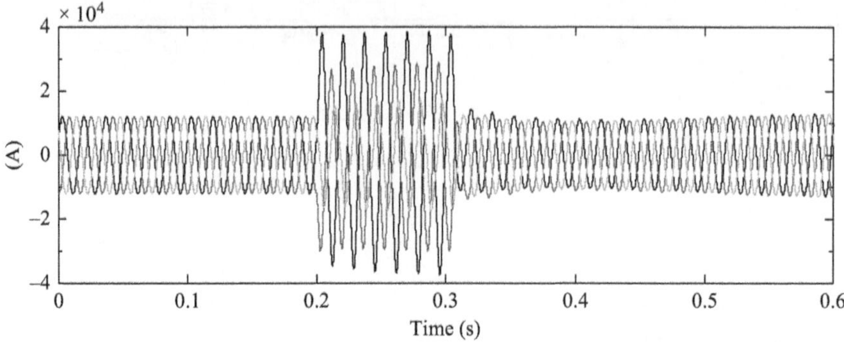

Figure 3.15   *Synchronous generator PowerPlant_10 phase currents, fault near bus B3 in Figure 3.13*

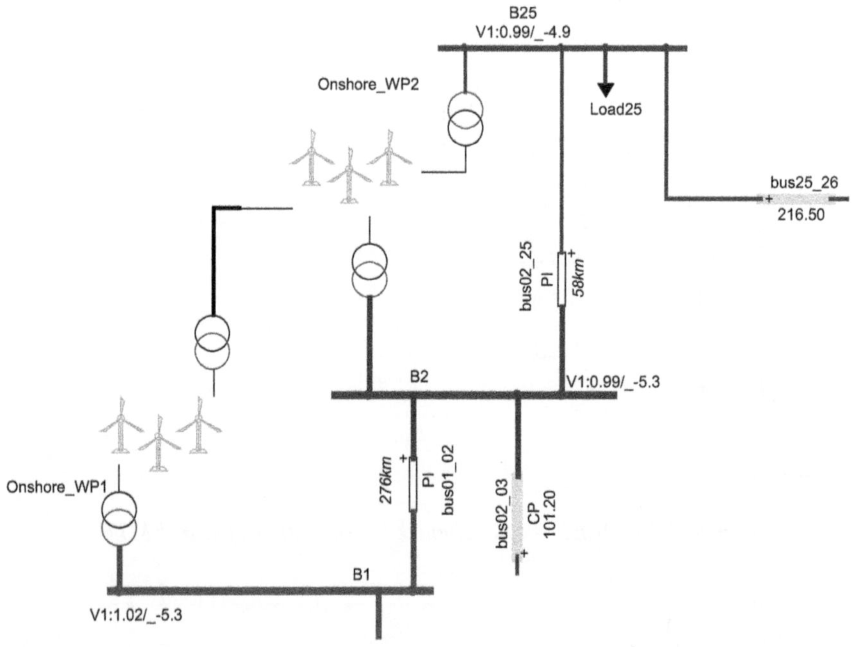

Figure 3.16   *Integration of wind generation into the IEEE-39 benchmark shown in Figure 3.13*

3.10.3   Wind generation

In this test case the IEEE-39 benchmark is modified to include wind generation. The wind parks are connected to the busses B1, B2, and B25 (Figures 3.16 and 3.17) for a total generation of 755 MW. The synchronous generators on buses B2 and B25

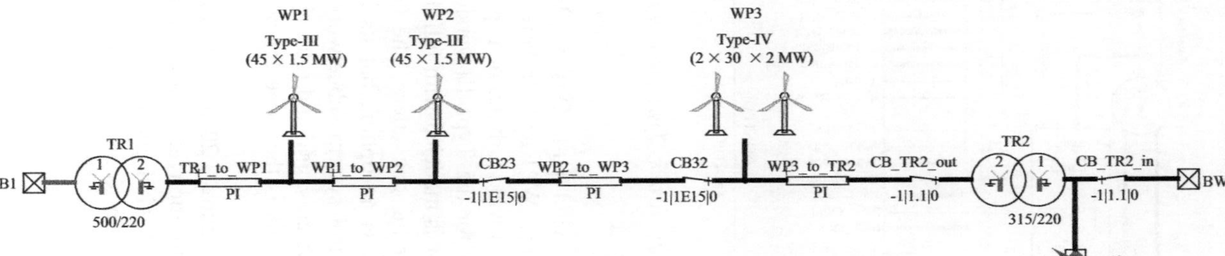

Figure 3.17   Wind park Onshore_WP1 subnetwork contents (see Figure 3.16)

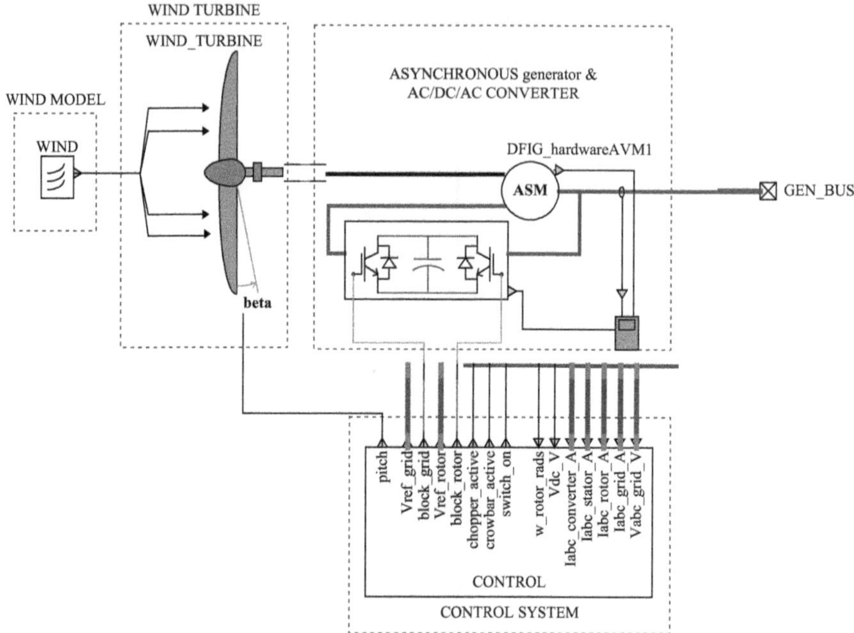

Figure 3.18   Wind generator subnetwork, Type III wind generator model

are now removed. The design of wind park Onshore_WP1 is presented in
Figure 3.17. Both wind parks (Onshore_WP1 and Onshore_WP2) contain aggre-
gated wind parks with a combination of wind turbines of Type III and Type IV. The
wind generator subnetwork contents for the Type III version are shown in Figure 3.18. The asynchronous generator is modeled with full details, but an average
value model is used in this case for the converters. The detailed (IGBT model
based) converter model is available as an option, but its increased accuracy is
unnecessary for this study. The control system subnetwork is presented in Figure 3.19. It includes the protection systems, grid and rotor side converter controls,
and pitch control. The setup of controls is the same for both average value and
detailed converter models.

The simulation results shown in Figures 3.20 and 3.21 are for a three-phase-to-
ground fault occurring at 1 s (see location in Figure 3.17) and cleared by the
breakers at 1.1 s. The initialization of wind generators is based on startup from
load-flow solution and initial connection to an ideal source. The ideal source is
switched off at 200 ms.

3.10.4   Geomagnetic disturbances

In this example, the IEEE-39 bus system in Figure 3.22 is used to study geo-
magnetic disturbances (GMDs). The Geoelectric Field (GEF) is modeled with a
dc voltage source injected in series with each transmission line. All transformers

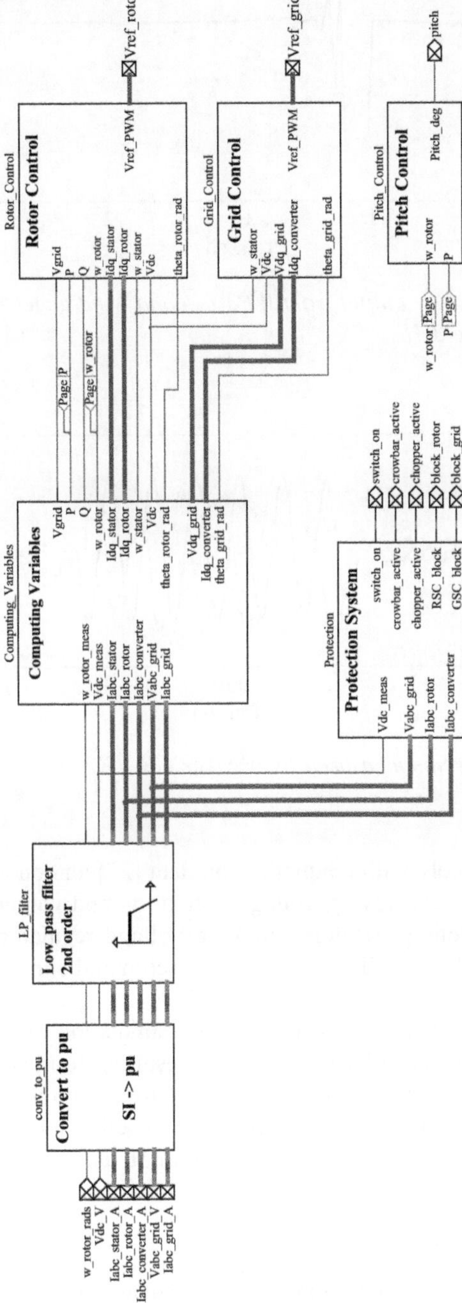

Figure 3.19　Type III wind generator control system (see subnetwork in Figure 3.18), top level subnetworks

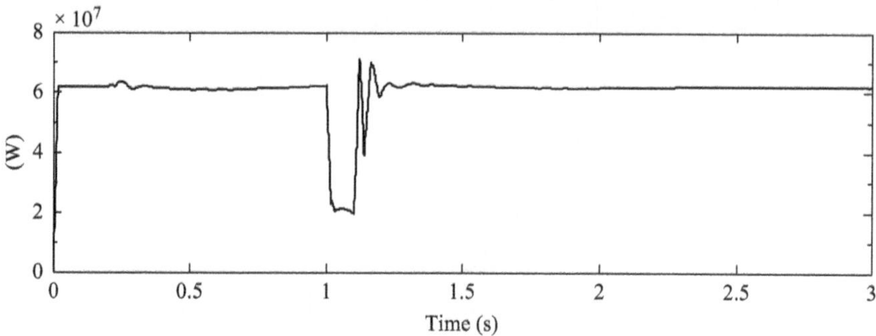

Figure 3.20   *Active power output from WP3 (second wind generator on the right) in Figure 3.18*

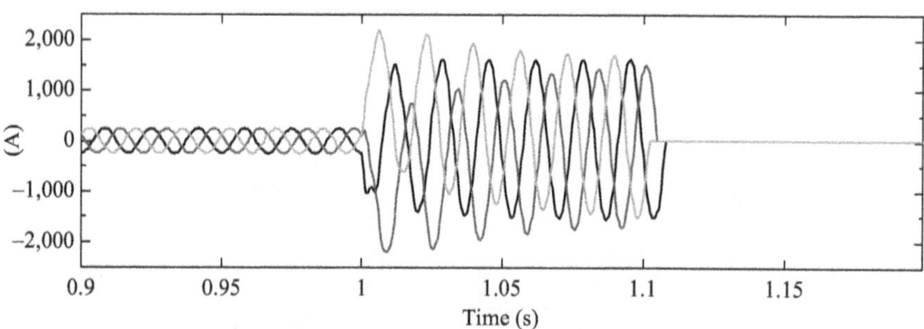

Figure 3.21   *Currents measured by breaker CB_TR2_out in Figure 3.17*

are modeled appropriately with magnetization data [27] and on load tap changer (OLTC) functionality for studying voltage regulation and voltage collapse conditions. The synchronous generator automatic voltage regulators include over-excitation limiters. Other details on this system setup and simulation results can be found in [27].

Such detailed modeling allows the accurate simulation of GMDs on power systems and predict voltage collapse conditions resulting from supplementary var consumptions by network transformers submitted to dc currents. In this type of analysis, the simulation interval must be extended to 600 s and highest accuracy is achieved using a numerical integration time step of 50 µs.

3.10.5   HVDC transmission

An example of HVDC transmission model simulated in EMTP is shown in Figure 3.23. The HVDC converters are modular multilevel converters (MMCs). The detailed simulation of such converter [28–30] requires modeling very large

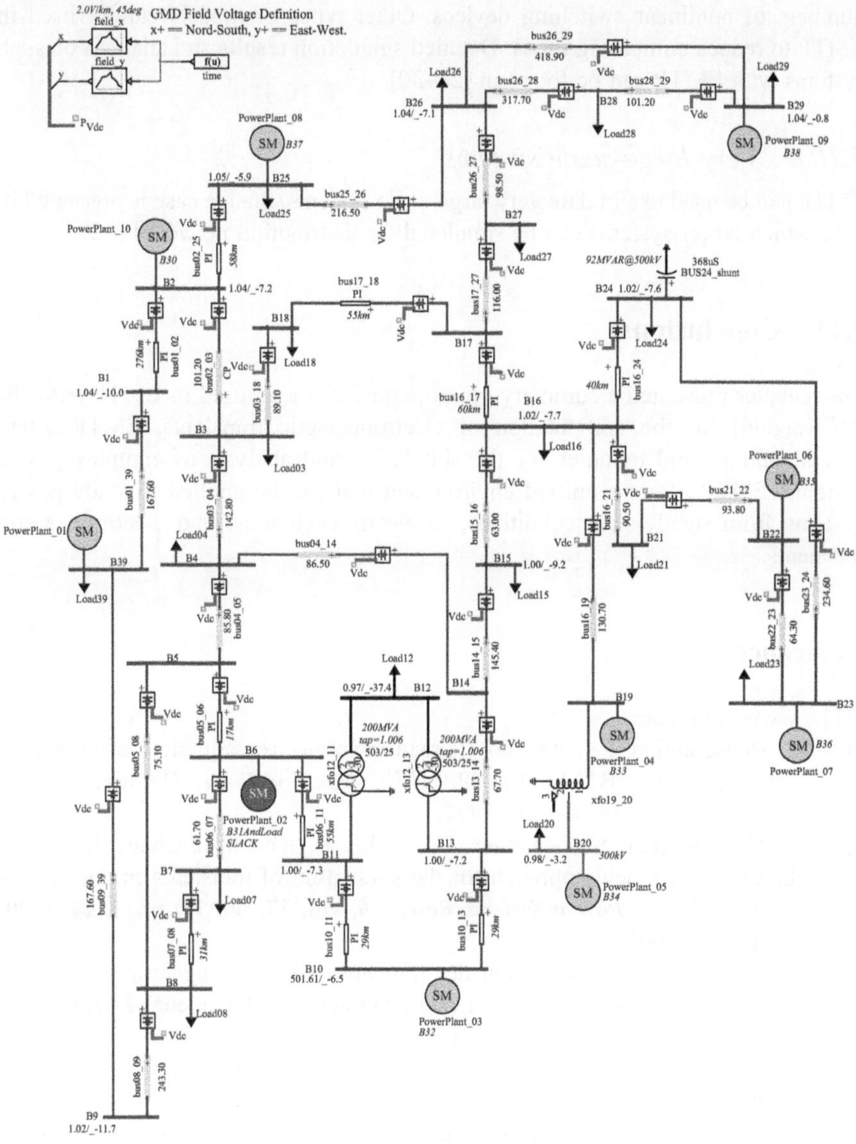

Figure 3.22   *IEEE-39 benchmark version modified for GMD studies in EMTP*

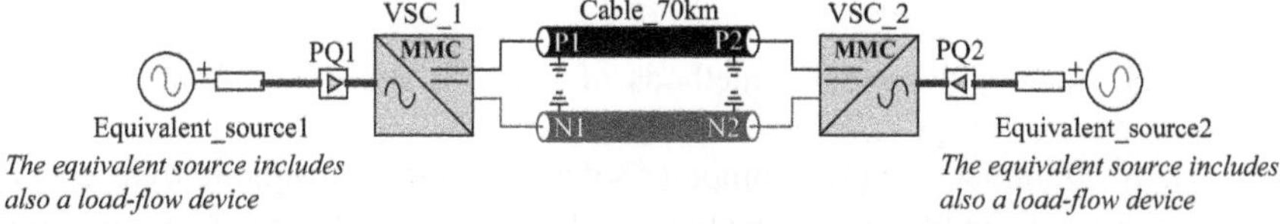

Figure 3.23   *MMC-HVDC transmission system*

numbers of nonlinear switching devices. Other types of models can be used in EMTP to reduce computing times. Detailed simulation results and analysis on such systems with EMTP can be found in [28–30].

3.10.6   Very large-scale systems

EMTP can be used to simulate very large-scale systems. Such a case is presented in [31]. Much larger systems can be simulated for distribution networks [32].

3.11   Conclusions

This chapter presented a summary on numerical methods used in EMTP (EMTP-RV version) for the computation of electromagnetic transients. EMTP offers several options and modules for the simulation and analysis of complex power systems. EMTP offers a unified environment that can be applied to study power systems from steady-state conditions, to electromechanical and electromagnetic transients.

References

[1]   www.emtp.com

[2]   J. Mahseredjian, F. Alvarado, "Creating an electromagnetic transients program in MATLAB®: MatEMTP," *IEEE Trans. on Power Delivery*, vol. 12, no. 1, pp. 380–388, January 1997.

[3]   J. Mahseredjian, S. Dennetière, L. Dubé, B. Khodabakhchian, L. Gérin-Lajoie, "On a new approach for the simulation of transients in power systems," *Electric Power Systems Research*, vol. 77, no. 11, pp. 1514–1520, September 2007.

[4]   J. Mahseredjian, "Simulation des transitoires électromagnétiques dans les réseaux électriques," Édition 'Les Techniques de l'Ingénieur', February 10, 2008, Dossier D4130. 2008, 12 pages.

[5]   B. Kulicke, "(Simulation program NETOMAC: Difference conductance method for continuous and discontinuous systems)," *Siemens Forschungs—und Entwicklungsberichte—Siemens Research and Development Reports*, vol. 10, no. 5, pp. 299–302, 1981.

[6]   J. R. Marti, J. Lin, "Suppression of numerical oscillations in the EMTP power systems," *IEEE Trans. on Power Systems*, vol. 4, no. 2, pp. 739–747, 1989.

[7]   F. H. Branin, "Computer methods of network analysis," *Proc. of IEEE*, vol. 55, no. 11, pp. 1787–1801, November 1967.

[8]   H. W. Dommel, "Digital computer solution of electromagnetic transients in single- and multiphase networks," *IEEE Trans. on Power Apparatus and Systems*, vol. 88, no. 4, pp. 734–741, April 1969.

[9]   L. W. Nagel, "SPICE2 A computer program to simulate semiconductor circuits," Memorandum No. UCB/ERL M520, May 9, 1975.

[10]   M. Tiberg, D. Bormann, B. Gustavsen, C. Heitz, O. Hoenecke, G. Muset, J. Mahseredjian, P. Werle, "Generic and automated simulation modeling based on measurements," *Proc. of International Conference on Power Systems Transients*, IPST 2007 in Lyon, France, June 3–7, 2007.

[11]   J. Mahseredjian, C. Dufour, U. Karaagac, J. Bélanger, "Simulation of power system transients using state-space grouping through nodal analysis," *Proc. of International Conference on Power Systems Transients*, IPST 2011 in Delft, Netherlands, June 14–17, 2011.

[12]   J. Mahseredjian, L. Dube, M. Zou, S. Dennetiere, G. Joos, "Simultaneous solution of control system equations in EMTP," *IEEE Trans. Power Systems*, vol. 21, no. 1, pp. 117–124, February 2006.

[13]   I. Kocar, J. Mahseredjian, U. Karaagac, G. Soykan, O. Saad, "Multiphase load flow solution of large scale distribution systems using the MANA approach," *IEEE Trans. on Power Delivery*, vol. 29, no. 2, pp. 908–915, April 2014.

[14]   T. A. Davis, E. P. Natarajan, "Algorithm 907: KLU, a direct sparse solver for circuit simulations problems," *ACM Transactions on Mathematical Software*, vol. 37, no. 3, pp. 36:1–36:17, September 2010.

[15]   S. A. Hutchinson, E. R. Keiter, R. J. Hoekstra, L. J. Waters, T. Russo, E. Rankin, S. D. Wix, C. Bogdan, "The XyceTM parallel electronic simulator—an overview," *IEEE International Symposium on Circuits and Systems*, Sydney (AU), May 2000.

[16]   E. P. Natarajan, "KLU A high performance sparse linear solver for circuit simulation problems," Master's Thesis, University of Florida, 2005.

[17]   H. W. Dommel, *EMTP Theory Book*, Portland, Oregon: Bonneville Power Administration, 1996.

[18]   U. Karaagac, J. Mahseredjian, O. Saad, "An efficient synchronous machine model for electromagnetic transients," *IEEE Trans. on Power Delivery*, vol. 26, no. 4, pp. 2456–2465, October 2011.

[19]   U. Karaagac, J. Mahseredjian, O. Saad, S. Dennetière, "Synchronous machine modeling precision and efficiency in electromagnetic transients," *IEEE Trans. on Power Delivery*, vol. 26, no. 2, pp. 1072–1082, April 2011.

[20]   U. Karaagac, J. Mahseredjian, I. Kocar, O. Saad, "An efficient voltage-behind-reactance formulation based synchronous machine model for electromagnetic transients," *IEEE Trans. on Power Delivery*, vol. 28, no. 3, pp. 1788–1795, April 2013.

[21]   J. R. Marti, "Accurate modeling of frequency-dependent transmission lines in electromagnetic transient simulations," *IEEE Transactions on Power Apparatus and Systems*, vol. 101, no. 1, pp. 147–155, January 1982.

[22]   A. Morched, B. Gustavsen, M. Tartibi, "A universal model for accurate calculation of electromagnetic transients on overhead lines and underground cables," *IEEE Trans. on Power Delivery*, vol. 14, no. 3, pp. 1032–1038, July 1999.

[23] I. Kocar, J. Mahseredjian, G. Olivier, "Improvement of numerically stability for the computation of transients in lines and cables," *IEEE Trans. on Power Delivery*, vol. 25, no. 2, pp. 1104–1111, April 2010.

[24] I. Kocar, J. Mahseredjian, G. Olivier, "Weighting method for transient analysis of underground cables," *IEEE Trans. on Power Delivery*, vol. 23, no. 3, pp. 1629–1635, July 2008.

[25] www.mathworks.com

[26] IEEE Task Force on Load Representation for Dynamic Performance. Load Representation for Dynamic Performance Analysis. *IEEE Transactions on Power Systems*, vol. 8, no. 2, pp. 472–482, May 1993.

[27] L. Gérin-Lajoie, J. Mahseredjian, S. Guillon, O. Saad, "Simulation of voltage collapse caused by GMDs—problems and solutions," *CIGRE Conference*, Paris, August 24–28, 2014.

[28] J. Peralta, H. Saad, S. Dennetière, J. Mahseredjian, S. Nguefeu, "Detailed and averaged models for a 401-level MMC-HVDC system," *IEEE Trans. on Power Delivery*, vol. 27, no. 3, pp. 1501–1508, July 2012.

[29] H. Saad, J. Peralta, S. Dennetière, J. Mahseredjian, *et al.*, "Dynamic averaged and simplified models for MMC-based HVDC transmission systems," *IEEE Trans. on Power Delivery*, vol. 28, no. 3, pp. 1723–1730, July 2013.

[30] H. Saad, S. Dennetière, J. Mahseredjian, *et al.*, "Modular multilevel converter models for electromagnetic transients," *IEEE Trans. on Power Delivery*, vol. 29, no. 3, pp. 1481–1489, June 2014.

[31] L. Gérin-Lajoie, J. Mahseredjian, "Simulation of an extra large network in EMTP: from electromagnetic to electromechanical transients," *Proc. of International Conference on Power Systems Transients*, IPST 2009 in Kyoto, Tokyo, June 2–6, 2009.

[32] V. Spitsa, R. Salcedo, R. Xuanchang, J. F. Martinez, R. E. Uosef, F. de Leon, D. Czarkowski, Z. Zabar, "Three-phase time–domain simulation of very large distribution networks," *IEEE Trans. on Power Delivery*, vol. 27, no. 2, pp. 677–687, 2012.

Chapter 4
PSCAD/EMTDC
D. Woodford, G. Irwin* and U.S. Gudmundsdottir[†]*

4.1 Introduction

During the early 1970s, Manitoba Hydro was building the Nelson River DC Transmission System to bring hydroelectric power from Northern Manitoba to Winnipeg. At that time, Hydro Quebec was also developing its northern hydro-electric resources and had created IREQ (Hydro Quebec Research Institute) to assist them in their endeavour. IREQ acquired an analogue dc simulator from English Electric and Manitoba Hydro made good use of it to study their Nelson River DC Transmission System. The controls for the dc simulator were the same as on Nelson River DC Bipole 1.

The difficulty with the analogue dc simulator as a study tool was the lengthy set-up and checking time. For a study that lasted 2 weeks, the first 9 days were devoted to set-up and checking, and most of the studies was carried out during the last three hours before the taxi came to drive the team to the airport to return back to Winnipeg.

In 1973, Manitoba Hydro received a copy of the Bonneville Power Administration's Electromagnetic Transients Program or EMTP [1], as it later became to be called. Dennis Woodford tried for 6 months to model a simple dc link on EMTP but gave up in despair. There was no control on modelling capability with EMTP in those days. So, starting from Hermann Dommel's classic paper on electromagnetic transients [2], Dennis Woodford wrote in FORTRAN4, the first code of EMTDC.[1] After the frustrations for trying to model dc transmission on EMTP, he was careful not to look at EMTP code. He did not want to build the limitations of EMTP at that time into EMTDC.

While working with English Electric in the United Kingdom in the late 1960s, Dennis Woodford made use of a digital controls modelling program known as KALDAS.[2] Then at Manitoba Hydro, in the early 1970s, IBM's CSMP[3] controls

---

*Electranix Co., Canada

[†]Dong Energy, Denmark

[1]EMTDC was the ElectroMagnetic Transients for DC and was developed at Manitoba Hydro as a simulation engine.

[2]'KALDAS, an algorithmically based digital simulation of analogue computation'. *Computer Journal*, vol. 9, no. 2, pp. 181–187 (1966).

[3]CSMP, Continuous System Modelling Program. *IBM Systems Journal*, vol. 6, no. 4, p. 242 (1967).

modelling software was used. Concepts from both these programs were designed to apply building blocks that emulated analogue computer functions. It made sense therefore to incorporate the concept of these programs into the EMTDC electromagnetic transients solution to interface with the network solution. The control functions were labelled CSMF or 'Continuous System Modelling Functions'. The EMTDC network solution and the controls building blocks of CSMF were successfully merged so that users could write their own controls system models in a FORTRAN subroutine labelled DSDYN or Digital System DYNamics. Node voltages and branch currents could be accessed by the CSMF functions in DSDYN and could control sources, switches and branch inductors, capacitors or resistors.

The very first code of EMTDC contained phasor network solutions as well as the controls. Eventually, the phasor solution was replaced by the electromagnetic transients solution.

The best way to model a large dc system was still not obvious. In the late 1970s, Dennis Woodford had a project meeting in Bismarck, North Dakota for a proposed 500 kV transmission line between Canada and the United States. The flight to Bismarck was at dawn, but it was fogged in. After several attempts, the plane landed. However, no more planes landed and Dennis was the only one who arrived for the meeting. The flight out was not until that evening. It was during the day in a quiet corner of Bismarck airport that the concept for modelling large dc systems in EMTDC was worked out.

At this time, a graduate student at the University of Manitoba, Ani Gole was investigating a project for Manitoba Hydro. He needed a good synchronous machine model for his study, and with Dr. Rob Menzies at the University of Manitoba, they teamed up with Dennis and produced one. Ani Gole's project led to his Ph.D. and he was the first graduate student to use the EMTDC.

Up until 1980 when Ani Gole and Rob Menzies joined the development of EMTDC, only Dennis Woodford had used the program. He did not publicize his efforts since every time he raised it for discussion, no support was offered. Engineers were quite negative about the possibility that dc transmission could be accurately simulated digitally. One engineer said he would only believe it is possible if it could model core saturation instability in dc transmission. It was felt that digital simulation of dc transmission was impossible to achieve. Without support, there was no chance of corporate funding for a project of 'dubious' outcome. Consequently, the development was bootlegged, which as it happens is a very efficient development method.

The challenge to simulate core saturation instability was considered seriously. This performance problem on the Nelson River HVDC Transmission had occurred on several occasions and could not be reproduced on the analogue dc simulator at IREQ. The simulator's high level of damping and losses were thought to inhibit the observation of core saturation instability. Dennis Woodford embarked on a project to model Bipoles One and Two of the Nelson River HVDC Transmission on EMTDC. Dr. Mohamed Rashwan, the dc engineer at Dorsey Station, was kind enough to provide data on the dc controls and protections that were subsequently programmed into the EMTDC model. Using recordings from known disturbances, the EMTDC simulation was tested and found to have good performance by comparison. John McNichol of Transmission Planning at Manitoba Hydro had system data and recordings of a core saturation instability event on the Nelson River HVDC Transmission that he was kind enough to share. When the system data was entered into the simulation model on EMTDC, and the fault that instigated the core saturation instability was applied, the EMTDC simulation immediately exhibited the elusive core saturation instability.

Even despite the success of simulating core saturation instability, there was still considerable reticence in accepting EMTDC as the replacement for analogue simulation. Even Dennis Woodford did not feel completely confident that EMTDC was simulating the right phenomenon. Another significant study was undertaken that involved a spurious 93 Hz sustained oscillation on Nelson River Bipole II that started up when its first 500 kV valve group was de-blocked. The EMTDC simulation could not reproduce the event. After some time, engineers from the Working Group (BBC, Siemens and AEG), who were the equipment suppliers of Bipole II, were in Winnipeg and a meeting was convened at Dorsey Station to review the problem. The key components of the EMTDC simulation model was reviewed with them and everything checked out satisfactorily. The response time of the dc current transducers was asked. The reply from the equipment suppliers was 'instantaneous'. It was countered that it can't be instantaneous. The Working Group engineers had some discussion among themselves, and it was obvious they didn't know and all they agreed on was that it was 'fast'.

During the modelling of Bipole II on EMTDC, this same question was asked of Mohamed Rashwan who had proposed the dc current transducer response time constant be modelled as 10 ms. So, it was reset to 1 ms in the model and the simulation repeated. Immediately the 93 Hz oscillation appeared. This increased confidence in the simulating capability of EMTDC for very complex power system models. It is safe to say that during the early 1980s, this was the largest dc simulation model ever produced.

Just because the simulation of the 93 Hz simulation was successful, it did not mean that the reason for it was obvious. It took some time before with the use of EMTDC, and it was determined to be a ground mode resonant frequency at 93 Hz on the dc transmission line reacting with a positive sequence 153 Hz ac system resonance generated between the rectifier ac filters and the Long Spruce generating station. Since the EMTDC simulation with a slower responding dc current transducer could not generate the offending 93 Hz oscillation, the immediate solution was to add a capacitor to the input to the analogue current control amplifier in the Bipole II dc controls to effectively slow down the current feedback signal to about a 10 ms response time. This solution stayed in effect for many years and the 93 Hz oscillation was never observed again.

In November 1981, the Manitoba HVDC Research Centre was incorporated as an independent non-profit company and EMTDC was given to them by Manitoba Hydro for further development and marketing, although Manitoba Hydro retained its intellectual property rights. The great contribution made by the Manitoba HVDC Research Centre was the development of the graphical user interface to EMTDC known today as PSCAD.[4] Garth Irwin took over development of EMTDC adding significant improvements to it and guided the development of PSCAD. Many talented and capable computer scientists and engineers continue to develop EMTDC and PSCAD to this day at the Manitoba HVDC Research Centre, now a fully owned subsidiary of Manitoba Hydro.

4.2 Capabilities of EMTDC

The development of EMTDC and PSCAD is guided by the needs of many users around the world. The following are some of the studies that can be conducted with PSCAD/EMTDC:

- Insulation coordination of ac and dc equipment, including steep front studies, lightning studies and circuit-breaker transient recovery voltages
- Ease of user-constructed models
- Studies of very large systems through hybrid simulation of PSCAD with $PSS^{®}E^{5}$ using E-TRAN[6]
- Designing power electronic systems and controls
- Incorporating the capabilities of $MATLAB^{®}/Simulink^{®7}$ directly into PSCAD
- Sub-synchronous oscillations, their damping and torsional interactions
- Effects of dc currents and geomagnetically induced currents on power systems, inrush effects and ferroresonance
- Distribution system design with custom power controllers and distributed generation
- Power quality analysis and improvement
- Design of modern transportation systems using power electronics (ships, rail, auto)
- Design, control coordination and system integration of wind and solar farms, diesel systems, energy storage
- Sub-synchronous control interactions (SSCI)
- Impedance scans of a power system network over a range of frequencies and contingencies
- Under government security, PSCAD has been used to study the impact of electromagnetic pulse on electric power systems

---

[4]PSCAD, Power System Computer Aided Design, the graphical user interface to EMTDC and the property of Manitoba Hydro and is developed by the Manitoba HVDC Research Centre (https://hvdc.ca/pscad/).
[5]$PSS^{®}E$ is an integrated, interactive program of Siemens PTI for simulating, analysing and optimizing power system performance.
[6]E-TRAN by Electranix Corporation for translating steady-state and dynamic data for $PSS^{®}E$ into PSCAD, for hybrid simulation between PSCAD and $PSS^{®}E$, and for multi-core processing of PSCAD.
[7]MATLAB/Simulink by MathWorks incorporates MATLAB, the language of technical computing and Simulink, for simulation and model-based design.

The development of frequency-dependent transmission lines and cables provides an important capability for PSCAD. Experts were enlisted to support this development and included Bjorn Gustaveson [3], L. Martin Wedepohl [4], Garth Irwin, John Nordstrom, Dr. Ani Gole and his graduate students. Improving the accuracy of frequency-dependent transmission lines and cables is an ongoing development for PSCAD.

The User's Guide on the use of PSCAD is available from the Manitoba HVDC Research Centre [5].

4.3   Interpolation between time steps

Simplified phasor modelling such as with power system stability programs is far from adequate when solid-state switches, circuit breakers, faults, complex control and protection systems require precise representation. This is achieved by electromagnetic transients programs with varying levels of accuracy depending on the solution method it applies. Solution methods include (1) forward only time-step progressions, whether fixed or variable in time steps where a switching or change of state can only occur at time points that are a multiple of the time step Δt or (2) interpolation between time steps.

PSCAD applies the principle of interpolation to deal with ensuring a switching or change of state event occurs at the exact instant in time. This is achieved by allowing the solution to 'back up' to the precise instant in time when the change of state occurs. The advantages of interpolation are as follows:

- Allows simulation to be run with a larger time step without affecting accuracy. The alternative of changing to smaller time steps when a switching instant is detected requires a retriangularization of the matrix and in general results in longer calculation times
- Results in correct theoretical harmonics generated by switching devices (including HVDC and FACTS controllers) because each switch device fires at the correct time
- Avoids voltage 'spikes' in VSC circuits due to incorrect back-diode turn-on times (unrealistic snubber circuits are required to control these spikes in fixed time-step programs)
- Avoids numerical instabilities that can occur due to arrangements of multiple switching devices in close proximity
- Results in more accurate models of non-linear devices (surge arresters), especially in energy calculations
- Models correctly represent the low-frequency damping and harmonics of switching devices interacting with sub-synchronous resonance effects in machines

The interpolation process applied in PSCAD ensures that any backing up in time is performed using solutions that are calculated by the same conductance matrix. As an example, an IGBT circuit switches when either the voltage or current

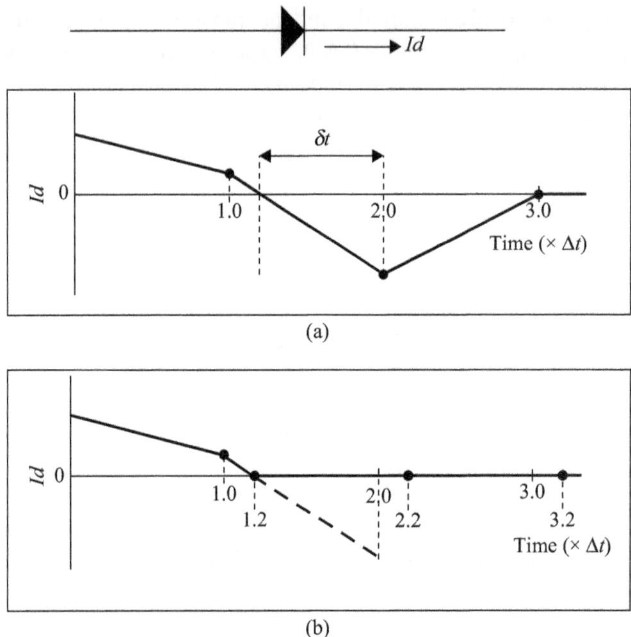

*Figure 4.1   Simulated diode switching off at a current zero (a) with fixed time step
and (b) with interpolation*

through it is zero, thus giving the correct losses the IGBT model and network it is
embedded in will provide. Other features in the interpolation are as follows:

- Large portions of a network separated by travelling wave transmission lines or
  cables are separated into sub-systems (so interpolations and switchings are
  isolated to only the required sub-network without having to manipulate the
  entire network).
- Control of chatter at inductive nodes and in capacitor loop currents is still
  controlled using the proven interpolated ½ step chatter removal process.
- The solution can be interpolated numerous times to accommodate multiple
  switches that occur in the same time step.

The process of interpolation is demonstrated by current through a simple
diode. The waveform in Figure 4.1(a) shows the current through the diode with a
standard fixed time-step switching algorithm. The current reverses at some time in
between time steps Δt and $2\Delta t$, but because of the discrete nature of the time step,
the impedance of the device can only be made infinite (i.e. the diode turns off) at
integrals of Δt. Here the simulated diode actually turns of at time $2\Delta t$, whereas in
reality the process should have occurred Δt time units earlier. The first recorded
instant of zero current is thus at $3\Delta t$.

Figure 4.1(b) shows the same device with the switching interpolated to the
correct instant. As before, the program calculates the solution at $t = \Delta t$ and $t = 2\Delta t$.

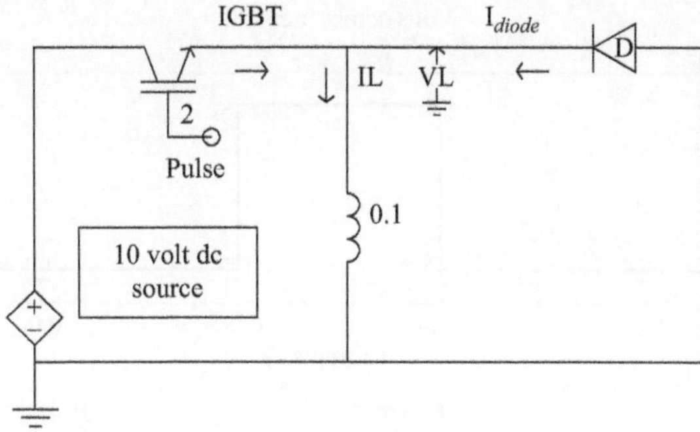

Figure 4.2   Network for simple test circuit with no snubber damping circuits

However, on noticing that at the latter time, the current has already crossed zero, it estimates the turn-off time to be $t = 1.2\Delta t$ based on a linear interpolation of the current within the switching interval. All the voltages and currents in the trapezoidal solution method are then also interpolated to this intermediate time in a linear fashion. The admittance matrix is then reformulated and the solution continues with the original time step, yielding the new solution one time step later at $t = 2.2\Delta t$. One additional interpolation step between $t = 1.2\Delta t$ and $2.2\Delta t$ yields the solution at $t = 2\Delta t$. The latter apparently cosmetic step is taken to put the solution back on the original time grid.

A simple circuit to illustrate the excellent performance of interpolation is shown in Figure 4.2. Here the IGBT is first turned on and current IL ramps up through the inductor. Then the IGBT is turned off, and at the same instance, the diode must turn on as the inductor current commutates to it. The commutation process is as shown in Figure 4.3.

The success of interpolation is evident in Figure 4.3 as no spikes in voltage and current occur. The simple test circuit of Figure 4.2 without snubbers on the IGBT or diode is an excellent test for any simulation algorithm.

Although a very simple example has been presented, the interpreted solution is being successfully applied to large power electronic simulations without known limitations and good speed of simulations [6].

4.4   User-built modelling

PSCAD/EMTDC is a platform for users to build their own models. It is usual for models to be incorporated into 'Components' that may consist of network and/or controls and perhaps with user-written descriptive code in FORTRAN or C. The actual process of creating a Component is not covered in this section. However, if a user-built Component is available and required for simulation, this is the process to be followed to include in the development of a simulation model for the development of a Component.

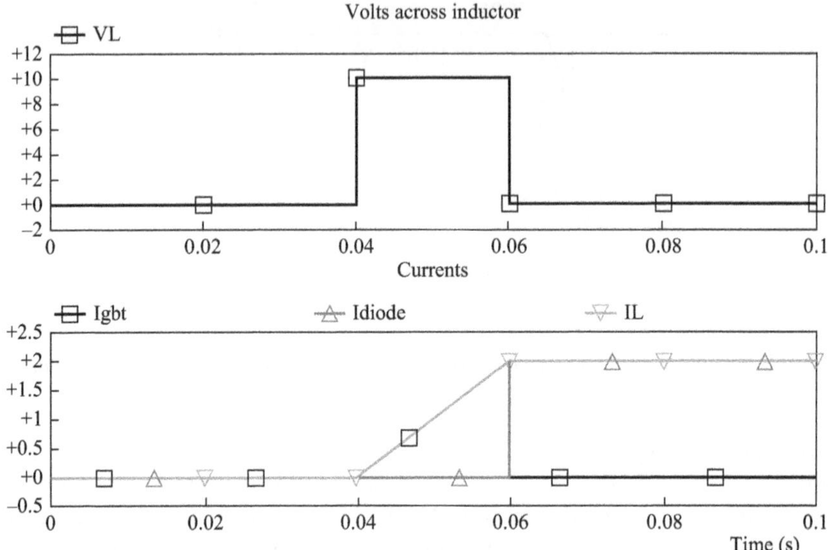

Figure 4.3   Correct solution to IGBT turning on and off for the simple test circuit

A Component is a graphical object in PSCAD created using the Component Wizard and is edited by the Component Workshop. When completed, it will have a unique Component definition name. The Component may be self-contained with user-written code included directly within it, or if extensive modelling has been applied, a file of one or more FORTRAN or C subroutines is added so that the Component consists of both the Definition and the external Subroutine(s).

A simulation model may be very large and unable to fit nicely on the Main Page of PSCAD. PSCAD allows large systems of controls and electric circuits to be assembled into pages. In fact, they can be assembled in pages within pages.

4.5   Interfacing to other programs

PSCAD has the ability to interface with many other programs such as MATLAB/ Simulink [7] and PSS®E [8] in particular.

4.5.1   Interfacing to MATLAB/Simulink

The following background on interfacing PSCAD to MATLAB/Simulink is provided in [5]:

> PSCAD provides the ability to utilize the functionality of MATLAB commands and toolboxes (including all the graphical commands) through a special interface. This is achieved by calling a special subroutine and from within a standard component in PSCAD.

Components that interface to MATLAB/Simulink are not offered as part of the PSCAD Master Library and must be specially developed for this purpose. Once the special component is developed, it is treated as a normal component in PSCAD. Of course MATLAB/Simulink must be installed on the computer with PSCAD in order to use the interface component. Instructions in writing a MATLAB or a MATLAB/Simulink interface component are presented in [5] where contact can be made for help if needed in preparing the interface component.

The MATLAB graphics functions are a very powerful addition to the PSCAD plots and graphical interface. Three-dimensional plots, active graphics and rotating images are possible and integrate seamlessly with the PSCAD graphical libraries.

An important factor to consider before attempting to interface with MATLAB from PSCAD: It will not function when using the free GNU FORTRAN 77 compiler.

4.5.2   *Interfacing with the E-TRAN translator*

Users of electromagnetic transients simulation programs such as PSCAD often face difficulties in obtaining data and developing cases suitable for their studies. Many utilities have the data available for their entire system in power flow programs, but a great deal of effort is required to re-enter the network data for use in EMT programs.

There are three basic challenges:

- Translation of circuit/network data (differences in per-unit systems, data entry etc.)
- The generation of network equivalents
- Initialization of machines, generators or sources in large interconnected networks

E-TRAN will directly create the necessary files for PSCAD from solved PSS/E power flow and transient stability files. The PSCAD files so translated are complete with correct data for all busses, loads, dc links, switched shunt devices, branches, generators, transformers, phase shifters etc., all correctly drawn and graphically interconnected. E-TRAN translates an entire circuit, a zone or area, or a central bus and 'N busses away'. Network drawings are grouped by location and voltage level, and can be created for a complete network or a portion of a network. The E-TRAN features are shown in Figure 4.4.

Further explanation follows.

4.5.2.1   Formation of network equivalents

The bus voltages and angles and power and reactive power flow information (from the solved power flow input file) are used to directly initialize sources/generators in PSCAD (and to initialize non-linear loads etc.). The initialization information can be used in simple voltage source representations of generators, or can be used to initialize detailed machines, governors, exciters, multi-mass systems etc. The resulting simulation in PSCAD exactly matches the solved load flow results for any circuit, and complex systems with any number of machines can be initialized without start-up oscillations.

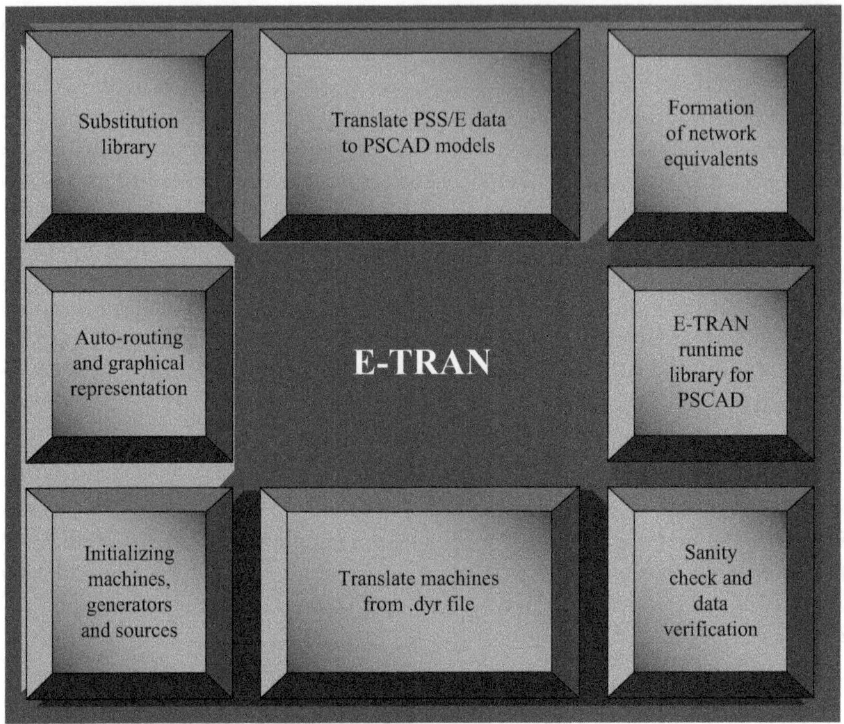

Figure 4.4   Illustration of the operations of E-TRAN

The algorithm used to automatically generate the circuit layout in PSCAD is based purely on busses and connection information (for the PSS/E power flow file is based on the graph theory).

4.5.2.2   Initialization of machines, generators and sources

E-TRAN includes a downloadable Runtime Library with the .psl extension for use in PSCAD. The E-TRAN library contains all models used in power flow programs including load models, sources, transmission lines, transformers, phase shifters etc. It has data entry based on the same per-unit system and data entry standards as used in power flow programs. PSCAD cases generated with E-TRAN can be freely distributed to other users without license fees.

E-TRAN also supports a Substitution Library (also with a .psl extension), which allows a user to maintain and automatically substitute its own more detailed models instead of simple power flow data. E-TRAN also reads the standard dynamic models in PSS®E.

E-TRAN generates warnings and error messages when it reads in PSS®E .raw input files. This will inform the user of areas of mismatched voltage levels, inconsistencies in the entry of data. The result messages are output to the E-TRAN log file.

The user identifies a portion of the network for a direct translation into PSCAD models (i.e. the 'kept' network), and E-TRAN creates a network equivalent of the rest of the network based on the available fundamental frequency impedance and power flow information. The equivalent is a multi-port representation that will be correct for steady state as well as for open-circuit and short-circuit conditions and contains Thevenin voltage sources to match PQ flow and represent generation in the equivalent network. The equivalents are generated based on the positive sequence data entered in PSS®E power flow files, and includes the effect of negative and zero sequence data if sequence files are available.

E-TRAN generates warnings and error messages when it reads in PSS®E power flow input files. This will inform the user of areas of mismatched voltage levels, inconsistencies in the entry of data. The result messages are output to the E-TRAN log file.

The E-TRAN library for PSCAD contains models which mirror the functionality of the device models contained in PSS/E such as constant power and constant reactive power loads. The E-TRAN program translates the results into PSCAD, and then the PSCAD differential equation solver is used during the run. The resulting solution will match the PSS®E power flow results to about three or four decimal points everywhere in the system.

E-TRAN generates a complex admittance matrix for the entire system which can be greater than 50,000 busses. It then calculates the power and reactive power mismatch of the entire system to see if the input power flow file is solved and decomposes using LDU sparse decomposition techniques to generate network equivalents.

4.5.2.3   E-TRAN substitution library

If a user has more detailed data available (e.g. geometrical data for transmission lines suitable for a frequency-dependent line model or a detailed core representation of a transformer with saturation data entered etc.) then the user can maintain this data in a Substitution Library and E-TRAN can directly write out this data instead of the simple data used by PSS®E. A device in the Substitution Library can also be a user-constructed component, in which case it can contain many sub-pages of electrical or control devices such as a detailed model of an HVDC link or SVC. An initialization feature can be used to define ±30 degree phase shifts (due to wye-delta transformers) and to accept the power flow steady-state voltage magnitude and phase angles. This can be used to initialize any device. The process that indicates how the substitution library is a part of the E-TRAN translation procedure is shown in Figure 4.5.

4.6   Operations in PSCAD

4.6.1   Basic operation in PSCAD

Components are either network components such as resistors, inductors, capacitors, switches, ac machines, transformers or power electronic devices, or measurement,

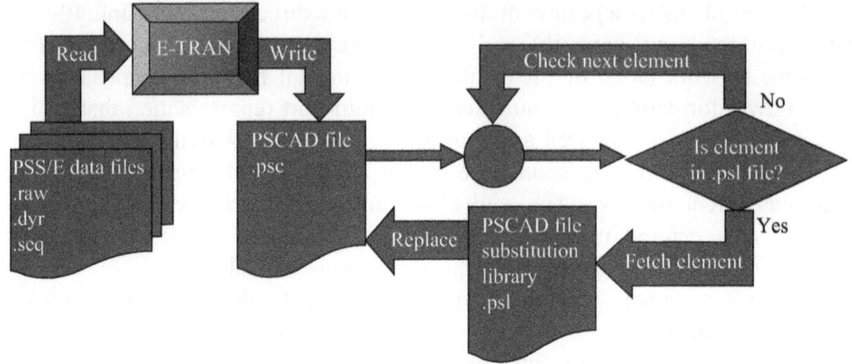

*Figure 4.5   State diagram showing at which stage of a translation E-TRAN checks
for detailed components of the substitution library*

control, monitoring functions, signal processing, outputs etc. Basic components are
available in a Master Library.

A valuable feature of PSCAD is the display of output results as the simulation
is running. Increased interaction is possible where the user can observe output
results either as the case is running or with the simulation in 'pause'.

With the help of 'fly-by' monitoring see exactly how the network and controls
are performing, and adjust or switch parameters online as desired. The fly-by help
is operational when the case is running or in 'pause' mode.

Initialization of power systems with complex and non-linear power electronic
equipment and their controls and protections is undertaken by a run-up to steady
state. A snapshot may be taken when steady state is reached, which becomes the
input data for a study requiring a fully initialized solution. Where E-TRAN is used,
it can extract initial conditions from a solved power flow case of the system and
input the initialized states of generators, motors, as well as models extracted from
the substitution library of E-TRAN. These fully initialized models from E-TRAN
stay fixed during start-up of a case until the steady state of all the complex nonlinear power electronic equipment has been reached and a snapshot taken.

PSCAD provides online plotting of user-selected quantities as well as post-run
plotting. Additional plotting programs are available for post-run plotting such as
Microsoft's Excel program, Livewire[8] and TOP.[9]

4.6.2   Hybrid simulation

The approach taken to implement the hybrid simulation with PSCAD is shown in
Figure 4.6, and has been implemented at this time with the PSS®E transient stability program. Hybrid simulation allows the simultaneous use of both PSS®E and

---

[8]Livewire 2 by Z Systems Inc. See also Manitoba HVDC Research Centre.
[9]TOP, The Output Processor, by Electrotek Concepts, Inc.

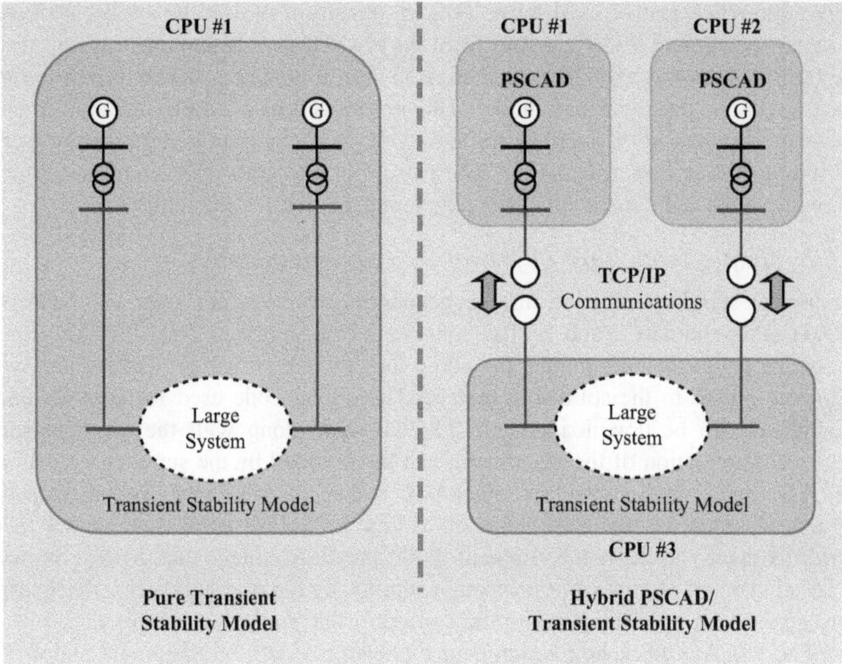

Figure 4.6   Hybrid simulation on multiple CPUs

PSCAD simulations, taking advantage of the strengths of each program for study capability that is not possible in either program. The applications for hybrid simulation become exciting, as many of the limitations of each type of simulation program will disappear. The PSS®E portion of the simulation models the full ac system, which includes all electromechanical modes of oscillation. Highly detailed PSCAD models for complex equipment are interfaced to the transient stability models currently used, and the resulting simulation will be more accurate and run faster than either program by itself. The combined PSS®E transient stability and PSCAD simulation is capable of modelling more detail using different simulation time steps, and run each portion of the network separately. There is also the capability of interconnecting two asynchronous electric power systems each modelled in PSS®E through a PSCAD network that includes an HVDC interconnector between the two asynchronous ac systems.

It is possible that each of the separate ac systems can be modelled with different versions of PSS®E, and each PSCAD system model can also be with a different PSCAD version.

As the hybrid simulation runs, the sub-systems communicate boundary electrical conditions with each other at the larger simulation time step. Voltage magnitude and angle/frequency information from the transient stability simulation is sent to the PSCAD simulation, where it is used to update equivalent voltage sources on the diagonals of a system equivalent that is auto-created based on the power flow admittance matrix. A discrete Fourier transform is used to extract positive sequence power and reactive power from the PSCAD simulation, and transmit it to the transient stability simulation, where it is used to update a dynamic power and reactive power generator/load model. This in turn acts as a current injection in the transients simulation to calculate the new voltage at the interfacing point by performing a power flow calculation. The two programs run synchronously in time, allowing faults to be made on either side to correctly affect the other.

4.6.3   Exact modelling of power system equipment

Equipment suppliers of wind turbine generators, inverters for solar wind farms, STATCOMs and SVCs, and HVDC converters can supply an exact PSCAD component model of their equipment including the digital controls, protections and on-off firing pulses to the solid-state valves. The digital code used for controls and protections can be downloaded into PSCAD and, along with the physical and electrical stimulation of the equipment, can be provided by the supplier for use in PSCAD, usually in the form of a 'black box' and accompanied by a non-disclosure agreement. This has the advantage of having a PSCAD simulation model that performs exactly as the real equipment. It has the disadvantage that there is limited or no adjustment of control or protection settings by the user and such models can only be used for applications with the consent of the equipment supplier.

The PSCAD black box component models provided by equipment suppliers must have the provision for storing state variables for support of snapshots and multiple instances. The multiple instance requirements allow copies to be generated in the simulation model for a study.

Only those who are independent of the equipment suppliers and have their confidence as well as the confidence of the equipment purchaser are in the position to undertake PSCAD studies with simulations that include multiple competitor vendor equipment.

4.6.4   Large and complex power system models

Large system PSCAD studies require a circuit representation which is able to accurately reproduce the detailed power electronic-based facilities of interest that also includes a reasonably accurate representation of electromechanical effects with generators, motors, governors and exciters. These studies include sub-synchronous resonance (SSR) studies, series capacitors studies, HVDC and SVC interaction studies and transient overvoltage (TOV) studies with lines using fast-reclosing and harmonic impedance frequency scans. To meet these study requirements, transient models are constructed using PSCAD and the E-TRAN program.

The study methodology to create a PSCAD electromagnetic transients model that includes both the correct frequency response and the correct electromechanical response is as follows:

- Start with a solved PSS®E power flow case. E-TRAN accepts .raw input files (all recent versions) and will translate this data into PSCAD. Studies can be undertaken with the electromagnetic transient area of interest including HVDC transmission located between two large ac asynchronous networks that are modelled in different versions of PSS®E.

- Determine how big an electrical and electromechanical system has to be included in the model by E-TRAN translation of successively larger systems. The harmonic impedance is computed using the PSCAD harmonic impedance component, and the size of the system translated should be increased until the harmonic impedance no longer changes appreciably in the transient frequency range of interest (typically 0–2 kHz).

- Develop detailed models of nearby facilities with power electronic controllers that can affect results such as HVDC transmission, SVCs and STATCOMs with custom controls, transmission lines and cables with detailed line geometries, wind and solar generators with exact supplier-provided models and store them in a PSCAD library. Detailed models are stored in the library/database according to the power flow 'from' bus 'to' bus and the circuit number. This allows E-TRAN to initialize and use the detailed models in the PSCAD case. The substitution library/database contains the key data and models that need to be backed up and saved. Eventually the substitution library will increase in size until detailed models for the full system have been entered and a study model can be quickly generated anywhere in the system modelled in PSS®E.

- Translate a case into PSCAD using E-TRAN. The selected circuit is auto-routed into PSCAD single-line diagram format. A multi-port system equivalent is auto-created for the portion of the network not represented in PSCAD. Machines, exciters, governors, stabilizers, min/max excitation limiters and compounding models are directly translated from the PSS®E transient stability file into PSCAD; all standard models have been written in PSCAD and tested. This process also fully initializes the machines and controls so they start at the correct power flow. If detailed models are found in the user-developed substitution library/database, they will be initialized and used and substituted in the PSCAD case – otherwise models will be created based on the PSS®E data. Transmission line data from the power flow is represented as a Bergeron model (if the line travel time is longer than the time step) or as a PI section (for short lines).

The harmonic impedance of the system is dominated by the overall system strength (i.e. the inductance) and the location of shunt capacitor banks and the capacitance of long transmission lines. The harmonic impedance has been observed to change for systems going up to 10 busses away from the study point. Successively larger systems are translated into PSCAD until the harmonic impedance as seen from the study point no longer changes. This process guarantees the integrity of the PSCAD model embedded in the larger network modelled in PSS®E.

4.7 Specialty studies with PSCAD

Modern electric power systems are becoming increasingly complex because of power electronic controllers proliferating through the network. Ensuring compatible performance between equipment with power electronic controllers is a task that PSCAD has been designed to address. Two common methods are global gain margin assessment [9] and use of the PSCAD multi-function optimization component control [10]. In addition SSR and control interaction effects are becoming increasingly important to investigate.

4.7.1   Global gain margin

The stability of a facility with power electronic controls is dependent on its gain margin. When there are multiple power electronic controllers in proximity to each other, individual controller gain margins may reduce resulting in a less robust power system. If access to the key control gain or gains in all local power system controllers is possible, the global gain margin analysis can be applied in PSCAD.

A global gain margin is defined as a setting on a scale between 0 and 1 which causes the ac system to reach the verge of instability when the gains on all fast acting control devices are simultaneously adjusted such that at a setting of 0, the gains of all devices are at their nominal values and at a setting of 1 and the gains of all devices are at their respective gain margin limits.

Deterministic solutions with PSCAD can be used to set individual controller gains such that the individual controller and system gain margins provide acceptable system damping to large and small signal system disturbances.

4.7.2   Multiple control function optimizations

The control parameters of one or more controllers are judiciously selected as being significant to the stability of the controllers and power system. The optimal solution by multiple runs converges to the best performance possible based on the measurement of robustness selected, such as the integral squared error method. In a PSCAD component, there is the choice of Nelder and Mead's non-linear Simplex, Hookes and Jeeves methods for multi-variable problems, and genetic algorithms for mixed integer problems.

4.7.3   Sub-synchronous resonance

PSCAD can undertake studies to evaluate the potential for the well-known phenomenon of sub-synchronous resonance (SSR). Such studies are required when a series capacitor is near gas/thermal turbines or generators with long shafts. They are also required when a thermal generator feeds radially into an HVDC converter.

Torsional amplification studies may also be required. Torsional amplification may occur due to the series capacitor located near the thermal generator, which can cause a sharp initial increase in the shaft torque when a fault clears as the energy in the series capacitor discharges into the generator.

It is recommended that three stages of analysis be undertaken with PSCAD for SSR investigation:

1. With series capacitors, a screening study is performed using harmonic impedance scanning methods to determine the extent of negative damping from the electric power system on the generator torsional modes. This is necessary to

efficiently cover the full set of contingency conditions and will identify generators which are affected significantly by the electrical resonance created by the series capacitor or DC links. This stage of work ignores the dynamic effect of SVCs, HVDC/VSC links and complex dynamic load models. This method does not provide a definitive answer, but is a screening study. This method is not useful for sub-synchronous torsional interactions with HVDC links.

2. For critical contingencies and suspect generators identified from Stage 1, small signal perturbation analysis is undertaken to determine more precisely the torsional impact of adding the series capacitors. This stage of work represents any nearby power electronic controllers, complex loads, generator exciters, governors, stabilizers and HVDC links. The perturbation analysis may be repeated with and without generator controls including exciter, governor and stabilizer to determine if these models have a large impact on the results. Power system stability levels of exciter, governor and stabilizer model are often not accurate at torsional frequencies. Mitigating and protection methods will be proposed and tested.

3. For critical cases identified from Stage 2 and for generators which have detailed shaft data available, a time domain large signal fault test will be performed – these results will correlate with Stage 2 results.

All three stages are applied using PSCAD with system data directly translated from solved PSS®E power flow and dynamics data using E-TRAN.

4.7.4   Sub-synchronous control interaction

Sub-synchronous control interaction (SSCI) studies are required when a series capacitor is near wind turbines or other power electronic controllers such as SVC, HVDC, STATCOM etc.

Studies and real-system experience have indicated that Type 3 doubly fed induction generator (DFIG) wind turbines may exhibit control instabilities when connected near or radial into series capacitors, resulting in damage to both series capacitors and wind turbines. The controls of Type 3 wind turbines can cause negative damping to the dominant resonant frequency oscillation of a series-compensated transmission line. Type 4 (full converter) wind turbines seem to be more tolerant of SSCI, although studies are still required to demonstrate acceptable performance.

SSCI studies are required to be performed to determine if unstable conditions are possible. Precise/accurate models of the wind turbine equipment and controls are required for the study (the best models use the actual C code from the real controls and protection hardware).

Mitigation methods are not trivial and can be expensive, including changes to the wind turbine controls, detection of SSCI oscillations and protective actions, series capacitor segment sizing, series capacitor parallel torsional bypass filters or other ac filters, thyristor-controlled series capacitor (TCSC) banks etc.

SSCI studies should be performed relatively early in the interconnection/study process, as it may have an impact on the turbines to be selected for the project, or on the series compensation scheme.

4.7.5   Harmonic frequency scan

Essential design and definition of ac filters for SVCs and some HVDC transmission converters require studies to generate the harmonic impedance of the ac system at the busbar to which the ac filters are to be located. PSCAD with system data translated by E-TRAN from PSS®E models is solved for a range of contingences. To obtain realistic results, impedances of dominant transmission lines and cables near to the busbar where the intended ac filters are to be located should be derived from frequency-dependent parameters. Specific details of large nearby loads should be included. For example, the frequency response of a large induction motor load is different to a power electronic converter such as a smelter, or of an electric arc furnace. Such detailed models not available from PSS®E files must be developed and placed in the E-TRAN substitution library.

Usually positive/negative sequence impedances as a function of frequency are adequate for ac filter design. The harmonic impedance study requires a case list that considers outages of many nearby lines, transformers, shunt capacitors and generators. Harmonic impedance plots are then computed for the base case and each of these contingencies. Impedance magnitude (ohms) and phase angle (degrees) are plotted as functions of frequency as well as resistance versus reactance over the range of, for example, 1–2,500 Hz. Many PSCAD cases must be run and the process can be automated and essential data compiled in Microsoft Excel spreadsheets to access for ac filter design.

4.8   Further development of PSCAD

4.8.1   Parallel processing

Study requirements are demanding larger and larger models to be used, so PSCAD simulation speed is a limiting factor with a single CPU. The PSCAD algorithm was initially designed for parallel processing, due to the natural time delay that occurs on transmission lines between substations since a 15 km transmission line has a 50 μs travelling time delay at the speed of light. The power system can be broken up into separate executables which can be run on the same computer or on different computers on a local area network wherever there is a travelling wave transmission line model. The inherent time delay in the travelling wave models is matched to the communication time delay/latency, so the overall simulation results are 100% identical and no approximations are made. Figure 4.7 indicates the basic configuration of parallel operation of PSCAD [11].

The results are dramatic, showing that modern computer networks have relatively little latency communication delays, and therefore a near linear speed up in solution time can be obtained if each parallel segment is of similar size. Different versions of PSCAD and different time steps can be paralleled and run on the separate CPUs or cores with different compilers.

As an example, a communication library and tools allowing each simulated subsystem to operate at different time steps enables a wind farm whose power electronics require small time steps to be simulated on one processor. The remaining part of

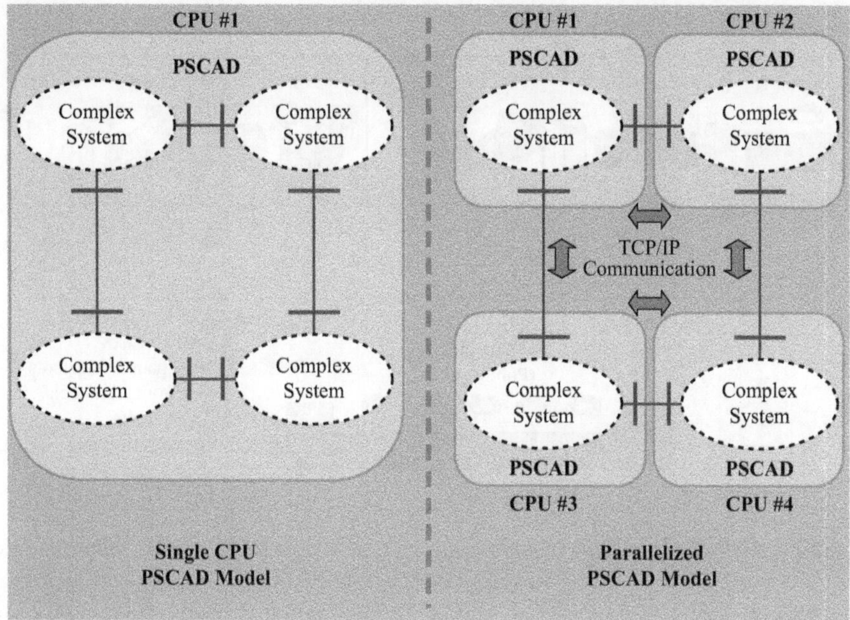

Figure 4.7   Parallel processing of PSCAD

the ac grid can be run at larger or normal time steps for a significant saving in simulation speed.

Another example application for the communication library is for black-boxing to hide confidential information. Manufacturers of power electronic equipment can prepare simulation models and release them as executables (not as binary libraries), thus hiding confidential details, ensuring the model is run at the designed-for time step and avoiding possible compatibility problems with future software compiler versions and versions of the programs.

4.8.2   Communications, security and management of large system studies

The parallel processing communication library has been implemented in several ways (with auto-selection of the best method), including TCP/IP (for communications across standard local area network cards), direct memory transfer (for communications between shared memory cores on the same CPU) or other custom interfaces such as Infiband, which allows inter-machine communications using ultra-low latency custom network cards.

The security of these tools including the auto-start mechanism which launches a program on a remote computer is based on the same industry standard capability inherent in the Windows operating systems. This becomes of importance for any of the highly confidential study involving models from many manufacturers. It is envisioned that each manufacturer would have a dedicated computer all connected

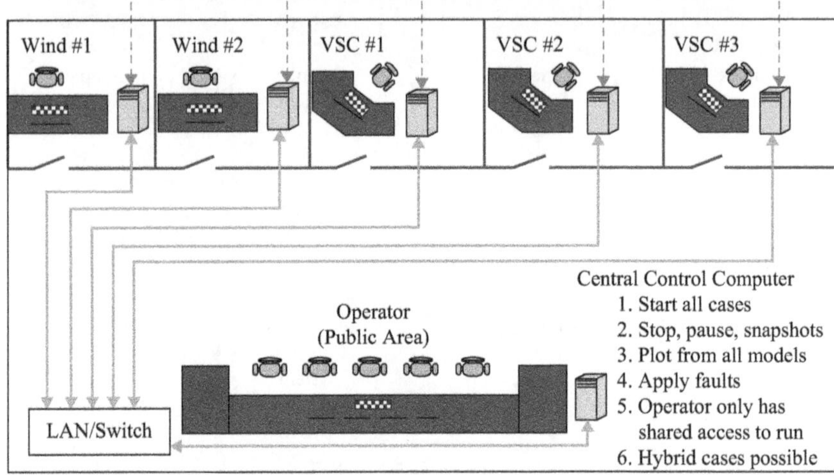

Figure 4.8   PSCAD/hybrid computation centre with security for multi-vendor models

via a local area network, on which they have installed their PSCAD cases. All executable PSCAD cases would be placed into a directory with share access given only to a single user as the Operator. The Operator would be the only person who could run all of the programs from each manufacturer and see the results on graphs on his/her computer. The Operator is able to control the entire simulation, apply faults, set the finish times, write output files etc. When the Operator starts the main case, it would auto-start the remote cases to establish communications every time step.

If a model or sub-system for a given manufacturer needs to be updated, then the manufacturer can connect in through a virtual private network (VPN) into their own computer, and update the models remotely. They also would be able to remote desktop into the remote computer and manually run PSCAD, thus seeing the full case details as they are simulated. In this case the interactive simulation would allow direct participation from the remote manufacturer but would not have access to other manufacturer's models. A concept for a secured PSCAD or hybrid computation centre for multi-vendor models is presented in Figure 4.8.

4.9   Application of PSCAD to cable transients

PSCAD has through the years been used for several types of system studies and for educational purposes. It is an important tool in the work of many power system engineers and is under constant improvements.

Modelling transmission cables has been an ongoing task for several years. As the electrical properties of underground cables differ considerably from that of overhead line (OHL), the modelling and studies of cables are considered very challenging. Not too much is known about long or many underground cables in a transmission system, as until today transmission systems have mainly been

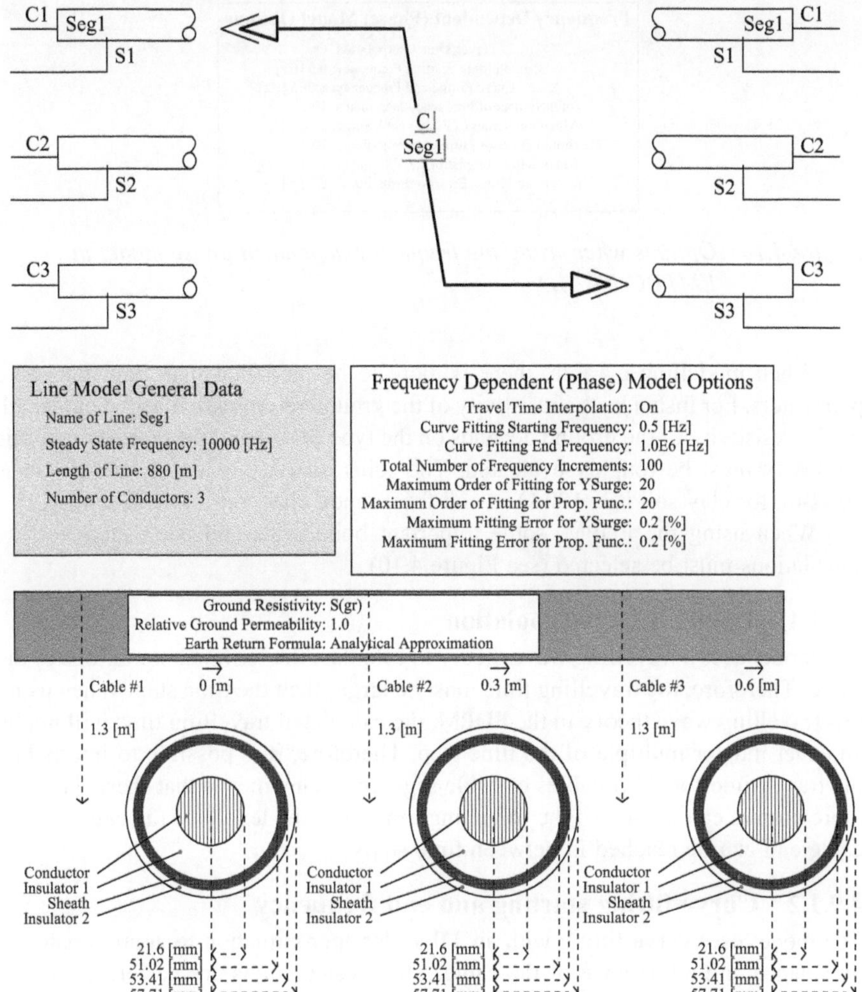

Figure 4.9   Simulation set-up in EMTDC/PSCAD for cable modelling

constructed of OHLs. Nevertheless, the modelling of cables has been an issue for several decades. This part introduces how transmission cables are normally modelled in EMTDC/PSCAD and how the modelling can be improved even further.

4.9.1   Simulation set-up

As explained previously, experts were enlisted to support the development of PSCAD. Bjørn Gustavsen was one of the experts who had his universal line model implemented in PSCAD, here named the frequency-dependent phase model (FDPM).

A layout of what the cable model looks like in EMTDC/PSCAD is shown in Figure 4.9, the upper part for the cable model and the lower part for the model segment cross section.

Frequency Dependent (Phase) Model Options

Travel Time Interpolation: On
Curve Fitting Starting Frequency: 0.5 [Hz]
Curve Fitting end Frequency: 1.0E6 [Hz]
Total Number of Frequency Increments: 100
Maximum Order of Fitting for YSurge: 20
Maximum Order of Fitting for Prop. Func.: 20
Maximum Fitting Error for YSurge: 0.2 [%]
Maximum Fitting Error for Prop. Func.: 0.2 [%]

Figure 4.10   Options when using the frequency-dependent phase model in EMTDC/PSCAD

When modelling a cable, there is more to be modelled than only the cable parameters. For instance, the resistivity of the ground return path must be modelled. As the resistivity of the ground depends on the type of soil and the temperature, this resistivity must be roughly evaluated. The soil resistivity of a very moist soil is 30 Ω-m, for clay soil it is 100 Ω-m and for a sandy clay soil it is 150 Ω-m [12].

When using the FDPM, some important boundaries and parameters for the calculations must be selected (see Figure 4.10).

4.9.1.1   Travel time interpolation

The model uses a travelling wave theory to calculate the terminal conditions of the cable. Therefore, the travelling time must be larger than the time step. When using this travelling wave theory in the FDPM, the calculated travelling time will not be an exact integer multiple of the time step. Therefore, it is possible to interpolate the travel time for short cables or cable segments. This means that there will be a more correct calculation of the travelling time from the length of the cable, as the cable end can be reached in between time steps.

4.9.1.2   Curve-fitting starting and end frequency

The model uses curve fitting with an Nth order approximation to approximate the impedance and admittance of the cable. These calculations are performed over a frequency range. The selection of this frequency range will affect the calculated characteristic impedance of the cable, $Z = \sqrt{(R + j\omega L)/(G + j\omega C)}$.

The selection of the curve-fitting starting frequency will affect the shunt conductance (G) of the line because this limits the DC surge impedance. This is because the solution at the minimum frequency is used for all frequencies from DC to the minimum. At low frequencies or DC, the surge impedance is $Z = \sqrt{R/G}$.

This means that at frequencies below the starting frequency, the shunt conductance has an effective value. If the lower limit of the frequency range is too high, this can result in a very large shunt conductance and shunt losses. Care must be taken not to choose a too small starting frequency as that can affect the accuracy of the curve fitting. This is because the specified maximum error is a percentage of the maximum, and the maximum surge impedance is larger for lower minimum frequency. PSCAD recommends a good starting frequency at approximately 0.5 Hz. The curve-fitting end frequency does not have as much effect on the calculations as the starting frequency. PSCAD states that this end frequency can usually be left at 1 MHz. This is because the time step used in the simulation uses the Nyquist criteria and places an upper limit on the frequencies. The cable model will therefore truncate any elements of the curve-fitted approximations, which are more than 1 decade above the Nyquist criteria.

4.9.1.3 Total number of frequency increments

It is possible to select between 100, 200, 500 and 1,000. This is the total number of calculation steps used in the frequency range for the curve fitting. An increased number of steps will result in a longer simulation time.

4.9.1.4 Maximum order of fitting for Y_{surge} and the propagation function

The frequency-dependent phase model calculates the terminal conditions of the cable by use of the propagation function and the surge admittance (see (4.1)).

$$Y_{surge} = \frac{\gamma}{R + j\omega L}$$

$$H = \exp(-\gamma \cdot l) \tag{4.1}$$

where $\gamma = \sqrt{(R + j\omega L)(G + j\omega C)}$; Y_{surge} is the characteristic admittance, Y_C; and H is the propagation function.

In order to find the surge admittance and the propagation function, the FDPM uses curve fitting. Usually the cable constants program will iterate and continuously increase the order of curve-fitted waveforms until the error is below a specified error for the fitting. On the other hand, for some real-time applications there is not enough time to continuously increase the curve-fitted waveforms until an acceptable accuracy is reached. Therefore, it is possible to set the maximum order of fitting for both the surge admittance and the propagation function.

4.9.1.5 Maximum fitting error for Y_{surge} and the propagation function

Choosing a large number here will result in a poorly fitted surge admittance and propagation function. This will then lead to an inaccurate calculation of the terminal conditions for the cable model. On the other hand, choosing a very small fitting error can lead to unstable simulations. It is recommended by PSCAD to use 0.2% for these values. This can be approximated by using the smallest value that does not lead to unstable simulation. The curve-fitting algorithm is based on a weighted least squares fitting. It is possible to specify different weighting factors for

- the frequency range from DC to the curve-fitting starting frequency;
- the given frequency at the curve-fitting starting frequency;
- from the lower limit to the curve-fitting end frequency.

Specifying a higher weighting factor results in the curve fitting at that frequency being more important. This will lead to placing poles and zeros so that the error at the given frequency is reduced. When the model boundaries and parameters have been implemented, the cable parameters need to be set for impedance calculations.

4.9.2   Parameters for cable constant calculations

The outcome of the simulations can only be as accurate as the input parameters to the program; therefore, care must be taken when implementing a model for the cable. In this section, an overview over how model parameters should be chosen is given, as well as an overview over the model layout.

4.9.2.1   Conductor

In most modelling software, it is only possible to model the conductor either as a solid conductor or as a hollow conductor. For stranded or segmental conductors the cross section is not solid and this is compensated by increasing the resistivity of the conductor. If the measured DC resistance per kilometre and the actual radius of the cable is known, then it is possible to correct the resistivity of the core by (4.2). ρ is the corrected resistivity, R_{DC} is the given DC resistance per kilometre of the conductor, r_1 is the core conductor radius and l is the length.

$$\rho' = R_{DC} \frac{r_1^2 \pi}{l} \tag{4.2}$$

When the DC resistance is not exactly known, it is possible to correct the core resistivity from the cross-sectional area and the given radius of the conductor. As shown in (4.2), the resistivity is given by the resistance of a uniform specimen of the material, the nominal cross-sectional area and the length of the material.

$$\rho = R_{DC} \frac{A}{l} \tag{4.3}$$

The new increased resistivity can be expressed as $\rho' = R_{DC} \frac{r^2 \pi}{l}$, where the resistance of a uniform specimen of the conductor is unchanged, as well as the conductor length. Equation (4.4) can therefore be used to calculate the increased resistivity for modelling the stranded conductor of the cable.

$$R_{DC} = \rho \cdot \frac{l}{A} = \rho' \cdot \frac{l}{r^2 \pi}$$

$$\Rightarrow \rho' = \rho \frac{l}{A} \cdot \frac{r^2 \pi}{l} \tag{4.4}$$

$$\Rightarrow \rho' = \rho \cdot \frac{r^2 \pi}{A}$$

4.9.2.2   Insulation and semiconductive layers – permittivity

It is normally not possible to model directly the semiconductive (SC) layers. The inner and outer SC layers have permittivity around 1,000 [13], compared to insulation permittivity of approximately 2.3, and a conductivity much lower than for both the core and screen conductors. Therefore, from the point of conduction, the SC layers can be neglected, while from the point of insulation they cannot. The effects of the SC layers are included by expanding the thickness of the insulation and increasing the relative permittivity, by assuming constant capacitance between the conductor and the metallic screen [14]. The capacitance between the conductor and the screen can be calculated using (4.5).

$$C = \epsilon \cdot \frac{2\pi l}{ln(b/a)} \tag{4.5}$$

When the corrected permittivity, ϵ', is calculated, the capacitance and the length of the cable are kept constant.

$$C = \epsilon \cdot \frac{2\pi l}{ln(b/a)} = \epsilon' \cdot \frac{2\pi l}{ln(r_2/r_1)}$$

$$\Rightarrow \epsilon' = \epsilon \cdot \frac{2\pi l}{ln(b/a)} \cdot \frac{ln(r_2/r_1)}{2\pi l} \tag{4.6}$$

$$\Rightarrow \epsilon' = \epsilon \cdot \frac{ln(r_2/r_1)}{ln(b/a)}$$

where ϵ is the known relative permittivity for the insulation, b is the outer radius of the insulation, a is the inner radius of the insulation, r_2 is the inner radius of the metal screen and r_1 is the outer radius of the core conductor.

4.9.2.3   Insulation and semiconductive layers – permeability

The permeability is related to the inductance, caused by the magnetic field from both the conductor and the metallic screen. Normally it is only possible to model coaxial conductors (both for the conductor and the screen). In reality, the conductor is often stranded and the metallic screen is made of a thin foil and wires that are helically wounded around the outer SC layer. The associated axial magnetic field will cause a solenoid effect and increase the total inductance [15]. By use of Ampéres circuital law, $\oint_C B \cdot dl = \mu_0 I$, the inductance of the conductors can be calculated from the stored inductive energy and the stored energy of the magnetic field, $\frac{1}{2}LI^2 = \int_V \frac{B}{2\mu}dV$. The magnetic field for the coaxial cable is calculated using (4.7).

$$0 \leq r \leq r_1 \; : B(r) = \frac{\mu_c I r}{2\pi r_1^2}$$

$$\tag{4.7}$$

$$r_1 \leq r \leq r_2 \; : B(r) = \frac{\mu_d I}{2\pi r}$$

where r_1 is the radius of the core conductor, r_2 is the radius of the outer SC layer, μ_c is the permeability of the core conductor and $\mu_d = 4\pi \cdot 10^{-7}$ is the given permeability of the insulation.

This magnetic field is used to calculate the inductance. The inductance is shown in (4.8).

$$0 \le r \le r_1 \; : L = \frac{\mu_c}{8\pi}$$

$$r_1 \le r \le r_2 \; : L = \frac{\mu_d}{2\pi} ln\left(\frac{r_2}{r_1}\right) \tag{4.8}$$

The flux density B_{sol} caused by the solenoid effect is given by the expression in (4.9) together with the associated inductance L.

$$B_{sol}(r) = \mu_d NI$$

$$L = \mu_d N^2 \pi \left(r_2^2 - r_1^2\right) \tag{4.9}$$

where N is the number of turns per metre of the cable.

In order to include the solenoid effect in the coaxial modes of propagation, the relative permeability of the main insulation is set larger than unity by the expression in (4.10).

$$\frac{\mu_c}{8\pi} + \frac{\mu_{d_{sol}}}{2\pi} ln\left(\frac{r_2}{r_1}\right) = \frac{\mu_c}{8\pi} + \frac{\mu_d}{2\pi} ln\left(\frac{r_2}{r_1}\right) + \mu_d N^2 \pi (r_2^2 - r_1^2)$$

$$\Rightarrow \mu_{d_{sol}} = \mu_d + \frac{\mu_d}{ln\left(\frac{r_2}{r_1}\right)} 2\pi^2 N^2 (r_2^2 - r_1^2) \tag{4.10}$$

where $\mu_{d_{sol}}$ is the corrected insulation permeability.

It should be noted that in EMTDC/PSCAD the relative permeability is used, so the corrected $\mu_{d_{sol}}$ should be divided by $\mu_0 = 4\pi \cdot 10^{-7}$.

4.9.2.4 Metallic screen

As is well known, the screen is constructed of two layers, a wired screen and a metallic tape (laminate layer), separated with a thin SC swelling tape. The wired screen and metallic tape are directly connected together both at each junction and cable ends and are therefore normally considered as a single conducting layer in cable modelling. It is a common practice when modelling the screen in EMT-based software to set the resistivity equal to the doubled wired screen resistivity [15]. This approximation has been implemented, because most of the current will be flowing in the wired screen. As shown in Figure 4.2 the wired screen does not fill the whole area around the outer SC layer, i.e. there are some spaces in between some of the wires of the screen. Therefore, the resistivity of the screen is increased. This is an assumption and must be considered when comparing simulations and field test results.

4.9.2.5 Outer insulation

There is no SC layer that needs to be taken into account when modelling the outer insulation, nor any solenoid effect. Therefore, when modelling the outer sheath, the permittivity is chosen as stated by the manufacturer and the permeability is chosen as unity. Normally the outer insulation is made of polyethylene with a permittivity of 2.3.

With an understanding of the frequency-dependent phase model in EMTDC/PSCAD and how to set modal boundaries and parameters, the model can be used for simulation studies and validation against field measurements.

4.9.3 Cable model improvements

It has been shown how cable modelling solely based on the above chapters will give some errors due to insufficient accuracy in representation of the metal screen layers and the proximity effect [16–18]. By correct use of sheath layer implementation in PSCAD and by including the proximity effect in the impedance modelling part of the EMTDC calculations, essential accuracy improvements can be made [19].

A typical layout of an XLPE HV cable is shown in Figure 4.11.

As shown in Figure 4.11, the metal screen consists of several layers. There are two conducting screen layers, the wired Cu screen and the Al foil. These two layers are separated by an SC swelling tape and connected to each other at cable ends and each cable junction. However, as shown in Figure 4.9, the PSCAD set-up contains only a single layer of metal screen.

Furthermore, when adjacent conductors carry current, the closeness of the conductors will cause the current distribution to be constrained to smaller regions. This phenomenon is called proximity effect. It was stated by Michael Faraday in 1831, that when a magnetic flux from a current carrying conductor starts changing, it will induce an electromotive force (emf) in an adjacent conductor. This emf will produce local currents in the adjacent conductor, which are perpendicular to the magnetic flux. These currents are called eddy currents. Because of the eddy currents, the overall distribution of current in the conductor will become non-homogeneous, or constrained to smaller regions. When AC currents flow in a conductor, the resulting magnetic flux will be time varying. When the frequency increases, the magnetic flux $\frac{d\varphi}{dt}$ will vary more, resulting in larger eddy currents of adjacent conductors. Therefore, the higher the frequency, the stronger the proximity effect.

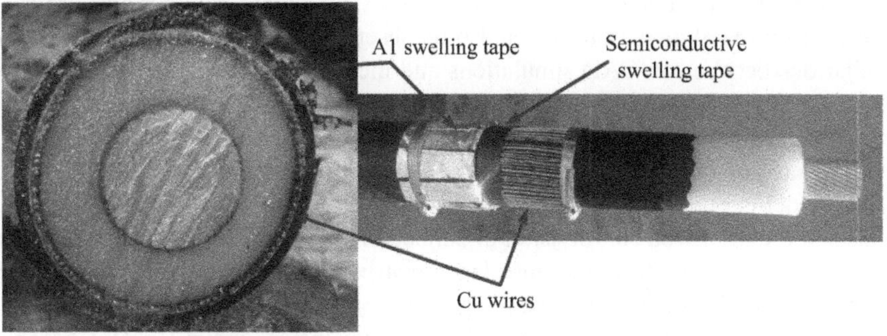

Figure 4.11   Layout of a typical 150 kV HV cable

As shown in Figure 4.11, the Cu wire part of the metal screen consists of many parallel conductors, having the current flowing in the same direction.

For taking the above into account, improvements of PSCAD cable modelling have been described in [19]. These include the following: (1) the physical layout of the screens is modelled more correctly and (2) both the correct physical screen layout is modelled and the proximity effect is taken into account.

Furthermore, these improvements have been validated against measurements when explicitly exciting an intersheath mode. As new simulation results are in very good agreement with field measurements of purely the intersheath mode on a single minor section, they are assumed validated for crossbonded cables. In order to verify this, these improvements are also to be validated against field measurements of longer crossbonded cables.

4.9.3.1   Model validation against measurements of a single major section

The single major section is a 2,510 m long 150 kV cable system having three minor sections. There are two crossbonding points but no grounding of the screen. In a comparison of field measurements with simulation results, for a non-improved cable model, it was observed that the deviation between measurements and simulations appear the transient currents starts to be divided between the metal screens of two of the three phases.

Figure 4.12 shows a comparison of simulation and field test results for a sending end current of the particular cable system.

As shown in Figure 4.12, the simulation results are very accurate when both layered screen and proximity effect are included. There is almost a perfect agreement between improved simulations and field measurement results.

Both simulation results are fully accurate until the transient currents start to be divided between the metal screens of two of the three phases, or at 20 µs. Until this point in time is reached, only the coaxial wave is measured which has already been proven to be correctly simulated by analytical calculations. After this time, however, the measured and non-improved simulated current waves become more and more out of phase as the current becomes affected by more crossbondings and reflected waves. In this set-up, there are no grounding points of the screen along the cable line, only at cable ends, and therefore only the crossbonding points are considered a source of deviation. It can also be observed how, despite deviation, some similarities between PSCAD simulations and measurements are shown, where the simulated wave appears to be less damped and delayed. This can be caused by inaccuracies in the real and imaginary parts of the characteristic admittance.

Because of correct metal screen impedances, the damping for the improved cable model is correct. The difference in impedances of the line constant calculations and the improved sub-division of conductors method is calculated to be up to 42% at 1 MHz. Correcting the impedance matrix will therefore result in a correct intersheath mode propagation damping, which is the cause for the much better simulation results. The damping at the higher frequencies (above 10 kHz), because of the proximity effect, is clearly seen by the fact that the high-frequency oscillations

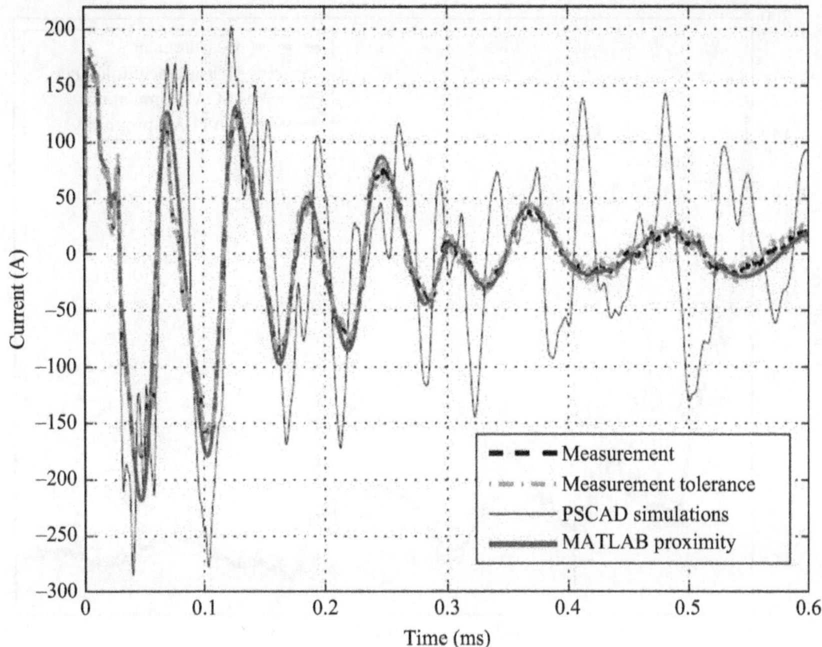

Figure 4.12 *Comparison for the measured and simulated core current at the sending end of the energized phase for the single major section cable line. The graph shows field measurements, universal line model simulations (PSCAD simulations) and simulations using both layered screen and proximity effect improvements (MATLAB proximity simulations)*

of the PSCAD simulations in Figure 4.12 are fully damped in the simulations with improved proximity and metal screen layer modelling. This does not help only with more accurate simulation results, but furthermore it allows for a larger time step giving faster simulations and less probability for an unstable model, which often is the case for long cables with many separately modelled minor sections and cross-bonding points.

4.9.3.2 Model validation against measurements of multiple major sections

The multiple major section is a 55 km long 150 kV cable system having 11 major sections with total of 22 crossbonding points and 10 grounding of the screens, apart from at cable ends. As for the single major section, when analysing the comparison of field measurement and simulation results of the multiple major sections, for a non-improved cable model, it was observed that the deviation between measurements and simulations appears when the intersheath mode starts flowing. Furthermore, the ground mode is equal to zero and will therefore not

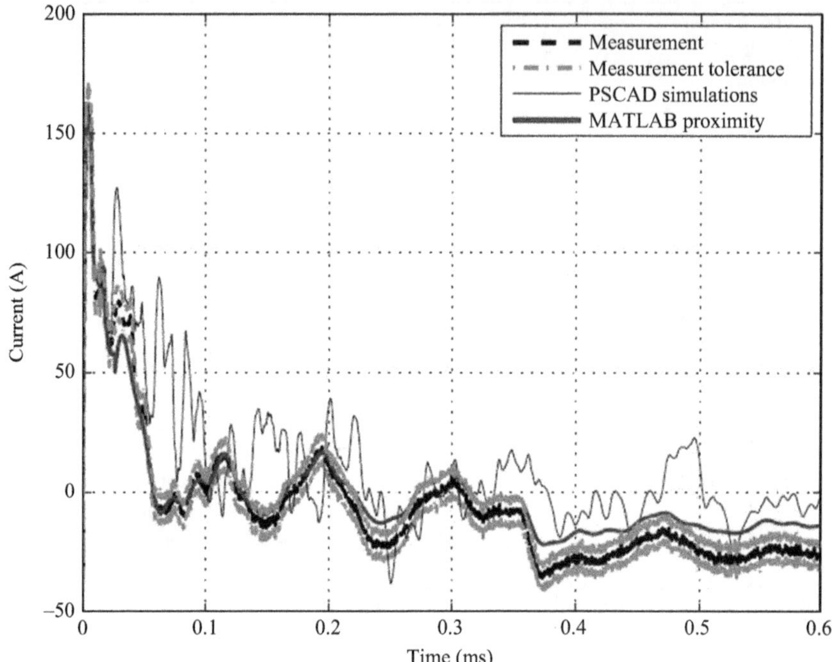

Figure 4.13   Comparison for the measured and simulated core current at the sending end of the energized phase for the multiple major section cable line. The graph shows field measurements, universal line model simulations (PSCAD simulations) and simulations using both layered screen and proximity effect improvements (MATLAB proximity simulations)

affect the simulation results. It is therefore assumed that improving only the simulations of intersheath mode will result in accurate simulation results.

Figure 4.13 shows a comparison of simulation and field test results for the sending end current.

As shown in Figure 4.13, the simulation results are very accurate when both layered screen and proximity effect are included. There is almost a perfect agreement between improved simulations and field measurement results.

The first intersheath reflection is measured at approximately 11 μs. Both simulation results are therefore fully accurate until this point in time is reached. After this time, however, the measured and non-improved simulated current waves become inaccurately damped as the current becomes affected by many cross-bondings and reflected waves. As for the single major section, despite deviation some similarities between PSCAD simulations and measurements are shown, where the simulated wave appears to be less damped. A correctly calculated $Z(\omega)$ matrix, when including the actual physical layout of the screen and the proximity effect, results in the correct damping of the current waves.

Correcting the impedance matrix results in a correct intersheath mode propagation damping, which is the cause for the much better simulation results. The damping at the higher frequencies (above 10 kHz), because of the proximity effect, is seen by the fact that the high-frequency oscillations of the PSCAD simulations in Figure 4.13 are fully damped in the MATLAB simulations in Figure 4.13. Only the modelling of the propagation characteristics of the intersheath mode has been improved and as the improved simulation results agree quite well with field measurements, the assumption of the ground mode not influencing the simulation results is correct.

4.9.4 *Summary for application of PSCAD to cable transients*

It has been shown how PSCAD can be used to make simulations of cable transients. For typical HV cables having two layers of metal screen, and a metal screen with parallel wires, it has, however, been observed that disagreement between cable simulations and field test results starts appearing due to crossbonding points. For non-crossbonded cables, the coaxial modal waves will dominate and when exclusively exciting the coaxial mode, the simulation results fall within the tolerance of the field measurements.

The presence of crossbonding points in long cables causes excitation of intersheath waves when the screens are shifted between cables. After the intersheath waves reach the measurement point, large deviations develop and the simulation model is therefore suggested to be insufficient for crossbonded cable systems. In order to validate this suggestion, a verification of improved simulations is shown. This is done not only for non-crossbonded field measurements with explicit excitation of the intersheath mode, but also against long crossbonded cables with several sections.

The model is therefore improved with respect to the screen layers. The actual physical layout of the screen is modelled more accurately by use of layered screen and resulting in a more accurate damping of the lower frequencies (below 10 kHz). The more accurate damping is obtained because when the correct physical layout of the screen is implemented in the model, the series impedance matrix $Z(\omega)$ becomes more correct. The terminal conditions of the cable depend on $Y_C(\omega) = \dfrac{Y(\omega)}{\sqrt{Y(\omega)Z(\omega)}}$ and $H(\omega) = e^{-\sqrt{Y(\omega)Z(\omega)}l}$. Therefore, when $Z(\omega)$ is more correct, the cable's simulated terminal conditions will become more correct.

Even though the simulations, when modelling the layered screen and field test results, agree quite well regarding the damping in the waveforms of the lower frequency component, all simulation results still have inadequate damping of higher frequency oscillations. Because of this, it is not sufficient to increase the damping at lower frequencies by introducing the layered screen. As the intersheath current propagates between the screens of adjacent cables, their propagation characteristics are also affected by proximity effects which are not taken into account by the simulation software. By expanding the improvements and including the proximity effect, more accurate simulation results at higher frequencies (above 10 kHz) are acquired. This is validated to give precise simulation results when explicitly exciting the intersheath mode, the origin of deviation between original simulations and field measurements.

In order to validate the improved cable model for transmission cables, the field measurements on a single major section with 2 crossbonding points and the field measurements on a cable with 11 major sections are performed. Comparison of these field measurement results and simulations including both layered screen and proximity effect does not give deviation outside of the tolerance of the field measurements. The improved cable model is therefore accurate and sufficient and delivers reliable model for simulating long-distance cables and a transmission system that is mostly or fully underground.

4.10   Conclusions

The development of EMTDC and its PSCAD operations interface with its associated software and its present capabilities are illustrated. The new simulation methods and tools to study complex power electronics applications in power systems are described. In addition, PSCAD has the ease in building user-defined models and can interface with other useful modelling and graphical software. Also, the latest application examples to cable transients are demonstrated.

The changing nature of electric power systems and associated equipment, including transportation, industrial requirements, and power electronic/control systems, are able to be studied and designed with PSCAD.

Very large electric grids are accommodated with parallel processing and hybrid simulation. Equipment supplier's confidential models of their equipment can be included in a model under secured conditions.

PSCAD continues to be aggressively developed to meet the requirements of the developing power system industry so much dependent on power electronics.

References

[1]   Scott-Meyer W. *EMTP Reference Manual*. Portland, OR: Bonneville Power Administration; January 1973.

[2]   Dommel H. W. 'Digital computer solution of electromagnetic transients in single and multiphase networks'. *IEEE Transactions on Power Apparatus and Systems*. April 1969; **PAS-88** (4):388–99.

[3]   Morched A., Gustavsen B., Semlyen A. 'A universal model for accurate calculation of electromagnetic transients on overhead lines and underground cables'. *IEEE Transactions on Power Delivery*. July 1999; **14** (3):1032–8.

[4]   Wedepohl L. M., Mohamed S. E. T. 'Multi-conductor transmission lines; theory of natural modes and Fourier integral applied to transient analysis'. *Proceedings of the IEE*. September 1969; **116** (9):1553–63.

[5]   Manitoba HVDC Research Centre (Canada). *User's Guide on the Use of PSCAD*. Winnipeg, 2004.

[6]   Gole A. M., Woodford S. A., Nordstrom J. E., Irwin G. D. 'A fully inter-polated controls library for electromagnetic transients simulation of power electronic systems'. *Proceedings of the International Conference on Power System Transients*; Rio de Janiero, Brazil, June 24–28, 1995.

[7]   MathWorks, 3, Apple Drive, Natick, MA 01760.

[8]   Siemens Power Technology International (USA), PSS/E User Guide, 400 State Street, P.O. Box 1058, Schenectady, NY 12301-1058, USA.

[9]   Woodford D. A. 'Electromagnetic design considerations for fast acting controllers'. *IEEE Transaction on Power Delivery*. 1996; **11** (3):1515–21.

[10]   Gole A. M., Filizadeh S., Wilson P. L. 'Inclusion of robustness into design using optimization-enabled transient simulation'. *IEEE Transactions on Power Delivery*. 2005; **20** (3):1991–7.

[11]   Irwin G. D., Amarasinghe C., Kroeker N., Woodford D. 'Parallel processing and hybrid simulation of HVDC/VSC PSCAD studies'. Paper B6.3 pre-sented to the ACDC 2012 10th International Conference on AC and DC Power Transmission, Birmingham, UK, December 2012.

[12]   Fluke. *Earth Ground Resistance – principles, testing methods and applications*. Available from http://www.newarkinone.thinkhost.com/brands/promos/Earth_Ground_Resistance.pdf [Accessed 19 November 2008].

[13]   Gustavsen B. 'Panel session on data for modeling system transients – insu-lated cables'. *IEEE Power Engineering Society Winter Meeting*, 2001. vol. 2, pp. 718–23.

[14]   Gustavsen B., Martinez J. A., Durbak D. 'Parameter determination for modeling system transients – part ii: insulated cables'. *IEEE Transactions on Power Delivery*. July 2005; **20** (3):1045–1050.

[15]   Gustavsen B. *A Study of Overvoltages in High Voltage Cables with Emphasis on Sheath Overvoltages*. Trondheim, Norway: Ph.D. Thesis, NTH, 1986.

[16]   Gudmundsdottir U. S., Gustavsen B., Bak C. L.,Wiechowski W., da Silva F. F. 'Field test and simulation of a 400 kV crossbonded cable system'. *IEEE Transactions on Power Delivery*. July 2011; **26** (3):1403–10.

[17]   Gudmundsdottir U. S., De Silva J., Bak C. L., Wiechowski W. *Double Layered Sheath in Accurate HV XLPE Cable Modeling'*. IEEE PES GM 2010, Minneapolis, MN, July 2010.

[18]   Gudmundsdottir U. S. 'Proximity effect in fast transient simulations of an underground transmission cable'. *Electric Power Systems Research*. 2014; **115**:50–6.

[19]   Gudmundsdottir U. S. *Modelling of Long High Voltage AC Cables in Transmission Systems*. Fredericia: Energinet.dk, Aalborg University: Ph.D. Thesis, 2010, ISBN 978-87-90707-73-6.

Chapter 5
XTAP
T. Noda[*]

5.1 Overview

In 1951, Central Research Institute of Electric Power Industry (CRIEPI) was founded by electric power companies in Japan. It conducts various research activities related to the electric power sector from various aspects. Electromagnetic transient (EMT) phenomena occurred in power systems are, of course, one of the subjects of its research activities. CRIEPI has been developing an electromagnetic transient analysis program called XTAP (eXpandable Transient Analysis Program) for its internal research purpose. One of the most important features of XTAP is that it uses a mathematically oscillation-free integration algorithm which never produces "fictitious" numerical oscillation. The widely used trapezoidal integration algorithm produces the numerical oscillation which does not exist in reality if an inductor current or a capacitor voltage is changed stepwise, and it has been an important problem requiring an effective solution. One solution is to use the backward Euler method twice with a halved time step size only at the moment a sudden change of a current or a voltage is detected, and this method, used by Electro-Magnetic Transients Program (EMTP), is known as the Critical Damping Adjustment (CDA). On the other hand, XTAP uses the 2-Stage Diagonally Implicit Runge-Kutta (2S-DIRK) method which is mathematically oscillation-free, for the numerical integration. It is mathematically guaranteed that the numerical oscillation is never observed in results obtained by a straightforward application of the 2S-DIRK method. Therefore, XTAP can be used for simulations involving switching without concern of the numerical oscillation, and it is used by electric power companies in Japan for simulations of power electronics applications such as HVDC, frequency conversion, FATCS, and integration of renewables. This chapter describes distinctive algorithms used in XTAP.

5.2 Numerical integration by the 2S-DIRK method

This section describes the integration algorithm used in XTAP by summarizing and revising [1]. In an EMT simulation, the circuit equations describing the power system of interest are solved at successive time steps with a small interval h, and the successive

[*]Central Research Institute of Electric Power Industry (CRIEPI), Japan

solutions give voltage and current waveforms at various points of the power system. In order to integrate the voltage-current relationships of dynamic components such as inductances and capacitances into the circuit equations, we have to convert the derivative terms appearing in the voltage-current relationships into algebraic relations defined at the successive time steps with h. For this conversion, an integration algorithm is used or discretization (the algebraic relations are obtained by "integrating" the derivative terms). EMTP uses the trapezoidal method for the integration [2]. The trapezoidal method is a simple one-step integration algorithm, which calculates the solution at the present time step only from values obtained at the previous time step. In spite of its simple formula, the trapezoidal method has a second-order accuracy and is A-stable [3]. When an integration method is said to be A-stable, it does not diverge for any time step size as long as the system to be simulated is stable. However, the trapezoidal method produces "fictitious" sustained numerical oscillation for a sudden change of a variable. This numerical oscillation appears when an inductor current or a capacitor voltage is suddenly changed, and this can be a serious drawback for EMT simulations that include power electronics converters utilizing semiconductor switches. To cope with this problem, the Critical Damping Adjustment (CDA) has been proposed and implemented in some versions of EMTP [4, 5]. In the CDA scheme, the trapezoidal method is basically used, and only at a moment a sudden change of a variable is detected the integration method is replaced by the backward Euler method with a halved time step. In most cases, an acceptable solution is obtained, since the backward Euler method does not produce sustained numerical oscillation. It should be noted that if the time step size of the backward Euler method is selected to the half of that of the trapezoidal method, the coefficients of the circuit equations do not change and thus additional LU decomposition is not required. In the practical implementations in old versions of EMTP, switching events, sudden changes of voltage and current source values, and changes of the operating points of nonlinear components represented by piece-wise linear curves are detected to switch the integration algorithm to the backward Euler method [5]. However, detecting a sudden change of a variable is not always feasible. For instance, when a sudden change of a voltage or a current is transmitted to another place by a transmission or distribution line represented by a traveling-wave-based model, detecting the sudden change at the receiving end is not an easy task (since the arriving instant is dependent on the traveling time of the line model). A voltage or current source controlled by a control system can produce a sudden change for instance due to a limiter operation, but detecting this is also difficult. Detecting all types of sudden changes by programming is not practical or can be a serious overhead of execution time even if such programming is done. An inherently "oscillation-free" integration method, therefore, is desired.

5.2.1 *The 2S-DIRK integration algorithm*

The 2S-DIRK integration algorithm is described using the differential equation:

$$\frac{dy}{dt} = f(t, y) \tag{5.1}$$

where t is time and f is function of t and the variable y. Assume that we are carrying out an EMT simulation using the 2S-DIRK method and have obtained the solution up to $t = t_{n-1}$. Now, we integrate (5.1) from $t = t_{n-1}$ to $t = t_n$ using the 2S-DIRK method which consists of two stages. The first stage calculates the solution at the intermediate time point which is in between $t = t_{n-1}$ and $t = t_n$:

$$\tilde{t}_n = t_{n-1} + \tilde{h}, \quad \tilde{h} = ah \tag{5.2}$$

where $a = 1 - 1/\sqrt{2}$ and h is the time step size used. If the values of y at $t = t_{n-1}$ and $t = \tilde{t}_n$ are respectively denoted by y_{n-1} and $\tilde{y}_n$, the first stage formula is

$$\tilde{y}_n = y_{n-1} + \tilde{h}f(\tilde{t}_n, \tilde{y}_n) \tag{5.3}$$

Before moving on to the second stage, $\tilde{y}_n$ is converted to $\tilde{y}_{n-1}$ using:

$$\tilde{y}_{n-1} = \alpha y_{n-1} + \beta \tilde{y}_n \tag{5.4}$$

where $\alpha = -\sqrt{2}$ and $\beta = 1 + \sqrt{2}$. The second stage calculates the solution y_n at $t = t_n$ by

$$y_n = \tilde{y}_{n-1} + \tilde{h}f(t_n, y_n) \tag{5.5}$$

It should be noted from the programming point of view that both (5.3) and (5.5) are in the same form as the backward Euler method, and code for the 2S-DIRK method can be written in such a way that the backward Euler method is applied twice.

The reason why the 2S-DIRK method has been selected, among many existing integration methods, for XTAP is explained as follows [1, 3, 6]. Power system circuits are generally large and have both very short and very long time constants, that is, they are large stiff circuits. To solve a large circuit, we have to save computer memory thus leading to a single step method rather than a multistep method. To solve a stiff circuit, we have to use an A-stable integration method that guarantees that the numerical solution obtained is always stable if the circuit equations of the given circuit satisfy the mathematical stability condition (if the given circuit is stable, its numerical solution is always stable). Implicit Runge-Kutta methods are single-step and A-stable, and thus, they are appropriate for the purpose. Since the method to be selected should be efficient, the number of stages should be two that is minimum and the following diagonality should be taken into account. When a general two-stage implicit Runge-Kutta method is used to solve a circuit resulting in n circuit equations, $2n$ simultaneous equations have to be solved to obtain the solution at each time step. This is because the two integration stages are coupled with each other. The coefficients of the two-stage implicit Runge-Kutta method can ingeniously be chosen so that the two stages are decoupled, and in this way the 2-Stage Diagonally Implicit Runge-Kutta (2S-DIRK) method has been obtained. The word "diagonally" comes from the fact that the two stages are decoupled by setting the "diagonal" coefficients of the integration formula to equal values. Using the 2S-DIRK method, two sets of n simultaneous equations are sequentially solved, rather than solving $2n$ simultaneous equations, thus resulting in more efficient simulations.

Since the 2S-DIRK method is considered as the "oscillation-free" substitute for the trapezoidal method in XTAP, here, the 2S-DIRK method is compared with the trapezoidal method quite briefly. Detailed comparisons with other integration methods are presented in section 5.2.3. From the algorithmic point of view, both the 2S-DIRK and the trapezoidal method are one-step methods which require values at the previous time step only to calculate the solution at the present time step. On the other hand, one of the major differences between the two, besides the issue of fictitious numerical oscillation, is that the 2S-DIRK method consists of two stages while the trapezoidal method consists of a single stage. The 2S-DIRK method thus requires additional calculations for the intermediate time step at $t = \tilde{t}_n$ as previously described. It should be noted that the 2S-DIRK method is more efficient compared with other two-stage implicit Runge-Kutta methods, since two sets of n simultaneous equations can be sequentially solved as mentioned above, but less efficient compared with the trapezoidal method due to the additional calculations at $t = \tilde{t}_n$.

5.2.2 Formulas for linear inductors and capacitors

Consider a linear inductor whose inductance is L. The relationship between the current i flowing through and the voltage v across the linear is given by

$$\frac{di}{dt} = \frac{1}{L}v \tag{5.6}$$

Applying the 2S-DIRK method described in section 5.2.1 to (5.6) gives the following formulas:

first stage: $$\tilde{i}_n = \frac{\tilde{h}}{L}\tilde{v}_n + i_{n-1} \tag{5.7}$$

variable conversion: $$\tilde{i}_{n-1} = \alpha i_{n-1} + \beta \tilde{i}_n \tag{5.8}$$

second stage: $$i_n = \frac{\tilde{h}}{L}v_n + \tilde{i}_{n-1} \tag{5.9}$$

Following section 5.2.1, subscripts are added to v and i to denote time steps and tilde symbols to indicate intermediate values. Both the first-stage formula (5.7) and the second-stage formula (5.9) lead to the equivalent circuit shown in Figure 5.1. The equivalent circuit consists of the conductance G with the current source J in parallel, and their values are given by

$$G = \frac{\tilde{h}}{L}, \quad J = i_{n-1} \tag{5.10}$$

for the first stage and by

$$G = \frac{\tilde{h}}{L}, \quad J = \tilde{i}_{n-1} \tag{5.11}$$

for the second stage. Note that the conductance values of the first and the second stage are the same, and this enables efficient computation as described below.

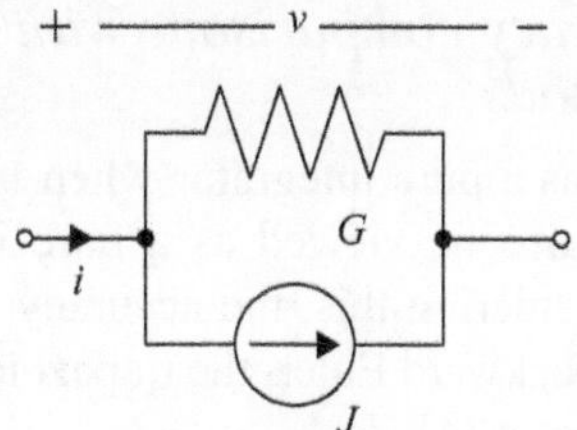

Figure 5.1   Equivalent circuit of linear inductors and linear capacitors

© 2009 IEEE. Reprinted with permission from Noda, T., Takenaka, K. and Inoue, T., 'Numerical integration by the 2-stage diagonally implicit Runge-Kutta method for electromagnetic transient simulations', *IEEE Transactions on Power Delivery*, vol. 24, no. 1, January 2009

Next, consider a linear capacitor whose capacitance is C. The relationship between the voltage v across and the current i through it is given by

$$\frac{dv}{dt} = \frac{1}{C} i \tag{5.12}$$

Applying the 2S-DIRK method described in section 5.2.1 to (5.12) gives the following formulas:

first stage: $\qquad \tilde{i}_n = \frac{C}{\tilde{h}} \tilde{v}_n - \frac{C}{\tilde{h}} v_{n-1} \tag{5.13}$

variable conversion: $\tilde{v}_{n-1} = \alpha v_{n-1} + \beta \tilde{v}_n \tag{5.14}$

second stage: $\qquad i_n = \frac{C}{\tilde{h}} v_n - \frac{C}{\tilde{h}} \tilde{v}_{n-1} \tag{5.15}$

Following section 5.2.1, subscripts are added to v and i to denote time steps and tilde symbols to indicate intermediate values. In the same way as the linear inductor case, both the first stage formula (5.13) and the second stage formula (5.15) lead to the equivalent circuit shown in Figure 5.1. The conductance value and the current source value of the equivalent circuit are given by

$$G = \frac{C}{\tilde{h}}, \quad J = -\frac{C}{\tilde{h}} v_{n-1} \tag{5.16}$$

for the first stage and by

$$G = \frac{C}{\tilde{h}}, \quad J = -\frac{C}{\tilde{h}} \tilde{v}_{n-1} \tag{5.17}$$

for the second stage. Again, in the same way as the linear inductor case, the conductance values of the first and the second stage are the same. This means that linear inductors and capacitors do not require refactorization of the large coefficient matrix obtained by the circuit equations describing the whole circuit.

For the 2S-DIRK formulas for nonlinear inductors and those for nonlinear capacitors, see [1].

5.2.3 Analytical accuracy comparisons with other integration methods

An inductor can be viewed as a pure integrator when its voltage is considered the input, and a capacitor can also be viewed as a pure integrator when its current is considered the input. Considering this, the accuracy of the 2S-DIRK method is compared with those of the backward Euler, the trapezoidal, and the Gear-Shichman method using the pure integrator [4].

The pure integrator is described by

$$\frac{dy}{dt} = x \tag{5.18}$$

where $x(t)$ is the input and $y(t)$ is the output. Applying the 2S-DIRK method described in section 5.2.1 to the equation above gives the following formula:

$$y_n = \left(1 - \frac{1}{\sqrt{2}}\right)x_n + \frac{1}{\sqrt{2}}x_{n-\frac{1}{\sqrt{2}}} + y_{n-1} \tag{5.19}$$

If the time step size h is normalized to unity, the Fourier transform of $x(n - m)$ is given by $X(j\omega)e^{-j\omega m}$ with $X(j\omega)$ being the frequency response, that is, the Fourier transform of $x(n)$. This brings (5.19) into the input-output frequency response:

$$H_{\text{2S-DIRK}}(j\omega) = \frac{\sqrt{2} - 1 + e^{-j\omega/\sqrt{2}}}{\sqrt{2}(1 - e^{-j\omega})} \tag{5.20}$$

Note that the angular frequency ω is also normalized so that the Nyquist frequency is $f_N = 1/2$. Equation (5.20) is the input-output frequency response of the pure integrator when the 2S-DIRK method is applied. In the same way, the input-output frequency response of the pure integrator is obtained using the backward Euler (BE), the trapezoidal (TR), and the Gear-Shichman (GS) method, respectively:

$$H_{\text{BE}}(j\omega) = \frac{1}{1 - e^{-j\omega}} \tag{5.21}$$

$$H_{\text{TR}}(j\omega) = \frac{1 + e^{-j\omega}}{2(1 - e^{-j\omega})} \tag{5.22}$$

$$H_{\text{GS}}(j\omega) = \frac{2}{3 - 4e^{-j\omega} + e^{-j2\omega}} \tag{5.23}$$

The three methods above are taken for comparisons due to the following reasons. The trapezoidal method has been used as the main integration method in the widely used EMTP programs. The backward Euler method is additionally used when the CDA scheme is activated in some versions of EMTP. The Gear-Shichman method, which is a two-step implicit integration method, is widely known and used in the field of electronic circuit analysis [7, 8].

Then, each of (5.20)–(5.23) is divided by $H(j\omega) = 1/(j\omega)$ that is the exact frequency response of the pure integrator, and they are plotted in Figure 5.2. Since they

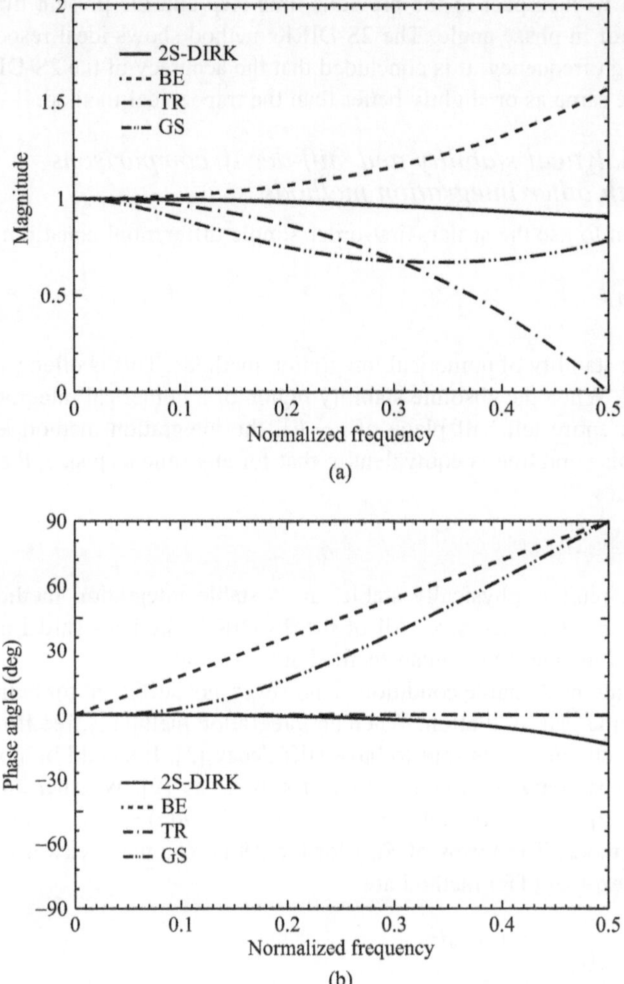

(a)

(b)

Figure 5.2   Comparison of the frequency responses of the pure integrator obtained by the 2S-DIRK, the backward Euler (BE), the trapezoidal (TR), and the Gear-Shichman (GS) method

© 2009 IEEE. Reprinted with permission from Noda, T., Takenaka, K. and Inoue, T., 'Numerical integration by the 2-stage diagonally implicit Runge-Kutta method for electromagnetic transient simulations', *IEEE Transactions on Power Delivery*, vol. 24, no. 1, January 2009

are divided by the exact frequency response, a response closer to unity in magnitude and zero in phase angle is more accurate. The 2S-DIRK and the trapezoidal method show better accuracy over the other two methods. In terms of magnitude, the 2S-DIRK method is better than the trapezoidal method. In terms of phase angle, on the other hand, the trapezoidal method is better. The trapezoidal method which is symmetrical with respect to the previous time step and the present time step produces no error in phase angle. The 2S-DIRK method shows ideal response up to a relatively high frequency. It is concluded that the accuracy of the 2S-DIRK method is almost the same as or slightly better than the trapezoidal method.

5.2.4 Analytical stability and stiff-decay comparisons with other integration methods

It is common to use the scalar, first-order simple differential equation:

$$\frac{dy}{dt} = \lambda y \tag{5.24}$$

to assess the stability of numerical integration methods. This is often called the test equation [3]. When the absolute stability region of a numerical integration method contains the entire left half plane of $z = h\lambda$, the integration method is said to be A-stable. This condition is equivalent to that for any time step size, the solution of (5.24) satisfies:

$$|y_n| \leq |y_{n-1}| \tag{5.25}$$

If a given circuit is physically stable, an A-stable integration method does not diverge for any time step size. All of the 2S-DIRK, the backward Euler, the trapezoidal, and the Gear-Shichman method are A-stable.

Satisfying the A-stable condition is, however, not sufficient for being free from sustained numerical oscillation. When an integration method is free from sustained numerical oscillation, it is said to have stiff decay [3]. It should be noted that the concept of S-stability is similar to that of stiff decay [9]. Whether an integration method has stiff decay or not can be assessed by $R(z) = y_n/y_{n-1}$ in the case of one-step methods. The forms of $R(z)$ for the 2S-DIRK, the backward Euler (BE), and the trapezoidal (TR) method are

$$R_{\text{2S-DIRK}}(z) = \frac{1 + \sqrt{2}az}{(1 - az)^2} \tag{5.26}$$

$$R_{\text{BE}}(z) = \frac{1}{1 - z} \tag{5.27}$$

$$R_{\text{TR}}(z) = \frac{2 + z}{2 - z} \tag{5.28}$$

where $z = h\lambda$. When the time constant of the test equation approaches zero, $\text{Re}\{\lambda\} \to -\infty$. In this situation, $R_{\text{2S-DIRK}}(z) \to 0$ and $R_{\text{BE}}(z) \to 0$ for any positive h. This means that:

$$|y_n| < |y_{n-1}| \tag{5.29}$$

is satisfied when the time constant is small compared with the time step size used. This is the condition that an integration method does not produce sustained numerical oscillation, that is, it has stiff decay. Therefore, the 2S-DIRK and the backward

Euler method have stiff decay. The Gear-Shichman method also has stiff decay, since it is one of the backward differentiation formulas (BDFs) that are designed to have stiff decay (see section 5.1.2 of [3]). In the case of the trapezoidal method, on the other hand, $R_{TR}(z) \rightarrow -1$ when the time constant approaches zero, thus leading to

$$y_n = -y_{n-1} \tag{5.30}$$

This equation itself indicates that the trapezoidal method produces sustained numerical oscillation and does not have stiff decay.

To sum up the discussion above, all of the 2S-DIRK, the backward Euler, the trapezoidal, and the Gear-Shichman method are A-stable and thus do not diverge for any time step size if the given circuit is physically stable (or, if the circuit equations given are mathematically stable). The 2S-DIRK, the backward Euler, and the Gear-Shichman method have stiff decay and do not produce sustained numerical oscillation even if an inductor current or a capacitor voltage are subjected to stepwise change (zero time constant). On the other hand, the trapezoidal method does not have stiff decay and produces sustained numerical oscillation under such a condition.

5.2.5   *Numerical comparisons with other integration methods*

In this section, five numerical examples are used to compare the 2S-DIRK method with the backward Euler, the trapezoidal, and the Gear-Shichman method, or with the CDA scheme.

5.2.5.1   Series RC circuit

Figure 5.3(a) shows a series RC circuit excited by a voltage source, while Figure 5.3(b) shows the voltage waveform generated by the voltage source E. The value of the resistance R is 100 Ω. In the first numerical test, the capacitance C is set to 1 μF, and the voltage v_C across the capacitance is calculated by the four integration methods mentioned above with a time step of 0.1 ms. The calculated results for this short time constant ($\tau = RC = 0.1$ ms) case are shown in Figure 5.4(a). In each waveform plot, the exact solution

$$v_C(t) = \frac{1}{h}\left[\left\{t - \tau\left(1 - e^{-t/\tau}\right)\right\} - \left\{t - h - \tau\left(1 - e^{-(t-h)/\tau}\right)\right\}u(t-h)\right] \tag{5.31}$$

where $u(t)$ is the unit step function, is superimposed by a dashed line. Figure 5.4(b) compares the deviations from the exact solution for the four integration methods. The backward Euler method exhibits the largest error and then the Gear-Shichman method follows. The 2S-DIRK and the trapezoidal method show similar errors, where the error of the former is slightly better than that of the latter. This result agrees with that of the frequency-domain comparison shown in Figure 5.2.

Next, the capacitance value is set to $C = 10$ μF. Figure 5.5(a) shows the calculated results for this long-time constant ($\tau = 1$ ms) case. Figure 5.5(b) shows their deviations from the exact solution (5.31). Again, the 2S-DIRK method exhibits the smallest error, and then, the trapezoidal, the Gear-Shichman, and the backward Euler method follow in this order.

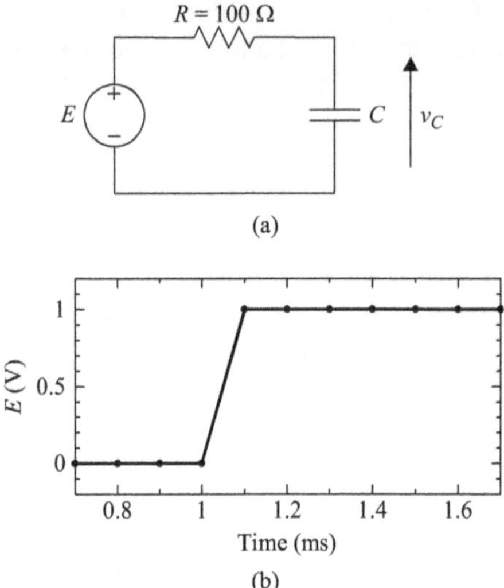

*Figure 5.3   Series RC circuit. (a) Shows the circuit diagram, and (b) the voltage
waveform generated by the voltage source E*

© 2009 IEEE. Reprinted with permission from Noda, T., Takenaka, K. and Inoue, T.,
'Numerical integration by the 2-stage diagonally implicit Runge-Kutta method for
electromagnetic transient simulations', *IEEE Transactions on Power Delivery*, vol. 24,
no. 1, January 2009

5.2.5.2   Series RL circuit

With this test circuit, the stiff-decay properties of the four integration methods are
compared. Figure 5.6(a) shows a series RL circuit excited by a current source, and
the waveform of the current source J is shown in Figure 5.6(b). The values of R
and L are 100 Ω and 10 mH, respectively. The current source initially injects 1 A and
then interrupts it at $t = 1$ ms. Since the magnetic energy stored in the inductor
cannot be released through the resistor, the time constant $t = L/R$ has no effect to
this current interruption. The voltage v_{RL} across the RL branch is calculated by
the four integration methods mentioned above with a time step of 0.1 ms, and the
results are shown in Figure 5.7. The exact solution:

$$v_{RL}(t) = \begin{cases} 100 & (t < 1 \text{ ms}) \\ -10^6(t - 10^{-3}) & (1 \text{ ms} < t < 1.1 \text{ ms}) \\ 0 & (t > 1.1 \text{ ms}) \end{cases} \tag{5.32}$$

is superimposed in each plot using a dashed line. It should be noted that the spike
voltage is observed in reality but does not initiate an oscillation in this circuit condition
without a capacitance. Both the 2S-DIRK and the backward Euler method correctly
reproduce the exact solution. The Gear-Shichman method gives an acceptable result,
but the peak value of the spike is larger than that of the exact solution and an unwanted
undershoot follows. Since these three methods have stiff decay as mentioned in
section 5.2.4, the current interruption does not initiate numerical oscillation. On the

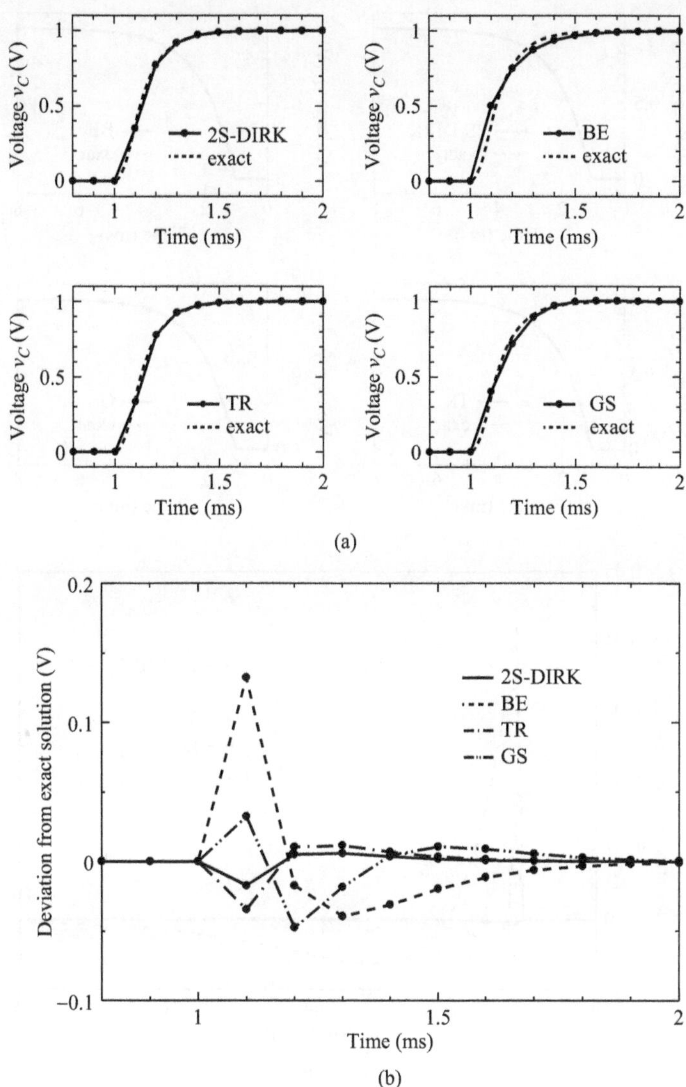

(a)

(b)

Figure 5.4 *Calculated capacitor voltage waveforms of the series RC circuit for the short-time constant ($\tau = 0.1$ ms) case. (a) Shows the calculated waveforms obtained by the four integration methods, and (b) comparison of deviations from the exact solution*

© 2009 IEEE. Reprinted with permission from Noda, T., Takenaka, K. and Inoue, T., 'Numerical integration by the 2-stage diagonally implicit Runge-Kutta method for electromagnetic transient simulations', *IEEE Transactions on Power Delivery*, vol. 24, no. 1, January 2009

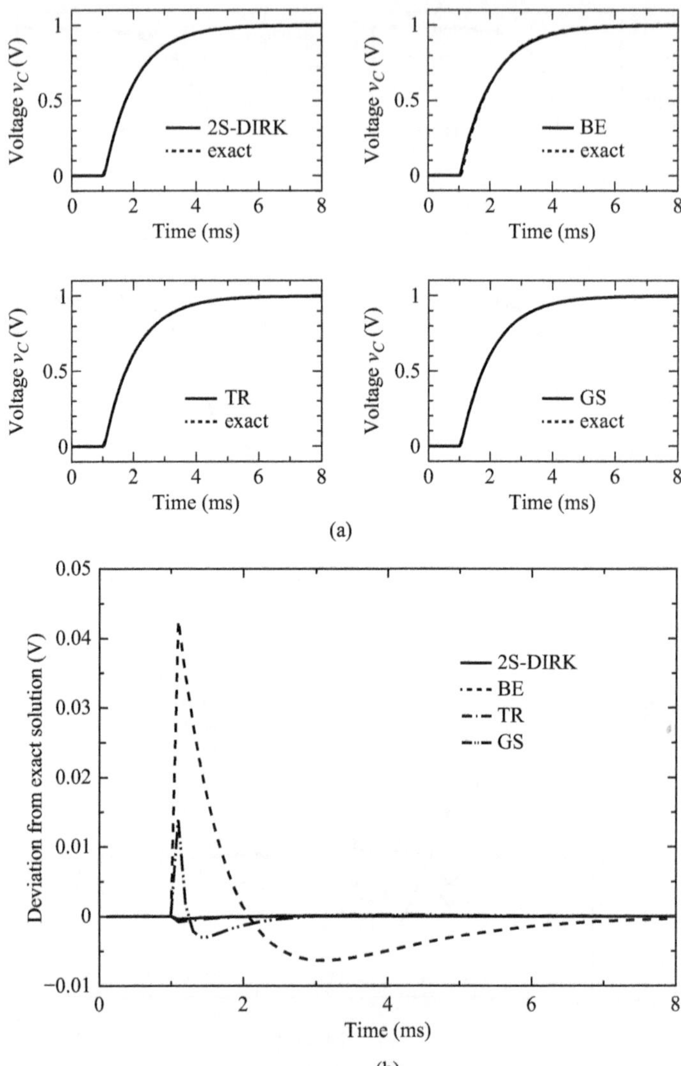

Figure 5.5   Calculated capacitor voltage waveforms of the series RC circuit for the long-time constant ($\tau = 1$ ms) case. (a) Shows the calculated waveforms obtained by the four integration methods, and (b) comparison of deviations from the exact solution

© 2009 IEEE. Reprinted with permission from Noda, T., Takenaka, K. and Inoue, T., 'Numerical integration by the 2-stage diagonally implicit Runge-Kutta method for electromagnetic transient simulations', *IEEE Transactions on Power Delivery*, vol. 24, no. 1, January 2009

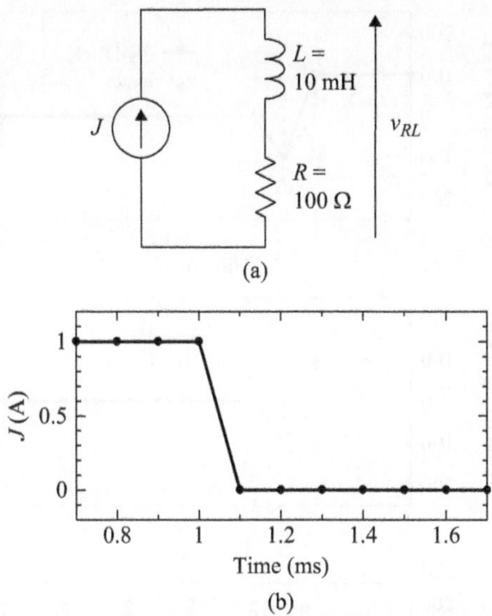

*Figure 5.6   Series RL circuit. (a) Shows the circuit diagram, and (b) the current
waveform generated by the current source J*

© 2009 IEEE. Reprinted with permission from Noda, T., Takenaka, K. and Inoue, T.,
'Numerical integration by the 2-stage diagonally implicit Runge-Kutta method for
electromagnetic transient simulations', *IEEE Transactions on Power Delivery*, vol. 24,
no. 1, January 2009

other hand, the result obtained by the trapezoidal method that does not have stiff decay
shows sustained numerical oscillation initiated by the current interruption.

5.2.5.3 Half-wave rectifier circuit

Figure 5.8 shows a half-wave rectifier circuit. The diode D is modeled by an ideal
switch with the following "diode" logic. It turns on (zero resistance) when the
voltage across it becomes positive and turns off (infinite resistance) when its current becomes negative. The representation of an ideal switch is described in
Appendix B of [1]. The simulation of this circuit is performed, with a time step of
0.1 ms, by the 2S-DIRK method and also by the trapezoidal method with and
without the CDA scheme, and the results are shown in Figure 5.9. The results by
the 2S-DIRK method and that by the trapezoidal method with the CDA scheme are
acceptable, while the result by the trapezoidal method without the CDA scheme is
spoiled by sustained numerical oscillation.

5.2.5.4 Distributed-parameter circuit

Figure 5.10 shows a test circuit in which a capacitor is excited by a voltage source
through a distributed-parameter circuit (transmission line). The distributed-parameter
circuit represents an underground cable, and Bergeron's model (lossless) [2]
is used for its simplified modeling. The capacitance value is 1,000 pF, and it

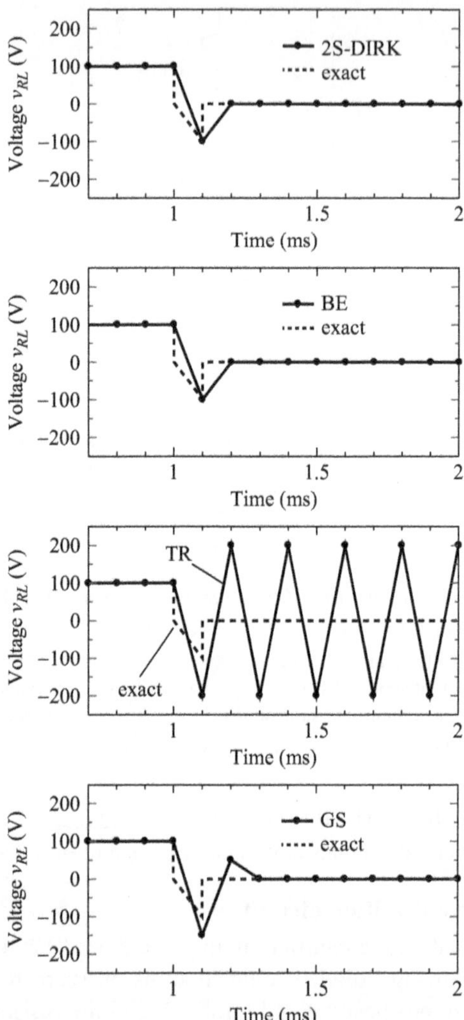

Figure 5.7   *Calculated waveforms of the resistor-inductor branch in the series RL circuit. (a) Shows the calculated waveforms obtained by the four integration methods, and (b) comparison of deviations from the exact solution*

© 2009 IEEE. Reprinted with permission from Noda, T., Takenaka, K. and Inoue, T., 'Numerical integration by the 2-stage diagonally implicit Runge-Kutta method for electromagnetic transient simulations', *IEEE Transactions on Power Delivery*, vol. 24, no. 1, January 2009

represents winding capacitance of a transformer. This circuit may be viewed as a simplified representation of a cable transmission system terminated by a transformer. The simulation is performed, with a time step of 1 μs, by the 2S-DIRK method and also by the trapezoidal method with the CDA scheme. The latter

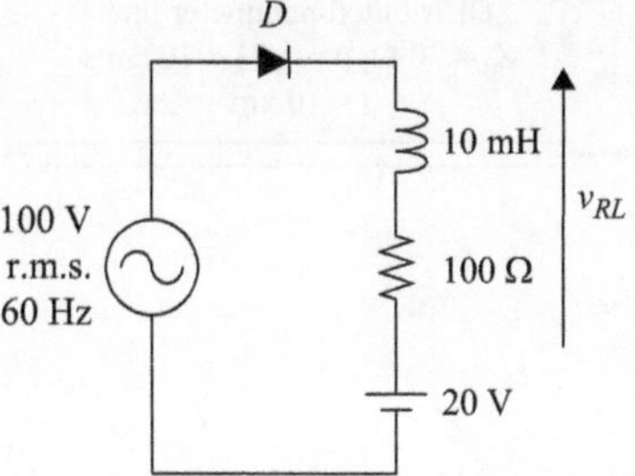

Figure 5.8   Half-wave rectifier circuit

© 2009 IEEE. Reprinted with permission from Noda, T., Takenaka, K. and Inoue, T., 'Numerical integration by the 2-stage diagonally implicit Runge-Kutta method for electromagnetic transient simulations', *IEEE Transactions on Power Delivery*, vol. 24, no. 1, January 2009

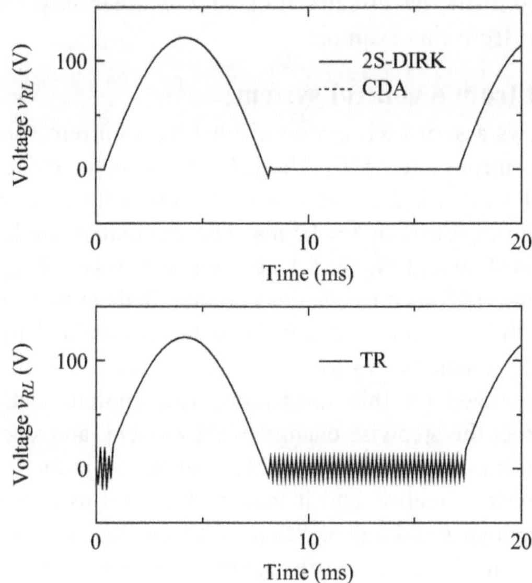

Figure 5.9   Simulation results of the half-wave rectifier circuit

© 2009 IEEE. Reprinted with permission from Noda, T., Takenaka, K. and Inoue, T., 'Numerical integration by the 2-stage diagonally implicit Runge-Kutta method for electromagnetic transient simulations', *IEEE Transactions on Power Delivery*, vol. 24, no. 1, January 2009

simulation was carried out by an existing version of EMTP. From the results shown in Figure 5.11, we can verify that the result obtained by the 2S-DIRK method is consistent with the circuit theory and thus correct, while that obtained by the trapezoidal method with the CDA scheme is spoiled by numerical oscillation. The implementation of the CDA scheme fails to detect the stepwise change of the voltage transmitted by the distributed-parameter circuit with a time

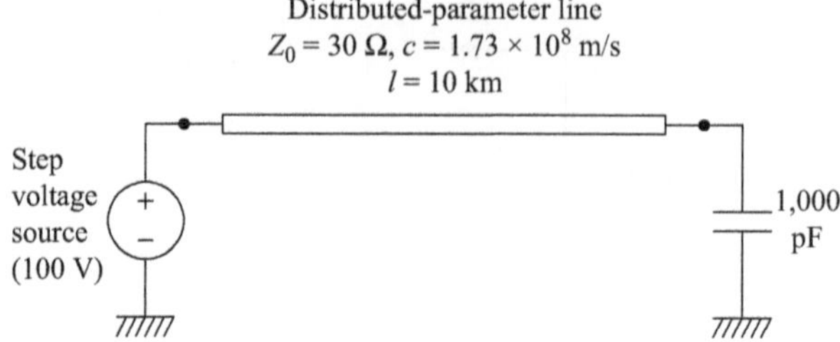

Figure 5.10   Test circuit in which a capacitor is excited through a distributed-parameter circuit

© 2009 IEEE. Reprinted with permission from Noda, T., Takenaka, K. and Inoue, T., 'Numerical integration by the 2-stage diagonally implicit Runge-Kutta method for electromagnetic transient simulations', *IEEE Transactions on Power Delivery*, vol. 24, no. 1, January 2009

delay. We may confirm the effectiveness of the inherently "oscillation-free" integration method from this example.

5.2.5.5   Output from a control system

Figure 5.12(a) shows a series RL circuit excited by a current source representing the output from a control system [10]. The voltage across the inductor is calculated, with a time step of 1 ms, for the case where the output from the control system is changed stepwise from 0 to 1 at $t = 10$ ms. The calculated result obtained by the 2S-DIRK method is shown in Figure 5.12(b), where a spike voltage is generated at the instant the current is forced to change stepwise. This is in accordance with the circuit theory and thus correct. The calculated result obtained by the trapezoidal method with the CDA scheme is shown in Figure 5.12(c). Again, an existing version of EMTP was used for this simulation. The implementation of the CDA scheme fails to detect the stepwise change of the current, and sustained numerical oscillation is produced, on the other hand. This sustained numerical oscillation, of course, does not exist in reality, and it may be too hard to expect for simulation beginners to distinguish this kind of "fictitious" oscillation from other kinds of "physical" oscillation. We can confirm again the effectiveness of the inherently "oscillation-free" integration method, that is, the 2S-DIRK method.

5.3   Solution by a robust and efficient iterative scheme

A power system often includes nonlinear circuit components such as surge arresters, windings with iron cores, and semiconductor switching devices. When the simulation circuit includes these nonlinear components that yield nonlinear constitutive relationships between their voltages and currents, the resultant circuit equations become also nonlinear. To carry out the EMT simulation of a nonlinear circuit, we have to solve, at each time step, a resistive nonlinear circuit which is derived by discretization of the dynamic circuit components included in the circuit. In XTAP, the dynamic circuit components such as inductors and capacitors are

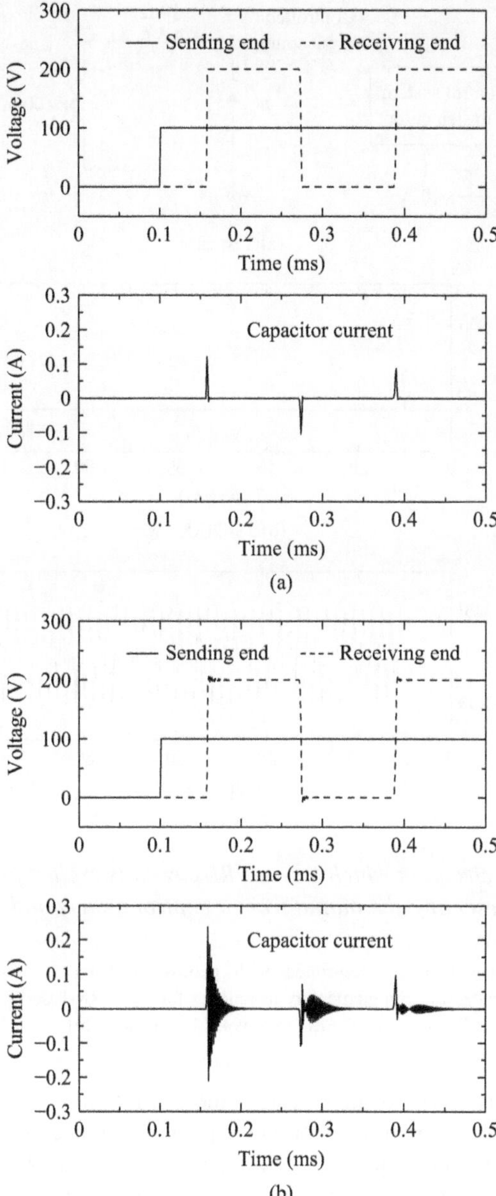

Figure 5.11   *Simulation results of the test circuit with a distributed-parameter circuit*

© 2009 IEEE. Reprinted with permission from Noda, T., Takenaka, K. and Inoue, T., 'Numerical integration by the 2-stage diagonally implicit Runge-Kutta method for electromagnetic transient simulations', *IEEE Transactions on Power Delivery*, vol. 24, no. 1, January 2009

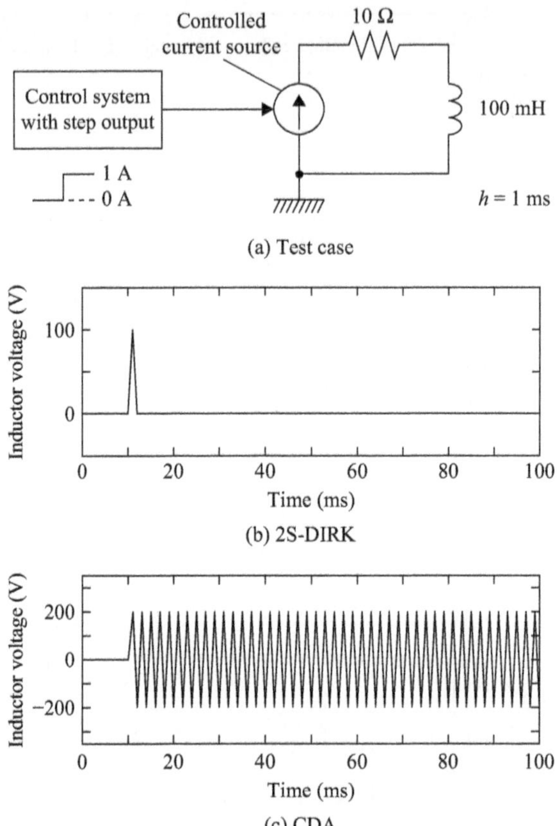

Figure 5.12   Test circuit in which a series RL circuit is excited by a current source representing the output from a control system and its simulation results

© 2014 Elsevier. Reprinted with permission from T. Noda, T. Kikuma and R. Yonezawa, 'Supplementary techniques for 2S-DIRK-based EMT simulations', *Electric Power Systems Research*, vol. 115, October 2014

discretized with respect to time using the 2S-DIRK method mentioned in section 5.2. In general, nonlinear circuit equations cannot be solved by a non-iterative method and thus require an iterative method such as the Newton-Raphson (NR) algorithm. In addition, the solution process using an iterative process is prone to convergence problems.

XTAP uses a robust and efficient iterative scheme for solving nonlinear circuit equations as its solution method for EMT simulations. It is described in this section by summarizing and revising [11].

The characteristics of nonlinear components are, in most cases, obtained by measurements, and they are represented by piece-wise linear curves. This is probably not true for electronic circuit simulations in which exponential formulas are often used for representing the characteristics of semiconductor devices. On the other hand, this is quite true for power system EMT simulations. Therefore, it is acceptable to assume that the characteristics of all nonlinear components are given by monotonically increasing piece-wise linear curves. As pointed out in [12], with the assumption above the standard NR algorithm shows good efficiency when it converges, but it is prone to get into an infinite loop thus resulting in non-convergence.

The iterative scheme used in XTAP combines the three different iterative algorithms, the standard NR, the biaxial NR, and the Katzenelson algorithm. The biaxial NR algorithm was developed especially for XTAP by extending the standard NR algorithm so as to select a better operating section (piece-wise linear section) by utilizing the information of both axes of the nonlinear characteristic rather than using the information of the horizontal axis only [11]. The Katzenelson algorithm [13–15] is famous for the fact that its convergence is mathematically guaranteed, but it is also known for its inefficiency. The iterative scheme used in XTAP first tries the standard NR algorithm that is most efficient but the least convergent. If it fails to converge, then the biaxial NR algorithm that is more convergent but less efficient is employed. In most cases, the iterative scheme converges to the solution at the standard or the biaxial NR stage. Just in case when the biaxial NR algorithm fails to converge, the Katzenelson algorithm whose convergence is guaranteed but least efficient is applied so as to ensure to obtain the solution. In this way, XTAP's iterative scheme always converges to the solution with a relatively small number of iteration steps.

5.3.1   Problem description

In most EMT simulation programs, first, each of the dynamic circuit components such as inductors and capacitors in the circuit is replaced by the parallel connection of a conductance and a current source shown in Figure 5.1 using a numerical integration method, and then, the circuit equations to be solved are formulated from the circuit obtained. This is the case where the program uses the nodal [16], the modified nodal [17], or the sparse tableau [18] approach for the formulation of circuit equations. Note that some programs use the state-space formulation [19], and the procedure described here cannot be applied to such state-space-based programs. The procedure mentioned above transforms a given nonlinear circuit including dynamic circuit components into a nonlinear resistive circuit. Therefore, the EMT simulation can be viewed as calculating the solutions of nonlinear resistive circuits obtained at successive time steps.

In general, the circuit equations of a nonlinear resistive circuit are obtained in the following form

$$f(x) = z \tag{5.33}$$

where x and z are column vectors of size N and $f(\cdot)$ is the function that gives N nonlinear relationships. The vector x contains unknown circuit variables, and z contains the contributions from independent voltage and current sources.

Equation (5.33) is composed of equations derived by Kirchhoff's current and voltage laws and Ohm's law. Linearizing (5.33) gives the recursive formula:

$$F^{(n-1)}x^{(n)} = y^{(n-1)} \quad (n = 1, 2, 3, \cdots) \tag{5.34}$$

where F, called the Jacobian of (5.33), is an N-by-N matrix given by

$$F^{(n-1)} = \left. \frac{\partial f(x)}{\partial x} \right|_{x=x^{(n-1)}} \tag{5.35}$$

and y is a column vector of size N containing the contributions from current sources representing the operating points (the intercepts of the $v - i$ characteristics) of nonlinear components in addition to those from independent voltage and current sources as in z. It should be noted that, in (5.35), each of nonlinear resistors in the circuit can be interpreted as the parallel connection of a conductance and a current source shown in Figure 5.1, where the conductance G and the current source J, respectively, represent the slope and the intercept of the present operating point in its nonlinear $v - i$ characteristic. The superscript n in parentheses indicates the iteration number. Starting with an appropriate initial solution $x^{(0)}$, successively solving (5.34) gives update solutions $x^{(n)}$ for $n = 1, 2, 3, \ldots$. Once $x^{(n)}$ has converged to the solution of the circuit, the iterative process is terminated, and then, the simulation moves on to the next time step.

In the iterative process mentioned above, the values of $F^{(n-1)}$ and $y^{(n-1)}$ in (5.34) are calculated using the solution $x^{(n-1)}$ obtained at the previous iteration step as follows. For each nonlinear component, its branch voltage and current calculated at the previous iteration step are obtained from $x^{(n-1)}$. Based on the branch voltage or current, the operating point of the nonlinear component is determined, and the values of G and J of the equivalent circuit shown in Figure 5.1 are obtained from its $v - i$ characteristic. Once the values of G's and J's of all nonlinear components have been obtained, $F^{(n-1)}$ and $y^{(n-1)}$ are reconstructed using the formulation method used. Most practical implementations follow the procedure mentioned here, that is, the procedure to directly build (5.34) utilizing the equivalent circuit shown in Figure 5.1 with a formulation method, rather than linearizing (5.33) using (5.35) [16].

5.3.2   Iterative methods

Using a simple illustrative circuit, the three iterative algorithms: (i) the standard NR, (ii) the biaxial NR, and (iii) the Katzenelson algorithm are compared in terms of convergence performance. Among these methods, the biaxial NR algorithm was developed for XTAP [11]. These three methods differ in how to obtain the values of $F^{(n-1)}$ and $y^{(n-1)}$ from $x^{(n-1)}$, that is, in how to obtain the values of G and J of each nonlinear component from its branch voltage and current calculated at the previous iteration step. Since it is assumed that the characteristic of each nonlinear component is represented by a piece-wise linear curve, obtaining G and J corresponds to choosing an operating section in the piece-wise linear curve.

Figure 5.13(a) shows the schematic diagram of the illustrative circuit. The linear resistor R and the nonlinear resistor R_N are connected in series to the dc voltage source e. The voltage-current characteristic of R_N, shown in Figure 5.13(b),

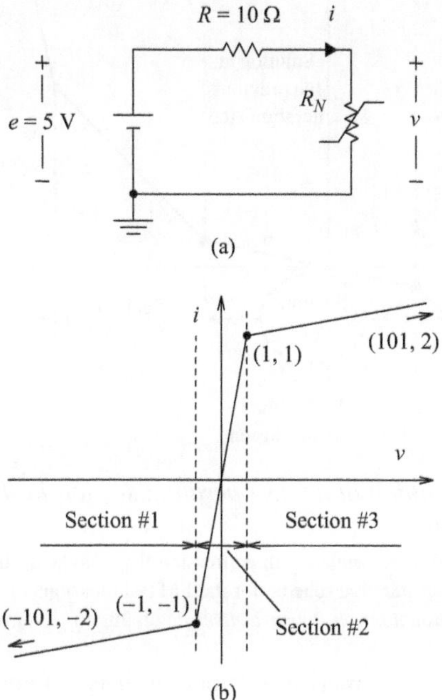

(a)

(b)

Figure 5.13   *Illustrative circuit for comparing (1) the standard NR, (2) the biaxial NR, and (3) the Katzenelson method in terms of convergence performance. (a) Schematic diagram, (b) voltage-current characteristic of the nonlinear resistor R_N*

© IEEE 2011. Reprinted with permission from Noda, T. and Kikuma, T., 'A robust and efficient iterative scheme for the EMT simulations of nonlinear circuits', *IEEE Transactions on Power Delivery*, vol. 26, no. 2, April 2011

consists of three linear sections. It is assumed that the characteristic outside the region defined by the breakpoints are linearly extrapolated, and the same applies to the other nonlinear characteristics presented in this section. In the following illustrations, the nonlinear resistor case is dealt with, but the same discussion applies to the nonlinear inductor and capacitor cases.

5.3.2.1   Standard NR algorithm

The standard NR algorithm chooses an operating section in the piece-wise linear characteristic of each nonlinear component in the following way. At the nth iteration step, the solution $x^{(n-1)}$ obtained at the previous iteration step gives the branch voltage of each nonlinear component calculated at the previous iteration step, and the section to which that branch voltage belongs is chosen as the operating section for the nth iteration step. A new operating section is determined based only on the branch voltage, and this procedure is illustrated in Figure 5.14. It is also possible to use the branch current to determine a new operating section. Anyhow, the standard NR algorithm determines a new operating section based on the information of either one of the two axes of the nonlinear characteristic.

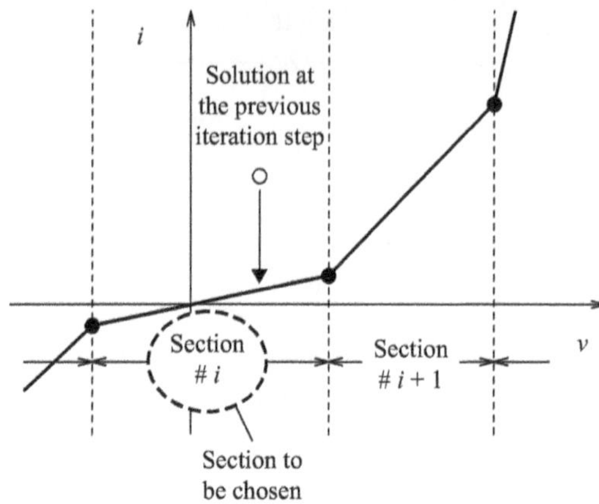

Figure 5.14   Determination of a new operating section in the standard NR algorithm

© IEEE 2011. Reprinted with permission from Noda, T. and Kikuma, T., 'A robust and efficient iterative scheme for the EMT simulations of nonlinear circuits', *IEEE Transactions on Power Delivery*, vol. 26, no. 2, April 2011

In most textbooks on numerical computing, it is common to describe the NR algorithm, which is called the "standard" NR algorithm for distinction in this book, in such a way that correction terms calculated by Jacobian matrices are successively added to an initial solution to obtain updated solutions. According to [16], it is common, however, in the field of circuit simulations to directly obtain updated solutions by successively solving (5.34). This is made possible by including the contributions from current sources representing the operating points of nonlinear components in the vector y in (5.34).

Using the illustrative circuit shown in Figure 5.13, the convergence performance of the standard NR algorithm is examined. In this examination, a new operating section is assumed to be determined based on the voltage axis. Figure 5.15 shows the solution locus during the iteration process when the standard NR algorithm is applied to solving the illustrative circuit. The initial solution assumed is designated by $x^{(0)}$. Due to this initial solution, the operating section at the first iteration step is section #3. The intersection of the straight line of section #3 and the straight line expressed by

$$e = Ri + v \tag{5.36}$$

is the solution at the first step, and it is designated by $x^{(1)}$ in Figure 5.15. The solution $x^{(1)}$ gives the branch voltage R_N at the first iteration step, and this determines the operating point at the second iteration step, that is, Point A. The operating section at the second iteration step is thus determined to section #1. The intersection of the straight line of section #1 and that of (5.36) is the solution at the second iteration step, and it is designated by $x^{(2)}$ in Figure 5.15. The solution $x^{(2)}$ gives the branch voltage of R_N at the second iteration step, and this determines the operating point at the third iteration step, that is, Point B. Since both the initial solution $x^{(0)}$ and

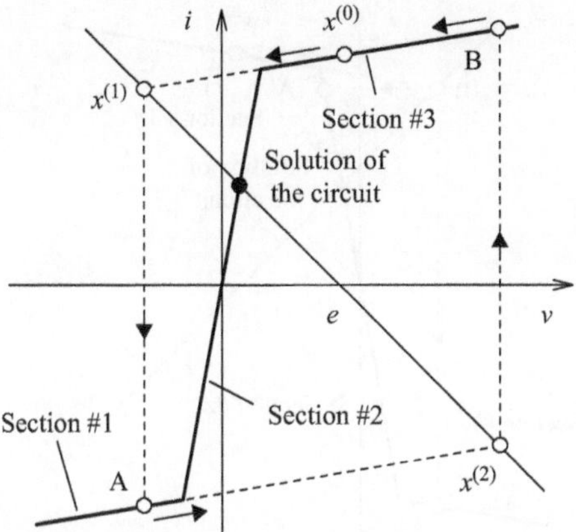

Figure 5.15 *Solution locus during the iteration process due to the standard NR*
algorithm

© IEEE 2011. Reprinted with permission from Noda, T. and Kikuma, T., 'A robust
and efficient iterative scheme for the EMT simulations of nonlinear circuits',
IEEE Transactions on Power Delivery, vol. 26, no. 2, April 2011

Point B are in section #3, the iteration steps following simply trace this cyclic locus. As a result, the solutions $x^{(n)}$ ($n = 3, 4, 5, \ldots$) following never reach the solution of the circuit marked by the solid dot in Figure 5.15. If a new operating section is determined based on the branch current, the iteration process converges to the solution of the circuit in this particular case. However, if the voltage-current characteristic were such that the voltage saturates with respect to the current, the iteration process would trace a cyclic locus and never reaches the solution of the circuit.

Since the process to find a new operating section can be implemented by a simple search algorithm, the computational complexity of the standard NR algorithm is quite small.

5.3.2.2   Biaxial NR algorithm

The biaxial NR algorithm that was developed especially for XTAP [11] adaptively alters which variable to use—the branch voltage or the branch current—for choosing a new operating section. The word "biaxial" comes from the fact that the algorithm utilizes the information of both axes of the nonlinear characteristic. At each iteration step, the biaxial NR algorithm first identifies the two candidates for the new operating section: one determined by the branch voltage and the other by the branch current. Then, the one that is closer to the operating section at the previous iteration step is selected as the new operating section.

Using the illustrative circuit shown in Figure 5.13, we examine the convergence performance of the biaxial NR algorithm. Figure 5.16 shows the solution locus when the biaxial NR algorithm is applied to solving the illustrative circuit. The initial solution $x^{(0)}$ is assumed at the same position as that of the standard NR algorithm case. The iteration process is the same as that of the standard NR algorithm, until the solution $x^{(1)}$ is obtained at the first iteration step. At the second

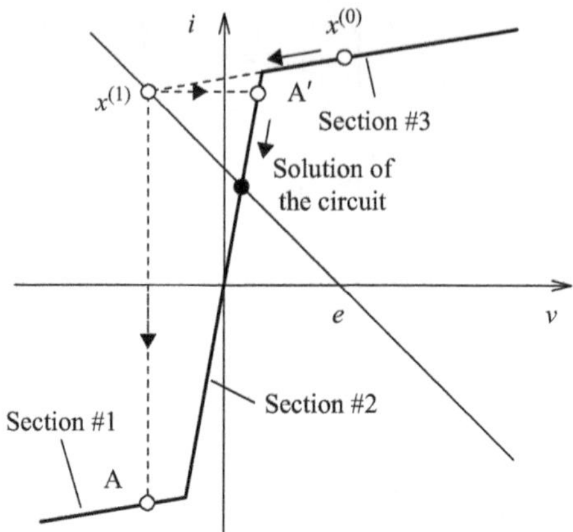

*Figure 5.16   Solution locus during the iteration process due to the biaxial NR
algorithm*

© IEEE 2011. Reprinted with permission from Noda, T. and Kikuma, T., 'A robust
and efficient iterative scheme for the EMT simulations of nonlinear circuits',
IEEE Transactions on Power Delivery, vol. 26, no. 2, April 2011

iteration step, the two candidates Point A and A' are identified. If the new operating section is chosen based on the branch voltage, Point A, that is, section #1 is
determined. If it is chosen based on the branch current, on the other hand, Point A',
that is, section #2 is determined. Since section #2 (Point A) is closer to the previous operating section, section #2 is finally selected as the new operating section
at the second iteration step. Next, the intersection of the straight line of section #2
and the straight line expressed by (5.36) is obtained as the solution at the second
iteration step. Since this is the solution of the circuit, the iteration process is terminated. Note that the biaxial NR algorithm is able to reach the solution of the
circuit, even if the voltage is saturated with respect to the current in the nonlinear
characteristic. It is clear from this example that the biaxial NR algorithm has much
improved probability of convergence to the solution of the circuit. It should also be
noted at the same time that its convergence is not mathematically guaranteed.

To find a new operating section, the biaxial NR algorithm has to repeat the
same search algorithm as that of the standard NR algorithm twice for both axes of
the nonlinear characteristic. However, the time required for such a simple search
algorithm is actually negligible, and the following fact is more essential. The
biaxial NR algorithm chooses an operating section that is closer to the previous
operating section out of the two candidates obtained by both axes. When a large
number of sections exist between the initial solution and the final solution to be
obtained, the number of sections to be jumped over by the biaxial NR algorithm is
thus reduced when compared with that by the standard NR algorithm. As a result,
the biaxial NR algorithm requires more iteration steps to reach the final solution,
since the number of sections to be jumped over is less. Therefore, the biaxial
NR algorithm is more demanding in terms of computational complexity when
compared with the standard NR algorithm.

5.3.2.3   Katzenelson algorithm

The Katzenelson algorithm is always able to reach the solution of the circuit with a finite number of iterations, when the characteristics of all nonlinear components are given in the form of monotonically increasing piece-wise linear curves [13]. And, its convergence is mathematically guaranteed [13–15]. The Katzenelson algorithm differs from the standard NR algorithm in how to choose a new operating section, just like the biaxial NR algorithm. In the case of the standard NR algorithm, even if the branch voltage or current determines a section which is far away from the previous one, that determined section is chosen as the new operating section. The Katzenelson algorithm, on the other hand, determines which direction to go based on the branch voltage or current but stops at the first breakpoint encountered, that is, at the boundary between the previous operating section and its neighboring section located in the determined direction. Then, that neighboring section is chosen as the new operating section, and the breakpoint is considered the updated solution. Therefore, the Katzenelson algorithm moves the operating section only to one of the neighboring sections and does not jump over multiple sections.

Using the illustrative circuit shown in Figure 5.13, the Katzenelson algorithm is briefly described here. Figure 5.17 shows the solution locus when the Katzenelson algorithm is applied to solving the illustrative circuit. The initial solution $x^{(0)}$ is assumed at the same position as that of the standard NR algorithm case. At the first iteration step, the intersection of the straight line of section #3 and that expressed by (5.36) is calculated as the tentative solution which is marked by Point C in Figure 5.17. Then, the tentative solution is modified to the first breakpoint encountered, that is, the breakpoint between sections #3 and #2 designated by $x^{(1)}$. Finally, section #2 is chosen as the new operating section. Since the solution of the

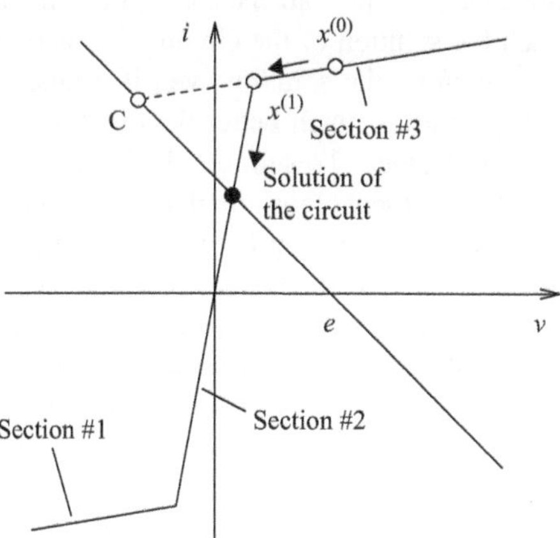

Figure 5.17   *Solution locus during the iteration process due to the Katzenelson algorithm*

© IEEE 2011. Reprinted with permission from Noda, T. and Kikuma, T., 'A robust and efficient iterative scheme for the EMT simulations of nonlinear circuits', *IEEE Transactions on Power Delivery*, vol. 26, no. 2, April 2011

circuit is in section #2, the Katzenelson algorithm reaches the final solution to be obtained at the next iteration step.

The Katzenelson algorithm is computationally quite demanding. When the tentative solution is modified to one of the neighboring breakpoints, interpolation is required. More importantly, the Katzenelson algorithm moves the operating section only to one of the neighboring section; in other words, it moves just one section per one iteration step. When the nonlinear characteristic has a large number of piecewise linear sections, the Katzenelson algorithm requires a significant number of iterations to reach the final solution to be obtained. Note that the biaxial NR algorithm can jump over multiple sections unlike the Katzenelson algorithm. Therefore, the Katzenelson algorithm requires further more computational complexity compared with the biaxial NR algorithm.

5.3.3   Iterative scheme used in XTAP

The iterative scheme used in XTAP combines the three iterative algorithms described above so as to realize robust and efficient iterations. Among the three algorithms, the standard NR algorithm is the most efficient if it converges, but its probability to converge to the solution of the circuit is the lowest, in other words, it is the least. Considering this property, the scheme first employs the standard NR algorithm. If it converges, the solution is obtained in the most efficient way. If it does not converge, then the biaxial NR algorithm which is more convergent but less efficient is employed. Due to its high probability of convergence, most cases converge at this stage. However, the convergence of the biaxial NR algorithm is not mathematically guaranteed. Just in case when it does not converge, the Katzenelson algorithm is finally employed to ensure the convergence. Since the convergence of the Katzenelson algorithm is mathematically guaranteed, the solution of the circuit is obtained with a finite number of iterations. As mentioned above, the Katzenelson algorithm requires a large number of iteration steps to reach the solution of the circuit. But, here, we have to make the following compromise: obtaining the solution even if it takes a significant time by using the Katzenelson algorithm is much better than getting into a cyclic locus and ending up with no solution obtained. To sum up, the iterative scheme starts with the most efficient but the least convergent algorithm that is the standard NR algorithm and successively changes it with a less efficient but more convergent algorithm if it does not converge. It finally employs the mathematically convergent Katzenelson algorithm to ensure the convergence. In this way, the iterative scheme used in XTAP always obtains the solution of the circuit with a relatively small number of iterations.

In a practical implementation, judging whether an iteration process is converging to the solution of the circuit or being in a cyclic locus is not an easy task. Thus, we specify a positive integer and assume that the process is not converging if the number of iterations exceeds that given positive integer. For the iterative scheme discussed, we specify the two positive integers N_B and N_K satisfying $N_B \leq N_K$. At first, the standard NR algorithm is employed, and if the number of iterations reaches N_B, the algorithm is changed with the biaxial NR algorithm. Then, if the number of iterations reaches N_K, the biaxial NR algorithm is changed with the Katzenelson

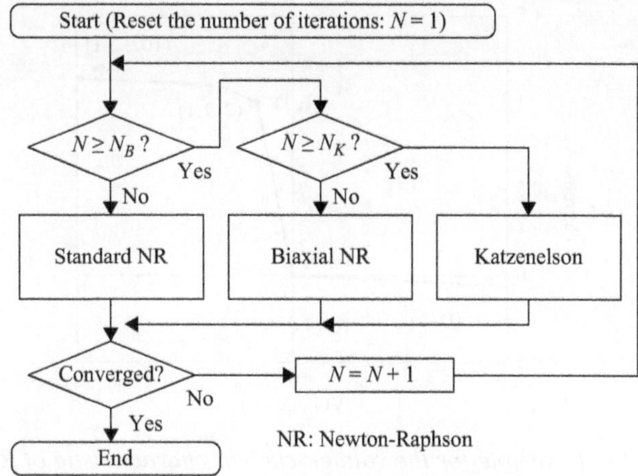

Figure 5.18   Flowchart of the iterative scheme used in XTAP

© IEEE 2011. Reprinted with permission from Noda, T. and Kikuma, T., 'A robust
and efficient iterative scheme for the EMT simulations of nonlinear circuits',
IEEE Transactions on Power Delivery, vol. 26, no. 2, April 2011

algorithm. The resultant flowchart of the iterative scheme is shown in Figure 5.18.
The values of N_B and N_K have to be specified by the user, and $N_B = 30$ and $N_K = 60$
are used as the default values in XTAP. To the best of this writer's knowledge, these
default values seem to be working generally well for most simulation cases.

5.3.4   Numerical examples

In this section, two more voltage-current characteristics for which the standard NR
algorithm does not converge are shown, and they are used to illustrate how the
iterative scheme used in XTAP converges to the solutions to be obtained. Then, an
actual inverter circuit for which the standard NR algorithm also does not converge
is presented to show the practical convergence performance of XTAP's iterative
scheme.

5.3.4.1   Voltage-current curves with cyclic loci

In the first example, we assume that the nonlinear resistor R_N of the illustrative
circuit shown in Figure 5.13 has the voltage-current characteristic shown in
Figure 5.19. Figure 5.20(a) shows the solution locus when the standard NR algorithm which determines the new operating section based on the branch voltage is
used. The initial solution $x^{(0)}$ is assumed to be at Point 0 shown in Figure 5.20(a).
The iteration process of the standard NR algorithm follows Points 1, 2, 3, 4, 5,
6, ..., and the solution is thus updated as $x^{(1)}, x^{(2)}, x^{(3)}, \ldots$. Following Point 6, the
iteration process obtains the solution $x^{(2)}$ at Point 3 that is the same solution as that
at the second iteration step. Therefore, the solution process gets into the infinite
loop formed by $6 \to 3 \to 4 \to 5 \to 6$, in which $x^{(2)}$ and $x^{(3)}$ alternate the solution.

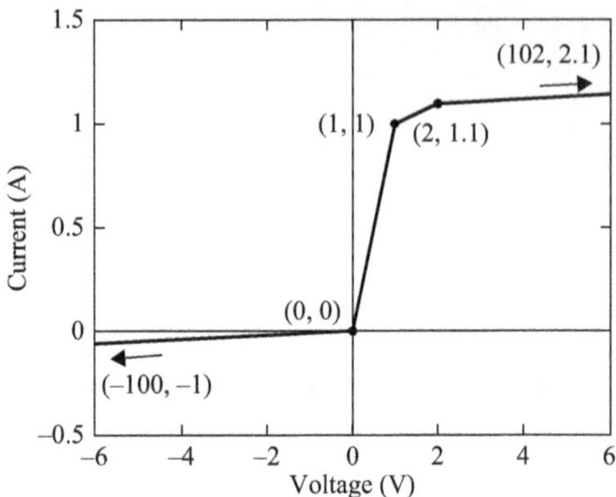

Figure 5.19   First example for the voltage-current characteristic of R_N

© IEEE 2011. Reprinted with permission from Noda, T. and Kikuma, T., 'A robust
and efficient iterative scheme for the EMT simulations of nonlinear circuits',
IEEE Transactions on Power Delivery, vol. 26, no. 2, April 2011

In summary, when the standard NR algorithm is applied, its iteration process first
goes around a cyclic locus and then gets into another cyclic locus which is an
infinite loop. When XTAP's iterative scheme is applied, on the other hand, the
first solution locus before the N_Bth iteration step is the same as that of the standard
NR algorithm mentioned above. At the N_Bth iteration step, the biaxial NR algorithm is employed and it reaches to the solution of the circuit with one or two
iteration steps, depending on the value of N_B specified, as shown in Figure 5.20(b).
If the solution at the $(N_B - 1)$-th iteration step is $x^{(1)}$ or $x^{(2)}$, it requires only one
more iteration step to reach the final solution. If the solution at the $(N_B - 1)$-th
iteration step is $x^{(3)}$, two more iteration steps are required. Figure 5.20(b) shows
these three different loci. Note that the biaxial NR algorithm effectively avoids
getting into the infinite loop.

   In the second example, we assume that the nonlinear resistor R_N has the
voltage-current characteristic shown in Figure 5.21. Figure 5.22(a) shows the
solution locus when the standard NR algorithm with its new operating sections
determined by the branch voltage is used. Starting from the initial solution $x^{(0)}$
designated by Point 0 in Figure 5.22(a), the iteration process follows Points 1,
2, 3, 4, 5, 6, 7, ..., and thus the solution is updated as $x^{(1)}$, $x^{(2)}$, $x^{(3)}$, $x^{(4)}$, ...
Since $x^{(4)}$ at Point 7 determines the operating section in which $x^{(0)}$ exists, the
iteration process gets into the infinite loop formed by $0 \rightarrow 1 \rightarrow 2 \rightarrow ... \rightarrow 6 \rightarrow$
$7 \rightarrow 0$. As a result, the iteration process of the standard NR algorithm keeps
going around the double infinite loop: $0 \rightarrow 1 \rightarrow 2 \rightarrow 3 \rightarrow 4$ and $5 \rightarrow 6 \rightarrow 7 \rightarrow 0$.
If XTAP's iterative scheme is applied, after the N_Bth iteration step, the biaxial
NR algorithm converges to the solution of the circuit with one or two iteration
steps, as shown in Figure 5.22(b). Again, it is verified that the biaxial NR
algorithm effectively avoids getting into the infinite loop, and the Katzenelson
algorithm is not used.

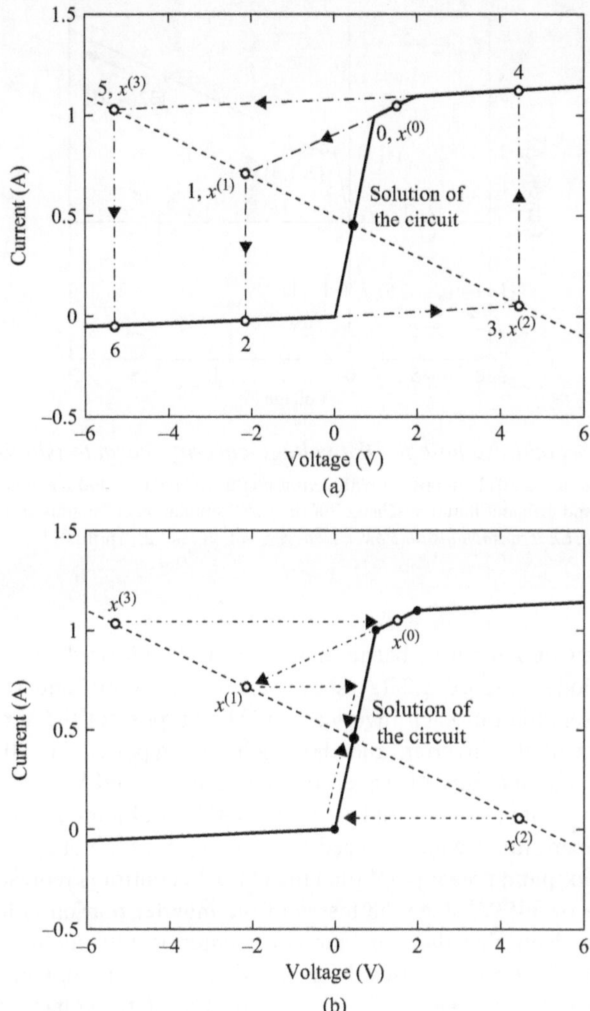

Figure 5.20   Solution locus during the iteration process for the first example of the voltage-current characteristic. (a) Shows the result obtained by the standard NR algorithm, (b) shows the result obtained by the iterative scheme used in XTAP

© IEEE 2011. Reprinted with permission from Noda, T. and Kikuma, T., 'A robust and efficient iterative scheme for the EMT simulations of nonlinear circuits', *IEEE Transactions on Power Delivery*, vol. 26, no. 2, April 2011

5.3.4.2   Actual inverter circuit

Figure 5.23 shows the inverter portion of an actual power conditioner for a photovoltaic (PV) power generation system. The power conditioner was developed as a prototype to achieve reduced losses by utilizing the improved efficiency of

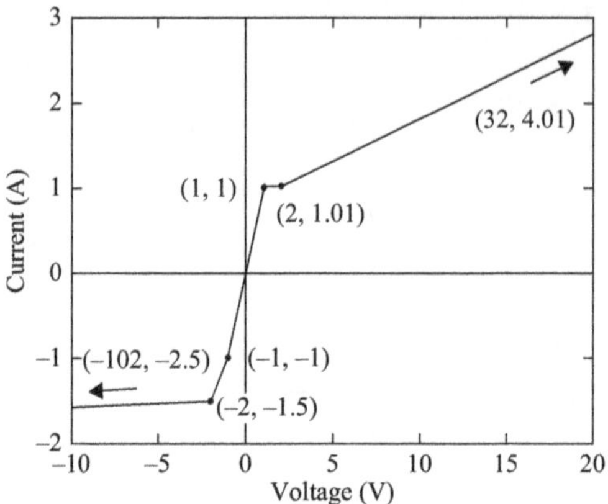

Figure 5.21   Second example for the voltage-current characteristic of R_N

© IEEE 2011. Reprinted with permission from Noda, T. and Kikuma, T., 'A robust and efficient iterative scheme for the EMT simulations of nonlinear circuits', *IEEE Transactions on Power Delivery*, vol. 26, no. 2, April 2011

silicon-carbide (SiC) Schottky barrier diodes (SBDs) compared with conventional silicon (Si) diodes. Figure 5.23(a) shows the main circuit, and (b) shows the detailed representation of each inverter arm. The purpose of the simulation is to evaluate losses of the inverter, and the results are reported in [20], where the calculated voltage and current waveforms agree quite well with measured ones including instants of turning on and off of the IGBTs. In Figure 5.23(a), the output from the chopper circuit which is used for stepping up the voltage of PV panels and also for maximum power point tracking (MPPT) control is represented by a dc voltage source of 365 V, since the losses of the inverter portion is focused on in this simulation. Note that the simulations presented in [20] include the chopper circuit. For each inverter arm, two discrete SiC SBDs are used, and their equivalent circuit model is shown in Figure 5.24(a). The static voltage-current characteristic of each SiC SBD is represented by the nonlinear resistor R_{st} whose characteristic is shown in Figure 5.25(a). The junction capacitance of the SiC SBD is represented by the nonlinear capacitor C_j whose charge-voltage characteristic is shown in Figure 5.25(b). The Si IGBT of each inverter arm is represented by Kraus's IGBT model as shown in Figure 5.24(b). For the details of Kraus's IGBT model, see [21]. The controlled voltage source V_{mos} represents the MOSFET (metal-oxide semiconductor field-effect transistor) portion of the IGBT, and the nonlinear capacitance of the oxide film by the nonlinear capacitance C_{ox}. Figure 5.25(c) shows the nonlinear charge-voltage characteristic of C_{ox}. The controlled current source I_Q represents the dynamics of the base current due to the variation of the base charge. The controlled current source I_{pc} represents the dynamics of the collector current mainly observed at the tail portion when the IGBT is turned off. D_1 and D_2 are nonlinear resistors representing ideal diode

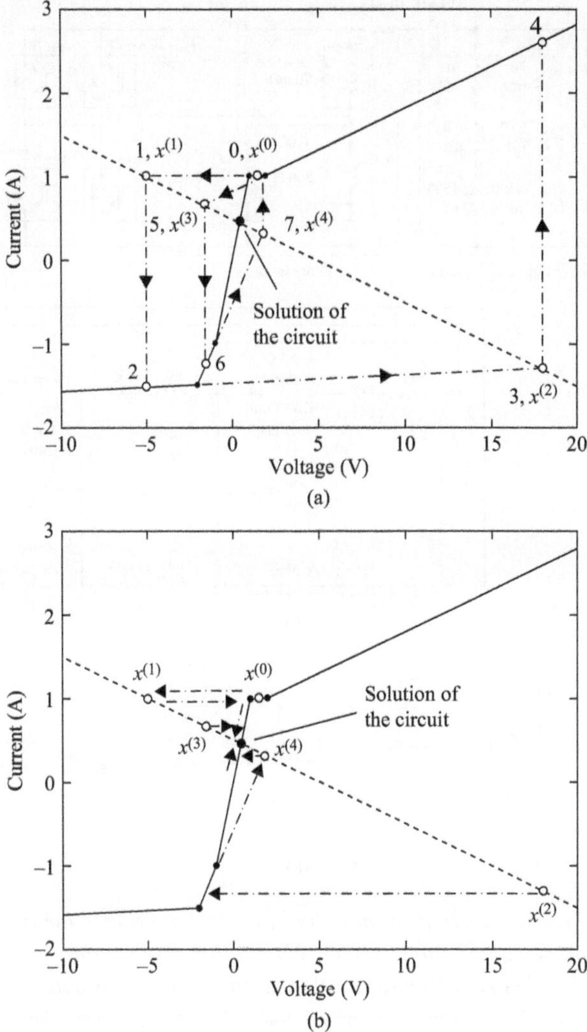

Figure 5.22   Solution locus during the iteration process for the second example of the voltage-current characteristic. (a) Shows the result obtained by the standard NR algorithm, (b) shows the result obtained by the iterative scheme used in XTAP

© IEEE 2011. Reprinted with permission from Noda, T. and Kikuma, T., 'A robust and efficient iterative scheme for the EMT simulations of nonlinear circuits', *IEEE Transactions on Power Delivery*, vol. 26, no. 2, April 2011

characteristics. The on- and off-resistance values of D_1 are 10^{-4} and 10^{14} Ω, and those of D_2 are 10^{-5} and 10^{13} Ω. Note that the nonlinear components are enclosed by dotted boxes in Figure 5.24. Since the simulation circuit involve many nonlinear components, its convergence is not easy.

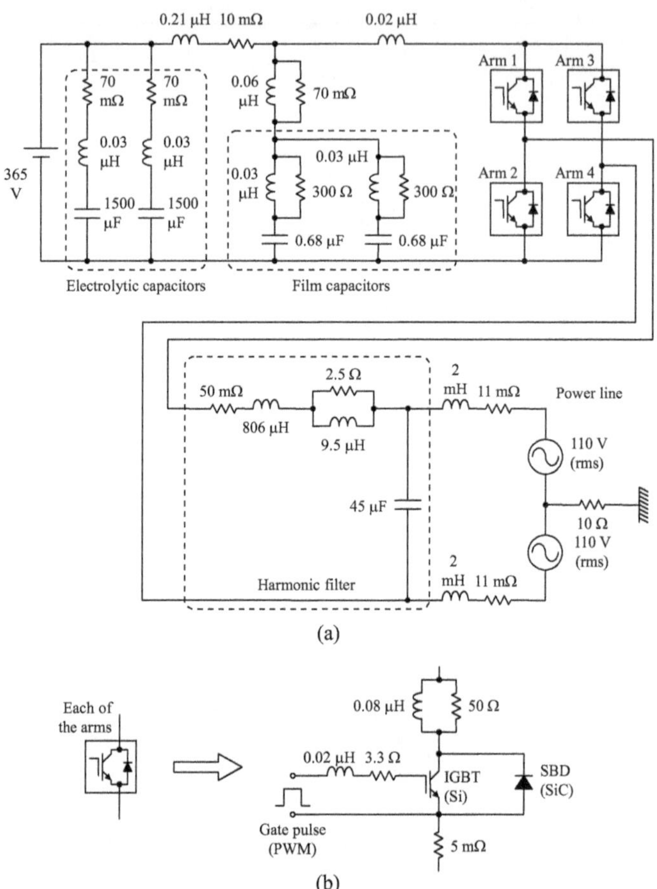

Figure 5.23   *Inverter portion of an actual power conditioner for a photovoltaic power generation system. The main circuit is shown in (a), and the detailed representation of each inverter arm is shown in (b)*

© IEEE 2011. Reprinted with permission from Noda, T. and Kikuma, T., 'A robust and efficient iterative scheme for the EMT simulations of nonlinear circuits', *IEEE Transactions on Power Delivery*, vol. 26, no. 2, April 2011

The EMT simulation of this inverter circuit is carried out with a time step of 10 ns. At each time step, the nonlinear circuit is solved iteratively. Figure 5.26 shows the calculated voltage and current waveforms of the inverter output. The transition of the iteration count with respect to time, when N_B and N_K are set to 100 and 200, respectively, is shown in Figure 5.27. By setting N_B and N_K to such large numbers, we can easily tell which algorithm has reached the solution of the circuit at each time step. As can be seen from Figure 5.27, at most time steps, the solution of the circuit is reached by the standard NR algorithm with about six iterations. At certain time steps, the iteration count exceeds 100, which indicates that the standard

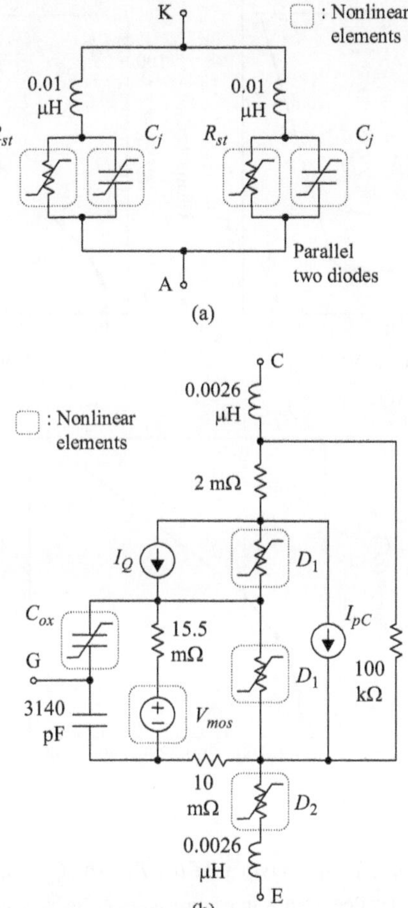

Figure 5.24 *Diode and IGBT models used in the simulation. The equivalent circuit model of the SiC SBD is shown in (a), and that of the Si IGBT is shown in (b)*

© IEEE 2011. Reprinted with permission from Noda, T. and Kikuma, T., 'A robust and efficient iterative scheme for the EMT simulations of nonlinear circuits', *IEEE Transactions on Power Delivery*, vol. 26, no. 2, April 2011

NR algorithm does not converge at those time steps, and the biaxial NR algorithm is employed. Then, the biaxial NR algorithm converges to the solution of the circuit with a small number of iterations. To identify what kind of instant the standard NR algorithm does not converge, the IGBT voltages of the lower-left and the upper-right arm are plotted with the same time scale as that of the iteration count in Figure 5.28. It is observed that the standard NR algorithm does not converge at instants when one of the IGBTs turns on or off. When the IGBTs do not turn on or off, the voltage and current distributions of the circuit do not change significantly,

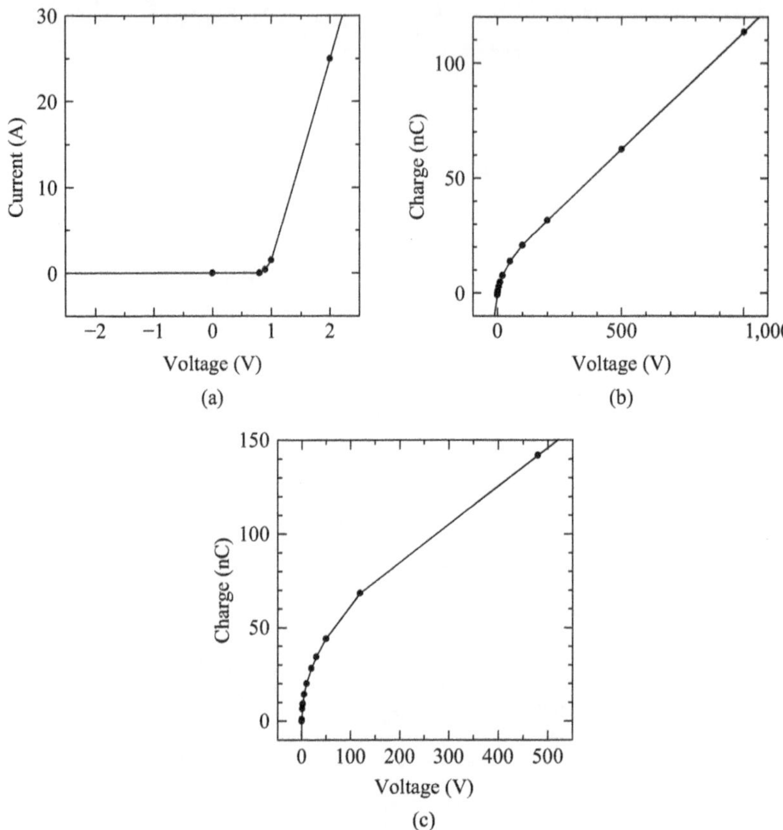

Figure 5.25   *Nonlinear characteristics of (a) R_{st}, (b) C_j, and (c) C_{ox}*

© IEEE 2011. Reprinted with permission from Noda, T. and Kikuma, T., 'A robust
and efficient iterative scheme for the EMT simulations of nonlinear circuits',
IEEE Transactions on Power Delivery, vol. 26, no. 2, April 2011

that is, the operating sections of nonlinear components do not largely move. When
one of the IGBTs turns on or off, on the other hand, the operating sections of some
nonlinear components largely move. If an infinite loop is formed in the iteration
process, then the standard NR algorithm does not converge. At the 100th ($=N_B$th)
iteration step, the biaxial NR algorithm is employed and it can get rid of the infinite
loop and reach the solution of the circuit. Like the two voltage-current characteristic examples, the Katzenelson algorithm is not used, since the iteration count does
not exceed 200 at any time step. From the result presented above, we can verify the
good convergence property of the biaxial NR algorithm. As mentioned before, N_B
and N_K have been set to large values (100 and 200) so that we can easily tell which
algorithm has reached the solution of the circuit in Figures 5.27 and 5.28(c).

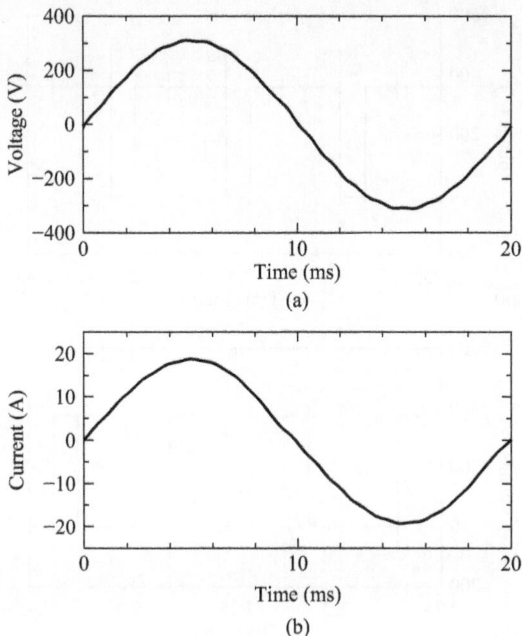

Figure 5.26 *Calculated voltage and current waveforms of the inverter output. (a) Shows the voltage, and (b) shows the current*

© IEEE 2011. Reprinted with permission from Noda, T. and Kikuma, T., 'A robust and efficient iterative scheme for the EMT simulations of nonlinear circuits', *IEEE Transactions on Power Delivery*, vol. 26, no. 2, April 2011

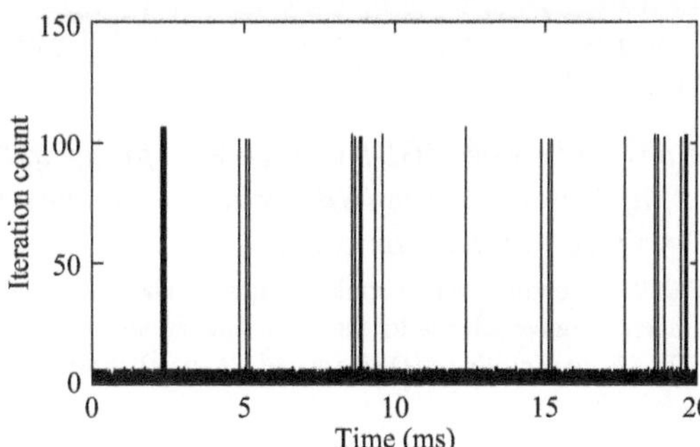

Figure 5.27 *Transition of the iteration count for one period*

© IEEE 2011. Reprinted with permission from Noda, T. and Kikuma, T., 'A robust and efficient iterative scheme for the EMT simulations of nonlinear circuits', *IEEE Transactions on Power Delivery*, vol. 26, no. 2, April 2011

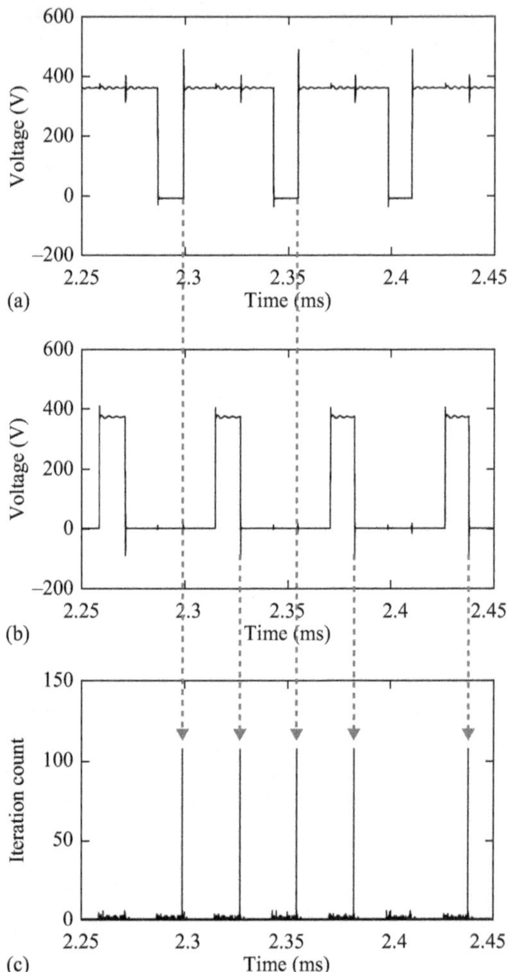

Figure 5.28   *Relationship between IGBT voltage waveforms and the iteration count. (a) Lower-left-arm IGBT voltage, (b) upper-right-arm IGBT voltage, (c) iteration count*

© IEEE 2011. Reprinted with permission from Noda, T. and Kikuma, T., 'A robust and efficient iterative scheme for the EMT simulations of nonlinear circuits', *IEEE Transactions on Power Delivery*, vol. 26, no. 2, April 2011

It should be noted that even if the default values $N_B = 30$ and $N_K = 60$ are used, identical simulation results can be obtained.

Figure 5.29 compares the inverter output voltage waveforms obtained by XTAP's iterative scheme and by the standard NR algorithm. XTAP's iteration scheme is able to reach the solution of the circuit at all-time steps, and the simulation has completed normally. On the other hand, the standard NR algorithm is not

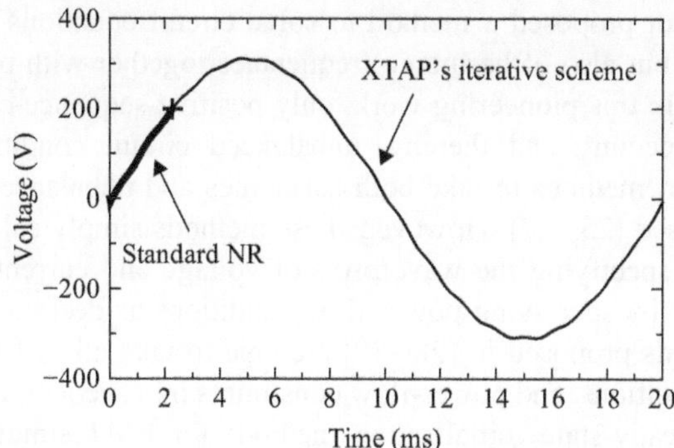

Figure 5.29   Comparison of the inverter output voltage waveforms obtained by XTAP's iterative scheme and by the standard NR algorithm

© IEEE 2011. Reprinted with permission from Noda, T. and Kikuma, T., 'A robust and efficient iterative scheme for the EMT simulations of nonlinear circuits', *IEEE Transactions on Power Delivery*, vol. 26, no. 2, April 2011

able to converge at a time step around $t = 2.3$ ms, which is marked by "+" in Figure 5.29, and the simulation has ended up with an abnormal termination.

5.4   Steady-state initialization method

The purposes of most EMT simulations are calculating EMT phenomena following disturbances occurring in steady-state conditions. A steady state of a given power system could be obtained by simply letting the simulation run from a zero initial condition until it reaches a periodic behavior, but this would need an enormous amount of time and may be impractical. In order to start an EMT simulation directly from a steady state, a function to calculate the steady state and to initialize the internal variables of the circuit components and those of the power system component models according to the steady-state solution obtained is required in an EMT simulation program. Such a method is called a steady-state initialization method. The steady-state initialization method used in XTAP is briefly described by summarizing [22].

So as to develop a practical steady-state initialization method, the following points should be noted:

- Since we are dealing with a power system, it is convenient to calculate a steady-state solution by specifying power-flow conditions.
- Due to unbalanced impedances of transmission lines and other power system components, voltages and currents are not perfectly balanced.
- The power system circuit for which we want to carry out an EMT simulation often includes distributed-parameter circuits for representing transmission lines.
- A steady-state solution by nature includes harmonics generated by nonlinear loads and power electronics devices.

Xia and Heydt proposed a method to solve circuit equations not only of the power frequency but also of harmonic frequencies together with power-flow constraints [23, 24]. In this pioneering work, only positive-sequence circuit equations are taken into account, and thereby, unbalanced circuit conditions cannot be represented. Later, methods to take both harmonics and unbalanced circuit conditions were proposed [25–27]. However, these methods simply calculate a steady-state solution by specifying the waveforms of voltage and current sources as the driving force, not by specifying power-flow conditions as desired by power engineers. The methods proposed in [28–30] are able to take all of harmonics, unbalanced circuit conditions, and power-flow constraints into account. If these methods are viewed as steady-state initialization methods for EMT simulations, there is unfortunately the following intrinsic drawback. These methods are based on the shooting method [31]. Since the shooting method assumes that the circuit equations are described by a set of ordinary differential equations (ODEs), distributed-parameter circuits which are described by partial differential equations (PDEs) cannot be included in the steady-state calculation [32].

Considering the discussion above, the three-stage method was proposed [22] and implemented in XTAP as its steady-state initialization function. The method consists of the following three stages:

Stage 1: Using power-flow conditions specified by the user, a conventional single-phase (positive-sequence) power-flow calculation is performed.

Stage 2: The generators and the substations connected to upper-voltage systems included in the simulation circuit are replaced by balanced three-phase sinusoidal voltage sources whose magnitudes and phase angles are obtained by the power-flow solution of Stage 1. Then, a three-phase power-frequency steady-state solution is calculated by solving the circuit equations in the frequency domain.

Stage 3: Based on the power-frequency steady-state solution obtained at Stage 2, the internal variables of the circuit components and those of the power system component models are initialized, and then, the EMT simulation is kept running for a certain number of cycles.

As an initial approximation, the conventional single-phase power-flow calculation at Stage 1 gives a balanced power-frequency solution. Then, at Stage 2, the three-phase power-frequency steady-state solution in the frequency domain takes into account unbalanced circuit conditions at the power frequency. Finally, the EMT simulation at Stage 3 generates harmonics due to nonlinear loads and power-electronics devices. In this way, we can obtain a steady-state solution taking harmonics and unbalanced circuit conditions into account by specifying power-flow conditions.

The three-stage method should not be considered a novel method, and actually, most users of EMT simulation programs have been manually carrying out these three stages in their routine simulations. For details of the three stage method, consult [22].

5.5   Object-oriented design of the simulation code

The name XTAP stands for eXpandable Transient Analysis Program. The word "eXpandable" comes from the fact that the simulation code of XTAP is written in the C++ programming language and its design philosophy accepts adding a new simulation model without substantial revision of the code. This topic is briefly introduced here by summarizing [33].

Power system engineers started using object-oriented methods for power system simulations in the late 1980s. Object-oriented methods are now used in various aspects of power system simulations. In [34], an object-oriented implementation of a load-flow program using the Objective-C language is illustrated. How power system components such as transmission lines, transformers, etc. are modeled using classes is explained, and it is concluded that the object-oriented approach reduces the efforts of programming. Similar results are obtained in [35–37] using the C++ language. In addition, [35] shows how to handle sparse matrices using C++, and [37] deals with large-scale networks including FACTS controllers. Object-oriented modeling of a power system for dynamic stability-type simulations is proposed in [38] and for obtaining the steady-state solution of a nonlinear power network in [39], both using C++. An object-oriented modeling methodology, called DCOM (Distribution Circuit Object Model), for a distribution analysis system is reported in [40]. As an application to other than programming, an object-oriented concept was used to develop a common database which generates input data files for different simulation programs ranging from load flow to electromagnetic transients [41]. The database stores parameters of power system components in various aspects and appropriately processes the parameters to generate an input file for a specific type of power system simulations. Reference [42] reports a development of graphical user interfaces for power system simulations using object-oriented programming. The development of a fault locating system using an object-oriented approach is described in [43].

Two important features of the object-oriented programming are (i) abstraction of data and their procedures using "classes" and (ii) derivation of a new class from an existing class using the concept of "inheritance." An efficient implementation of an electromagnetic transient program is reported in [44] based on a partly object-oriented approach using Fortran 95. The approach uses (i) for optimizing the execution speed.

The core simulation code of XTAP is based on the following object-oriented design philosophy utilizing the C++ programming language. The object-oriented methodology first defines a base class called "Branch" from which all kinds of circuit components are derived using inheritance. This is illustrated in Figure 5.30. The Branch class has data and their procedures which are common to all circuit components but no specific model parameters and behaviors are defined. Using inheritance, specific model parameters and behaviors are added to the Branch class, and as a result a specific circuit component is derived. From the Branch class, not only simple components like resistors, inductors, capacitors, etc. but also complex components such as a frequency-dependent line model are derived. In order to

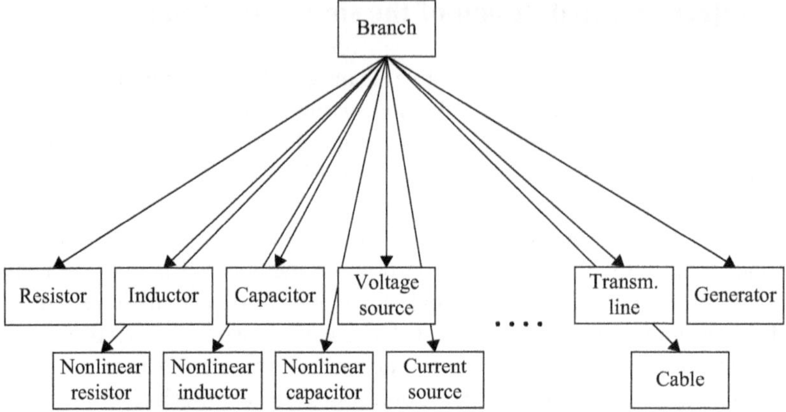

Figure 5.30   Class hierarchy of the circuit components in the core simulation code of XTAP

© IPST 2011. Reprinted with permission from T. Noda and K. Takenaka, 'A practical steady-state initialization method for electromagnetic transient simulations,' *Proc. of IPST 2011*, Paper # 11IPST097, Delft, The Netherlands, 2011

carry out an electromagnetic transient simulation, each circuit component has to be initialized preceding the time step loop and then updated at each time step (if the component is nonlinear, then updated at each iteration step). The design methodology significantly simplifies these processes. Since all circuit components are derived from the common base class Branch, all circuit components can be put into an array of the Branch class. Once all components of a circuit are stored in the array, the circuit solver can issue an initialization or updating message to each element of the array. When issuing these messages, the circuit solver does not have to specify how to initialize or how to update and even does not have to know what each element in the array is. On the other hand, each element knows how to initialize or update itself, since the initialization and updating processes are defined when each component has been derived from the Branch class. This enables separately writing the code of the circuit solver and that of circuit components and thus significantly simplifies the entire code structure. Code maintainability is also enhanced. Especially, adding a new component or a user-defined component becomes an easy task. When adding a new component, the circuit solver does not have to be revised and only the code for deriving the new component from the Branch class is required. For details, consult [33].

References

[1]   T. Noda, K. Takenaka, and T. Inoue, "Numerical integration by the 2-stage diagonally implicit Runge-Kutta method for electromagnetic transient simulations," *IEEE Trans., Power Delivery*, vol. 24, no. 1, pp. 390–399, Jan. 2009.

[2]   H. W. Dommel, "Digital computer solution of electromagnetic transients in single- and multi-phase networks," *IEEE Trans. Power Apparatus and Systems*, vol. PAS-88, no. 4, pp. 388–399, Apr. 1969.

[3]   U. M. Ascher and L. R. Petzold, "*Computer Methods for Ordinary Differential Equations and Differential-Algebraic Equations,*" SIAM, Philadelphia, PA, USA, 1998.

[4]   J. R. Marti and J. Lin, "Suppression of numerical oscillations in the EMTP," *IEEE Trans., Power Systems*, vol. 4, no. 2, pp. 739–747, May 1989.

[5]   J. Lin and J. R. Marti, "Implementation of the CDA procedure in the EMTP," *IEEE Trans., Power Systems*, vol. 5, no. 2, pp. 394–402, May 1990.

[6]   R. Alexander, "Diagonally implicit Runge-Kutta methods for stiff O.D.E.'s," *SIAM Journal on Numerical Analysis*, vol. 14, no. 6, pp. 1006–1021, Dec. 1977.

[7]   C. W. Gear, *Numerical Initial Value Problems in Ordinally Differential Equations*, Prentice-Hall, Inc., NJ, USA, 1971.

[8]   H. Shichman, "Integration system of a nonlinear network-analysis program," *IEEE Trans., Circuit Theory*, vol. CT-17, no. 3, pp. 378–386, Aug. 1970.

[9]   A. Prothero and A. Robinson, "On the stability and accuracy of one-step methods for solving stiff systems of ordinary differential equations," *Math. Comp.*, vol. 28, pp. 145–162, 1974.

[10]   T. Noda, T. Kikuma, and R. Yonezawa, "Supplementary techniques for 2S-DIRK-based EMT simulations," *Electric Power Systems Research*, Elsevier, Vol. 115, pp. 87–93, Oct. 2014. See http://www.sciencedirect.com/science/article/pii/S0378779614001515.

[11]   T. Noda and T. Kikuma, "A robust and efficient iterative scheme for the EMT simulations of nonlinear circuits," *IEEE Trans., Power Delivery*, vol. 26, no. 2, pp. 1030–1038, Apr. 2011.

[12]   L. O. Chua, "Efficient computer algorithms for piecewise-linear analysis of resistive nonlinear networks," *IEEE Trans., Circuit Theory*, vol. CT-18, no. 1, pp. 73–85, Jan. 1971.

[13]   J. Katzenelson, "An algorithm for solving nonlinear resistor networks," *Bell Syst. Tech. J.*, vol. 44, no. 8, pp. 1605–1620, Oct. 1965. See http://onlinelibrary.wiley.com/doi/10.1002/j.1538-7305.1965.tb04195.x/abstract.

[14]   T. Fujisawa and E. S. Kuh, "Piecewise-linear theory of nonlinear networks," *SIAM J. Appl. Math.*, vol. 22, no. 2, pp. 307–328, Mar. 1972.

[15]   T. Ohtsuki, T. Fujisawa, and S. Kumagai, "Existence theorems and a solution algorithm for piecewise-linear resistor networks," *SIAM J. Math. Anal.*, vol. 8, no. 1, pp. 69–99, Feb. 1977.

[16]   L. O. Chua and P.-M. Lin, *Computer-Aided Analysis of Electronic Circuits—Algorithms and Computational Techniques*, Prentice-Hall, Englewood Cliffs, NJ, USA, 1975.

[17]   C.-W. Ho, A. E. Ruehli, and P. A. Brennan, "The modified nodal approach to network analysis," *IEEE Trans., Circuit and Systems*, vol. CAS-22, no. 6, pp. 504–509, Jun. 1975.

[18]   G. D. Hachtel, R. K. Brayton, and F. G. Gustavson, "The sparse tableau approach to network analysis and design," *IEEE Trans., Circuit Theory*, vol. CT-18, no. 1, pp. 101–113, Jan. 1971.

[19] R. A. Rohrer, *Circuit Theory: An introduction to the state variable approach*, McGraw-Hill, Englewood Cliffs, NJ, USA, 1975.

[20] T. Kikuma, N. Okada, M. Takasaki, A. Kuzumaki, and K. Kodani, "Simulation analysis of SiC diode applied inverter," *Proc., 2008 IEEJ Industry Applications Society Conference*, pp. I 347–350, 2008 (in Japanese).

[21] R. Kraus, P. Turkes, and J. Sigg, "Physics-based models of power semiconductor devices for the circuit simulator SPICE," *Proc., 29th Annual IEEE Power Electronics Specialists Conference Record*, vol. 2, pp. 1726–1731, May 1998.

[22] T. Noda and K. Takenaka, "A practical steady-state initialization method for electromagnetic transient simulations," *Proc. of IPST 2011*, Paper # 11IPST097, Delft, The Netherlands, 2011.

[23] D. Xia and G. T. Heydt, "Harmonic power flow studies Part I—Formulation and solution," *IEEE Trans., Power Apparatus and Systems*, vol. 101, no. 6, pp. 1257–1265, Jun. 1982.

[24] D. Xia and G. T. Heydt, "Harmonic power flow studies Part II—Implementation and practical application," *IEEE Trans., Power Apparatus and Systems*, vol. 101, no. 6, pp. 1266–1270, Jun. 1982.

[25] A. Semlyen and A. Medina, "Computation of the periodic steady state in systems with nonlinear components using a hybrid time and frequency domain methodology," *IEEE Trans., Power Systems*, vol. 10, no. 3, pp. 1498–1504, Aug. 1995.

[26] E. Acha and M. Madrigal, *Power Systems Harmonics: Computer Modelling and Analysis*, 1st edition, John Wiley & Sons, Ltd, West Sussex, England Jun. 4, 2001.

[27] T. Noda, A. Semlyen, and R. Iravani, "Entirely harmonic domain calculation of multiphase nonsinusoidal steady state," *IEEE Trans., Power Delivery*, vol. 19, no. 3, pp. 1368–1377, Jul. 2004.

[28] A. Semlyen and M. Shlash, "Principles of modular harmonic power flow methodology," *IEE Proceedings—Generation, Transmission and Distribution*, vol. 147, no. 1, pp. 1–6, Jan. 2000.

[29] N. Garcia and E. Acha, "Periodic steady-state analysis of large-scale electric systems using Poincare map and parallel processing," *IEEE Trans., Power Systems*, vol. 19, no. 4, pp. 1784–1793, Nov. 2004.

[30] K. L. Lian and T. Noda, "A time-domain harmonic power-flow algorithm for obtaining nonsinusoidal steady-state solutions," *IEEE Trans., Power Delivery*, vol. 25, no. 3, pp. 1888–1898, Jul. 2010.

[31] T. J. Aprille and T. N. Trick, "Steady state analysis of nonlinear circuits with periodic inputs," *Proceedings of the IEEE*, vol. 60, no. 1, pp. 108–114, Jan. 1972. See http://ieeexplore.ieee.org/xpl/login.jsp?tp=&arnumber=1450493 &url=http%3A%2F%2Fieeexplore.ieee.org%2Fxpls%2Fabs_all.jsp%3Far number%3D1450493

[32] K. S. Kundert, "Steady-state methods for simulating analog circuits," Ph.D. Thesis, University of California, Berkley, Technical Report No. UCB/ERL M89/63, 1989.

[33] T. Noda, "Object oriented design of a transient analysis program," *Proc. of IPST 2011*, Paper # 07IPST075, Lyon, France, 2007.

[34] A. F. Neyer, F. F. Wu, and K. Imhof, "Object-oriented programming for flexible software: Example of load flow," *IEEE Trans. on Power Systems*, vol. 5, no. 3, pp. 689–696, Aug. 1990.

[35] B. Hakavik and A. T. Holen, "Power system modeling and sparse matrix operations using object-oriented programming," *IEEE Trans. on Power Systems*, vol. 9, no. 2, pp. 1045–1051, May 1994.

[36] E. Z. Zhou, "Object-oriented programming, C++ and power system simulation," *IEEE Trans. on Power Systems*, vol. 11, no. 1, pp. 206–215, Feb. 1996.

[37] C. R. Fuerte-Esquivel, E. Acha, S. G. Tan, and J. J. Rico, "Efficient object oriented power systems software for the analysis of large-scale networks containing FACTS-controlled branches," *IEEE Trans. on Power Systems*, vol. 13, no. 2, pp. 464–472, May 1998.

[38] A. Manzoni, A. S. e Silva, and I. C. Decker, "Power system dynamics simulation using object-oriented programming," *IEEE Trans. on Power Systems*, vol. 14, no. 1, pp. 249–255, Feb. 1999.

[39] A. Medina, A. Ramos-Paz, R. Mora-Juarez, and C. R. Fuerte-Esquivel, "Object oriented programming techniques applied to conventional and fast time domain algorithms for the steady state solution of nonlinear electric networks," *Proc. 2004 IEEE Power Engineering Society General Meeting*, pp. 342–346, Jun. 6–10, 2004.

[40] J. Zhu and D. L. Lubkeman, "Object-oriented development of software systems for power system simulations," *IEEE Trans. on Power Systems*, vol. 12, no. 2, pp. 1002–1007, May 1997.

[41] D. G. Flinn and R. C. Dugan, "A database for diverse power system simulation applications," *IEEE Trans. on Power Systems*, vol. 7, no. 2, pp. 784–790, May 1992.

[42] M. Foley, W. Mitchell, and A. Faustini, "An object based graphical user interface for power systems," *IEEE Trans. on Power Systems*, vol. 8, no. 1, pp. 97–104, Feb. 1993.

[43] M. Kezunovic and Y. Liao, "Development of a fault locating system using object-oriented programming," *Proc. 2001 IEEE ower Engineering Society Winter Meeting*, pp. 763–768, Jan. 28–Feb. 1, 2001.

[44] J. Mahseredjian, B. Bressac, A. Xemard, L. Gerin-Lajoie, and P.-J. Lagace, "A Fortran-95 implementation of EMTP algorithms," *Proc. of IPST 2001*, Paper # 01IPST118, Rio de Janeiro, Brazil, 2001.

Chapter 6

Numerical electromagnetic analysis using the FDTD method

Y. Baba[*]

6.1 Introduction

Numerical electromagnetic analysis (NEA) methods, which include the method of moments (MoM) [1], the finite-element method (FEM) [2], the partial-element equivalent-circuit (PEEC) method [3], the hybrid electromagnetic/circuit model (HEM) [4], the transmission-line-matrix (TLM) method [5], and the finite-difference time-domain (FDTD) method [6], have been widely used in analyzing lightning electromagnetic pulses (LEMPs) and surges (e.g., [7–11]). One of the advantages of NEA, in comparison with circuit simulation methods, is that it allows a self-consistent full-wave solution for both the transient current distribution in a 3D conductor system and resultant electromagnetic fields, although they are computationally expensive.

Among NEA methods, the FDTD method has been most frequently used in LEMP calculations. The first peer-reviewed paper, in which the FDTD method was used in a lightning electromagnetic field simulation, was published in 2003 by Baba and Rakov [12], and it was first applied to the surge analysis of grounding electrodes by Tanabe [13] in 2001. The amount of published material on applications of the FDTD method to lightning electromagnetic field and surge simulations is quite large. Interest in using the FDTD method continues to grow because of the availability of numerical codes and increased computer capabilities.

In this chapter, applications of the FDTD method to lightning electromagnetic field and surge simulations are reviewed. The structure of this chapter is as follows. In section 6.2, the fundamental theory of the FDTD method is briefly described, and advantages and disadvantages of the FDTD method in comparison with other representative NEA methods are identified. In section 6.3, seven representations of the lightning return-stroke channel used to date are described. In section 6.4, representative applications of the FDTD method are reviewed. They include simulations of (i) lightning electromagnetic fields at close and far distances [12, 14–30], (ii) lightning surges on overhead power transmission lines

[*]Doshisha University, Japan

and towers [31–41], (iii) lightning surges on overhead power distribution lines [42–50], (iv) lightning electromagnetic environment in power substations [51], (v) lightning electromagnetic environment in airborne vehicles [52, 53], (vi) lightning surges and electromagnetic environment in buildings [54–58], and (vii) surges on grounding electrodes [13, 59–65].

6.2   FDTD method

6.2.1   Fundamentals

The FDTD method involves the space and time discretization of the working space and the finite-difference approximation to Maxwell's equations. For the analysis of the electromagnetic response of a structure in an unbounded space using the FDTD method, an absorbing boundary condition such as Liao's condition [66] or perfectly matched layers [67], which suppress unwanted reflections, should be applied.

The method requires the whole working space to be divided into cubic or rectangular cells. Cells for specifying or computing electric field (electric field cells) and magnetic field cells are placed relative to each other as shown in Figure 6.1. Generally, the cell size (Δx, Δy, and Δz) should not exceed one-tenth of the wavelength corresponding to the highest frequency in the excitation. Time-updating equations for electric- and magnetic-field components, E_x, E_y, E_z, H_x, H_y, and H_z, are derived and as shown below.

Ampere's law is given as

$$\nabla \times \boldsymbol{H}^{n-\frac{1}{2}} = \varepsilon \frac{\partial \boldsymbol{E}^{n-\frac{1}{2}}}{\partial t} + \boldsymbol{J}^{n-\frac{1}{2}} = \varepsilon \frac{\partial \boldsymbol{E}^{n-\frac{1}{2}}}{\partial t} + \sigma \boldsymbol{E}^{n-\frac{1}{2}} \tag{6.1}$$

where $\boldsymbol{H}$ is the magnetic field intensity vector, $\boldsymbol{E}$ is the electric field intensity vector, $\boldsymbol{J}$ is the conduction-current-density vector, ε is the electric permittivity, σ is

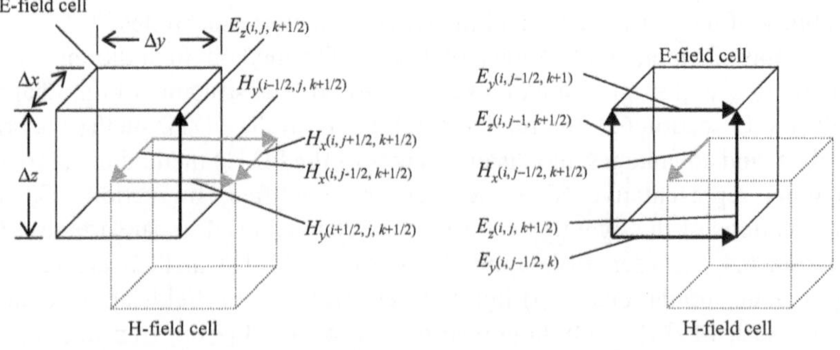

Figure 6.1   Placement of electric-field and magnetic-field cells for solving discretized Maxwell's equations using the FDTD method

the electric conductivity, t is the time, and $n - 1/2$ is the present time step. If the time-derivative term in (6.1) is approximated by its central finite difference, (6.1) can be expressed as

$$\varepsilon \frac{\partial E^{n-\frac{1}{2}}}{\partial t} + \sigma E^{n-\frac{1}{2}} \approx \varepsilon \frac{E^n - E^{n-1}}{\Delta t} + \sigma \frac{E^n + E^{n-1}}{2} \approx \nabla \times H^{n-\frac{1}{2}} \tag{6.2}$$

where Δt is the time increment. Note that $E^{n-1/2}$ in the second term of (6.2) is approximated by its average value, $(E^n + E^{n-1})/2$. Rearranging (6.2), the updating equation for electric field at time step n from its one time-step previous value E^{n-1} and half time-step previous magnetic-field curl value $\nabla \times H^{n-1/2}$ is obtained as follows:

$$E^n = \left(\frac{1 - \dfrac{\sigma \Delta t}{2\varepsilon}}{1 + \dfrac{\sigma \Delta t}{2\varepsilon}} \right) E^{n-1} + \left(\frac{\dfrac{\Delta t}{\varepsilon}}{1 + \dfrac{\sigma \Delta t}{2\varepsilon}} \right) \nabla \times H^{n-\frac{1}{2}} \tag{6.3}$$

From (6.3), the updating equation for E_z^n at the space point $(i + 1/2, j, k)$ (see, e.g., Figure 6.1) is expressed as

$$E_z^n\left(i,j,k+\frac{1}{2}\right) = \frac{1 - \dfrac{\sigma(i,j,k+1/2)\Delta t}{2\varepsilon(i,j,k+1/2)}}{1 + \dfrac{\sigma(i,j,k+1/2)\Delta t}{2\varepsilon(i,j,k+1/2)}} E_z^{n-1}\left(i,j,k+\frac{1}{2}\right)$$

$$+ \frac{\dfrac{\Delta t}{\varepsilon(i,j,k+1/2)}}{1 + \dfrac{\sigma(i,j,k+1/2)\Delta t}{2\varepsilon(i,j,k+1/2)}} \left[\frac{\partial H_y^{n-\frac{1}{2}}\left(i,j,k+\frac{1}{2}\right)}{\partial x} - \frac{\partial H_x^{n-\frac{1}{2}}\left(i,j,k+\frac{1}{2}\right)}{\partial y} \right]$$

$$\approx \frac{1 - \dfrac{\sigma(i,j,k+1/2)\Delta t}{2\varepsilon(i,j,k+1/2)}}{1 + \dfrac{\sigma(i,j,k+1/2)\Delta t}{2\varepsilon(i,j,k+1/2)}} E_z^{n-1}\left(i,j,k+\frac{1}{2}\right) + \frac{\dfrac{\Delta t}{\varepsilon(i,j,k+1/2)}}{1 + \dfrac{\sigma(i,j,k+1/2)\Delta t}{2\varepsilon(i,j,k+1/2)}} \frac{1}{\Delta x \Delta y}$$

$$\times \left[\begin{array}{l} H_y^{n-\frac{1}{2}}\left(i+\frac{1}{2},j,k+\frac{1}{2}\right)\Delta y - H_y^{n-\frac{1}{2}}\left(i-\frac{1}{2},j,k+\frac{1}{2}\right)\Delta y \\[2mm] -H_x^{n-\frac{1}{2}}\left(i,j+\frac{1}{2},k+\frac{1}{2}\right)\Delta x + H_x^{n-\frac{1}{2}}\left(i,j-\frac{1}{2},k+\frac{1}{2}\right)\Delta x \end{array} \right]$$

$$\tag{6.4}$$

where the spatial-derivative terms in (6.4) are approximated by their central finite differences. Updating equations for E_x^n and E_y^n can be derived in the same manner.

Faraday's law is written as

$$\nabla \times \boldsymbol{E}^n = -\mu \frac{\partial \boldsymbol{H}^n}{\partial t} \tag{6.5}$$

where n is the present time step and μ is the magnetic permeability. If the time-derivative term in (6.5) is approximated by its central finite difference, (6.5) is expressed as

$$\mu \frac{\partial \boldsymbol{H}^n}{\partial t} \approx \mu \frac{\boldsymbol{H}^{n+\frac{1}{2}} - \boldsymbol{H}^{n-\frac{1}{2}}}{\Delta t} \approx -\nabla \times \boldsymbol{E}^n \tag{6.6}$$

Rearranging (6.6), we obtain the updating equation for magnetic field at time step $n + 1/2$ from its one time-step previous value $\boldsymbol{H}^{n-1/2}$ and half time-step previous electric-field curl value $\nabla \times \boldsymbol{E}^n$ as follows:

$$\boldsymbol{H}^{n+1/2} = \boldsymbol{H}^{n-1/2} - \frac{\Delta t}{\mu} \nabla \times \boldsymbol{E}^n \tag{6.7}$$

From (6.7), the updating equation for $H_x^{n+1/2}$ at the location $(i, j - 1/2, k + 1/2)$ (see, e.g., Figure 6.1) is expressed as follows:

$$
\begin{aligned}
H_x^{n+\frac{1}{2}}\left(i,j - \frac{1}{2}, k + \frac{1}{2}\right) &= H_x^{n-\frac{1}{2}}\left(i,j - \frac{1}{2}, k + \frac{1}{2}\right) + \frac{\Delta t}{\mu(i,j - 1/2, k + 1/2)} \\
&\times \left[-\frac{\partial E_z^n\left(i,j - \frac{1}{2}, k + \frac{1}{2}\right)}{\partial y} + \frac{\partial E_y^n\left(i,j - \frac{1}{2}, k + \frac{1}{2}\right)}{\partial z} \right] \\
&\approx H_x^{n-\frac{1}{2}}\left(i,j - \frac{1}{2}, k + \frac{1}{2}\right) + \frac{\Delta t}{\mu(i,j - 1/2, k + 1/2)} \frac{1}{\Delta y \Delta z} \\
&\times \left[\begin{aligned} &-E_z^n\left(i,j, k + \frac{1}{2}\right)\Delta z + E_z^n\left(i,j - 1, k + \frac{1}{2}\right)\Delta z \\ &+E_y^n\left(i,j - \frac{1}{2}, k + 1\right)\Delta y - E_y^n\left(i,j - \frac{1}{2}, k\right)\Delta y \end{aligned} \right]
\end{aligned}
\tag{6.8}
$$

where the spatial-derivative terms in (6.8) are approximated by their central finite differences. Updating equations for $H_y^{n+1/2}$ and $H_z^{n+1/2}$ can be derived in the same manner.

By updating E_x^n, E_y^n, E_z^n, $H_x^{n+1/2}$, $H_y^{n+1/2}$, and $H_z^{n+1/2}$ at every point, transient electric and magnetic fields throughout the working volume are obtained.

6.2.2   Advantages and disadvantages

Advantages of the FDTD method in comparison with other NEA methods are summarized as follows:

1. It is based on a simple procedure in electric and magnetic field computations, and therefore its programming is relatively easy.
2. It is capable of treating complex geometries and inhomogeneities.
3. It is capable of incorporating nonlinear effects and components.
4. It can handle wideband quantities from one run with the help of a time-to-frequency transforming tool.

    Disadvantages are as follows:

1. It is computationally expensive compared to other methods such as the MoM.
2. It cannot deal with oblique boundaries that are not aligned with the Cartesian grid.
3. It would require a complex procedure for incorporating dispersive materials.

6.3   Representations of lightning return-stroke channels and excitations

6.3.1   Lightning return-stroke channels

Seven types of representation of lightning return-stroke channels are used in electromagnetic pulse and surge computations:

1. a perfectly conducting/resistive wire in air above ground;
2. a wire loaded by additional distributed series inductance in air above ground;
3. a wire surrounded by a dielectric medium (other than air) that occupies the entire half space above ground (this fictitious configuration is used only for finding current distribution, which is then applied to a vertical wire in air above ground for calculating electromagnetic fields);
4. a wire coated by a dielectric material in air above ground;
5. a wire coated by a fictitious material having high relative permittivity and high relative permeability in air above ground;
6. two parallel wires having additional distributed shunt capacitance in air (this fictitious configuration is used only for finding current distribution, which is then applied to a vertical wire in air above ground for calculating electromagnetic fields); and
7. a phased-current-source array in air above ground, each current source being activated successively by the arrival of lightning return-stroke wavefront progressing upward at a specified speed.

The return-stroke speed, along with the current peak, largely determines the radiation field initial peak (e.g., [68]). The characteristic impedance of the lightning return-stroke channel influences the magnitude of lightning current and/or the current reflection coefficient at the top of the strike object when a lumped voltage source is employed. It is therefore desirable that the return-stroke speed and the characteristic impedance of simulated lightning channel agree with observations that can be summarized as follows:

1. Typical values of return-stroke speed are in the range from $c/3$ to $c/2$ [69], as observed using optical techniques, where c is the speed of light.
2. The equivalent impedance of the lightning return-stroke channel is expected to be in the range from 0.6 to 2.5 kΩ [70].

Type 1 was used, for example, by Baba and Rakov [12] in their FDTD simulation on electromagnetic fields due to a lightning strike to flat ground. Note that this lightning-channel representation was first used by Podgorski and Landt [71] in their simulation of lightning current in a tall structure with the MoM in the time domain. The speed of the current wave propagating along a vertical perfectly conducting/resistive wire is nearly equal to the speed of light, which is two to three times larger than typical measured values of return-stroke wavefront speed ($c/3$ to $c/2$). This discrepancy is the main deficiency of this representation. The characteristic impedance of the channel-representing vertical wire of radius 3 cm is estimated to be around 0.6 kΩ at a height of 500 m above ground (it varies with height above ground). This is right at the lower bound of its expected range of variation (0.6 to 2.5 kΩ). Note that a current wave suffers attenuation (distortion) as it propagates along a vertical wire even if it has no ohmic losses [31]. Further attenuation can be achieved by loading the wire by distributed series resistance.

Type 2 was used, for example, by Baba and Rakov [7] in their FDTD simulation of current along a vertical lightning channel. Note that this lightning-channel representation was first used by Kato *et al.* [72] in their simulation on lightning current in a tall structure and its associated electromagnetic fields with the MoM in the time domain. The speed of the current wave propagating along a vertical wire loaded by additional distributed series inductance of 17 and 6.3 μH/m in air is $c/3$ and $c/2$, respectively, if the natural inductance of vertical wire is assumed to be $L_0 = 2.1$ μH/m (as estimated by Rakov [73] for a 3-cm-radius wire at a height of 500 m above ground). The corresponding characteristic impedance ranges from 1.2 to 1.8 kΩ (0.6 k$\Omega \times [(17 + 2.1)/2.1]^{1/2} = 1.8$ kΩ, and 0.6 k$\Omega \times [(6.3 + 2.1)/2.1]^{1/2} = 1.2$ kΩ) for the speed ranging from $c/3$ to $c/2$. The characteristic impedance of the inductance-loaded wire is within the range of values of the expected equivalent impedance of the lightning return-stroke channel. Note that additional inductance has no physical meaning and is invoked only to reduce the speed of current wave propagating along the wire to a value lower than the speed of light. The use of this representation allows one to calculate both the distribution of current along the channel-representing wire and remote electromagnetic fields in a single, self-consistent procedure. Bonyadi-ram *et al.* [74] have incorporated additional distributed series inductance that increases with increasing height in order to simulate the optically observed reduction in return-stroke speed with increasing height (e.g., [75]).

Type 3 was used, for example, by Baba and Rakov [7] in their FDTD simulation of current along a vertical lightning channel. Note that this lightning-channel representation was first used by Moini *et al.* [76] in their simulation on lightning electromagnetic fields with the MoM in the time domain. The artificial dielectric medium was used only for finding current distribution along the lightning channel, which was then removed for calculating electromagnetic fields in air. When the relative permittivity is 9 or 4, the speed is $c/3$ or $c/2$, respectively. The corresponding characteristic impedance ranges from 0.2 to 0.3 kΩ (0.6 k$\Omega/\sqrt{9} = 0.2$ kΩ, and 0.6 k$\Omega/\sqrt{4} = 0.3$ kΩ) for the speed ranging from $c/3$ to $c/2$. These characteristic impedance values are smaller than the expected ones (0.6 to 2.5 kΩ).

Type 4 was used, for example, by Baba and Rakov [7] in their FDTD simulation of current along a vertical lightning channel. Note that this lightning-channel representation was first used by Kato *et al.* [77] in their simulation of lightning electromagnetic fields with the MoM in the frequency domain. Baba and Rakov [7] represented the lightning channel by a vertical perfectly conducting wire, which had a radius of 0.23 m and was placed along the axis of a dielectric rectangular parallelepiped of relative permittivity 9 and cross section 4 m $\times$ 4 m. This dielectric parallelepiped was surrounded by air. The speed of the current wave propagating along the wire was about $0.74c$. Such a representation allows one to calculate both the distribution of current along the wire and the remote electromagnetic fields in a single, self-consistent procedure, while that of a vertical wire surrounded by an artificial dielectric medium occupying the entire half space (type 3 described above) requires two steps to achieve the same objective. However, the electromagnetic fields produced by a dielectric-coated wire in air might be influenced by the presence of coating.

Type 5 was first used by Miyazaki and Ishii [78] in their FDTD simulation of electromagnetic fields due to a lightning strike to a tall structure. Although the exact values of relative permittivity and relative permeability of the coating were not given in [78], the speed of the current wave propagating along the wire was about $0.5c$. Similar to type 4, this representation allows one to calculate both the distribution of current along the wire and the remote electromagnetic fields in a single, self-consistent procedure. For the same speed of current wave, the characteristic impedance value for this channel representation is higher than that for type 4, since both relative permittivity and permeability are set at higher values in the type 5 representation.

Type 6 has not been used in LEMP and surge simulations with the FDTD method to date. It was, however, used by Bonyadi-ram *et al.* [79] in their simulation with the MoM in the time domain. The speed of the current wave propagating along two parallel wires having additional distributed shunt capacitance in air is $0.43c$ when the additional capacitance is 50 pF/m. In this model, each of the wires has a radius of 2 cm, and the separation between the wires is 30 m. Similar to type 3 described above, this representation employs a fictitious configuration for finding a reasonable distribution of current along the lightning channel, and then this current distribution is applied to the actual configuration (vertical wire in air above ground).

Type 7 was used by Baba and Rakov [12] in their FDTD calculations of lightning electromagnetic fields. This representation can be employed for simulation of "engineering" lightning return-stroke models. Each current source of the phased-current-source array is activated successively by the arrival of lightning return-stroke wavefront that progresses upward at a specified speed. Although the impedance of this channel model is infinity, appropriate reflection coefficients at the top and bottom of the structure and at the lightning attachment point can be specified to account for the presence of tall strike object and upward-connecting leader (e.g., [80, 81]).

Among the above seven types, types 2 and 5 appear to be best in terms of the resultant return-stroke wavefront speed, the characteristic impedance, and the procedure for current and field computations. Type 7 is also useful since the return-stroke wavefront speed and the current attenuation with height are controlled easily with a simple mathematical expression: "engineering" return-stroke model such as the transmission-line (TL) model [82], the traveling-current-source (TCS) model [83], or the modified transmission-line model with linear current decay model with linear current decay (MTLL) model [68].

6.3.2   Excitations

Methods of excitation of the lightning channel used in electromagnetic pulse and surge computations include:

1.   closing a charged vertical wire at its bottom end with a specified impedance (or circuit);
2.   a lumped voltage source (same as a delta-gap electric-field source);
3.   a lumped current source; and
4.   a phased-current-source array.

Type 1 was used, for example, by Baba and Rakov [7] in their FDTD simulation of currents along a vertical lightning channel. Note that this representation was first used by Podgorski and Landt [71] in their simulation of lightning currents with the MoM in the time domain. Baba and Rakov [7] represented a leader/return-stroke sequence by a pre-charged vertical perfectly conducting wire connected via a nonlinear resistor to flat ground. In their model, closing a charged vertical wire in a specified circuit simulates the lightning return-stroke process.

Type 2 representation was used, for example, by Baba and Rakov [7] in their FDTD simulation of lightning currents, but it was first used by Moini *et al.* [84] in their simulation of lightning-induced voltages with the MoM in the time domain. This type of source generates a specified electric field, which is independent of current flowing through the source. Since it has zero internal impedance, its presence in series with the lightning channel and strike object does not disturb any transient processes in them. If necessary, one could insert a lumped resistor in series with the voltage source to adjust the impedance seen by waves entering the channel from the strike object to a value consistent with the expected equivalent impedance of the lightning channel.

Type 3 was used, for example, by Noda *et al.* [33]. However, in contrast with a lumped voltage source, a lumped current source inserted at the attachment point is justified only when reflected waves returning to the current source are negligible. This is the case for a branchless subsequent lightning strike terminating on flat ground, in which case an upward connecting leader is usually neglected and the return-stroke current wave is assumed to propagate upward from the ground surface. The primary reason for the use of a lumped current source at the channel base is a desire to use directly the channel-base current, known from measurements for both natural and triggered lightning, as an input parameter of the model. When one employs a lumped ideal current source at the attachment point in analyzing lightning strikes to a tall grounded object, the lightning channel, owing to the infinitely large impedance of the ideal current source, is electrically isolated from the strike object, so that current waves reflected from ground cannot be directly transmitted to the lightning channel (only electromagnetic coupling is possible). Since this is physically unreasonable, a series ideal current source is not suitable for modeling of lightning strikes to tall grounded objects [80].

Features of type 4 excitation are described in section 6.3.1 for type 7 representation of lightning return-stroke channel.

6.4 Applications

In this section, representative applications of the FDTD method are reviewed. These include simulations of (i) lightning electromagnetic fields at close and far distances [12, 14–30], (ii) lightning surges on overhead power transmission lines and towers [31–41], (iii) lightning surges on overhead power distribution lines [42–50], (vi) lightning electromagnetic environment in power substations [51], (v) lightning electromagnetic environment in airborne vehicles [52, 53], (vi) lightning surges and electromagnetic environment in buildings [54–58], and (vii) surges on grounding electrodes [13, 59–65].

6.4.1 *Lightning electromagnetic fields at close and far distances*

The FDTD method has been widely used in analyzing electromagnetic fields due to lightning strikes to flat ground. When the lightning channel is vertical and the configuration is cylindrically symmetrical, simulations are often performed in the 2D cylindrical coordinate system (e.g., [14, 15, 18–20, 23, 25, 29, 30]), since

2D simulations are computationally much less expensive than the corresponding 3D simulations. Since the FDTD method yields reasonably accurate results for lossy ground, it has been used to test the validity of approximate expressions for horizontal electric fields [14, 23, 24] such as the Cooray-Rubinstein formula [85]. Also, underground electric and magnetic fields have been analyzed by Mimouni *et al.* [29]. Furthermore, the influence of horizontally stratified ground has been studied by Shoory *et al.* [20] and Zhang *et al.* [27], and the influence of rough ground surface has been investigated by Li *et al.* [26, 28].

Baba and Rakov [17] have calculated, using the 3D FDTD method, the vertical electric field E_z and azimuthal magnetic field H_φ due to lightning strikes in the presence and in the absence of building at the field point, as shown in Figure 6.2. Strikes to both flat ground and tall objects of height $h = 100$, 200, or 500 m were considered, and the strike object and the lightning current were represented by a vertical array of current sources specified using the "engineering" TL model as described by Baba and Rakov [86]. The magnitude of H_φ is not much influenced by the presence of either building at the field point or strike object, while the magnitude of E_z is significantly influenced by either of them. In the case of a lightning strike to flat ground, the magnitude of E_z at the top of the building (at the center point of its flat roof) of plan area $S_b = 40$ m $\times$ 40 m and height $h_b = 20$, 50, or 100 m located at a horizontal distance d ranging from 100 to 500 m from the lightning channel is about 1.5, 2, or 3 times, respectively, greater than that at the same horizontal distance on the ground surface in the absence of building, as shown in Table 6.1. The magnitude of E_z at ground level in the immediate vicinity of the building is reduced relative to the case of no building, with this shielding effect becoming negligible at horizontal distances from the building exceeding twice the height of the building. The ratios for lightning strikes to a tall object of height 100, 200, or 500 m are almost the same as those in the case of lightning strike to flat ground, although the corresponding results are not shown here. On the basis of these results, Nag *et al.* [87] employed an electric field enhancement factor of 1.4 for their antenna installed on the roof of a five-story building.

Berenger [15] has analyzed E_z radiated by a lightning strike at a distance of more than 5,000 km on the flat Earth surface using the 2D cylindrical FDTD method. He represents lightning channel by a 1 km vertical current dipole, and uses a 1,000-km-long domain that moved horizontally and was divided into 0.5 km $\times$ 0.5 km square cells. Oikawa *et al.* [21] have carried out a similar simulation of E_z using the 3D FDTD method. They represent lightning channel as a 10 m vertical current source, simulate a realistic bumpy ground geometry with 10 m $\times$ 10 m $\times$ 10 m cubic cells, and employ a 7 km $\times$ 4 km $\times$ 7 km moving domain within a 7 km $\times$ 28 km $\times$ 7 km working volume. Haddad *et al.* [22] computed vertical electric field waveforms at distances ranging from 100 to 400 km using an FDTD model of VLF propagation in the Earth-ionosphere waveguide [16].

Using the 2D cylindrical FDTD method, Baba and Rakov [18] have calculated E_z, horizontal (radial) electric field E_h, and H_φ produced on the ground surface by lightning strikes to a 160 m and a 553 m high conical strike objects representing the

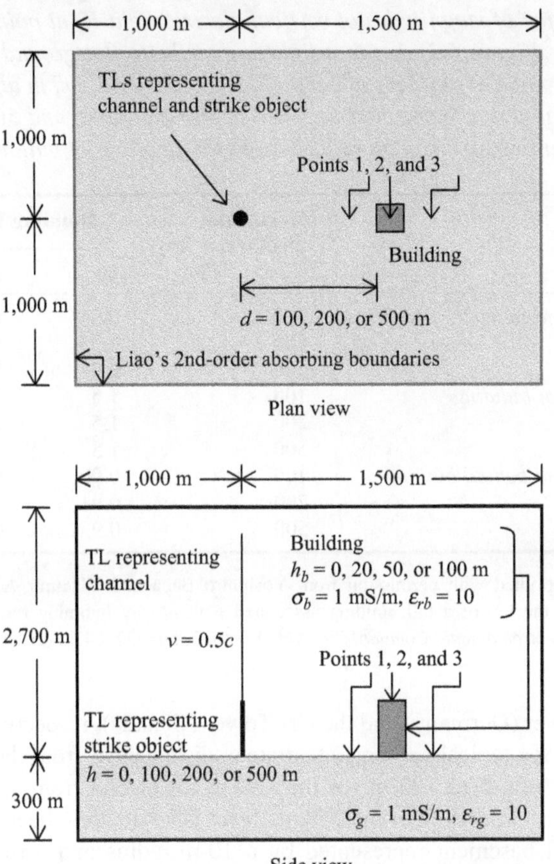

Figure 6.2   *Building of height $h_b = 0$ (for the case of no building), 20, 50, or 100 m and plan area $S_b = 40$ m × 40 m located at distance d = 100, 200, or 500 m from a tall grounded object of height h = 0 (for the case of lightning strike to flat ground), 100, 200, or 500 m struck by lightning. The strike object and the lightning channel are represented by a vertical array of current sources specified using the "engineering" TL model described by Baba and Rakov [86]. The current propagation speed along the lightning channel is set to v = 0.5c. The working volume of 2,000 m × 2,500 m × 3,000 m is divided into 10 m × 10 m × 10 m cubic cells and surrounded by six planes of Liao's second-order absorbing boundary condition [66] to avoid reflections there*

© 2007 IEEE. Reprinted with permission from Yoshihiro Baba and Vladimir A. Rakov, 'Electromagnetic fields at the top of a tall building associated with nearby lightning return strokes', *IEEE Transactions on Electromagnetic Compatibility*, vol. 49, no. 3, pp. 632–643, August 2007, Figure 1

Table 6.1 Ratios of magnitudes of vertical electric field E_z at points 1, 2, and 3 (see Figure 6.2) due to a lightning strike to flat ground ($h = 0$) in the presence of building of height $h_b = 20$, 50, and 100 m at the field point, located at a horizontal distance of $d = 100$, 200, and 500 m from the lightning strike point and those in the absence of building ($h_b = 0$)

Field point	Horizontal distance, d (m)	Building height, h_b (m)		
		20	50	100
Point 1, *on ground in front of building*	100	0.98	0.94	0.88
	200	0.95	0.88	0.77
	500	0.96	0.82	0.71
Point 2, *on roof of building*	100	1.5	2.1	2.7
	200	1.5	2.2	3.0
	500	1.5	2.2	3.2
Point 3, *on ground behind building*	100	0.90	0.69	0.46
	200	0.91	0.72	0.51
	500	0.93	0.75	0.57

© 2007 IEEE. Reprinted with permission from Yoshihiro Baba and Vladimir A. Rakov, 'Electromagnetic fields at the top of a tall building associated with nearby lightning return strokes', *IEEE Transactions on Electromagnetic Compatibility*, vol. 49, no. 3, pp. 632–643, August 2007, Table I.

Peissenberg tower (Germany) and the CN Tower (Canada), respectively. The fields are computed for a typical subsequent stroke at distances d' from the bottom of the object ranging from 5 to 100 m for the 160 m tower and from 10 to 300 m for the 553 m tower. Grounding of the 160 m object is assumed to be accomplished by its underground basement represented by a 10-m-radius and 8-m-long perfectly conducting cylinder with or without a reference ground plane located 2 m below. The reference ground plane simulates, to some extent, a higher conducting ground layer that is expected to exist below the water table. The configuration without reference ground plane actually means that this plane is present, but is located at an infinitely large depth. Grounding of the 553 m object is modeled in a similar manner but in the absence of reference ground plane only. In all cases considered, waveforms of E_h and H_φ are not much influenced by the presence of strike object, while waveforms of E_z are. Waveforms of E_z are essentially unipolar (as they are in the absence of strike object) when the ground conductivity σ is 10 mS/m (the equivalent transient impedance is several ohm) or greater. For the 160 m tower and for $\sigma = 1$ and 0.1 mS/m, waveforms of E_z become bipolar (exhibit polarity change) at $d' \leq 10$ m and $d' \leq 50$ m, respectively, regardless of the presence of the reference ground plane. The corresponding equivalent transient grounding impedances are about 30 and 50 Ω in the absence of the reference ground plane and smaller than 10 Ω in the presence of the reference ground plane. The source of opposite polarity E_z is the potential rise at the object base (at the air/ground interface) relative to the reference ground plane. For a given grounding electrode geometry, the strength of this source increases with decreasing σ, provided that the grounding impedance is linear. Potential rises at the strike object base for $\sigma = 1$ and 0.1 mS/m are some hundreds of kilovolts, which is sufficient to produce electrical breakdown from relatively sharp edges of the basement over a distance of several meters (or more) along the ground surface. The resultant ground surface arcs will serve to reduce the equivalent grounding impedance and, hence, potential rise. Therefore, the polarity change of E_z near the Peissenberg tower, for which σ is probably about 1 mS/m, should be a rare phenomenon, if it occurs at all. The equivalent transient grounding impedance of the cylindrical basement is similar to that of a hemispherical grounding electrode of the same radius. For the 160 m tower and for hemispherical grounding electrode, the transient grounding impedance is higher than its dc grounding resistance for $\sigma = 10$ and 1 mS/m, but lower for $\sigma = 0.1$ mS/m. For the 553 m tower, the transient grounding impedance of hemispherical electrode is equal to or larger than its dc resistance for all values of σ considered. Mimouni *et al.* [19] have performed a similar simulation using the 2D cylindrical FDTD method.

Baba and Rakov [30] have examined, using the 2D cylindrical FDTD method for the configuration shown in Figure 6.3, the effect of upward-extending wire, used for artificial lightning initiation from natural thunderclouds, and corona space charge emanated from this wire on the close ground-level electric field (prior to lightning initiation). The wire current and charge transfer are also esti-mated. The lightning-triggering wire is assumed to be perfectly conducting and extending vertically upward with a constant speed of $v = 150$ m/s. Due to the limitations of the FDTD method, the wire radius is set to $r = 0.27$ m, larger than the actual radius (0.1 mm), but the results are not expected to be much influenced by this assumption. The corona space charge emanated from the wire surface is represented by a conducting cylindrical sheath of outer radius $r = 2$, 4, 8, and 16 m, coaxial with the wire (dynamics of corona discharge are not considered here). It has been found that the results are insensitive to the value of corona-sheath conductivity ranging from 10^{-8} S/m to infinity. The corona-space-charge layer at ground is simulated by perfectly conducting cylindrical tubes placed on the ground surface, coaxial with the upward-extending conductor. The quasi-static electric field between the thundercloud charge source and the ground is simulated by creating a quasi-uniform upward-directed electric field of 43 kV/m between two parallel conducting plates limiting the FDTD computational domain. The simulated corona space charge at ground reduces the electric field at the ground surface to 5.5 kV/m, a typical value at the time of rocket launch. The upward-directed electric field E_z on the ground surface in the vicinity of triggering wire decreases with increasing the altitude of the wire top. When the wire-top altitude is 200 m, the reduction of E_z at horizontal distance $d = 60$ m is about 17%, 26%, 31%, 40%, and 52% relative to the background value of 5.5 kV/m for $r = 0.27$, 2, 4, 8, and 16 m, respectively, while the corresponding reduction of E_z at $d = 360$ m in all cases is only 1% or less (see Figure 6.4 for $r = 4$ m). The calculated results for $r \approx 4$ to 16 m agree reasonably well with E_z variations measured at $d = 60$ and 350 m from the triggering wire by Biagi *et al.* [88]. This indicates that the electric-field reduction in the vicinity of triggering wire, prior to lightning initiation, is

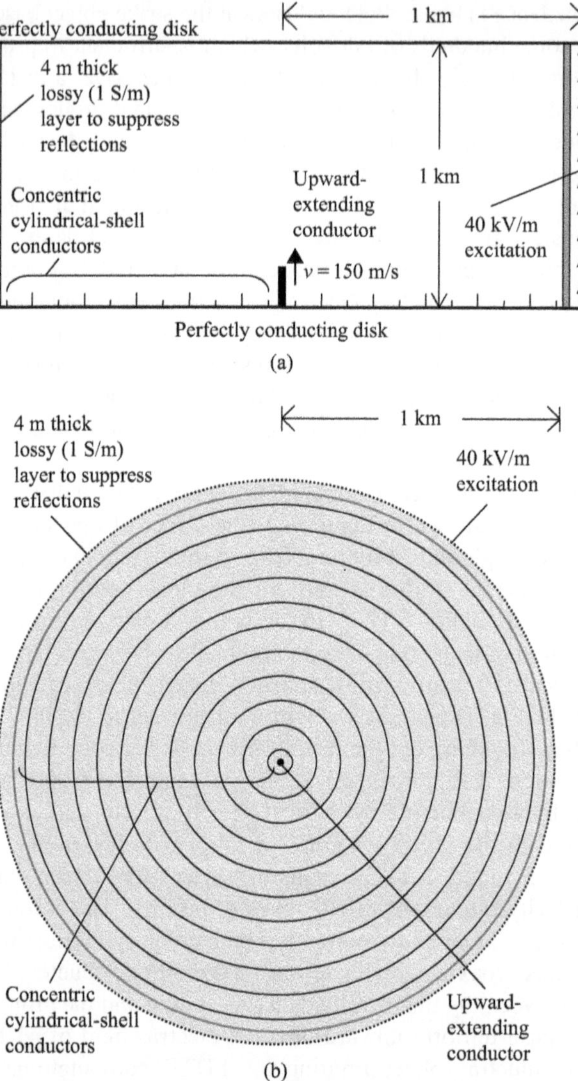

Figure 6.3   *(a) Side view and (b) plan view of the configuration analyzed using the FDTD method, including a grounded conductor of radius 0.27, 2, 4, 8, or 16 m, which extends upward at a constant speed v = 150 m/s in a quasi-static upward-directed electric field. The latter is formed between two perfectly conducting disks of 1 km radius and 1 km apart. Perfectly conducting cylindrical tubes on the ground surface, coaxial with the upward-extending conductor, simulate the presence of corona-space-charge layer at ground*

© 2011 AGU. Reprinted with permission from Yoshihiro Baba and Vladimir A. Rakov, 'Simulation of corona at lightning-triggering wire: Current, charge transfer, and the field-reduction effect', *Journal of Geophysical Research*, vol. 116, no. D21, doi: 10.1029/2011JD016341, November 2011, Figure 3

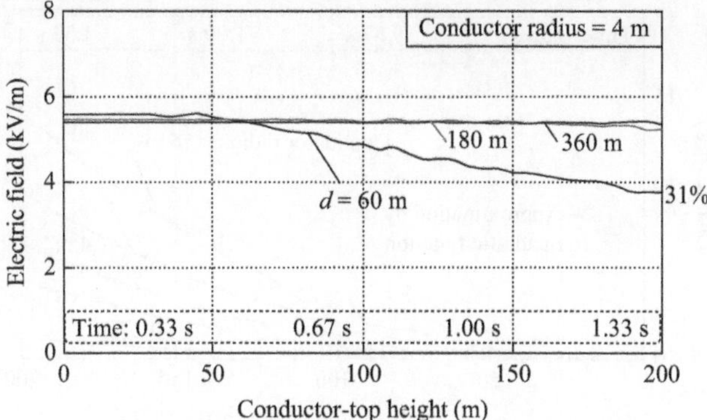

Figure 6.4   *FDTD-calculated variations of upward-directed electric field E_z on the ground surface as a function of conductor-top height (or time) at $d = 60$, 180, and 360 m for conductor radius 4 m. The conductor extends upward at a speed of $v = 150$ m/s up to an altitude of 200 m. The pre-launch electric field on the ground surface is 5.5 kV/m and 43 kV/m aloft. The percent electric field change at $d = 60$ m for $h = 200$ m is indicated on the right of the panel*

© 2011 AGU. Reprinted with permission from Yoshihiro Baba and Vladimir A. Rakov, 'Simulation of corona at lightning-triggering wire: Current, charge transfer, and the field-reduction effect', *Journal of Geophysical Research*, vol. 116, no. D21, doi: 10.1029/2011JD016341, November 2011, Figure 7(c)

primarily caused by the presence of corona space charge emanated from the wire to a radius of about 4 m or more, as opposed to the presence of wire alone. The total charge transfer from ground to the wire, the top of which is at an altitude of 200 m, is 1.2, 4.5, 6.6, 9.5, and 14 mC for $r = 0.27$, 2, 4, 8, and 16 m, respectively, as shown in Figure 6.5. The corresponding currents flowing in the wire are 2.1, 7.9, 11, 15, and 22 mA, as shown in Figure 6.6. The model-predicted charges and currents are consistent with limited measurements available in the literature computed for $r = 2$ to 4 m, smaller than the values based on the field reduction calculations, but still of the order of meters. The radial electric field near the top of 200 m high cylindrical conductor can exceed 400 kV/m, which is sufficient for positive streamer propagation, when the conductor radius is up to 8 m, confirming corona-sheath radii of the order of meters inferred from the field reduction and wire charge/current analyses.

6.4.2   Lightning surges on overhead power transmission lines and towers

The 3D FDTD method has been widely used in analyzing lightning surges on overhead power transmission line conductors and towers. Noda [33, 35] and Noda *et al.* [34] have analyzed insulator voltages of a 500 kV double-circuit transmission line struck by lightning. The lightning channel is represented by a

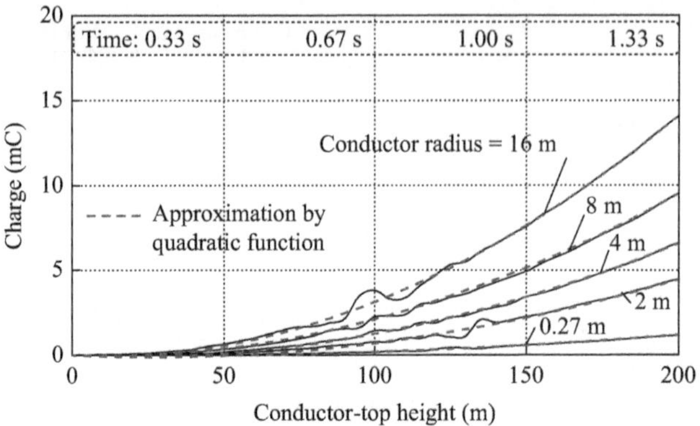

Figure 6.5   *FDTD-calculated variations of total charge on the surface of an upward-extending conductor (representing both the triggering wire and its corona sheath) as a function of conductor-top height (or time) for different conductor radii, 0.27, 2, 4, 8, and 16 m. The apparent irregularities (superimposed oscillations) are related to the presence of conducting tubes simulating the effect of corona-space-charge layer at ground. Broken-line curves are approximations by quadratic functions used for evaluating corresponding currents shown in Figure 6.6*

© 2011 AGU. Reprinted with permission from Yoshihiro Baba and Vladimir A. Rakov, 'Simulation of corona at lightning-triggering wire: Current, charge transfer, and the field-reduction effect', *Journal of Geophysical Research*, vol. 116, no. D21, doi: 10.1029/2011JD016341, November 2011, Figure 8

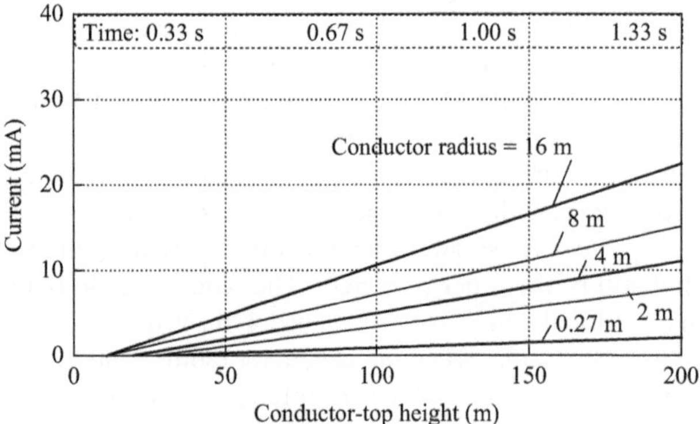

Figure 6.6   *Currents flowing in the upward-extending conductor as a function of conductor-top height (or time) for different conductor radii, 0.27, 2, 4, 8, and 16 m, estimated as time derivatives of quadratic-function-approximated variations of charge shown in Figure 6.5*

© 2011 AGU. Reprinted with permission from Yoshihiro Baba and Vladimir A. Rakov, 'Simulation of corona at lightning-triggering wire: Current, charge transfer, and the field-reduction effect', *Journal of Geophysical Research*, vol. 116, no. D21, doi: 10.1029/2011JD016341, November 2011, Figure 9 vertical wire loaded by additional distributed series inductance [33, 34]. Yao *et al.* [40] have analyzed lightning currents in a 500-kV single-circuit tower struck by lightning and transient magnetic fields around the tower, with lightning represented by a lumped current source. Takami *et al.* [41] have studied lightning surges propagating on inclined ground wires and phase conductors from a double-circuit tower struck by lightning to the substation, and have shown that the surge voltages computed for realistic inclined power line conductors are significantly different from those computed for horizontal conductors.

Baba and Rakov [32] have analyzed, using the 3D FDTD method, the characteristics of current wave propagating along perfectly conducting conical structures (see Figure 6.7). A conical conductor is excited at its apex by a lumped current source in Figure 6.7(a), and it is excited at its base in Figure 6.7(b). They

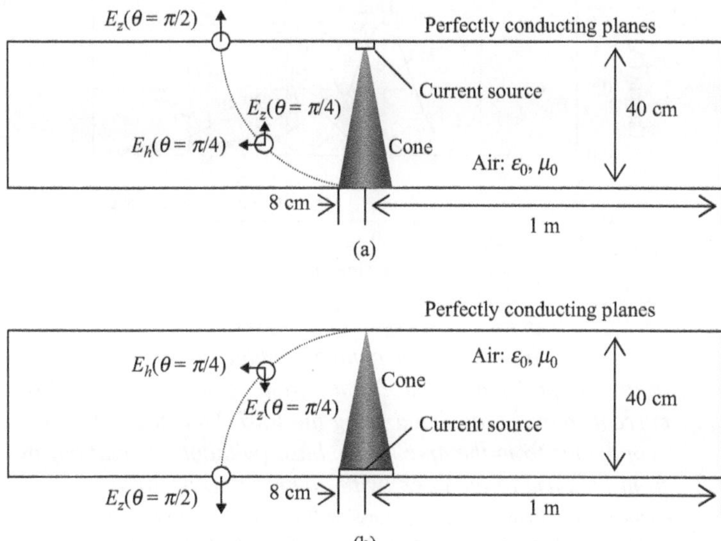

Figure 6.7   *(a) A perfectly conducting cone in air excited at its apex by a current source having a height of 1 cm and a cross-sectional area of 1.5 cm × 1.5 cm, and (b) a similar cone excited at its base by a current source having a height of 1 cm and an approximately circular cross-sectional area with a radius of 8.5 cm, analyzed using the FDTD method. The current source produces a Gaussian pulse having an amplitude of 1 A and a half-peak width of 0.33 ns. The working volume of 2 m × 2 m × 0.4 m, which is divided into 0.5 cm × 0.5 cm × 1 cm cells, is surrounded by six perfectly conducting planes*

© 2005 IEEE. Reprinted with permission from Yoshihiro Baba and Vladimir A. Rakov, 'On the interpretation of ground reflections observed in small-scale experiments: Simulating lightning strikes to towers', *IEEE Transactions on Electromagnetic Compatibility*, vol. 47, no. 3, pp. 533–542, August 2005, Figure 1

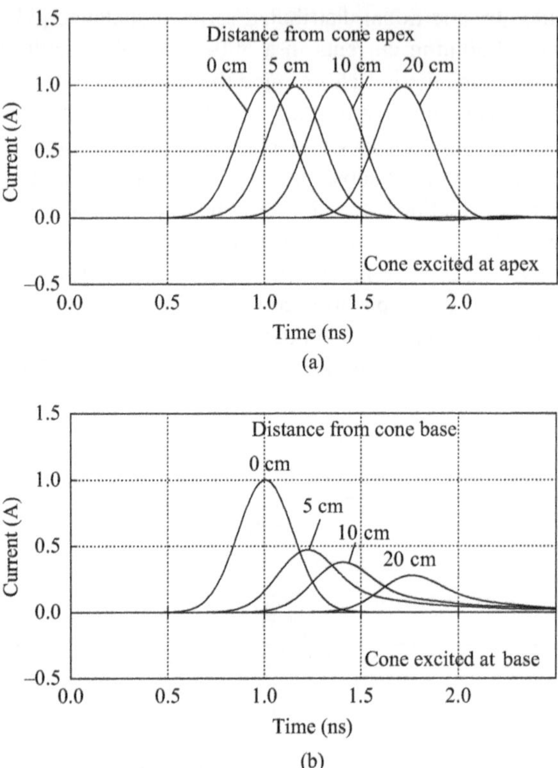

Figure 6.8   Current waveforms for the conical conductor (see Figure 6.7) excited at its (a) apex and (b) base at different vertical distances from the current source, calculated using the FDTD method. The current pulse propagates from the apex to the base (without attenuation) in (a) and from the base to the apex (with attenuation) in (b)

© 2005 IEEE. Reprinted with permission from Yoshihiro Baba and Vladimir A. Rakov, 'On the interpretation of ground reflections observed in small-scale experiments: Simulating lightning strikes to towers', *IEEE Transactions on Electromagnetic Compatibility*, vol. 47, no. 3, pp. 533–542, August 2005, Figure 2

have found that a current wave suffers no attenuation when it propagates downward from the apex of the conical conductor to its base, but it attenuates significantly when it propagates upward from the base of the conical conductor to its apex, as seen in Figure 6.8. The current reflection coefficient at the base of the conical conductor is close to 1, so that the equivalent grounding impedance of the conducting plane is close to zero. Our analysis suggests that the relatively high grounding impedance (60 Ω) of conducting plane inferred from the small-scale experiments [89] is an engineering approximation to the neglected attenuation of upward propagating waves. When the dependence of cone's waveguiding properties on the direction of propagation is taken into account, the results of small-scale experiments simulating lightning strikes to towers can be interpreted without invoking the fictitious grounding impedance of conducting plane. Representation of a vertical strike object by a uniform transmission line terminated in a fictitious high grounding impedance (60 Ω or so) appears to be justified in computing lightning-generated magnetic fields and relatively distant electric fields, but may be inadequate for calculating electric fields in the immediate vicinity of the object.

Baba and Rakov [31] have also shown that a current wave suffers attenuation as it propagates along a vertical perfect conductor of uniform, nonzero thickness (e.g., cylinder) above perfectly conducting ground excited at its bottom by a lumped source. The associated electromagnetic field structure is non-TEM, particularly near the source region. They decompose the "total" current in the conductor into two components that they refer to as the "incident" and "scattered" currents. The "incident" current serves as a reference (no attenuation) and is specified disregarding the interaction of resultant electric and magnetic fields with the conductor, while the "scattered" current, found here using the 3D FDTD method, can be viewed as a correction to account for that interaction. The scattered current modifies the incident current, so that the resultant total current pulse appears attenuated. The attenuation of the total current pulse is accompanied by the lengthening of its tail, such that the total charge transfer is independent of height. Takami *et al.* [36] have shown experimentally that a current wave suffers attenuation and distortion when it propagates along a vertical uniform conductor, both well reproduced by their 3D FDTD simulation.

Thang *et al.* [38] have analyzed, using the 3D FDTD method, lightning surges propagating along overhead wires with corona discharge, and compared FDTD-computed results with the experiments of Inoue [90] and Wagner *et al.* [91]. In Inoue's experiment, a 12.65 mm radius, 1.4 km long overhead wire was employed. The radial progression of corona streamers from the wire is represented as the radial expansion of cylindrical conducting region, the conductivity of which is 40 μS/m, as shown in Figure 6.9 [37]. The critical electric field on the surface of the 12.65 mm radius wire for corona initiation is set to $E_0 = 2.4$ MV/m. The critical background electric field for streamer propagation is set to $E_{cp} = 0.5$ MV/m for positive voltage application and $E_{cn} = 1.5$ MV/m for negative voltage application. Figure 6.10 shows the FDTD-computed waveforms (note wavefront distortion and attenuation at later times) of surge voltages at three different distances from the energized end of the wire, which agree reasonably well with the corresponding measured waveforms. Also, the FDTD-computed waveforms of surge voltages induced on a nearby parallel bundled conductor agree fairly well with the corresponding measured waveforms, although they are not shown here. In the FDTD simulation, the conductor system is accommodated in a working volume of 60 m $\times$ 1460 m $\times$ 80 m, which is divided nonuniformly into rectangular cells. The side length in y direction (conductor-axis direction) of all the cells is 1 m (constant). Cell sides along x and z axes are not constant: 5.5 cm in the vicinity (220 cm $\times$ 220 cm) of the wire, increasing gradually (to 10, 20, and 100 cm) beyond that region.

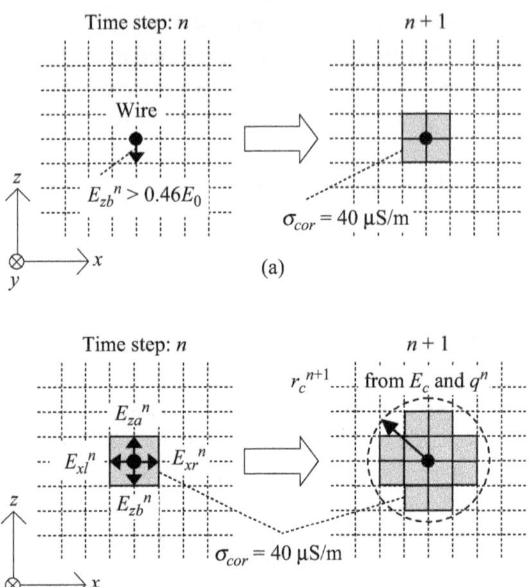

Figure 6.9   FDTD representations of (a) inception of corona discharge at the wire surface and (b) radial expansion of corona discharge

© 2012 IEEE. Reprinted with permission from Tran Huu Thang, Yoshihiro Baba, Naoto Nagaoka; Akihiro Ametani, Jun Takami, Shigemitsu Okabe, and Vladimir A. Rakov, 'A simplified model of corona discharge on overhead wire for FDTD computations', *IEEE Transactions on Electromagnetic Compatibility*, vol. 54, no. 3, pp. 585–593, June 2012, Figure 3

Thang *et al.* [39] have analyzed, using the 3D FDTD method, transient voltages across insulators of a 60 m high transmission line whose overhead ground wire is struck by lightning. The progression of corona streamers from the ground wire is represented as the radial expansion of cylindrical conducting region around the wire. The lightning channel is represented by a vertical phased-current-source array, along which a current wave propagates upward at speed 130 m/μs. On the basis of the computed results, the role of corona discharge at the ground wire on transient insulator voltages is examined. The insulator voltages are reduced by corona discharge on the ground wire. The reduction of insulator-voltage peak due to the ground-wire corona is not very significant: the upper-, middle-, and lower-phase-voltage peaks are reduced by 15%, 14%, and 13% for a positive-stroke with 50 kA peak and 3 μs risetime current, and those for negative stroke with the same current waveform parameters are reduced by 10%, 9%, and 8%, respectively. In the FDTD simulation, the conductor system is accommodated in a working volume of 400 m × 500 m × 750 m, which is divided nonuniformly into rectangular cells: 9 cm in the vicinity of the ground wire and each of the phase conductors, increasing gradually (to 50, 100, and 500 cm) beyond that region.

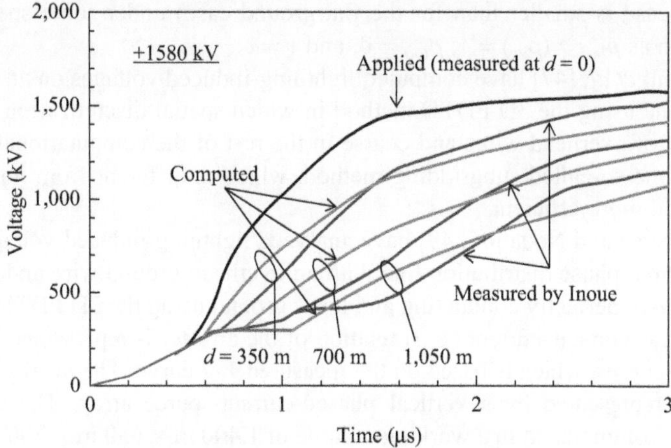

Figure 6.10 *FDTD-computed (for $\sigma_{cor} = 40\ \mu S/m$ and $E_0 = 2.4\ MV/m$) and measured waveforms of surge voltage at d = 0, 350, 700, and 1,050 m from the energized end of the 12.65 mm radius, 1.4 km long horizontal wire located 22.2 m above ground of conductivity 10 mS/m. The applied voltage is positive and $E_{cp} = 0.5\ MV/m$. Applied voltage peak is 1,580 kV*

© 2012 IEEE. Reprinted with permission from Tran Huu Thang, Yoshihiro Baba, Naoto Nagaoka, Akihiro Ametani, Jun Takami, Shigemitsu Okabe, and Vladimir A. Rakov, 'FDTD simulation of lightning surges on overhead wires in the presence of corona discharge', *IEEE Transactions on Electromagnetic Compatibility*, vol. 54, no. 6, pp. 1234–1243, December 2012, Figure 2(a)

6.4.3 Lightning surges on overhead power distribution lines

Baba and Rakov [42] have analyzed, using the 3D FDTD method, transient voltages on an overhead single wire above lossy ground due to nearby lightning strikes to flat ground and to a 100 m high object. The nearby lightning channel with or without the strike object is represented by a vertical phased-current-source array. The conductor system is accommodated in a working volume of 1,400 m × 600 m × 850 m, which is divided uniformly into cubic cells of 5 m × 5 m × 5 m. They have examined the ratios of magnitudes of lightning-induced voltages for the cases of strikes to a 100 m high object and to flat ground as a function of distance from the lightning channel, d, current reflection coefficients at the top of the strike object, ρ_{top}, and at the bottom of the strike object, ρ_{bot}, the current reflection coefficient at the channel base (in the case of strikes to flat ground), ρ_{gr}, and the return-stroke speed, v. The ratio of magnitudes of lightning-induced voltages for tall-object and flat-ground cases increases with increasing d, decreasing ρ_{bot} (<1), decreasing ρ_{top} (<0, except for the case of $\rho_{bot} = 0$), and decreasing v ($<c$). The ratio is larger than unity (strike object serves to enhance the induced voltage) for $d = 40$ to 200 m and realistic conditions such as $\rho_{bot} = (\rho_{gr}) = 1$, $\rho_{top} = -0.5$, and $v = c/3$, but becomes smaller than unity (lightning-induced voltage for the

tall-object case is smaller than for the flat-ground case) under some special conditions such as $\rho_{bot} = (\rho_{gr}) = 1$, $\rho_{top} = 0$, and $v = c$.

Sumitani *et al.* [47] have computed lightning-induced voltages on an overhead two-wire line using the 3D FDTD method in which spatial discretization is fine in the vicinity of overhead wires and coarse in the rest of the computational domain. They used the so-called subgridding method, which is different from nonuniform gridding and more efficient.

Tatematsu and Noda [45, 49] have analyzed lightning-induced voltages on an overhead three-phase distribution line with an overhead ground wire and lightning arresters above perfectly conducting and lossy ground using the 3D FDTD method. The nonlinear voltage–current (V–I) relation of the arrester is represented by piecewise linear curves, which is based on the measured V–I curve. The nearby lightning channel is represented by a vertical phased-current-source array. The conductor system is accommodated in a working volume of 1,400 m $\times$ 650 m $\times$ 700 m, which is divided nonuniformly into rectangular cells: 2 cm in the vicinity of the horizontal ground wires, increasing gradually (to 200 cm) beyond that region.

Thang *et al.* [48] have computed, using the 3D FDTD method, lightning-induced voltages at different points along a 5 mm radius, 1 km long single overhead wire shown in Figure 6.11. Corona space charge around the wire is considered. FDTD computations are performed using a 3D nonuniform grid. The progression of corona streamers from the wire is represented as the radial expansion of cylindrical weakly conducting (40 μS/m) region around the wire. The magnitudes of FDTD-computed lightning-induced voltages in the presence of corona discharge are larger than those computed without considering corona, as shown in Figure 6.12. The observed trend, although less pronounced, is in agreement with that reported by Nucci *et al.* [92]. When corona is considered, there is a tendency for induced-voltage risetimes to increase. Figure 6.13 shows the variation with time of corona radius at different points along the wire and the variation of corona radius along the wire at time 5 μs, computed for the case of perfectly conducting ground. The maximum radius of corona region around the wire for stroke location A (see Figure 6.11) and 35 kA current peak is 19.8 cm. It follows from Figure 6.13 that the presence of lightning-induced corona on the wire makes the transmission line (formed by the wire and its image) nonuniform. Note that corona-radius variation is step-like due to the size of rectangular cells employed in the FDTD computations. It appears that the distributed impedance discontinuity, associated with corona development on the wire, is the primary reason for higher induced-voltage peaks and longer voltage risetimes compared to the case without corona.

Ren *et al.* [43], using the 2D cylindrical FDTD method instead of approximate expressions, evaluated the electric fields over lossy ground, and calculated the lightning-induced voltages on an overhead two-wire distribution line using the FDTD-computed fields with Agrawal *et al.*'s field-to-wire coupling model [93]. One of the reasons for using the two-step or hybrid approach is apparently related to a difficulty of representing closely spaced overhead thin wires in the 3D uniform-grid FDTD method. Zhang *et al.* [50] have used the same approach for investigating the influence of horizontally stratified ground on lightning-induced voltages.

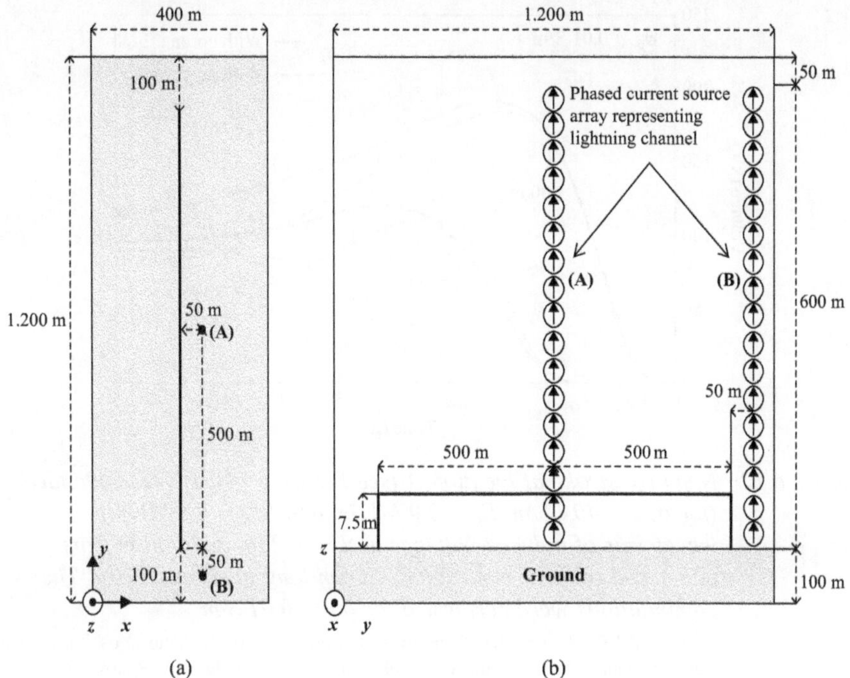

Figure 6.11   (a) *Plan (xy plane) and (b) side (yz plane) views of a 5 mm radius,
1 km long overhead horizontal wire located 7.5 m above ground.
Corona discharge is assumed to occur only on the horizontal wire.
Ground strike points are shown in (a), and simulated lighting
channels are shown in (b)*

© 2014 IEEE. Reprinted with permission from Tran Huu Thang, Yoshihiro Baba,
Naoto Nagaoka, Akihiro Ametani, Naoki Itamoto, and Vladimir A. Rakov, 'FDTD
simulations of corona effect on lightning-induced voltages', *IEEE Transactions on
Electromagnetic Compatibility*, vol. 56, no. 1, pp. 168–176, February 2014, Figure 1

Yang *et al.* [46] have employed a different two-step approach for evaluating
lightning-induced voltages on an overhead single wire above lossy ground. The
first step is a 2D cylindrical FDTD computation of electric and magnetic fields
generated by a nearby lightning strike that illuminate the 3D volume accommodating the overhead wire. The second step is a 3D FDTD computation of
lightning-induced voltages on the overhead wire illuminated by incident electromagnetic fields from the boundary of the 3D working volume.

Matsuura *et al.* [44] have analyzed transient voltages due to direct lightning
strikes to the overhead ground wire of three-phase distribution line over lossy
ground using the 3D FDTD method. The working volume is divided nonuniformly.
The lightning channel attached to the ground wire is represented by a vertical
perfectly conducting wire, excited by a lumped current source inserted between the
simulated lightning channel and the ground wire.

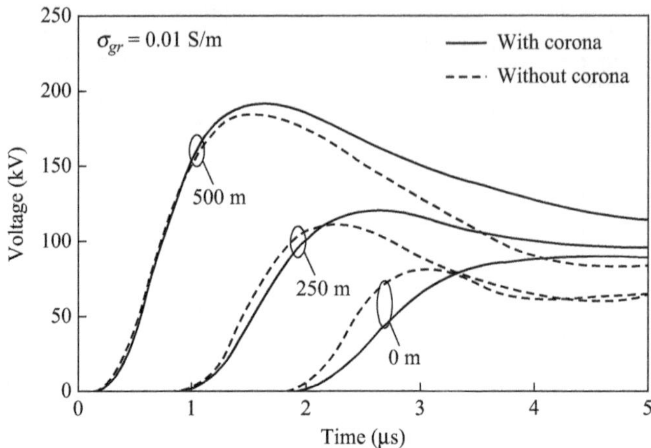

Figure 6.12   *Negative stroke at location A (see Figure 6.11): FDTD-computed*
(for σ_{cor} = 40 μS/m, E_0 = 2.9 MV/m, and E_{cn} = 1.5 MV/m)
waveforms of induced voltages at d = 0, 250, and 500 m from
either end of the 5 mm radius, 1.0 km long horizontal wire. The
computations were performed for σ_{gr} = 0.01 S/m

© 2014 IEEE. Reprinted with permission from Tran Huu Thang, Yoshihiro Baba, Naoto Nagaoka, Akihiro Ametani, Naoki Itamoto, and Vladimir A. Rakov, 'FDTD simulations of corona effect on lightning-induced voltages', *IEEE Transactions on Electromagnetic Compatibility*, vol. 56, no. 1, pp. 168–176, February 2014, Figure 4(b)

6.4.4   *Lightning electromagnetic environment in power substation*

Oliveira and Sobrinho [51], using the 3D FDTD method, have analyzed electromagnetic environment in an air-insulated power substation struck by lightning. Electric potentials and step and touch voltages are calculated for a realistic configuration. The lightning channel is not considered, and the lightning current is appeared to be injected by a lumped current source.

6.4.5   *Lightning electromagnetic environment in airborne vehicles*

Apra *et al.* [52], using the 3D FDTD method, have analyzed transient magnetic fields and voltages induced on cables in an airborne vehicle struck by lightning. The lightning channel is represented by a vertical perfectly conducting wire. A lumped current source appears to be used as an exciter. The airborne vehicle, which is partially made of carbon fiber composite of conductivity 1.5×10^4 S/m, is 22 m in length, 30 m in wing span, and 9 m in height. It is represented with cubic cells of 0.1 m × 0.1 m × 0.1 m. Perrin *et al.* [53] carried out a similar 3D FDTD computation for estimating currents induced on cables in an airborne vehicle struck by lightning. The representation of lightning channel is not given in [53]. The airborne vehicle, which is partially made of carbon fiber composite of conductivity 1,000 S/m, is represented with cubic cells of 5 cm × 5 cm × 5 cm.

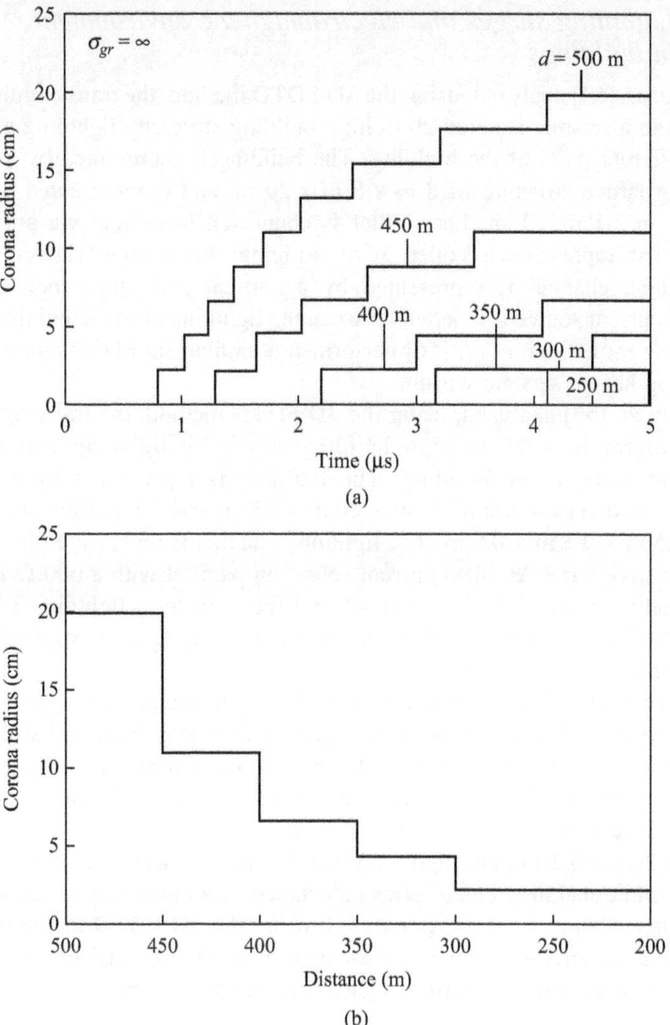

Figure 6.13 *Negative stroke at location A: (a) Time variation of corona radius at d = 500, 450, 400, 350, 300, and 250 m from either end of the 5 mm radius, 1.0 km long horizontal wire located above perfectly conducting ground, and (b) corona radius as a function of distance from either end of the wire at time 5 μs*

© 2014 IEEE. Reprinted with permission from Tran Huu Thang, Yoshihiro Baba, Naoto Nagaoka, Akihiro Ametani, Naoki Itamoto, and Vladimir A. Rakov, 'FDTD simulations of corona effect on lightning-induced voltages', *IEEE Transactions on Electromagnetic Compatibility*, vol. 56, no. 1, pp. 168–176, February 2014, Figure 6

6.4.6   Lightning surges and electromagnetic environment in buildings

Nagaoka *et al.* [54] analyzed, using the 3D FDTD method, the transient distribution of current in a seismic-isolated 20 m high building struck by lightning and potentials at different parts of the building. The building is represented by a perfectly conducting lattice structure of 8 m × 8 m × 20 m, and is represented with cubic cells of 1 m × 1 m × 1 m. Each pillar is connected to ground via a 1 or 2 µH inductor that represents a coiled wire bridging the seismic rubber insulator. The lightning channel is represented by a vertical perfectly conducting wire. A lumped current source is inserted between the lightning channel and the center of the building roof. Chen *et al.* [56] performed a similar 3D FDTD simulation, but the building has no seismic wiring.

Ishii *et al.* [57] analyzed, using the 3D FDTD method, the transient distribution of current in a 25 m high building struck by lightning and potentials at different parts of the building. The building is represented by a perfectly conducting lattice structure of 20 m × 20 m × 25 m, and is represented with cubic cells of 0.5 m × 0.5 m × 0.5 m. The lightning channel is represented by a vertical 1 Ω/m resistive wire. An ideal current source in parallel with a 600 Ω resistor is inserted between the lightning channel and the 5 m long lightning rod on the building roof. Surge protective devices for suppressing lightning overvoltages are incorporated in the simulation.

Chamié *et al.* [58] analyzed, using the 3D FDTD method, voltages induced on electrical wirings of nine buildings of height 24 m, due to lightning strikes to the nearby 50 m high radio base station. Each building is represented by a perfectly conducting lattice structure of 24 m × 12.8 m × 24 m with walls and floors of zero conductivity and relative permittivity 7.5. It is represented with cubic cells of 0.2 m × 0.2 m × 0.2 m. The lightning channel is represented by a vertical wire loaded by additional distributed series inductance. A lumped current source is used as an exciter. Oshiro *et al.* [55] analyzed, using the 3D FDTD method, voltages induced on an electrical wiring in a house by nearby lightning strikes. The lightning channel is represented by a vertical phased-current-source array.

6.4.7   Surges on grounding electrodes

Tanabe [13] has first applied the 3D FDTD method to analyzing the surge response of a vertical grounding electrode of 0.5 m × 0.5 m × 3 m, and shown that FDTD-computed response agrees well with the corresponding measured one. Tsumura *et al.* [60] analyzed the surge response of horizontal grounding electrode using the 3D FDTD method, and used the FDTD-computed response to test the validity of a simplified equivalent circuit model of horizontal grounding electrode.

Baba *et al.* [59] have proposed a procedure for representing thin cylindrical wires buried in lossy ground, which is implemented in 3D FDTD surge simulations. Xiong *et al.* [62, 63] have proposed procedures for representing flat thin wires buried in lossy ground, which is used in 3D FDTD surge simulations. Here "thin" means the radius of the thickness of the conductor is smaller than the side length of cell employed. Xiong *et al.* [64] have also carried out surge simulations for a symmetrical vertical grounding electrode using the 2D cylindrical FDTD method.

Ala *et al.* [61] considered the soil ionization around a grounding electrode in their 3D FDTD computations. The ionization model is based on the dynamic soil-resistivity model of Liew and Darveniza [94]. In the model, the resistivity of each soil-representing cell is controlled by the instantaneous value of the electric field there and time. Otani *et al.* [65] have applied the soil ionization model to analyzing different grounding electrodes, and compared 3D FDTD-computed responses with corresponding measured one.

6.5   Summary

In this chapter, the fundamental theory of the FDTD method has been reviewed, and advantages and disadvantages of the FDTD method in comparison with other representative NEA methods have been identified. Also, seven representations of the lightning return-stroke channel used to date have been described. Furthermore, the representative applications of the FDTD method have been reviewed. They include simulations of (i) lightning electromagnetic fields at close and far distances, (ii) lightning surges on overhead power transmission lines and towers, (iii) lightning surges on overhead power distribution and telecommunication lines, (iv) lightning electromagnetic environment in power substations, (v) lightning electromagnetic environment in airborne vehicles, (vi) lightning surges and electromagnetic environment in buildings, and (vii) surges on grounding electrodes.

References

[1]   Harrington R. F. *Field Computation by Moment Methods*, Macmillan, New York, 1968.

[2]   Sadiku M. N. O. 'A simple introduction to finite element analysis of electromagnetic problems'. *IEEE Transactions on Education.* 1989, vol. 32, no. 2, pp. 85–93.

[3]   A. Ruehli, 'Equivalent circuit models for three-dimensional multiconductor systems'. *IEEE Transactions on Microwave Theory and Techniques.* 1974, vol. 22, no. 3, pp. 216–221.

[4]   Visacro S., Soares A. Jr. 'HEM: A model for simulation of lightning related engineering problems'. *IEEE Transactions on Power Delivery.* 2005, vol. 20, no. 2, pp. 1206–1207.

[5]   Johns P. B., Beurle R. B. 'Numerical solutions of 2-dimensional scattering problems using a transmission-line matrix'. *Proceedings of the IEE.* 1971, vol. 118, no. 9, pp. 1203–1208.

[6]   Yee K. S. 'Numerical solution of initial boundary value problems involving Maxwell's equations in isotropic media'. *IEEE Transactions on Antennas and Propagation.* 1966, vol. 14, no. 3, pp. 302–307.

[7]   Baba Y., Rakov V. A. 'Electromagnetic models of the lightning return stroke'. *Journal of Geophysical Research*. 2007, vol. 112, no. D4, doi:10.1029/2006JD007222, 17 pp.

[8]   Baba Y., Rakov V. A. 'Applications of electromagnetic models of the lightning return stroke'. *IEEE Transactions on Power Delivery*. 2008, vol. 23, no. 2, pp. 800–811.

[9]   Baba Y., Rakov V. A. 'Electric and magnetic fields predicted by different electromagnetic models of the lightning return stroke versus measured fields'. *IEEE Transactions on Electromagnetic Compatibility*. 2009, vol. 51, no. 3, pp. 479–483.

[10]   Baba Y., Rakov V. A. 'Electromagnetic models of lightning return strokes'. In *Lightning Electromagnetics* (V. Cooray, Ed.). IET. 2012, London, UK. pp. 263–313.

[11]   CIGRE Working Group C4.501. *Guide for Numerical Electromagnetic Analysis Methods: Application to Surge Phenomena and Comparison with Circuit Theory-based Approach*. Technical Brochure. 2013, no. 543, ISBN: 978-2-85873-237-1.

[12]   Baba Y., Rakov V. A. 'On the transmission line model for lightning return stroke representation'. *Geophysical Research Letters*. 2013, vol. 30, no. 24, 4 pp.

[13]   Tanabe K. 'Novel method for analyzing dynamic behavior of grounding systems based on the finite-difference time-domain method'. *IEEE Power Engineering Review*. 2001, vol. 21, no. 9, pp. 55–57.

[14]   Yang C., Zhou B. 'Calculation methods of electromagnetic fields very close to lightning'. *IEEE Transactions on Electromagnetic Compatibility*. 2004, vol. 46, no. 1, pp. 133–141.

[15]   Berenger J.-P. 'Long range propagation of lightning pulses using the FDTD method'. *IEEE Transactions on Electromagnetic Compatibility*. 2005, vol. 47, no. 4, pp. 1008–1011.

[16]   Hu W., Cummer S. A. 'An FDTD model for low and high altitude lightning-generated EM fields'. *IEEE Transactions on Antennas and Propagation*. 2006, vol. 54, no. 5, pp. 1513–1522.

[17]   Baba Y., Rakov V. A. 'Electromagnetic fields at the top of a tall building associated with nearby lightning return strokes'. *IEEE Transactions on Electromagnetic Compatibility*. 2007, vol. 49, no. 3, pp. 632–643.

[18]   Baba Y., Rakov V. A. 'Influence of strike object grounding on close lightning electric fields'. *Journal of Geophysical Research*. 2008, vol. 113, no. D12, doi:10.1029/2008JD009811, 18 pp.

[19]   Mimouni A., Rachidi F., Azzouz Z. 'A finite-difference time-domain approach for the evaluation of electromagnetic fields radiated by lightning strikes to tall structures'. *Journal of Electrostatics*. 2008, vol. 66, no. 9, pp. 504–513.

[20]   Shoory A., Mimouni A., Rachidi F., Cooray V., Moini R., Sadeghi S. H. H. 'Validity of simplified approaches for the evaluation of lightning electromagnetic fields above a horizontally stratified ground'. *IEEE Transactions on Electromagnetic Compatibility*. 2010, vol. 52, no. 3, pp. 657–663.

[21]   Oikawa T., Sonoda J., Sato M., Honma N., Ikegawa Y. 'Analysis of lightning electromagnetic field on large-scale terrain model using three-dimensional MW-FDTD parallel computation' (in Japanese). *IEEJ Transactions on Fundamentals and Materials*. 2012, vol. 132, no. 1, pp. 44–50.

[22]   Haddad M., Rakov V. A., Cummer S. 'New measurements of lightning electric fields in Florida: Waveform characteristics, interaction with the ionosphere, and peak current estimates'. *Journal of Geophysical Research*. 2012, vol. 117, no. D10, doi:10.1029/2011JD017196, 26 pp.

[23]   Zhang Q., Li D., Fan Y., Zhang Y., Gao J. 'Examination of the Cooray-Rubinstein (C-R) formula for a mixed propagation path by using FDTD'. *Journal of Geophysical Research*. 2012, vol. 117, no. D15, doi:10.1029/2011JD017331, 7 pp.

[24]   Khosravi-Farsani M., Moini R., Sadeghi S. H. H., Rachidi F. 'On the validity of approximate formulas for the evaluation of the lightning electromagnetic fields in the presence of a lossy ground'. *IEEE Transactions on Electromagnetic Compatibility*. 2013, vol. 55, no. 2, pp. 362–370.

[25]   Zhang S., Wu Q. 'A new perspective to the attributes of lightning electromagnetic field above the ground'. *Journal of Electrostatics*. 2013, vol. 71, no. 4, pp. 703–711.

[26]   Li D., Zhang Q., Liu T., Wang Z. 'Validation of the Cooray-Rubinstein (C-R) formula for a rough ground surface by using three-dimensional (3-D) FDTD'. *Journal of Geophysical Research*. 2013, vol. 118, no. 22, doi:10.1002/2013JD020078, 6 pp.

[27]   Zhang Q., Tang V., Gao J., Zhang L., Li D. 'The influence of the horizontally stratified conducting ground on the lightning-induced voltages'. *IEEE Transactions on Electromagnetic Compatibility*. 2014, vol. 56, no. 1, pp. 143–148.

[28]   Li D., Zhang Q., Wang Z., Liu T. 'Computation of lightning horizontal field over the two-dimensional rough ground by using the three-dimensional FDTD'. *IEEE Transactions on Electromagnetic Compatibility*. 2014, vol. 56, no. 1, pp. 143–148.

[29]   Mimouni A., Rachidi F., Rubinstein M. 'Electromagnetic fields of a lightning return stroke in presence of a stratified ground'. *IEEE Transactions on Electromagnetic Compatibility*. 2014, vol. 56, no. 2, pp. 413–418.

[30]   Baba Y., Rakov V. A. 'Simulation of corona at lightning-triggering wire: Current, charge transfer, and the field-reduction effect'. *Journal of Geophysical Research*. 2011, vol. 116, no. D21, doi:10.1029/2011JD016341, 12 pp.

[31]   Baba Y., Rakov V. A. 'On the mechanism of attenuation of current waves propagating along a vertical perfectly conducting wire above ground: Application to lightning'. *IEEE Transactions on Electromagnetic Compatibility*. 2005, vol. 47, no. 3, pp. 521–532.

[32]   Baba Y., Rakov V. A. 'On the interpretation of ground reflections observed in small-scale experiments: Simulating lightning strikes to towers'. *IEEE Transactions on Electromagnetic Compatibility*. 2005, vol. 47, no. 3, pp. 533–542.

[33]   Noda T. 'A tower model for lightning overvoltage studies based on the result of an FDTD simulation' (in Japanese). *IEEJ Transactions on Power and Energy*. 2007, vol. 127, no. 2, pp. 379–388.

[34]   Noda T., Tatematsu A., Yokoyama S. 'Improvements of an FDTD-based surge simulation code and its application to the lightning overvoltage calculation of a transmission tower'. *Electric Power Systems Research*. 2007, vol. 77, no. 11, pp. 1495–1500.

[35]   Noda T. 'A numerical simulation of transient electromagnetic fields for obtaining the step response of a transmission tower using the FDTD method'. *IEEE Transactions on Power Delivery*. 2008, vol. 23, no. 2, pp. 1262–1263.

[36]   Takami J., Tsuboi T., Okabe S. 'Measured distortion of current waves and electrical potentials with propagation of a spherical wave in an electromagnetic field'. *IEEE Transactions on Electromagnetic Compatibility*. 2010, vol. 52, no. 3, pp. 753–756.

[37]   Thang T. H., Baba Y., Nagaoka N., Ametani A., Takami J., Okabe S., Rakov V. A. 'A simplified model of corona discharge on overhead wire for FDTD computations'. *IEEE Transactions on Electromagnetic Compatibility*. 2012, vol. 54, no. 3, pp. 585–593.

[38]   Thang T. H., Baba Y., Nagaoka N., Ametani A., Takami J., Okabe S., Rakov V. A. 'FDTD simulation of lightning surges on overhead wires in the presence of corona discharge'. *IEEE Transactions on Electromagnetic Compatibility*. 2012, vol. 54, no. 6, pp. 1234–1243.

[39]   Thang T. H., Baba Y., Nagaoka N., Ametani A., Itamoto N., Rakov V. A. 'FDTD simulation of insulator voltages at a lightning-struck tower considering ground-wire corona'. *IEEE Transactions on Power Delivery*. 2013, vol. 28, no. 3, pp. 1635–1642.

[40]   Yao C., Wu H., Mi Y., Ma Y., Shen Y., Wang L. 'Finite difference time domain simulation of lightning transient electromagnetic fields on transmission lines'. *IEEE Transactions on Dielectrics and Electrical Insulation*. 2013, vol. 20, no. 4, pp. 1239–1246.

[41]   Takami J., Tsuboi T., Yamamoto K., Okabe S., Baba Y., Ametani A. 'Lightning surge response of a double-circuit transmission tower with incoming lines to a substation through FDTD simulation'. *IEEE Transactions on Dielectrics and Insulation*. 2014, vol. 21, no. 1, pp. 96–104.

[42]   Baba Y., Rakov V. A. 'Voltages induced on an overhead wire by lightning strikes to a nearby tall grounded object'. *IEEE Transactions on Electromagnetic Compatibility*. 2006, vol. 48, no. 1, pp. 212–224.

[43]   Ren H.-M., Zhou B.-H., Rakov V. A., Shi L.-H., Gao C., Yang J.-H. 'Analysis of lightning-induced voltages on overhead lines using a 2-D FDTD method and Agrawal coupling model'. *IEEE Transactions on Electromagnetic Compatibility*. 2008, vol. 50, no. 3, pp. 651–659.

[44]   Matsuura S., Tatematsu A., Noda T., Yokoyama S. 'A simulation study of lightning surge characteristics of a distribution line using the FDTD method'. *IEEJ Transactions on Power and Energy*. 2009, vol. 129, no. 10, pp. 1225–1232.

[45] Tatematsu A., Noda T. 'Development of a technique for representing lightning arresters in the surge simulations based on the FDTD method and its application to the calculation of lightning-induced voltages on a distribution line' (in Japanese). *IEEJ Transactions on Power and Energy*. 2010, vol. 130, no. 3, pp. 373–382.

[46] Yang B., Zhou B.-H., Gao C., Shi L.-H., Chen B., Chen H.-L. 'Using a two-step finite-difference time-domain method to analyze lightning-induced voltages on transmission lines'. *IEEE Transactions on Electromagnetic Compatibility*. 2011, vol. 53, no. 1, pp. 256–260.

[47] Sumitani H., Takeshima T., Baba Y., Nagaoka N., Ametani A., Takami J., Okabe S., Rakov V. A. '3-D FDTD computation of lightning-induced voltages on an overhead two-wire distribution line'. *IEEE Transactions on Electromagnetic Compatibility*. 2012, vol. 54, no. 5, pp. 1161–1168.

[48] Thang T. H., Baba Y., Nagaoka N., Ametani A., Itamoto N., Rakov V. A. 'FDTD simulations of corona effect on lightning-induced voltages'. *IEEE Transactions on Electromagnetic Compatibility*. 2014, vol. 56, no. 1, pp. 168–176.

[49] Tatematsu A., Noda T. 'Three-dimensional FDTD calculation of lightning-induced voltages on a multiphase distribution line with the lightning arresters and an overhead shielding wire'. *IEEE Transactions on Electromagnetic Compatibility*. 2014, vol. 56, no. 1, pp. 159–167.

[50] Zhang Q., Tang V., Gao J., Zhang L., Li D. 'The Influence of the horizontally stratified conducting ground on the lightning-.induced voltages'. *IEEE Transactions on Electromagnetic Compatibility*. 2014, vol. 56, no. 2, pp. 435–443.

[51] Oliveira R. Melo e Silva de, Sobrinho C. L. da Silva Souza. 'Computational environment for simulating lightning strokes in a power substation by finite-difference time-domain method'. *IEEE Transactions on Electromagnetic Compatibility*. 2009, vol. 51, no. 4, pp. 995–1000.

[52] Apra M., D'Amore M., Gigliotti K., Sarto M. S., Volpi V. 'Lightning indirect effects certification of a transport aircraft by numerical simulation'. *IEEE Transactions on Electromagnetic Compatibility*. 2008, vol. 50, no. 3, pp. 513–523.

[53] Perrin E., Guiffaut C., Reineix A., Tristant F. 'Using a design-of-experiment technique to consider the wire harness load impedances in the FDTD model of an aircraft struck by lightning'. *IEEE Transactions on Electromagnetic Compatibility*. 2013, vol. 55, no. 4, pp. 747–753.

[54] Nagaoka N., Morita H., Baba Y., Ametani A. 'Numerical simulations of lightning surge responses in a seismic isolated building by FDTD and EMTP'. *IEEJ Transactions on Power and Energy*. 2008, vol. 128, no. 2, pp. 473–478.

[55] Oshiro R., Isa S., Kaneko E. 'A simulation study of induced surge on low voltage distribution system nearby the lightning point' (in Japanese). *IEEJ Transactions on Power and Energy*. 2010, vol. 130, no. 7, pp. 687–693.

[56] Chen J., Zhou B., Zhao F., Qiu S. 'Finite-difference time-domain analysis of the electromagnetic environment in a reinforced concrete structure when

struck by lightning'. *IEEE Transactions on Electromagnetic Compatibility*. 2010, vol. 52, no. 4, pp. 914–920.

[57] Ishii M., Miyabe K., Tatematsu A. 'Induced voltages and currents on electrical wirings in building directly hit by lightning'. *Electric Power Systems Research*. 2012, vol. 85, pp. 2–6.

[58] Filho Ricardo H.T. Chamié, Carvalho Lorena F.P., Machado Péricles L., Oliveira Rodrigo M.S. de. 'Analysis of voltages induced on power outlets due to atmospheric discharges on radio base stations'. *Applied Mathematical Modelling*. 2013, vol. 37, no. 9, pp. 6530–6542.

[59] Baba Y., Nagaoka N., Ametani A. 'Modeling of thin wires in a lossy medium for FDTD simulations'. *IEEE Transactions on Electromagnetic Compatibility*. 2005, vol. 47, no. 1, pp. 54–60.

[60] Tsumura M., Baba Y., Nagaoka N., Ametani A. 'FDTD simulation of a horizontal grounding electrode and modeling of its equivalent circuit'. *IEEE Transactions on Electromagnetic Compatibility*. 2006, vol. 48, no. 4, pp. 817–825.

[61] Ala G., Buccheri P. L., Romano P., Viola F. 'Finite difference time domain simulation of earth electrodes soil ionisation under lightning surge condition'. *IET Science, Measurement & Technology*. 2008, vol. 2, no. 3, pp. 134–145.

[62] Xiong R., Chen B., Mao Y.-F., Deng W., Wu Q., Qiu Y.-Y. 'FDTD modeling of the earthing conductor in the transient grounding resistance analysis'. *IEEE Trans. Antennas and Wireless Propagation Letters*. 2012, vol. 11, pp. 957–960.

[63] Xiong R., Chen B., Fang D. 'An algorithm for the FDTD modeling of flat electrodes in grounding systems'. *IEEE Transactions on Antennas and Propagation*. 2014, vol. 62, no. 1, pp. 345–353.

[64] Xiong R., Chen B., Zhou B. H., Gao C. 'Optimized programs for shaped conductive backfill material of grounding systems based on the FDTD simulations'. *IEEE Transactions on Power Delivery*. 2014, vol. 29, no. 4, pp. 1744–1751.

[65] Otani K., Shiraki Y., Baba Y., Nagaoka N., Ametani A., Itamoto N. 'FDTD surge analysis of grounding electrodes considering soil ionization'. *Electric Power Systems Research*. 2014, vol. 113, pp. 171–179.

[66] Liao Z. P., Wong H. L., Yang B. P., Yuan Y. F. 'A transmission boundary for transient wave analysis'. *Science Sinica*. 1984, vol. A27, no. 10, pp. 1063–1076.

[67] Berenger J. P. 'A perfectly matched layer for the absorption of electromagnetic waves'. *Journal of Computational Physics*. 1994, vol. 114, pp. 185–200.

[68] Rakov V. A., Dulzon A. A. 'Calculated electromagnetic fields of lightning return stroke' (in Russian). *Tekh. Elektrodinam*. 1987, vol. 1, pp. 87–89.

[69] Rakov V. A. 'Lightning return stroke speed'. *Journal of Lightning Research*. 2007, vol. 1, pp. 80–89.

[70] Gorin B. N., Shkilev A. V. 'Measurements of lightning currents at the Ostankino tower' (in Russian). *Electrichestrvo*. 1984, vol. 8, pp. 64, 65.

[71] Podgorski A. S., Landt J. A. 'Three dimensional time domain modelling of lightning'. *IEEE Trans. Power Delivery*. 1987, vol. 2, no. 3, pp. 931–938.

[72] Kato S., Narita T., Yamada T., Zaima E. 'Simulation of electromagnetic field in lightning to tall tower'. Paper presented at 11th International Symposium on High Voltage Engineering, no. 467, London, UK, August 1999.

[73] Rakov V. A. 'Some inferences on the propagation mechanisms of dart leaders and return strokes'. *Journal of Geophysical Research*. 1998, vol. 103, no. D2, pp. 1879–1887.

[74] Bonyadi-ram S., Moini R., Sadeghi S. H. H., Rakov V. A. 'On representation of lightning return stroke as a lossy monopole antenna with inductive loading'. *IEEE Transactions on Electromagnetic Compatibility*. 2008, vol. 50, no. 1, pp. 118–127.

[75] Idone V. P., Orville R. E. 'Lightning return stroke velocities in the Thunderstorm Research International Program (TRIP)'. *Journal of Geophysical Research*. 1982, vol. 87, pp. 4903–4915.

[76] Moini R., Kordi B., Rafi G. Z., Rakov V. A. 'A new lightning return stroke model based on antenna theory'. *Journal of Geophysical Research*. 2000, vol. 105, no. D24, pp. 29693–29702.

[77] Kato S., Takinami T., Hirai T., Okabe S. 'A study of lightning channel model in numerical electromagnetic field computation' (in Japanese). Paper presented at 2001 IEEJ National Convention, no. 7-140, Nagoya, Japan, March 2001.

[78] Miyazaki S., Ishii M. 'Reproduction of electromagnetic fields associated with lightning return stroke to a high structure using FDTD method' (in Japanese). Paper presented at 2004 IEEJ National Convention, no. 7-065, p. 98, Kanagawa, Japan, March 2004.

[79] Bonyadi-ram S., Moini R., Sadeghi S. H. H., Rakov V. A. 'Incorporation of distributed capacitive loads in the antenna theory model of lightning return stroke'. Paper presented at 16th International Zurich Symposium on Electromagnetic Compatibility, pp. 213–218, Zurich, March 2005.

[80] Baba Y., Rakov V. A. 'On the use of lumped sources in lightning return stroke models'. *Journal of Geophysical Research*. 2005, vol. 110, no. D3, D03101, doi:10.1029/2004JD005202, 10 pp.

[81] Baba Y., Rakov V. A. 'Influences of the presence of a tall grounded strike object and an upward connecting leader on lightning currents and electromagnetic fields'. *IEEE Transactions on Electromagnetic Compatibility*. 2007, vol. 49, no. 4, pp. 886–892.

[82] Uman M. A., McLain D. K., Krider E. P. 'The electromagnetic radiation from a finite antenna'. *American Journal of Physics*. 1975, vol. 43, pp. 33–38.

[83] Heidler F. 'Traveling current source model for LEMP calculation'. Paper presented at 6th International Symposium on Electromagnetic Compatibility, Zurich, Switzerland, pp. 157–162, 1985.

[84] Moini R., Kordi B., Abedi M. 'Evaluation of LEMP effects on complex wire structures located above a perfectly conducting ground using electric field

integral equation in time domain'. *IEEE Transactions on Electromagnetic Compatibility*. 1998, vol. 40, no. 2, pp. 154–162.

[85] Rubinstein M. 'An approximate formula for the calculation of the horizontal electric field from lightning at close, intermediate, and long range'. *IEEE Transactions on Electromagnetic Compatibility*. 1996, vol. 38, no. 3, pp. 531–535.

[86] Baba Y., Rakov V. A. 'On the use of lumped sources in lightning return stroke models'. *Journal of Geophysical Research*. 2005, vol. 110, no. D3, D03101, doi:10.1029/2004JD005202.

[87] Nag A., Rakov V. A., Cramer J. A. 'Remote measurements of currents in cloud lightning discharges'. *IEEE Transactions on Electromagnetic Compatibility*. 2011, vol. 53, no. 2, pp. 407–415.

[88] Biagi C. J., Uman M. A., Gopalakrishnan J., Hill J. D., Rakov V. A., Ngin T., Jordan D. M. 'Determination of the electric field intensity and space charge density versus height prior to triggered lightning'. *Journal of Geophysical Research*. 2011, vol. 116, no. D15, doi:10.1029/2011JD015710, 15 pp.

[89] Chisholm W. A., Janischewskyj W. 'Lightning surge response of ground electrodes'. *IEEE Transactions on Power Delivery*. 1989, vol. 4, no. 2, pp. 1329–1337.

[90] Inoue A. 'Study on propagation characteristics of high-voltage traveling waves with corona discharge' (in Japanese). CRIEPI Report, 1983, no. 114.

[91] Wagner C. F., Gross I. W., Lloyd B. L. 'High-voltage impulse tests on transmission lines'. *AIEE Transactions on Power Apparatus and Systems*. 1954, vol. 73, pp. 196–210.

[92] Nucci C. A., Guerrieri S., Correia de Barros M. T., Rachidi F. 'Influence of corona on the voltages induced by nearby lightning on overhead distribution lines'. *IEEE Transactions on Power Delivery*. 2000, vol. 15, no. 4, pp. 1265–1273.

[93] Agrawal A. K., Price H. J., Gurbaxani H. H. 'Transient response of multi-conductor transmission lines excited by a nonuniform electromagnetic field'. *IEEE Transactions on Electromagnetic Compatibility*. 1980, vol. 22, no. 2, pp. 119–129.

[94] Liew A. C., Darveniza M. 'Dynamic model of impulse characteristics of concentrated earth'. *Proceedings of IEE*. 1974, vol. 121, no. 2, pp. 123–135.

Chapter 7

Numerical electromagnetic analysis with the PEEC method

*Peerawut Yutthagowith**

The conventional full-wave formulation of the partial element equivalent circuit (PEEC) method was derived from the mixed-potential integral equation (MPIE) for the free space by A. E. Ruehli [1] in 1974. The final results are interpreted Maxwell's equations to a circuit domain by inductance, capacitance, and resistance extraction including retardation in theory. In that time, the retardation is neglected because the minimum wavelength corresponding to the considered maximum frequency is longer than dimensions of an interested domain. His development is based on the concept of inductance and capacitance extraction, which was firstly introduced by Rosa [2] in 1908, and further developed by Grover [3] in 1946 and Hoer and Love [4] in 1965. Despite the fact that the PEEC method has an exact field theoretical basis, it was not primarily developed for the computation of electromagnetic fields. Instead, the circuit designer has a tool at hand to analyze the parasitic effects on connecting structures of circuits without leaving the familiar area of a network theory. The original detailed description of the approach can be found in [1]. This method was considerably developed further in the 1990s where retardation, external field excitation, and the treatment of a dielectric material were investigated [5].

During the last 20 years, an interest, research efforts, and development of the PEEC method have increased significantly in many areas such as:

- Inductance calculations [6]
- EM radiation from printed circuit boards [7]
- Transmission line modeling [8]
- Noise effect modeling [9]
- Scattering problems [10]
- Power electronic circuits [11, 12]
- Antenna analysis [13]
- Lightning-induced effects in electronic devices [14] and in power systems [15]

*King Mongkut's Institute of Technology Ladkrabang, Thailand

- Lightning protection systems [16]
- Surge over-voltage phenomena due to lightning strike in power systems [17]
- Inductance effects in chip design [18]
- Electromagnetic coupling effect on a measuring system [19, 20]
- Ground potential rise due to lightning strike in power systems [21]

The PEEC approach [1] is an integral equation (IE)–based technique and a numerical method for modeling the electromagnetic behavior of arbitrary three-dimensional electrical interconnection structures. It is based on the mixed potential integral equation (MPIE) from which an equivalent circuit is extracted. After discretizing the structure, introducing evaluation functions for current and charge densities and applying the Galerkin method, the PEEC system equations are obtained in a general form. With the introduction of generalized partial elements, the circuit interpretation results directly in PEEC models. The main difference of the PEEC from the method of moment (MOM) is the possibility to extract equivalent circuits from the integral equations. Unlike differential equation–based methods, the PEEC method provides fewer unknowns, since it does not require discretization of the whole space of interests. Though the resultant matrices are dense, recent fast solvers have greatly improved the solution time for PEEC simulations [22]. Like the other IE techniques, spurious resonances may occur [23], most likely resulting from poor geometric meshing. The formulations of the PEEC methods can be carried out either in the frequency or in the time domain.

In the case of frequency domain formulation, the time response is obtained by using the inverse Fourier or Laplace transform. In the case of the time-domain formulation, nonlinear elements are handled easily. However, the solution is quite sensitive, i.e., it may result in an instability problem. Potentials and currents are required to be memorized every time step. Therefore, computation time efficiency is not much different from that in the frequency domain formulation.

The first advantage of this method is that it can incorporate electrical components based on a circuit theory, such as resistors, inductors, capacitors, transmission lines, cables, transformers, switches, and so on. The second one is that in the frequency domain formulation the composition matrix in this method depends on the configuration of a considered system and a medium, and does not depend on sources. The third one is the main advantage of the PEEC method. A potential and a current on a conductor is calculated directly from a node potential and an element current, respectively. Therefore, post-processing for calculating scalar potentials and currents is not required.

In the PEEC method, the MPIE is interpreted as Kirchhoff's voltage law applied to a current element, and the continuity equation or the charge conservation equation is applied via Kirchhoff's current law to a potential element. The equivalent circuit is extracted from three-dimensional geometries of a considered structure.

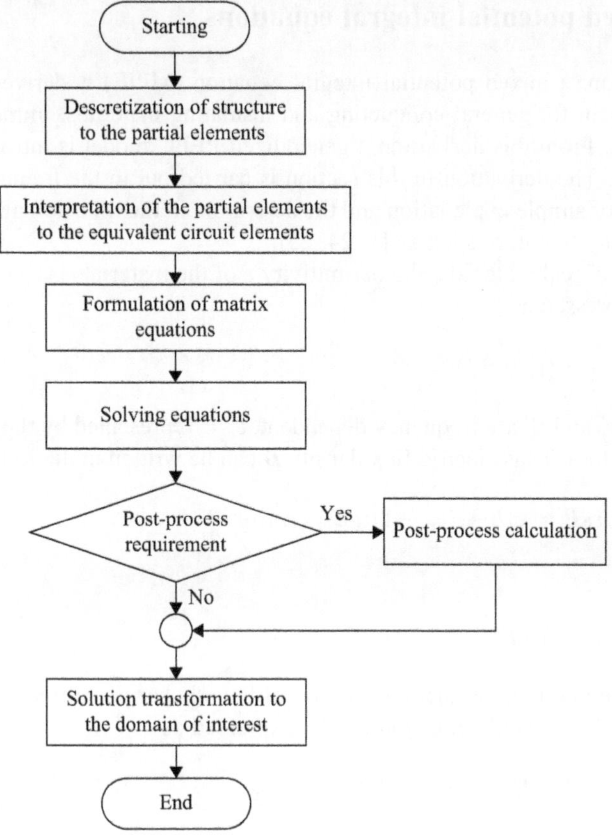

Figure 7.1   Procedures in the simulation of PEEC models

The procedure of a PEEC model simulation is illustrated in Figure 7.1, and is described in the following procedures:

–   Discretization of the structure to the partial elements (current elements and potential elements)
–   Interpretation of the partial elements as equivalent circuit elements
–   Matrix formulation with the MPIE and the continuity equation
–   Matrix solution by solving the circuit equations to obtain currents and potentials in the element structure (in the time domain or in the frequency domain)
–   Post-processing of currents and potentials to obtain field variables
–   Solution transformation to the domain of interest

7.1   Mixed potential integral equations

In this section, a mixed potential integral equation (MPIE) is derived in the frequency domain for general conducting and insulating materials, immersed in free space media. From this derivation a generalized PEEC model is introduced in the next section. The derivation in this section is carried out in the frequency domain for the sake of simple explanation and the derivation in the time domain can also be found in many literature such as [1, 24, 25].

In general real materials, the permittivity ε of the materials is a complex value which is expressed as

$$\varepsilon = \varepsilon_r \varepsilon_0 = (\varepsilon_r' - j\varepsilon_r'')\varepsilon_0 \tag{7.1}$$

where both ε_r' and ε_r'' are frequency dependent. ε_r'' is represented by the dielectric or polarization loss. Thus electric flux density $\boldsymbol{D}$ can be written in the following form:

$$\boldsymbol{D} = \varepsilon_r \varepsilon_0 \boldsymbol{E} = \varepsilon_0 \boldsymbol{E} + (\varepsilon_r - 1)\varepsilon_0 \boldsymbol{E} = \varepsilon_0 \boldsymbol{E} + \boldsymbol{P} \tag{7.2}$$

with

$$\boldsymbol{P} = (\varepsilon_r - 1)\varepsilon_0 \boldsymbol{E} \tag{7.3}$$

where $\boldsymbol{P}$ is represented by the polarization of bound charges in free space.

The third Maxwell's equation can be expressed as

$$\nabla \cdot \boldsymbol{D} = \varepsilon \nabla \cdot \boldsymbol{E} = \varepsilon_0 \nabla \cdot \boldsymbol{E} + \nabla \cdot \boldsymbol{P} = \rho_f \tag{7.4}$$

with

$$\nabla \cdot \boldsymbol{P} = \rho_b \tag{7.5}$$

where ρ_f is density of free charge and ρ_b is density of bound charge. Therefore, from (7.4) and (7.5) the relation between electric field intensity and charges can be expressed as

$$\nabla \cdot \boldsymbol{E} = \frac{\rho_f}{\varepsilon} = \frac{\rho_f + \rho_b}{\varepsilon_0} = \frac{\rho_t}{\varepsilon_0} \tag{7.6}$$

where ρ_t is total charge density.

It means that both bound and free charges are sources of the electric fields in free space. Further, the second Maxwell's equation can be written in the form:

$$\nabla \times \boldsymbol{H} = \boldsymbol{J}_C + \boldsymbol{J}_D = \sigma \boldsymbol{E} + j\omega\varepsilon\boldsymbol{E}$$
$$= \sigma \boldsymbol{E} + j\omega\varepsilon_0(\varepsilon_r - 1)\boldsymbol{E} + j\omega\varepsilon_0\boldsymbol{E} \tag{7.7}$$

where $\boldsymbol{J}_C = \sigma\boldsymbol{E}$ is conduction current density, $\boldsymbol{J}_D = j\omega\varepsilon\boldsymbol{E}$ is displacement current density in the medium, $\boldsymbol{J}_P = j\omega\varepsilon(\varepsilon_r - 1)\boldsymbol{E}$ is polarization current density,

$J_{D0} = j\omega\varepsilon_{0r}E$ is displacement current density in free space, and $J = J_C + J_P$ is total current density based on moving charges.

Substituting (7.1) into (7.7) yields:

$$\nabla \times H = \left(\sigma + \omega\varepsilon_0\varepsilon_r''\right)E + j\omega\varepsilon_0\left(\varepsilon_r' - 1\right)E + j\omega\varepsilon_0 E \tag{7.8}$$

The equivalent conductivity σ_e including polarization loss is defined as

$$\sigma_e = \sigma + \omega\varepsilon_0\varepsilon_r'' \tag{7.9}$$

The total electric field strength at a point r on the material is calculated by (7.10).

$$E = E^i + E^s \tag{7.10}$$

where E^i is an incident electric field and E^s is a scattering electric field reacted to E^i.

The scattering field can be written in the form of the auxiliary potentials as the following equation.

$$E^s = -j\omega A - \nabla\phi \tag{7.11}$$

The magnetic vector potential A is introduced according to (7.11) and using Lorenz gauge for free space by taking into account conduction and polarization currents as well as free and bound charges. The decoupled wave equations of A and ϕ in the region, which is linear, isotopic, and homogeneous media, can be expressed as

$$\nabla^2 A\varepsilon - \gamma^2 A = -\mu_0(\sigma_e + j\omega\varepsilon_0(\varepsilon_r' - 1))E = -\mu_0 J \tag{7.12}$$

$$\nabla^2\phi - \gamma^2\phi = -\frac{\rho_t}{\varepsilon_0} \tag{7.13}$$

where $\gamma^2 = -\omega^2\mu_0\varepsilon_0$ or $\gamma = j\omega\sqrt{\mu_0\varepsilon_0}$

Substituting (7.12) and (7.13) into (7.11), the MPIE, which will be used in the next section, can be written in the following form:

$$E^i = \frac{J}{\sigma_e + j\omega\varepsilon_0\left(\varepsilon_r' - 1\right)} + j\omega\mu_0\int_{v'} g(r,r')J(r',\omega)dv' + \nabla\phi(r,\omega) \tag{7.14}$$

where $g(r,r') = \frac{e^{-\gamma|r-r'|}}{4\pi|r-r'|}$ and $\phi = \frac{1}{\varepsilon_0}\int_{v'} g(r,r')\rho_v(r',\omega)dv'$

In the case of ideal conductors ($\varepsilon_r = 1$), (7.14) becomes as (7.15):

$$E^i = \frac{J}{\sigma} + j\omega\mu_0\int_{v'} g(r,r')J(r',\omega)dv' + \nabla\phi(r,\omega) \tag{7.15}$$

For the material immersed in the lossy dielectric space, the free space permittivity in (7.14) and (7.15) can be replaced by complex permittivity as the following form:

$$\varepsilon^C = \varepsilon - j\sigma_s/\omega \tag{7.16}$$

where ε and σ_s are permittivity and conductivity of the lossy dielectric space, respectively.

7.2   Formulation of the generalized PEEC models

7.2.1   Derivation of the generalized PEEC method

The derivation of the PEEC method starts from the MPIE which is explained in the previous section. The whole structure is discretized into two groups which are current elements, and charge or potential elements. The current elements and potential elements are interleaved each other as illustrated in Figures 7.2 and 7.3. Assume that there are N current volume elements and M potential elements.

Considering a current element, of which the definition is illustrated in Figure 7.4, its current density can be approximated by a pulse basis function weighted by an unknown coefficient as illustrated in Figure 7.3. All current densities can be written in the following form:

$$J(r) = \sum_{n=1}^{N} b_n^c(r) I_n \tag{7.17}$$

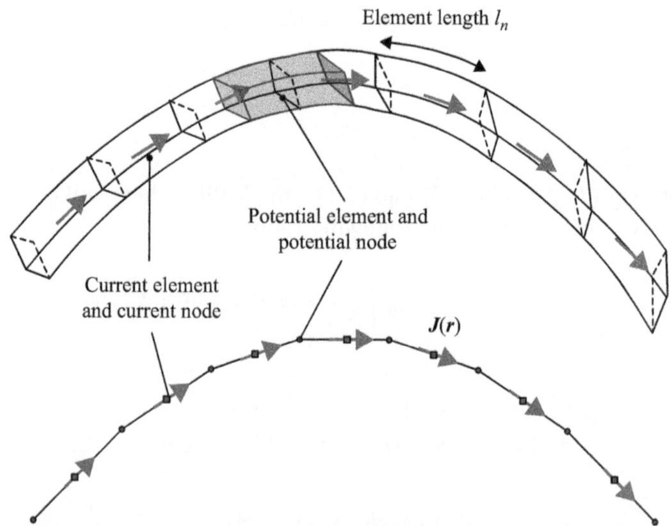

Figure 7.2   Discretization of structure in the PEEC models

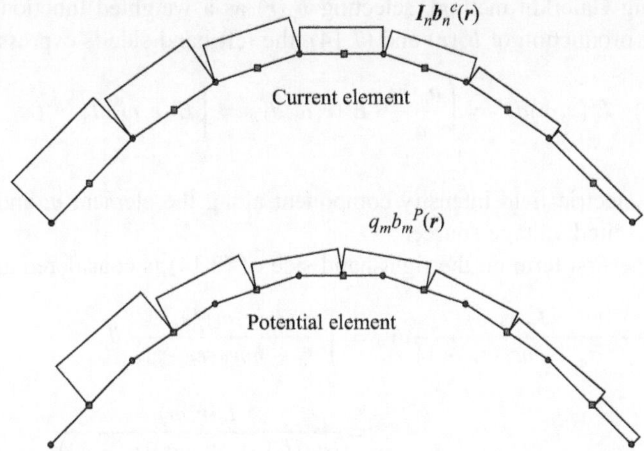

Figure 7.3   Pulse basis functions on current elements and potential element

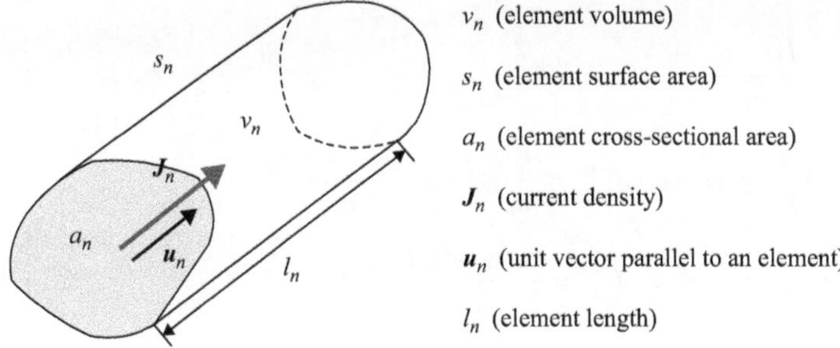

Figure 7.4   Current element n in the PEEC model

with

$$b_n^c(r) = \begin{cases} \dfrac{u_n(r)}{a_n} & r \in v_i \\ 0 & \text{otherwise} \end{cases}$$

The charge density can also be defined in the similar form:

$$\rho(r) = \sum_{m=1}^{M} b_m^P(r) q_m$$

$$b_m^P(r) = \begin{cases} \dfrac{1}{s_m} & r \in v_i \\ 0 & \text{otherwise} \end{cases}$$

$$(7.18)$$

where b_n^c and b_m^P is basis functions employed in PEEC models for current elements and potential elements, respectively, as illustrated in Figures 7.2, 7.3, and 7.4.

Applying Galerkin method, selecting $b_n^c(r)$ as a weighted function, and considering dot production of $b_n^c(r)$ and (7.14), the left-hand side is expressed as

$$\int_{v'} b_n^c(r) \cdot E^i(r,\omega)dv' = \int_{v_n} \frac{u_n(r)}{a_n} \cdot E^i(r,\omega)dv_n = \int_{l_n} E^i(r,\omega)dl_n = U_n^i \qquad (7.19)$$

where E^i is electric field intensity component along the element n, and U_n^i is the equivalent excited voltage source.

Then, the first term on the right-hand side of (7.14) is considered as

$$\int_{v_n} \frac{u_n(r)}{a_n} \cdot \frac{J_n(r,\omega)}{\sigma_e + j\omega\varepsilon_0(\varepsilon_r - 1)} dv_n = \int_{l_n} \frac{I_n(r,\omega)/a_n}{\sigma_e + j\omega\varepsilon_0(\varepsilon_r - 1)} dl_n$$

$$= \frac{I_n(r,\omega)}{(a_n/l_n)(\sigma_e + j\omega\varepsilon_0(\varepsilon_r - 1))} \qquad (7.20)$$

Equation (7.20) can be interpreted as a voltage U_n^1 between both ends and the relation of the current and the voltage can be expressed as

$$U_n^1 = \frac{I(r,\omega)}{Y_n^1}$$

$$Y_n = \frac{1}{Z_n} = G_n + j\omega C_n^+ = \frac{\sigma_e l_n}{a_n} + j\omega \frac{\varepsilon_0(\varepsilon_r - 1)l_n}{a_n} \qquad (7.21)$$

$$G_n = \frac{1}{R_n} = \frac{\sigma_e l_n}{a_n}, \quad C_n^+ = \frac{\varepsilon_0(\varepsilon_r - 1)l_n}{a_n}$$

From (7.20) and (7.21), the first term on the right-hand side can be interpreted as a resistor R_n paralleled with an excess capacitor C_n^+.

Now, the second term on the right-hand side of (7.14), the magnetic vector potential term, is under consideration. In a similar manner, the second term is derived as

$$\int_{v_n} \frac{u_n(r)}{a_n} \cdot \left(j\omega\mu \int_{v'} g(r,r')J(r',\omega)dv' \right) dv_n$$

$$\approx \sum_{i=1}^{N} \left(\frac{j\omega\mu\cos\theta_{in}}{a_i a_n} \int_{v_i}\int_{v_n} g(r,r')dv_n dv_i \right) I_i(r',\omega)$$

$$= \sum_{i=1}^{N} (j\omega L_{in})I_i(r',\omega) \qquad (7.22)$$

where $L_{in} = \frac{\mu\cos\theta_{in}}{a_i a_n} \int_{v_i}\int_{v_n} g(r,r')dv_n dv_i$

The complex coefficient L_{in} in (7.22) has a unit of inductance, in particular, partial self ($i = n$) and partial mutual ($i \neq n$) inductances.

The last term on the right-hand side refers to the electric scalar potential. Similarly, the last term is derived as

$$\nabla\phi(\boldsymbol{r}, \omega) = \frac{\nabla}{\varepsilon} \int_{v'} g(\boldsymbol{r}, \boldsymbol{r}')\rho_v(\boldsymbol{r}', \omega)dv'$$

$$\int_{v_n} \frac{\boldsymbol{u}_n(\boldsymbol{r})}{a_n} \cdot \nabla\phi(\boldsymbol{r}, \omega)dv_n \approx \underbrace{\frac{1}{a_n l_n} \int_{v_n} \phi(\boldsymbol{r} + \boldsymbol{l}_n/2, \omega)dv_n}_{\phi_{n+}} - \underbrace{\frac{1}{a_n l_n} \int_{v_n} \phi(\boldsymbol{r} - \boldsymbol{l}_n/2, \omega)dv_n}_{\phi_{n-}}$$

$$(7.23)$$

The last terms on the right-hand side of (7.14) represents a voltage drop between both ends of a current element. The voltage drop can be calculated from the difference of the average value of potentials ϕ_{n+} and ϕ_{n-}, which are on the both ends of the element. This is the reason for introducing a second group of the M elements called potential elements, which shift by a half of the discretization length to the current elements as illustrated in Figure 7.5. The potential elements are surface elements, since the charge is distributed on the surfaces only.

The potentials can be calculated on the potential elements in the similar manner as a calculation of series impedance:

$$\phi_m \approx \sum_{i=1}^{M} \left(\frac{1}{\varepsilon_0 s_i s_m} \iint_{s_i s_m} g(\boldsymbol{r}, \boldsymbol{r}')ds_m ds_i \right) q_i(\boldsymbol{r}', \omega)$$

$$= \sum_{i=1}^{M} (P_{im}) \, q_i(\boldsymbol{r}', \omega) \tag{7.24}$$

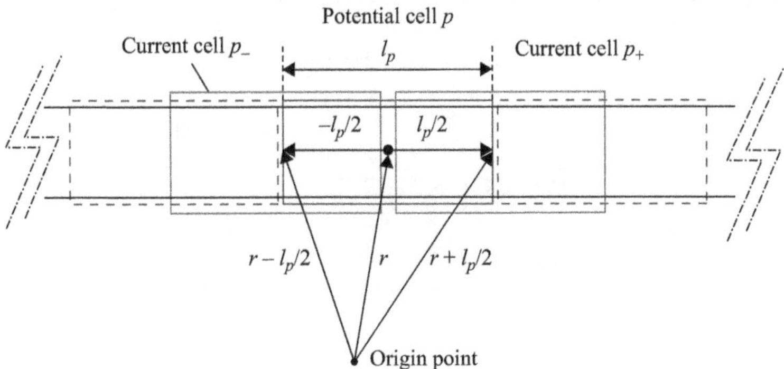

Figure 7.5   Current elements and potential elements in the PEEC model

where s_i is a surface area of a potential element i, q_i is an unknown total charge on a surface of an element i, and $P_{im} = \frac{1}{\varepsilon_0 s_i s_m} \int_{s_i} \int_{s_m} g(r, r') ds_m ds_i$.

The coefficient P_{im} in (7.24) is generalized as a potential coefficient in particular self ($i = m$) coefficient or mutual ($i \neq m$) potential coefficient.

Applying the above (7.22) and (7.23) for N current elements, N equations are obtained in the following form:

$$U_n^i = Z_n I_n + j\omega \sum_{k=1}^{N} L_{kn} I_k + \phi_{n+} - \phi_{n-} \tag{7.25}$$

Then, applying the continuity equation to M potential elements, M equations are obtained in the form:

$$\left(\sum_{j=1}^{n_m} I_{m-j} - I_{ms} \right) + j\omega q_m = 0 \tag{7.26}$$

where n_m is the number of current elements connected to the potential element m, I_{m-j} is a current away from a potential element m to a potential element j, and I_{ms} is an external current source applied to the potential element m.

Figure 7.6 shows the current elements interleaved with a potential element for clarifying explanation.

The assignment of the oriented currents in the current element n is from a potential element $n-$ to a potential element $n+$, and a connectivity matrix A, which has N (the number of current elements) row and M (the number of potential elements) column, is defined as

$$a_{nm} = \begin{cases} +1 & \text{current } I_n \text{ along current element } n \text{ is away from potential element } m \\ -1 & \text{current } I_n \text{ along current element } n \text{ is to potential element } m \\ 0 & \text{no connection} \end{cases}$$

$$\tag{7.27}$$

where a_{mn} is an element of A at row n and column m.

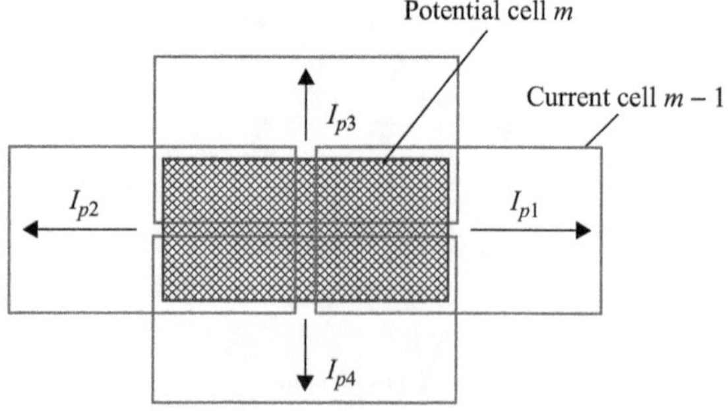

Figure 7.6   Potential element m with four interleaved current elements (2D discretization)

Whole equations can be written in the form of matrix equations as follows:

$$\begin{bmatrix} -AP & Z+j\omega L \\ j\omega 1 & A^T \end{bmatrix} \begin{bmatrix} q \\ I \end{bmatrix} = \begin{bmatrix} U_S \\ I_S \end{bmatrix} \tag{7.28}$$

The solutions of (7.28) are charges and currents along a conductor. The potentials on the potential elements can be calculated by

$$\Phi = Pq \tag{7.29}$$

From (7.28) and (7.29), another form of equations, which give the solution of potentials and currents along the conductor, can be expressed as

$$\begin{bmatrix} -A & Z+j\omega L \\ j\omega P^{-1} & A^T \end{bmatrix} \begin{bmatrix} \Phi \\ I \end{bmatrix} = \begin{bmatrix} U_S \\ I_S \end{bmatrix} \tag{7.30}$$

where the parameters in the system matrix are defined as follows:

Z is a diagonal matrix (N, N) of local impedances calculated by (7.21).
L is a matrix (N, N) of generalized partial inductances calculated by (7.22).
P is a matrix (M, M) of generalized potential coefficients calculated by (7.24).
A is a connectivity matrix (N, M) calculated by (7.27).
I is a vector $(N, 1)$ of currents calculated by (7.28) or (7.30).
q is a vector $(M, 1)$ of charges calculated by (7.28) and (7.29).
Φ is a vector $(M, 1)$ of potentials calculated by (7.28) and (7.29) or (7.30).
U_S is a vector $(N, 1)$ of excited voltage sources calculated by (7.19) or input external lump sources.
I_S is a vector $(M, 1)$ of excited current sources.

The system matrix in (7.30) is equivalent to the system matrix which constructed by modified nodal analysis (MNA) approach. It is noted that for solving (7.30) in the case of large amount of unknowns is not effective when it is compared with (7.28) and (7.29), because the matrix P has to be inversed.

7.2.2   Circuit interpretation of the PEEC method

The circuit interpretation of the system (7.30) derived in the previous section and written in a dense matrix form will be developed in this section.

Each of the N equations in (7.25) can be interpreted as a Kirchhoff's voltage law (KVL) illustrated in Figure 7.7. Formally, the usual symbols of the circuit theory for resistances, inductances, capacitances, and independent and controlled voltage sources are used regardless of the complex values of the generalized partial elements.

The remaining M equations are related to the potential elements and to (7.26). Each of the continuity equations in (7.26) can be interpreted as Kirchhoff's current law (KCL). Definition of transverse current I_{Tm} at each potential element is introduced as the following:

$$j\omega q_m = I_{Tm} \tag{7.31}$$

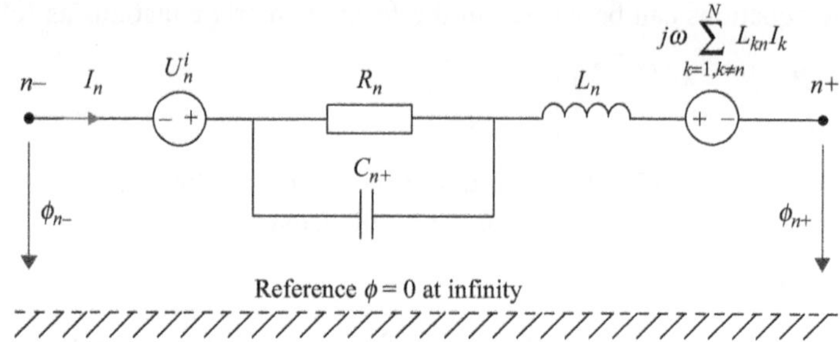

Figure 7.7   *Current interpretation via KVL*

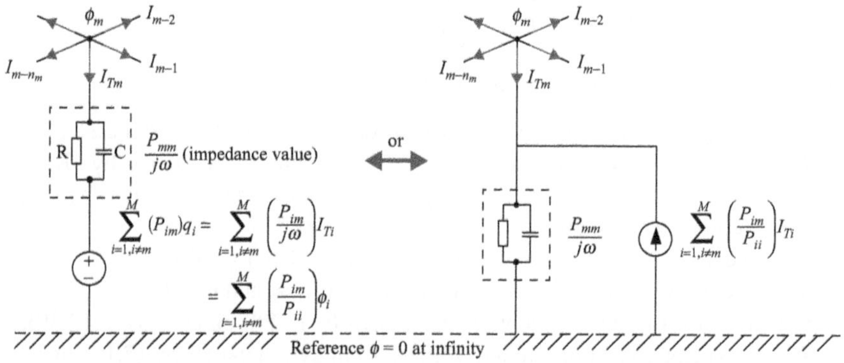

Figure 7.8   *Equivalent circuit model of a potential element*

Substituting (7.31) into (7.26) and rearranging, the other form can be obtained as

$$I_{Ti} = \frac{j\omega}{P_{ii}}\phi_i - \sum_{m=1,m\neq i}^{M} \frac{P_{im}}{P_{im}} I_{Tm} \tag{7.32}$$

or

$$\phi_i = \phi_i^Y - \sum_{m=1,m\neq i}^{M} \frac{P_{im}}{P_{im}} \phi_m^Y \tag{7.33}$$

Equations (7.32) and (7.33) can be interpreted as an equivalent circuit composed of shunt capacitor and resistor, and a controlled voltage or a current source as illustrated in Figure 7.8. It is noted that the shunt resistor in Figure 7.8 is interpretation of the PEEC structure in lossy media. The shunt resistor is disappeared in the case of free space.

7.2.3   *Discretization of PEEC elements*

The level of the discretization or the sizes of the volume and surface elements are determined by the highest frequency in the PEEC model of interest. A rule of thumb for modeling is that the size of the volume and surface elements must never exceed one tenth of the shortest wavelength to ensure a correct representation of an actual waveform. Time and memory limitations are one aspect which is in consideration, because the number of partial elements used in a PEEC model is directly proportional to the discretization. Thus, the calculation time of the partial elements and the computational time for solution increase for an over-discretized problem. On the other hand, the element size, which is not small enough, will lead to the solution being erroneous and/or instable.

The discretization starts with the placement of nodes (on a potential element) in the structure. The current is in between the nodes, in the current elements, while each node charge is approximated with one surface element. The calculations of general 1D, 2D, and 3D partial elements are explained in section 7.3.3.

7.2.4   PEEC models for a plane half space

The PEEC models given in section 7.2.1 are a free space model based on the free space Green's function. The PEEC models for a plane half space is necessary for some configurations of which the structure is located above a large ground plane. The Green function for a plane half space can be derived in the simple form by using the image method as illustrated in Figure 7.9. The Green function of the plane half space is given as summation of the Green functions of free space and its image. The generalized partial inductances of the half space are calculated by

$$L_{in}^{h} = \frac{(\boldsymbol{u}_i \cdot \boldsymbol{u}_n)\mu}{a_i a_n} \iint\limits_{v_i v_n} g(\boldsymbol{r}, \boldsymbol{r}')dv_n dv_i + \frac{(\boldsymbol{u}_{i(im)} \cdot \boldsymbol{u}_n)\mu}{a_i a_n} \iint\limits_{v_i v_n} g(\boldsymbol{r}, \boldsymbol{r}' - 2z'\boldsymbol{e}_z)dv_n dv_i$$

$$(7.34)$$

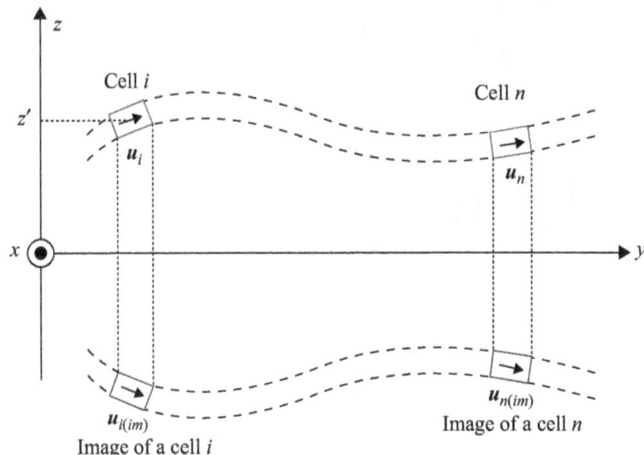

Figure 7.9   PEEC models in a plane half space

Similarly, the potential coefficient of the half space can be calculated as

$$P_{im}^h = \frac{1}{\varepsilon_0 s_i s_m} \int\int_{s_i s_m} g(\boldsymbol{r}, \boldsymbol{r}') ds_m ds_i - \frac{1}{\varepsilon_0 s_i s_m} \int\int_{s_i s_m} g(\boldsymbol{r}, \boldsymbol{r}' - 2z' \boldsymbol{e}_z) ds_m ds_i \tag{7.35}$$

where $\boldsymbol{u}_i$ and $\boldsymbol{u}_{i(im)}$ are unit vectors along element i and its image, respectively. $\boldsymbol{e}_z$ is a unit vector in z direction.

7.3   Some approximate aspects of PEEC models

7.3.1   Center-to-center retardation approximation

In (7.22) and (7.24) for calculating a partial element in the frequency domain the distributed retardation of Green's function included in the integral leads to heavy computation because the partial elements have to be calculated at every frequency step. The approximation of constant retardation between the center to the center of elements can simplify the formulation of (7.22) and (7.24) by moving the retardation term out of the integral. The result is expressed by

$$L_{in}(j\omega) = \frac{\mu \cos\theta_{in}}{a_i a_n} \int\int_{v_i v_n} g(\boldsymbol{r}, \boldsymbol{r}') dv_n dv_i$$

$$\approx \left(\frac{\mu \cos\theta_{in}}{a_i a_n} \int\int_{v_i v_n} \frac{1}{4\pi |\boldsymbol{r} - \boldsymbol{r}'|} dv_n dv_i \right) \times e^{-\gamma d_{in}}$$

$$= L_{in}(0) \times e^{-\gamma d_{in}} \tag{7.36}$$

with

$$L_{in}(0) = \frac{\mu \cos\theta_{in}}{a_i a_n} \int\int_{v_i v_n} \frac{1}{4\pi |\boldsymbol{r} - \boldsymbol{r}'|} dv_n dv_i \tag{7.37}$$

and

$$P_{im}(j\omega) \approx \left(\frac{1}{\varepsilon s_i s_m} \int\int_{s_i s_m} \frac{1}{4\pi |\boldsymbol{r} - \boldsymbol{r}'|} ds_m ds_i \right) \times e^{-\gamma d_{im}}$$

$$= P_{im}(0) \times e^{-\gamma d_{im}} \tag{7.38}$$

with

$$P_{im}(0) = \frac{1}{\varepsilon s_i s_m} \int\int_{s_i s_m} \frac{1}{4\pi |\boldsymbol{r} - \boldsymbol{r}'|} ds_m ds_i \tag{7.39}$$

The retardation term depending on frequency can be written in another form with propagation time τ and attenuation constant α:

$$e^{-\gamma d_{in}} = e^{-(\alpha+j\beta)d_{in}} = e^{-\alpha d_{in}+j\frac{\omega d_{in}}{v}} = e^{-\alpha d_{in}} \times e^{-j\omega\tau_{in}} \tag{7.40}$$

with d_{in} as a center-to-center distance, α as attenuation constant, $\tau_{in} = d_{in}/v$ as a center-to-center propagation time, and v is the speed of an electromagnetic wave in the media.

It means that partial elements at a different frequency step can be calculated by multiplication of the partial element at zero frequency and a retardation term depending on the frequency. However, the approximation technique is applicable when the element length is much shorter than the minimum wavelength of an applied source. The element length, which is not short enough, may lead to an instability problem in the calculated results. The reason for the instability problem can be explained by the damped characteristic of partial element values at a high frequency where it cannot be found in the approximated retardation. Figure 7.10

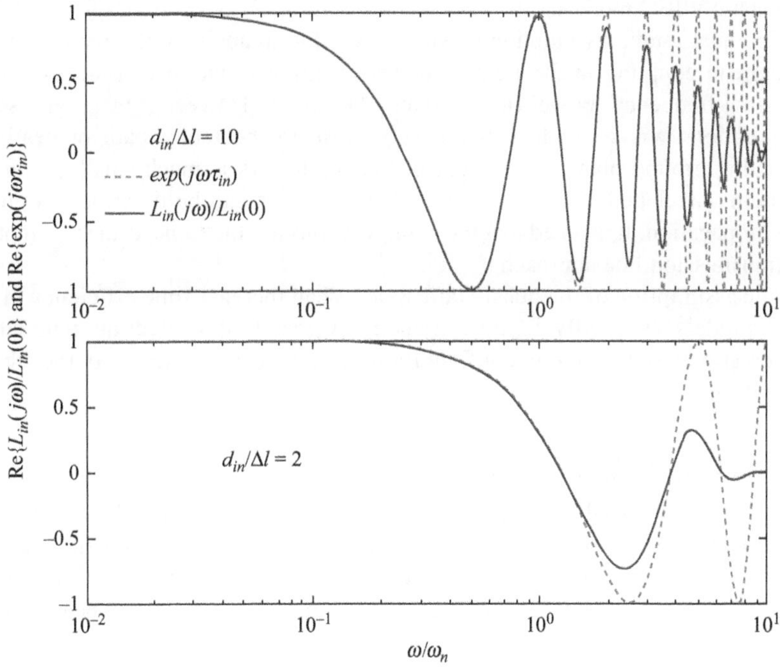

Figure 7.10   Frequency behaviors of exp(jωτ$_{in}$) and L$_{in}$(jω)/L$_{in}$(0) with various distance d$_{in}$ [26]

shows comparison of normalized calculated values based on distributed retardation and assumption of constant retardation. The ω_m is corresponding to

$$\omega_m = 2\pi f_m = \frac{c}{10\Delta l} \qquad (7.41)$$

where c is the light velocity, and Δl is element length.

The details of the reason for the instability problems can be found in [26].

7.3.2　Quasi-static PEEC models

The full-wave PEEC model including retardation has been explained in the previous section. A quasi-static PEEC model neglects the retardation. The structure has to be electrically small compared with the minimum wave length of an applied source, i.e.,

$$d << \lambda \qquad (7.42)$$

where d is the maximum distance between two elements and λ is a minimum wave length of an applied source.

The retardation term in (7.36) and (7.38) can be approximated as unity in the quasi-static PEEC models.

The quasi-static assumption is widely used for an analysis of a parasitic electromagnetic coupling in the field of power electronics due to consideration of a quite low-frequency range (lower than 100 kHz). However, there are some configurations where conductors are very close to the return path or nearby a conducting ground plane. The height of the conductors is much smaller than the minimum wavelength. The quasi-static PEEC models can be described as quasi-TEM propagation. It is noted that the configurations are the same as the assumption of a transmission line approach.

The assumption of the quasi-static models can increase time efficiency in the PEEC models especially in time domain. A result of neglecting retardation is overestimated when the configuration is not corresponding to the above assumption.

7.3.3　Partial element calculation

A calculation of partial elements is a step for a PEEC simulation. The partial element calculations are based on the mathematical formulae in section 7.2.1. In this step, the following calculations of partial inductances, partial coefficients of potential, element resistances, and excess capacitances of dielectric elements are performed.

Reduction of dimensionality of a PEEC element from a three-dimensional structure to a two- or single-dimensional structure is an effective way for increasing efficiency or reducing computational time. For example, a two-dimensional discretization can be used as a representative model in a high-frequency range based on an approximation that currents and charges are distributed on the surface of a metallic structure. Especially, for some structures composed of cylindrical wires such as transmission towers, transmission lines, and a building structure, a thin wire approximation (single-dimensional structure) can be used. Those are typical cases used for an analysis of transient surge phenomena in electrical power systems.

Equations (7.22), (7.24), (7.37), and (7.39) of the partial elements can be solved using a numerical integration technique with general geometries of the structure. The numerical integration technique is time consuming and also introducing errors in the partial element values. To avoid these problems, exact analytical expressions have been developed to calculate the partial elements for a number of basic geometries [27, 28].

7.3.3.1 Partial inductance calculation

By using the center-to-center retardation, partial elements at frequency zero can be calculated in an integral form. The formula of a partial inductance for a three-dimension structure is repeated again here for clarity.

$$L_{in}(0) = \frac{\mu \cos\theta_{in}}{4\pi a_i a_n} \iint\limits_{v_i v_n} \frac{1}{|r - r'|} dv_n dv_i$$

The formula for two- and single-dimensional structures can be derived from the above formula. The final result is given here for two- and single-dimensional structures in (7.43) and (7.44), respectively.

$$L_{in}(0) = \frac{\mu \cos\theta_{in}}{4\pi w_i w_n} \iint\limits_{s_i s_n} \frac{1}{|r - r'|} ds_n ds_i \tag{7.43}$$

where w_i and w_n are widths of elements i and n, and s_i and s_n are surface areas of elements i and n.

$$L_{in}(0) = \frac{\mu \cos\theta_{in}}{4\pi} \iint\limits_{l_i l_n} \frac{1}{|r - r'|} dl_n dl_i \tag{7.44}$$

where l_i and l_n are lengths of elements i and n.

To achieve a satisfactory accuracy and fast evaluation of the self-partial inductance, basic geometries have been defined. The most important basic geometry is the rectangular conductor illustrated in Figure 7.11, where the length l_n is in the direction of the current flow. The formula of the self-inductance of a

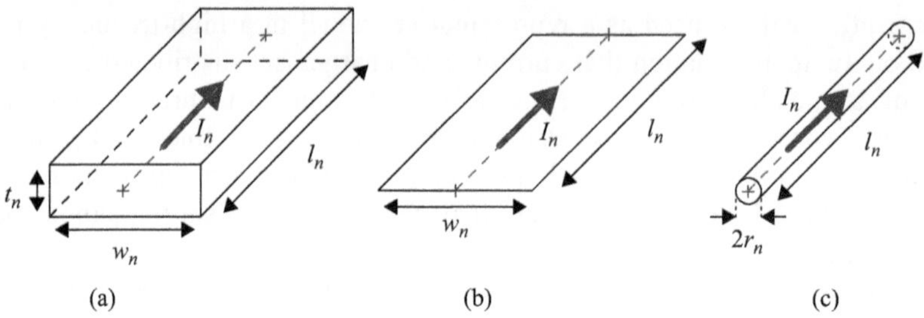

Figure 7.11   Current elements for (a) 3D, (b) 2D, and (c) 1D (thin wire structure)

three-dimensional rectangular conductor illustrated in Figure 7.11(a) can be calculated as follows [27]:

$$
\begin{aligned}
\frac{L_{ii}(0)}{l_i} = \frac{2\mu}{\pi} \Bigg\{ &\frac{w^2}{24u}\left[ln\left(\frac{1+A_2}{w}\right) - A_5 \right] + \frac{1}{24uw}\left[ln(w+A_2) - A_6 \right] \\
&+ \frac{w^2}{60u}(A_4 - A_3) + \frac{w^2}{24}\left[ln\left(\frac{u+A_3}{w}\right) - A_7 \right] \\
&+ \frac{w^2}{60u}(w - A_2) + \frac{1}{20u}(A_2 - A_4) + \frac{u}{4}A_5 \\
&- \frac{u^2}{6w}tan^{-1}\left(\frac{w}{uA_4}\right) + \frac{u}{4w}A_6 - \frac{w}{6}tan^{-1}\left(\frac{u}{wA_4}\right) + \frac{A_7}{4} \\
&- \frac{1}{6w}tan^{-1}\left(\frac{uw}{A_4}\right) + \frac{1}{24w^2}\left[ln(u+A_1) - A_7 \right] \\
&+ \frac{u}{20w^2}(A_1 - A_4) + \frac{1}{60w^2u}(1 - A_2) \\
&+ \frac{1}{60w^2u}(A_4 - A_1) + \frac{u}{20}(A_3 - A_4) \\
&+ \frac{u^3}{24w^2}\left[ln\left(\frac{1+A_1}{u}\right) - A_5 \right] + \frac{u^3}{24w}\left[ln\left(\frac{w+A_3}{u}\right) - A_6 \right] \\
&+ \frac{u^3}{60w^2}\left[(A_4 - A_1) - (u - A_5) \right] \Bigg\}
\end{aligned}
$$

$$\tag{7.45}$$

where $u = l_n/W_n$, $w = t_n/W_n$, $A_1 = \sqrt{1 + u^2}$, $A_2 = \sqrt{1 + w^2}$, $A_3 = \sqrt{w^2 + u^2}$, $A_4 = \sqrt{1 + w^2 + u^2}$, $A_5 = ln\left(\frac{1 + A_4}{A_3}\right)$, $A_6 = ln\left(\frac{w + A_4}{A_1}\right)$, and $A_7 = ln\left(\frac{u + A_4}{A_2}\right)$.

The formula of the self-inductance of a two-dimensional rectangular conductor as illustrated in Figure 7.11(b) can be calculated as the following [27]:

$$\frac{L_{ii}(0)}{l_i} = \frac{\mu}{6\pi}\left\{ 3ln\left(u + \sqrt{u^2 + 1}\right) + u^2 + 1/u \right.$$

$$\left. + 3uln\left(1/u + \sqrt{(1/u)^2 + 1}\right) - \left(u^{4/3} + (1/u)^{2/3}\right)^{3/2} \right\} \tag{7.46}$$

The formula of the self-inductance of a thin wire conductor as illustrated in Figure 7.11(c) can be calculated by

$$\frac{L_{ii}(0)}{l_i} = \frac{\mu}{2\pi}\left\{ ln\left(\frac{l_i}{r} + \sqrt{\left(\frac{l_i}{r}\right)^2 + 1}\right) + \frac{r}{l_i} - \sqrt{\left(\frac{r}{l_i}\right)^2 + 1} \right\} \tag{7.47}$$

The analytical formula of the mutual inductance for coplanar and plane-perpendicular cases can be found in [27, 28]. It seems there is no analytical formula for a mutual inductance calculation in arbitrary configuration. Therefore, numerical integration is employed for the calculation.

7.3.3.2 Partial potential coefficient calculation

A calculation of a partial potential coefficient can be carried out similarly to those in the previous section. The integration on the surface of a structure is given by

$$P_{im}(0) = \frac{1}{4\pi\varepsilon S_i S_m} \iint\limits_{S_i S_m} \frac{1}{|r - r'|} ds_m ds_i$$

For a single-dimensional structure or a thin wire structure the formula is reduced to

$$P_{im}(0) = \frac{1}{4\pi\varepsilon l_i l_m} \iint\limits_{l_i l_m} \frac{1}{|r - r'|} dl_m dl_i \tag{7.48}$$

The formula of the self-potential coefficient of a conductor surface illustrated in Figure 7.11(b) can be found in [28]. The following formula is a modified version of the formula which is used for calculating a partial inductance.

$$P_{ii}(0) = \frac{1}{6\pi\varepsilon l_i}\left\{ 3ln\left(u + \sqrt{u^2 + 1}\right) + u^2 + 1/u \right.$$

$$\left. + 3uln\left(1/u + \sqrt{(1/u)^2 + 1}\right) - \left(u^{4/3} + (1/u)^{2/3}\right)^{3/2} \right\} \tag{7.49}$$

The equation of the self-potential coefficient of a thin wire conductor as illustrated in Figure 7.11(c) can be calculated in the similar manner as employed for calculating the self-inductance. The formula is given as

$$P_{ii}(0) = \frac{1}{2\pi\varepsilon l_i}\left\{ ln\left(\frac{l_i}{r} + \sqrt{\left(\frac{l_i}{r}\right)^2 + 1}\right) + \frac{r}{l_i} - \sqrt{\left(\frac{r}{l_i}\right)^2 + 1} \right\} \qquad (7.50)$$

For a mutual potential coefficient calculation, the analytical formula for coplanar and plane-perpendicular cases can be found in [28]. No analytical formula for mutual potential coefficient calculation in arbitrary configuration can be found. Therefore, numerical integration is employed for the calculation of the mutual potential coefficient.

7.3.3.3 Partial resistance calculation

The partial resistance in a PEEC model is calculated by the volume element discretization and the resistance formula is given a simple form.

$$R_i = \frac{l_i}{a_i \sigma_i} \qquad (7.51)$$

where l_i is the length of the volume element in the current direction, a_i is the cross-section normal to the current direction, and σ_i is the conductivity of the volume element material.

The resistance in the PEEC model accounts for losses in the conductors and can be neglected for lossless PEEC models.

7.3.3.4 Excess capacitance

When there is a dielectric material, an excess capacitance, C_n^+, of the dielectric element is to be calculated as defined in (7.21). The calculation is straightforward and requires no further investigation.

7.4 Matrix formulation and solution

This section describes some approaches for solving a system equation which is formulated from the extraction of the currents in the current elements and the charge at the potential elements in section 7.2. Focus is on the frequency domain solution of PEEC models, and the time-domain solution is explained in brief.

For the solution of PEECs in the time and frequency domains a nodal analysis (an admittance method) and a modified nodal analysis (MNA) method are presented. The MNA is adopted by using a stamp inspection technique to increase time efficiency of the PEEC method. The admittance method produces a minimal but dense system matrix to obtain the potentials on the structure. The MNA solves both voltages and currents on a structure and therefore produces a larger, and sparser system matrix. The MNA method is widely used in modern circuit analysis software because of its full-spectrum properties and flexibility to include additional circuit elements. The choice between the two methods depends on a given problem.

7.4.1 Frequency domain circuit equations and the solution

The potential of each potential element and the current flowing in each current element can be calculated from the equations employing a nodal analysis (admittance method) or an MNA approach in the frequency domain at each frequency over the considered frequency range. The frequency step in the PEEC method depends on the maximum observation time and the frequency range specific to the circuit condition and a measured result investigated. The length of a small element shall be much smaller than the minimum wavelength which is inversely proportional to the maximum frequency.

The time solutions of the PEEC methods in the frequency domain are obtained by using the inverse Fourier transform. It is well known that a frequency transform method is suffered by aliasing phenomenon, Gibb's oscillation and truncation error occurred in the time domain as a result of employing conventional Fourier transform and conventional inverse Fourier transform such as discrete Fourier transform (DFT), fast Fourier transform, and their discrete inverse Fourier transforms (DIFTs). To mitigate those problems by using DIFT and inverse fast Fourier transform (IFFT), it is necessary to reduce sizes of a frequency step and partial element, and also increase the number of frequency step and enlarge a size of matrix equations. It leads to a large CPU time.

The modified Fourier transform (MFT) and the modified Fourier transform (MIFT) are developed to avoid those problems [29]. By employing MFT and MIFT, the results in the time domain are released from Gibb's oscillation in a short time period and truncation error in a large time period.

7.4.1.1 Admittance method formulation

The full-wave solution of PEEC models in the frequency domain requires retardation which is included in a partial inductance and partial potential coefficient. Let's consider (7.30) formulated based on an MNA method by adding an external admittance matrix Y_L, which is related to a node potential vector Φ and an additional admittance current vector I_L as

$$I_L = Y_L \Phi \tag{7.52}$$

and

$$\begin{bmatrix} -A & Z + j\omega L \\ j\omega P^{-1} + Y_L & A^T \end{bmatrix} \begin{bmatrix} \Phi \\ I \end{bmatrix} = \begin{bmatrix} U_S \\ I_S \end{bmatrix} \tag{7.53}$$

Equation (7.53) can be written in the separate form as

$$-A\Phi + (Z + j\omega L)I = U_s \tag{7.54}$$

$$(j\omega P^{-1} + Y_L)\phi + A^T I = I_s \tag{7.55}$$

Substituting (7.55) into (7.54), the admittance method equation is formulated as

$$\left\{ \left(j\omega P^{-1} + Y_L \right) + A^T (Z + j\omega L)^{-1} A \right\} \phi = I_s - A^T (Z + j\omega L)^{-1} U_s \tag{7.56}$$

Or, a single expression,

$$Y\Phi = I'_s \tag{7.57}$$

where

$$Y = \left(j\omega P^{-1} + Y_L \right) + A^T (Z + j\omega L)^{-1} A \tag{7.58}$$

and

$$I'_s = I_s - A^T (Z + j\omega L)^{-1} U_s \tag{7.59}$$

The current I flowing into the current element can be calculated by substitution of Φ calculated by (7.57) into (7.54). The result is expressed as

$$I = (Z + j\omega L)^{-1} (U_s + A\Phi) \tag{7.60}$$

As mentioned above, the admittance method equation is densed and smaller than the MNA equation. It seems to be more practical for solving but the formulation of the admittance method equation is, in fact, suffered by a few times of matrix inversion and matrix multiplication. The iterative method for solving the equation is also not effective. For a large system composed of thousand unknowns, the MNA method is more effective in practice.

7.4.1.2   Modified nodal analysis formulation with stamp inspection

In an MNA method, all the node potentials and currents flowing in the current elements are calculated at the same time. The created system matrix is larger than that of the admittance method, but it is also more sparsed than that of the admittance method. A sparse matrix solver can be used. The solutions of (7.28) are charges and currents. To calculate potential, (7.29) has to be employed. The system equation (7.28) is sparser than equation (7.30), but the implementation of lump elements such as a resistor, an inductor, a capacitor, with (7.28) is not straightforward. Therefore, (7.30) is more widely used in practice than system equations (7.38) and (7.29). However, the inversion of a potential coefficient matrix P is still necessary for formulating system equation (7.30). In the case of very large amount of unknowns, the inversion is heavy time consuming. To avoid the matrix inversion, another kind of matrix formulation based on an MNA method with a stamp inspection technique is introduced in this section.

The MNA method described here is derived from (7.30). The matrix properties are found as

$$P^{-1} = FS^{-1} \tag{7.61}$$

$$FS^{-1} = \left(S^{-1}\right)^T F \tag{7.62}$$

where F and S having dimension $M \times M$ are pseudo capacitance matrices and their definitions of those elements are as follows:

$$f_{nm} = \begin{cases} 1/p_{nn} & \text{if } m = n \\ 0 & \text{otherwise} \end{cases} \tag{7.63}$$

where f_{mn} is an element of F at row n and column m.

$$s_{nm} = \begin{cases} 0 & \text{if } m = n \\ p_{nm}/p_{nn} & \text{otherwise} \end{cases} \tag{7.64}$$

where s_{mn} is an element of S at row n and column m.

Implementing the properties in (7.61) and (7.62), the following system equation is given:

$$\begin{bmatrix} -A & Z + j\omega L \\ j\omega F + S^T Y_L & S^T A^T \end{bmatrix} \begin{bmatrix} \Phi \\ I \end{bmatrix} = \begin{bmatrix} U_S \\ S^T I_S \end{bmatrix} \tag{7.65}$$

Note that the result of the equation has no inversion of matrix on the second row of (7.65), but there are three additional multiplications of matrices. The stamp inspection approach is more effective in a calculation time than that of matrix multiplication. Because the matrices Y_L, A^T, and I_S are sparsed, the terms of $S^T Y_L$, $S^T A^T$, and $S^T I_S$ can be calculated by using the stamp inspection technique and also implemented readily in computer codes.

7.4.2   *Time-domain circuit equations and the solution*

Time-domain equations for the solution of a PEEC model can be derived from the system equations in the frequency domain. Let's consider (7.66) and (7.67) displaying the induced voltage on an element due to the magnetic field coupling with the element. The difference in the time-domain approach from that in the frequency domain is the inclusion of the time retardation that is expressed as a complex part in the partial mutual coupling for the frequency domain models as given in (7.66), while in the time-domain PEEC models, the retardation is written as a finite time delay as shown in (7.67).

$$\phi_{in}(\omega) = j\omega L_{in} I_n e^{-j\omega\tau_{in}} \tag{7.66}$$

$$\phi_{in}(t) = L_{in}\frac{dI_n(t - \tau_{in})}{dt} \tag{7.67}$$

where τ_{in} is time delay between element i and element n.

A time step employed in the time-domain method must be smaller than the smallest time delay and an interpolation technique is employed to calculate a current and a potential, which is not on discrete time points.

7.4.2.1   Time integration methods

The solution of time-domain (TD) PEEC models involves the construction and solution of the corresponding TD equations as detailed above. The stability measures explained in the next section can produce stable PEEC models in combination with a well-suited time integration technique for the type of equations arising for PEEC models. Commonly used time integration techniques are the:

-   Forward Euler
-   Backward Euler
-   Trapezoid
-   Lobatto III-C [30]

The backward Euler (BE) and Lobatto III-C are preferred over the forward Euler and trapezoid [30] because of stability.

7.4.2.2   Modified nodal analysis and admittance method formulations

Exclusion of retardation or quasi-static assumption in the time domain will increase the computational time efficiency of the PEEC method, because the past histories of potentials and currents are memorized only at the previous time step. The inversion of an impedance matrix is required only once and also the model of reduction (MOR) [31, 32] can be employed to reduce the equation size. For quasi-static time-domain PEEC models, the time-domain circuit equations can be derived from (7.53) or (7.65). Employing (7.53) or (7.65) in the formulation does not affect much in time efficiency.

Firstly the MNA method is considered. From the derivation based on the MNA method, the formulation of an admittance method can be derived.

Let's consider (7.53), neglecting excess capacitance from dielectric elements for simplicity in explanation. The excess capacitance, if necessary, can be derived in a similar manner by using an equivalent resistance and an additional voltage source from the previous time step. The approach is a well-known technique employed in a circuit simulator for reducing the size of a system equation. The formulation neglecting excess capacitance in the time domain is written as

$$\begin{bmatrix} -A & R+L\dfrac{d}{dt} \\ P^{-1}\dfrac{d}{dt}+Y_L & A^T \end{bmatrix} \begin{bmatrix} \Phi \\ I \end{bmatrix} = \begin{bmatrix} U_S \\ I_S \end{bmatrix} \tag{7.68}$$

Discretizing the matrix (7.68) in the time domain by using the BE scheme, the following equation is obtained.

$$\begin{bmatrix} -A & R+L\dfrac{1}{\Delta t} \\ P^{-1}\dfrac{1}{\Delta t}+Y_L & A^T \end{bmatrix} \begin{bmatrix} \Phi^n \\ I^n \end{bmatrix} = \begin{bmatrix} U_S^n+L\dfrac{1}{\Delta t}I^{n-1} \\ I_S^n+P^{-1}\dfrac{1}{\Delta t}\Phi^{n-1} \end{bmatrix} \tag{7.69}$$

The above equation can be rewritten to a single equation to calculate the current PEEC node potentials, Φ^n, in the following form.

$$\Phi^n = \left(\frac{1}{\Delta t}P^{-1} + Y_L + A^T\left(R+L\frac{1}{\Delta t}\right)^{-1}A \right)^{-1}$$

$$\times \left(I_S^n + \frac{1}{\Delta t}P^{-1}\Phi^{n-1} - A^T\left(R+L\frac{1}{\Delta t}\right)^{-1}\left(U_S^n+L\frac{1}{\Delta t}I^{n-1}\right) \right) \tag{7.70}$$

The element currents are calculated using post-processing by

$$I^n = \left(R+L\frac{1}{\Delta t}\right)^{-1} \times \left(U_S^n - L\frac{1}{\Delta t}I^{n-1} + A\Phi^n \right) \tag{7.71}$$

For full-wave time-domain PEEC models, accounting for the individual retardation between the current element and the potential element, potentials and currents at many time steps to be stored, and it requires heavy storage and computational time resources. The system of ordinary differential equations (ODEs) describing a quasi-static PEEC model is transformed into a system of neutral delay differential equations (NDDEs) [33, 34] for the full-wave case. The full wave formulation requires the separation of the self (instantaneous) and mutual (retarded) couplings in the previous matrix formulations, and results in the following equations.

$$L = L_S + L_M$$

$$P = P_S + P_M$$

For the MNA method, the subscripts S and M indicate self and mutual terms, respectively. The resulting system of the NDDEs becomes:

$$C_0 y(t) + G_0 y(t) + C_1 y'(t-\tau) + G_1 y(t-\tau) = B\,u(t, t-\tau), \quad t > t_0$$

$$y(t) = g(t), \quad t \le t_0 \tag{7.72}$$

The matrices C_0, G_0, $y(t)$, C_1, G_1, and B in the above equation are real valued and unrelated to previous matrix notations. The retarded couplings must be interpreted in an operator sense. The separation of the self-term and the mutual term results in an extended right-hand side in the corresponding equation system, because the mutual coupling is considered to be known for they are related to past current and potential values. This requires reevaluation at each time step. The time-domain modeling of PEECs is further investigated in [35].

7.4.2.3   PEEC model order reduction

The basic idea of a model order reduction (MOR) technique is to reduce the size of a system described by circuit equations by describing only the dominant behavior of the original system [31]. Within quasi-static PEEC modeling, the PRIMA algorithm has been used with a satisfactory result as discussed in [32].

However, in a typical case in lightning surge analysis, there are few thousand elements and a few hundred calculation points. Over 80 percent in the process is spent for the formulation of the impedance matrix and inversion. Therefore, the MOR is not an effective way to increase time efficiency of the method in this case.

As it is well known, the transmission line approach assuming TEM mode propagation is still effective for the horizontal conductor close to a ground plane because the electric field parallel to the line is not dominant. The PEEC method transforms the MPIE to the circuit domain, circuit elements and also a transmission line model can be implemented readily with the method. Fortunately, in the case of a lightning surge analysis in electrical power systems, most of typical elements can be represented as horizontal wires close to a ground plane. Therefore, adopting the transmission line model with the PEEC method is an effective way to increase time efficiency. In section 7.7.1, the full-wave PEEC method and the PEEC method neglecting a retardation effect in the time domain adopting the transmission line model is investigated for the lightning surge analyses.

7.5   Stability of PEEC models

The instability of the discretized electric field integral equation (EFIE) and/or MPIE formulations for the solution in the time domain is well known [30]. An appropriate PEEC model with center-to-center retardation is described by a linear, ordinary delay differential equation (LODDE) system of which instability is also well known and depends on the delay. Reference [30] shows that by using a numerical method such as backward Euler (BE) or Lobatto III-C, one can obtain a stable time-domain solution for a stable PEEC model. In fact, the instability is observed in the solution of a PEEC model in the time domain and also in the frequency domain. The instability is observed as spurious resonances of large amplitudes.

There are some measures to mitigate an instability problem in PEEC models. In this section two measures are explained. The other measures for mitigation of instability can be found in [26].

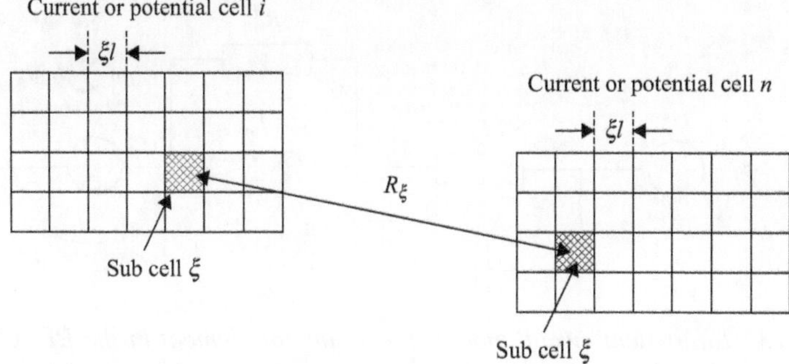

Figure 7.12   Extended current element in the PEEC model

7.5.1   +PEEC formulation

The most important addition to stabilize PEEC models is the +PEEC formulation detailed in [36]. The +PEEC formulation suggests a refined calculation of the partial elements as illustrated in Figure 7.12 to obtain a better phase response in the +PEEC model than in the basic PEEC model. The formulation utilizes a partitioning of the current and potential elements for both self and partial mutual element (an inductance and a potential coefficient). This results in a new partial inductance calculation according to

$$L_{in}^{+}(j\omega) = \frac{\mu \cos \theta_{in}}{a_i a_n} \sum_{\nu=1}^{N_n} \sum_{\mu=1}^{N_i} e^{-\gamma d_{\nu\mu}} \int\limits_{v_\nu} \int\limits_{v_\mu} g(\mathbf{r},\mathbf{r}')dv_\nu dv_\mu \tag{7.73}$$

The maximum frequency f_{max} is changed to the extended frequency f_e which is related to f_{max} and the maximum number of subdivision k.

$$f_e = kf_{max} \tag{7.74}$$

The extended partial coefficient can be calculated in the similar manner as the extended inductance. This approach gives a more stable PEEC model without introducing additional number of unknowns, but a number of frequency steps have to be increased.

7.5.2   Parallel damping resistors

To further stabilize PEEC models in the extended frequency range, a parallel damping resistor (R_n^d) can be used in conjunction with the +PEEC formulation. The basic PEEC model including a parallel damping resistor is shown in Figure 7.13.

The inclusion of the parallel resistance, R_n^d in Figure 7.13, introduces losses in the PEEC model that can improve the stability without increasing the number of

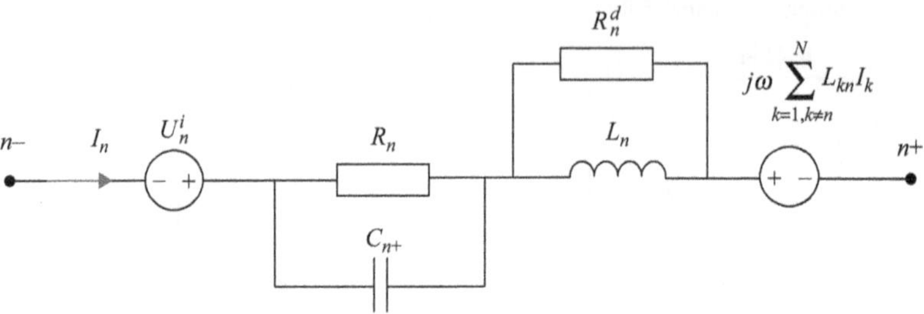

Figure 7.13   Equivalent circuit models for a current element in the PEEC model including a damping resistor

unknowns and a number of frequency steps. The value of the parallel damping resistance is calculated for each element according to

$$R_n^d = k\omega_{max}L_n \tag{7.75}$$

where $\omega_{max} = 2\pi f_{max}$ and $k = 10$ to 100 depending on the problem [36].

However, the parallel resistance leads to over-damped results if a constant K, which is evaluated case by case, is not selected suitably.

7.6   Electromagnetic field calculation by the PEEC model

Electromagnetic fields generated from a small thin wire and calculated by the PEEC method are discussed in this section. Only the longitudinal current (flowing in the conductor) I is enough to calculate electromagnetic fields in the conventional method [37]. To confirm the existence of a transverse current I_T, which is produced by charges on the conductor and can be derived from the continuity equation in the PEEC model, another approach of electromagnetic field calculation is proposed. This comparison of results is calculated by the proposed method and the conventional method can be found in [15]. Good agreement between them is found.

A charge element (a potential element) and a current element in PEEC models are interleaved each other. For simplicity, an element of a line charge or a line current is assumed laying out at origin as illustrated in Figure 7.14. From the distribution of the longitudinal and transversal currents along the conductor, the scalar potential ϕ and magnetic vector potential A from a small element are calculated by (7.76) and (7.77).

$$\phi = \frac{I_{Ti}}{4\pi l_i(\sigma + j\omega\varepsilon)} \, ln\left(\frac{\sqrt{(z + l_i/2)^2 + \rho^2} + z + l_i/2}{\sqrt{(z - l_i/2)^2 + \rho^2} + z - l_i/2}\right) e^{-\gamma R} \tag{7.76}$$

$$A = \frac{j\omega\mu I_i}{4\pi} \, ln\left(\frac{\sqrt{(z + l_i/2)^2 + \rho^2} + z + l_i/2}{\sqrt{(z - l_i/2)^2 + \rho^2} + z - l_i/2}\right) e^{-\gamma R}\boldsymbol{u}_z \tag{7.77}$$

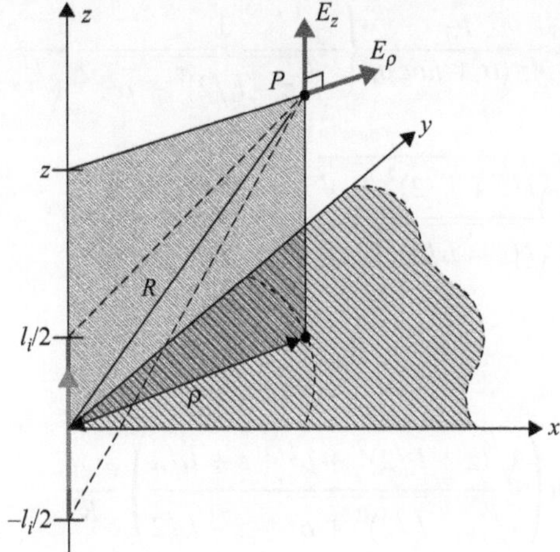

Figure 7.14   Small element in a considered media

Total electric field can be calculated by contribution of a charge or a scalar potential, and contribution of a moving charge (current) or a magnetic vector potential is given in the following equation.

$$E = E_c + E_A \tag{7.78}$$

with

$$E_c = -\nabla\phi \tag{7.79}$$

and

$$E_A = -j\omega A \tag{7.80}$$

From the configuration illustrated in Figure 7.14, (7.78), (7.79), and (7.80), the electric fields can be expressed in the following form:

$$E_c = E_{c\rho}u_\rho + E_{cz}u_z \tag{7.81}$$

$$E_A = E_{Az}u_z \tag{7.82}$$

with

$$E_{c\rho} = -\frac{\partial\phi}{\partial\rho} = \frac{I_{Ti}}{4\pi(\sigma + j\omega\varepsilon)l_i}\left\{\frac{1}{\rho}\left(\frac{z + l_i/2}{\sqrt{(z + l_i/2)^2 + \rho^2}} - \frac{z - l_i/2}{\sqrt{(z - l_i/2)^2 + \rho^2}}\right)\right.$$
$$\left. + \gamma\ln\left(\frac{\sqrt{(z + l_i/2)^2 + \rho^2} + z + l_i/2}{\sqrt{(z - l_i/2)^2 + \rho^2} + z - l_i/2}\right)\frac{\rho}{R}\right\}e^{-\gamma R} \tag{7.83}$$

$$E_{cz} = -\frac{\partial \phi}{\partial z} = \frac{I_{Ti}}{4\pi(\sigma + j\omega\varepsilon)l_i} \left\{ \frac{1}{\sqrt{(z - l_i/2)^2 + \rho^2}} - \frac{1}{\sqrt{(z + l_i/2)^2 + \rho^2}} \right.$$

$$\left. + \Gamma ln \left(\frac{\sqrt{(z + l_i/2)^2 + \rho^2} + z + l_i/2}{\sqrt{(z - l_i/2)^2 + \rho^2} + z - l_i/2} \right) \frac{z}{R} \right\} e^{-\gamma R}$$

$$(7.84)$$

and

$$E_{Az} = -\frac{j\omega I_i}{4\pi} ln \left(\frac{\sqrt{(z + l_i/2)^2 + \rho^2} + z + l_i/2}{\sqrt{(z - l_i/2)^2 + \rho^2} + z - l_i/2} \right) \frac{e^{-\gamma R}}{R} \qquad (7.85)$$

where u_z and u_ρ are unit vectors corresponding to z and ρ directions as illustrated in Figure 7.14.

Assuming $R \gg l_i$, (7.80) to (7.82) can be simplified as

$$E_{c\rho} = \frac{I_T e^{-\gamma R}}{4\pi(\sigma + j\omega\varepsilon)\, l_i} \left\{ \frac{4l_i}{4R^2 - l_i^2} + \Gamma ln \left(\frac{2R + l_i}{2R - l_i} \right) \right\} \frac{\rho}{R} \qquad (7.86)$$

$$E_{cz} = \frac{I_T e^{-\gamma R}}{4\pi(\sigma + j\omega\varepsilon)l_i} \left\{ \frac{4l_i}{4R^2 - l_i^2} + \Gamma ln \left(\frac{2R + l_i}{2R - l_i} \right) \right\} \frac{z}{R} \qquad (7.87)$$

and

$$E_{Az} = -\frac{j\omega I_i}{4\pi} ln \left(\frac{2R + l_i/2}{2R - l_i/2} \right) \frac{e^{-\gamma R}}{R} \qquad (7.88)$$

The verification of the formulae in (7.86) to (7.88) under the assumption of $R \gg l_i$ is confirmed by the calculated results in [15]. It is not applicable to an electric field calculation in the case of the considered point close to the conductor surface.

Magnetic field intensity is also calculated by

$$\boldsymbol{H} = \frac{1}{\mu} \nabla \times \boldsymbol{A} \qquad (7.89)$$

$$H_\phi \approx -\frac{1}{\mu} \left(\frac{\partial A_z}{\partial \rho} \right) = \frac{I_L e^{-\gamma R}}{4\pi} \left\{ \frac{4l_i}{4R^2 - l_i^2} + \Gamma ln \left(\frac{2R + l_i}{2R - l_i} \right) \right\} \frac{\rho}{R} \qquad (7.90)$$

The total fields are obtained by the summation of the fields produced by all elements including the elements from the image method in the case of the plane half space.

7.7 Application examples

In this section, the applications of the partial element equivalent circuit (PEEC) method in the time and frequency domains for calculating tower surge responses and insulator voltages of an actual transmission tower, and transient performance of grounding systems are presented. Incorporation of nonlinear elements such as a flashover model and a soil ionization model and a transmission line model [38] with the PEEC method in the time domain is presented. Neglecting of electromagnetic retardation effect in the PEEC method in the time domain and incorporation of a transmission line model are able to increase efficiency of the method. Comparisons of the simulation results by the proposed methods in the time and frequency domains with other numerical methods and with available experimental data show satisfactory agreement.

7.7.1 Surge characteristics of transmission towers

Experiment on an actual transmission tower was selected as a test case. The PEEC methods are applied to determining the tower surge response of a 500-kV double-circuit transmission line and voltages across insulators and are compared with experimental results reported by Ishii et al. [39].

In the actual tower case, the towers composed of slant elements and crossarms including overhead ground wires and phase wires are considered in the simulation. The configuration of the experiment on an actual system of a 500-kV double circuit transmission line is illustrated in Figure 7.15(a). The simulated system is composed

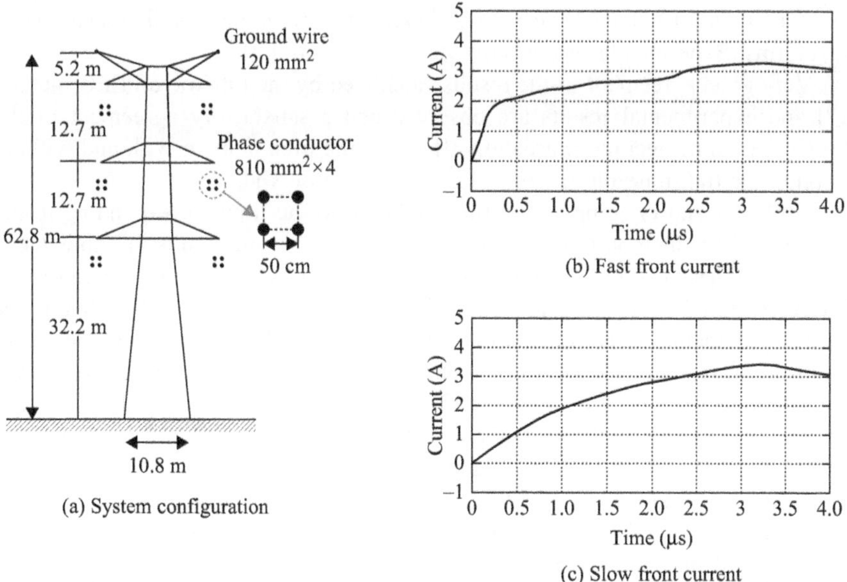

Figure 7.15   System configuration and applied current waveforms

of three towers with separation distances of 450 m and 560 m between towers, two overhead ground wires, and six-phase wires. A pulse current is applied to the top of the middle tower.

To measure the insulator voltages, two current pulses characterized by different risetimes were applied. The first pulse (Figure 7.15(b)) has about 0.2 μs risetime (fast front current) and the second (Figure 7.15(c)) has a time to crest of 3 μs. The peak amplitude of the two pulses is 3.4 A. To measure voltages across insulator strings 10-kΩ resistive voltage dividers were employed.

The full-wave PEEC simulation in the frequency domain involves 256 frequencies up to 5 MHz and 5-m element length. The tower is composed of four main poles of which elements have 0.2 m radius; slant and horizontal elements have 0.1 m radius. The crossarms are composed of 0.2-m radius elements. To increase the efficiency of the PEEC method with incorporation of the transmission line model, the overhead conductors are modeled using a lossless eight-conductor transmission line (six-phase wires and two ground wires). The same conditions are employed in the PEEC simulation neglecting retardation effect in the time domain with a 16-ns time step.

In a case of incorporation with a transmission line model, the towers and parts of the transmission lines close to the tower are modeled by the PEEC method for taking into account the retardation of electromagnetic fields around the tower by self and mutual impedances of the elements. The total number of elements without the use of the transmission line model is 2,488 while the total number of elements when using the transmission line model is reduced to 604.

In the simulation, the tower-footing resistance is represented by a 17-Ω resistance by connecting four 68-Ω resistors at the bottom of four main poles of the tower. Figures 7.16 and 7.17 show a comparison between measured and simulated waveforms. Experimental waveforms were taken from [39].

Very good agreement of the results calculated by the full-wave PEEC method [17] and experimental results are observed and a satisfactory agreement of the PEEC method neglecting retardation effects in the time domain with and without adopting the transmission line model [38] is also observed.

Table 7.1 shows computation times relative to the time corresponding to the full-wave PEEC method in the frequency domain. From the results in Table 7.1, it can be seen that the computation efficiency of the PEEC method neglecting retardation effect in the time domain is much better than that of the PEEC method in the frequency domain. Furthermore, the incorporation with the transmission line model for the representation of overhead conductors is very effective to increase the computational efficiency of the method without losing accuracy. In the test cases, the computation time of the time-domain PEEC method is less than 1 percent of the computation time of the frequency-domain PEEC method.

Next, the accuracy of the PEEC method neglecting retardation effect in the time domain is tested with a model of a nonlinear element. To demonstrate the application of the method incorporated with a flashover model [40] in analyzing voltages across insulator, numerical analyses on a double-circuit 500-kV transmission line have been carried out by considering one-phase back-flashover for the

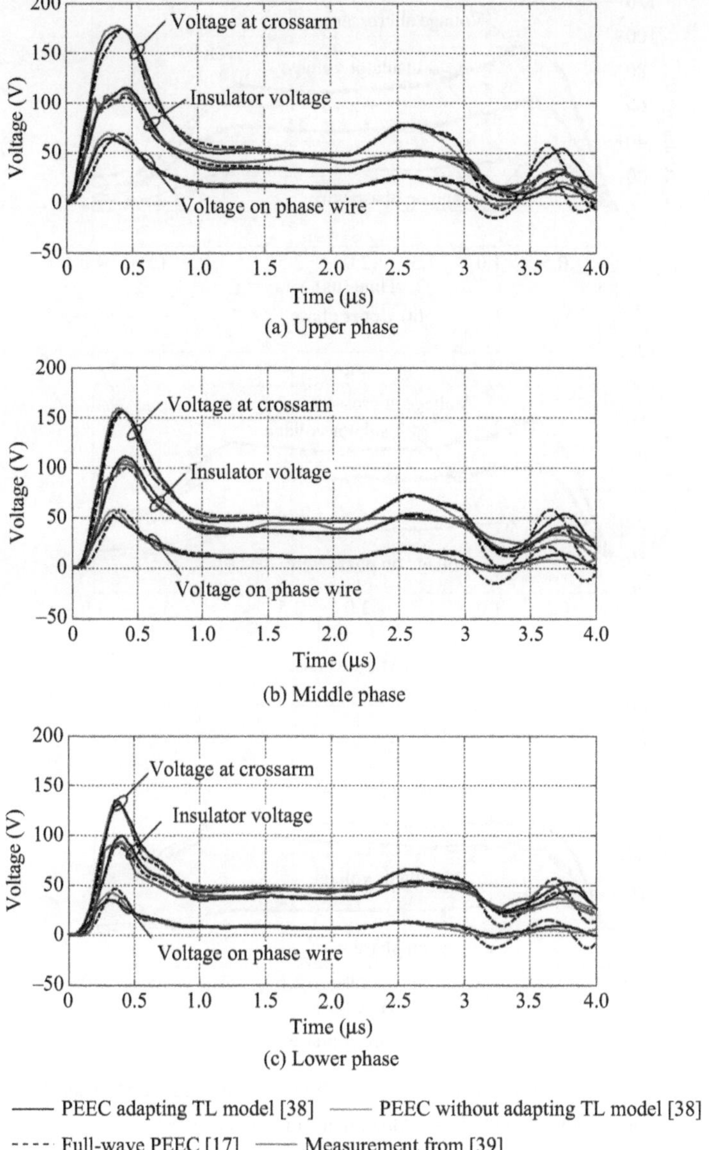

Figure 7.16   Comparison between simulated and measured waveforms, fast front current injected

sake of demonstration of nonlinear resistance corporation and comparison with thin wire time-domain (TWTD) simulation.

The accuracy of TWTD code in analyzing a full-scale transmission tower was investigated by comparison with the results obtained by the Numerical Electromagnetics Code (NEC-2) [31] and experimental results. The difference was found

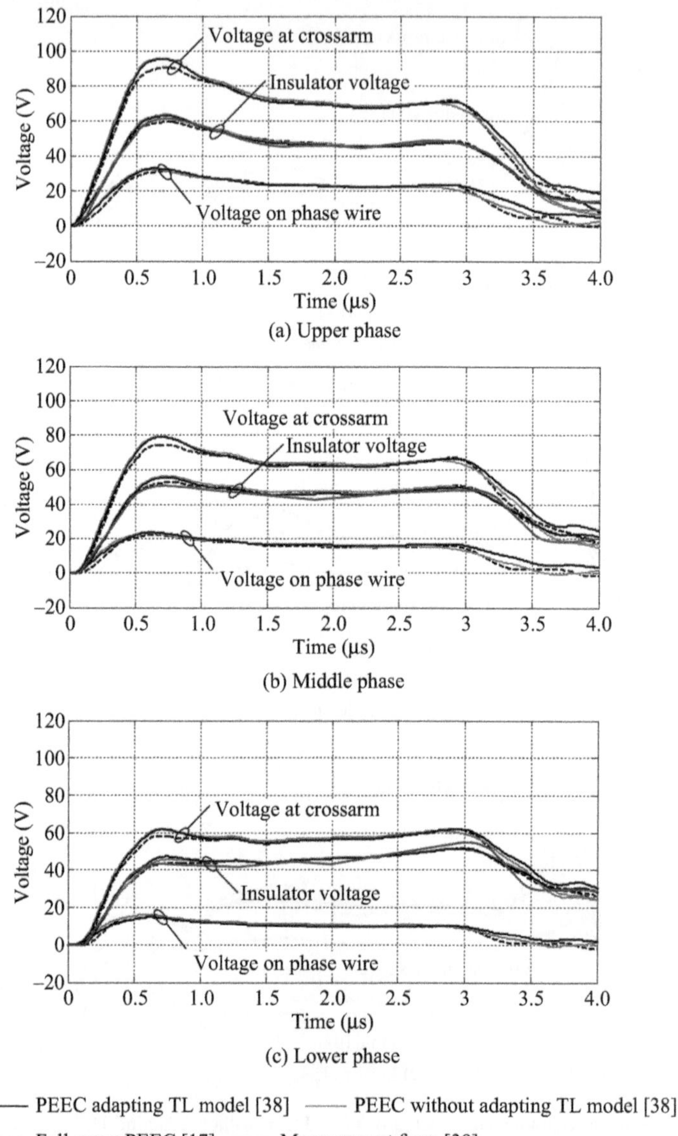

Figure 7.17   Comparison between simulated and measured waveforms, slow front current injected

to be less than 5 percent. The TWTD code has also been applied to analyzing insulator voltages incorporated with Motoyama's flashover model [40] by Mozumi et al. in [41]. The same model has been analyzed by the PEEC method.

Figure 7.18 illustrates the structure to be analyzed, which simulates a 500-kV twin-circuit transmission tower. It includes four crossarms and slanted bars, as well

Table 7.1   Comparison of computation time base relative to computation time of full-wave PEEC method

PEEC in the time domain	Formulation of an impedance matrix and inversion (p.u.)	Calculation of result by multiplication (p.u.)	Total (p.u.)
Without adopting transmission line model	5.56×10^{-3}	5.00×10^{-4}	6.06×10^{-3}
With adopting transmission line model	3.33×10^{-4}	6.66×10^{-5}	4.00×10^{-4}

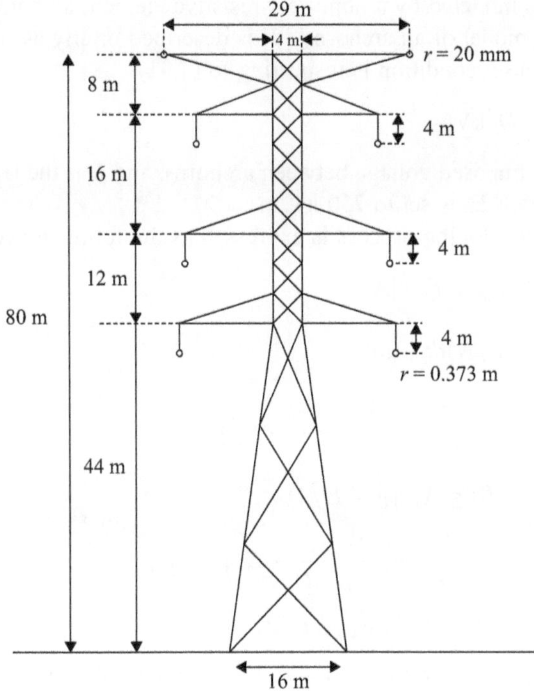

Figure 7.18   Structure of a tower subject to analysis

as four main poles. Each main pole has a variable radius of 0.1 m at the top and 0.35 m at the bottom. Each main pole is connected to a perfectly conducting ground through 40-Ω resistors to realize a total tower footing resistance of 10 Ω. The radius of elements composing crossarms and slanted bars are set to 0.1 m. A conductor of 0.1 m in radius and 400 m in length, simulating a lightning channel, is stretched vertically from the tower top. A pulse generator, the internal impedance of which is 5 kΩ, is inserted between the tower top and the lower end of the simulated lightning channel. Two parallel shield wires of 20 mm in radius and 800 m in length are stretched horizontally, and they are attached to the edges of shield-wire crossarm at their centers. A phase wire, which is actually composed of six bundle-conductors with radius 14.25 mm, is modeled by an equivalent single conductor of 0.373 m in radius. To save the computation time, only one three-phase circuit is considered with two shield wires, and the other circuit was removed from the model. The phase wires are in parallel with shield wires, and their length is 800 m. Each phase wire is suspended 4 m under the edge of the relevant crossarm. The ends of shield wires and phase wires are open. These terminating conditions do not affect the phenomena at the tower during the time of the first 2.6 μs.

For the numerical analysis, the conductors of the system are divided into cylindrical segments of about 4 m in length. The time increment is set to 20 ns. In these simulations, a 1.0 μs ramp current of 150 kA in magnitude is injected, and a back-flashover is modeled by a nonlinear resistive element as stated below.

A flashover model of an archorn [40] is described briefly as follows.

The leader onset condition is respecting to (7.91)

$$V(t) \geq E_0 \cdot D \, [\text{kV}] \tag{7.91}$$

where $V(t)$ is the imposed voltage between archorns, and D is the length of a gap in meter. The constant E_0 is set to 750 kV/m.

The leader-developing process is expressed as the following equations.

$$I_L(t) = 2K_0 v_{LAVE}(t) \quad [\text{A}] \tag{7.92}$$

$$X_{LAVE} = \int v_{LAVE}(t)dt \quad [\text{m}] \tag{7.93}$$

$$v_{LAVE} = \begin{cases} K_{10} \cdot \left(\dfrac{V(t)}{D - 2X_{LAVE}(t)} - E_0 \right) \\ \qquad (0 \leq X_{LAVE} \leq D/4) \\ K_{11} \cdot \left(\dfrac{V(t)}{D - 2X_{LAVE}(t)} - E' \right) + v'_{LAVE} \\ \qquad (D/4 \leq X_{LAVE} \leq D/2) \end{cases} \quad [\text{m/s}] \tag{7.94}$$

where I_L is the leader current; X_{LAVE} is the average value of leader-developing length; and v_{LAVE} is that of leader-developing velocity. The constants K_0, K_{10}, and K_{11} are set to 410 μC/m, 2.5 m^2/V s, and 0.42 m^2/V s, respectively. E' is the value of $V(t)/(D - 2X_{LAVE})$ when X_{LAVE} is $D/4$, and v'_{LAVE} is that of v_{LAVE} for $X_{LAVE} = D/4$.

The breakdown occurs when X_{LAVE} attains $D/2$. If the imposed voltage $V(t)$ is less than $E_0(D - 2X_{LAVE})$ during a leader-developing process, the leader is considered to stop its development.

The dynamic characteristic of an arcing horn explained above is represented by a nonlinear resistive element [40, 41], inserted between the edge of the relevant crossarm and the phase conductor. The instantaneous resistance value of this resistive element is determined as the ratio of $V(t)$ to $I_L(t)$. Before the leader onset, the resistance value is set to 1 MΩ, and it is set to 0.1 Ω after the breakdown.

Figure 7.19 shows comparison of the PEEC-computed results and the TWTD-computed results of back-flashover voltages on the upper phase, middle phase, and lower phase, respectively. It can be seen that the PEEC results are in good agreement with TWTD ones.

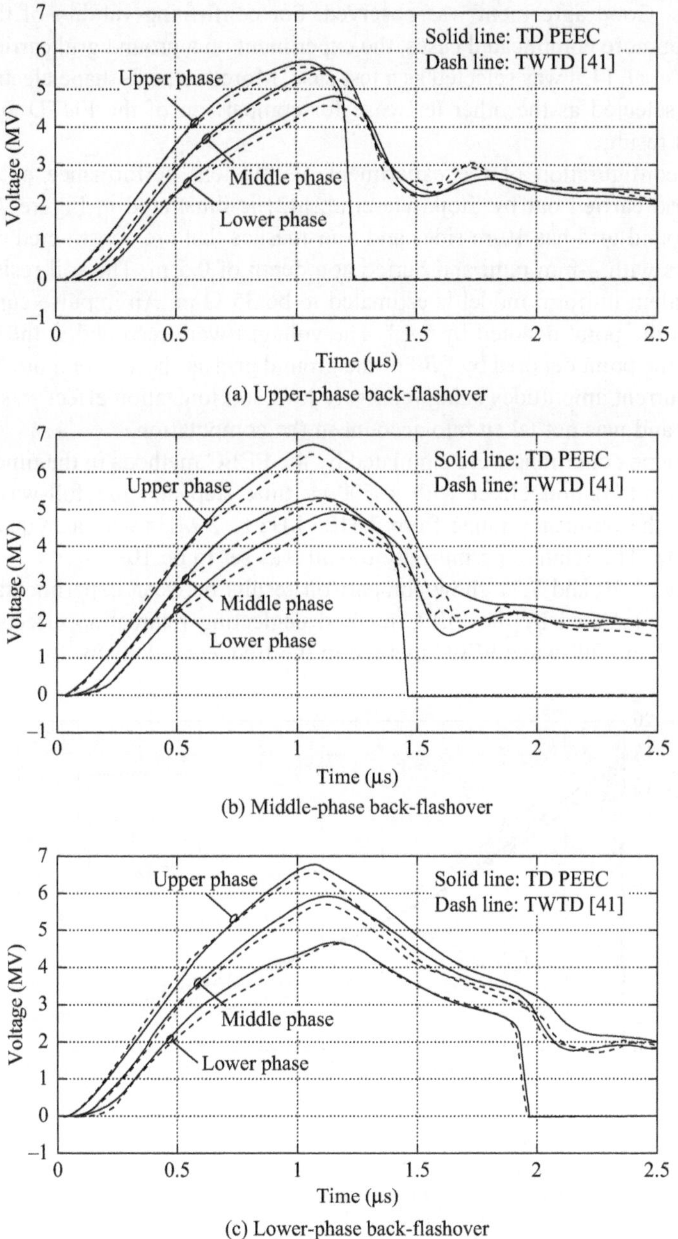

(a) Upper-phase back-flashover

(b) Middle-phase back-flashover

(c) Lower-phase back-flashover

Figure 7.19 *Waveforms of archorn voltages computed by the PEEC or by TWTD methods, 150-kA, 1.0-μs ramp current injection*

7.7.2   *Surge characteristics of grounding systems*

In this subsection, the PEEC methods either including or neglecting retardation effect are employed for analyzing transient performance of grounding electrodes. In [38], the PEEC methods have been validated with simple structures of grounding electrodes. Good agreement was observed. For confirming validity of the PEEC methods in more complicated cases, the experiment on a ground grid carried out by Stojkovic et al. [42] was selected as a test case. Moreover, a T-shape electrode [43] was also selected as the other test case for comparison of the FDTD and PEEC computed results.

The configuration of the experiment of transient performance of a square ground grid carried out by Stojkovic et al. [42] is illustrated in Figure 7.20. The square ground grid has 10-m sides and four meshes that are constructed of copper conductors with 4-mm radii and buried at a depth of 0.5 m. The soil resistivity of the equivalent uniform model is estimated to be 35 Ω·m. An impulse current was applied to the point denoted by "*A.*" The voltages were recorded at the injection point and the point denoted by "*B*" of the ground grid as shown in Figure 7.20. The injected current amplitudes being relatively low, soil ionization effect was unlikely to appear and was not taken into account in the computation.

The same conditions were simulated by the PEEC methods in the time domain neglecting retardation effect with a 500-ns time step and the full-wave PEEC method in the frequency range from 7.8125 kHz to 2 MHz with a frequency step 7.8125 kHz. The relative permittivity of soil was set to be 10.

Figures 7.20 and 7.21 show comparison results between experimental results and those calculated by the PEEC method neglecting retardation effects in time domain and the full-wave PEEC method in the frequency domain.

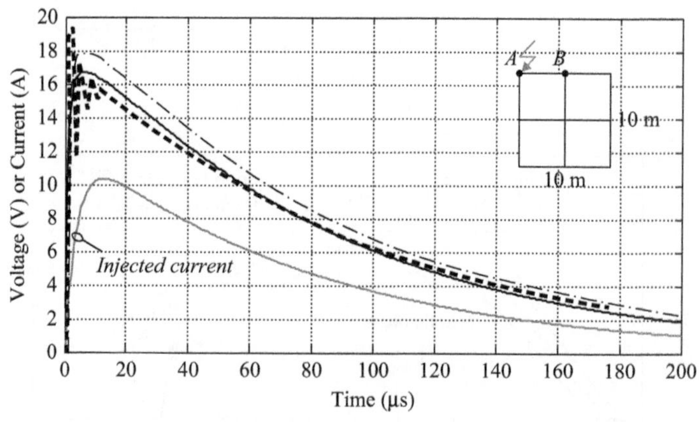

---- Measurement [32]
—— Computed by the full-wave PEEC method
—·— Computed by the PEEC method neglecting retardation effect

Figure 7.20   Comparison of measured and PEEC-computed transient potentials at point A of the square ground grid

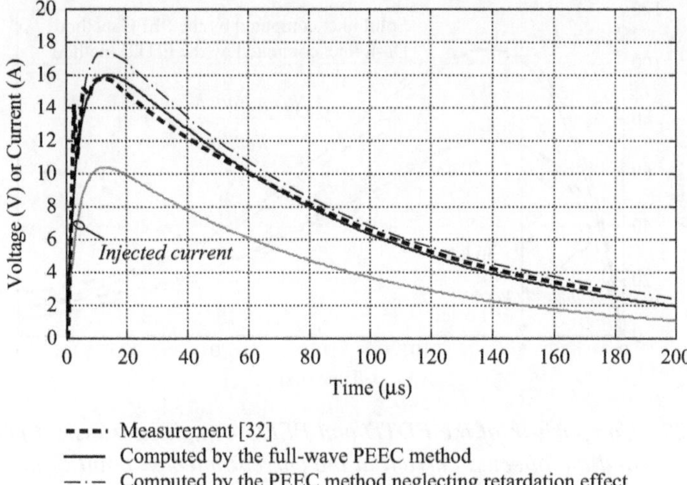

Figure 7.21   Comparison of measured and PEEC-computed transient potentials at point B of the square ground grid

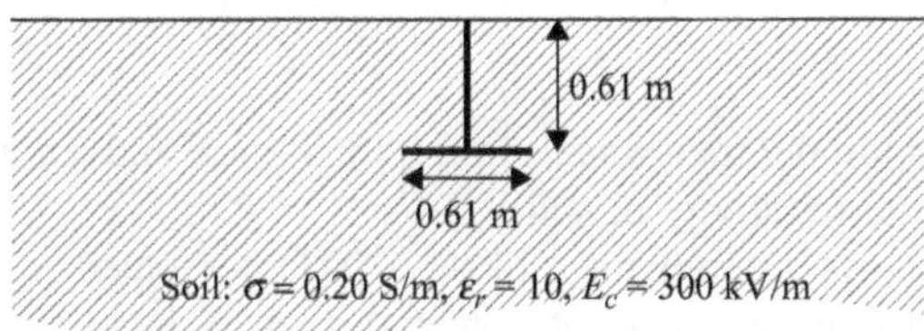

Figure 7.22   Configuration of the T-shape grounding electrode and soil parameters

In addition, we compare our results computed by the PEEC method neglecting the retardation effect with the FDTD-computed result in [43]. The T-shape electrode being in consideration is composed of perfect conductors with 14-mm radii as shown in Figure 7.22. The conductivity and relative permittivity of soil are set to be 0.02 S/m and 10, respectively. The same configuration and conditions were simulated by the PEEC method neglecting retardation in the time domain with a 200-ns time step. The current was injected at the top end of the electrode. In this case, the injected current is high enough to initiate soil ionization. Therefore, soil ionization model was included in the simulation.

During the soil ionization process a longitudinal current along the elements still flows inside the conductors, and thus it is estimated that the series self and mutual impedances of the elements are not changed and only the shunt admittances of the elements are affected by the soil ionization. During the soil ionization process, the radii of the elements are varied when the transverse electric field on the surface of the element is over the critical electric field strength of soil (E_c). In the

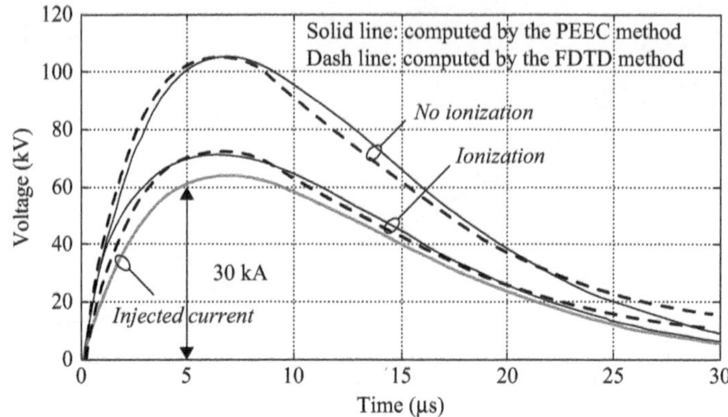

Figure 7.23   Comparison of the FDTD and PEEC-computed transient potentials and the injected current at the current injected point of the T-shape grounding electrode

simulation E_c was set to be 300 kV/m. The transverse electric field on the boundary of the ionized zone can be calculated by the following equation [44].

$$\frac{I_{Ti}}{2\pi r_i l_i} = \frac{E_c}{\rho} \tag{7.95}$$

where I_{Ti}, r_i, and l_i are the transverse current, the radius, and length of ith element, respectively. ρ is the resistivity of soil.

Figure 7.23 shows a comparison of the FDTD-computed results [43] and the PEEC computed results of transient potential of a T-shape electrode when soil ionization is neglected and included in the simulation. The figure shows a satisfactory agreement of the calculated results by the PEEC method with the FDTD-computed results collected from [43].

In fact the PEEC method neglecting the retardation effect is based on an assumption of quasi-static electromagnetic fields. The dimensions of the considered system should be smaller than the minimum wavelength of the applied source. A satisfactory agreement is still observed not only for the amplitudes but also for the wave shapes in all the cases.

References

[1]   Ruehli A. E. "Equivalent circuit models for three-dimensional multiconductor systems." *IEEE Trans. Microw. Theory Tech.* 1974, vol. MTT-22(3), pp. 216–21.

[2]   Rosa E. B. "The self and mutual inductance of linear conductors." *Bulletin of the National Bureau of Standards*. 1908, vol. 4(2), pp. 301–44.

[3]   Grover F. *Inductance Calculations: Working Formulas and Tables*. Van Nostrand; New York 1946.

[4]   Hoer C., Love C. "Exact inductance equations for rectangular conductors with applications to more complicated geometries." *Journal of Research of*

the National Bureau of Standards C, Engineering and Instrumentation. 1965, vol. 69C(2), pp. 127–37.

[5]   Ruelhi A. E., Heeb H. "Circuit models for three-dimensional geometrices including dielectrics." *IEEE Trans. Microw. Theory Tech.* 1992, vol. 40(7), pp. 1507–16.

[6]   Ruehli A. E., Paul C., Garrett J. "Inductance calculations using partial inductances and macromodels." *Proceedings of the Int. Symposium on EMC.* Atlanta, GA, USA, Aug. 1995, pp. 23–28.

[7]   Ruelhi A. E., Cangellaris A. C. "Application of the partial element equivalent circuit (PEEC) method to realistic printed circuit board problem." *Proceedings of the IEEE Int. Symposium on EMC.* Denver, CO, USA, Jul. 1998, pp. 182–97.

[8]   Ruehli A. E., Pinello W. P., Cangellaris A. C. "Comparison of differential and common mode response for short transmission line using PEEC models." *Proceedings of the IEEE Topical Meeting on Electrical Performance and Electronic Packaging.* Napa, CA, USA, Oct. 1996, pp. 169–71.

[9]   Cangellaris A. C., Pinello W., Ruehli A. E. "Circuit-based description and modeling of electromagnetic noise effects in packaged low-power electronics." *Proceedings of the Int. Conference on Computer Design.* Austin, TX, USA, Aug. 1997, pp. 136–42.

[10]   Ruehli A. E., Garrett J., Paul C. R. "Circuit models for 3D structures with incident fields." *Proceedings of the IEEE Int. Symposium on EMC.* Dallas, TX, USA, Aug. 1993, pp. 28–31.

[11]   Antonini G. "Full wave analysis of power electronic systems through a PEEC state variable method." *Proceedings of the IEEE Int. Symposium on Industrial Electronics.* Italy, Jul. 2002, pp. 1386–91.

[12]   Antonini G., Cristina S. "EMI design of power drive system cages using the PEEC method and genetic optimization techniques." *Proceedings of the IEEE Int. Symposium on EMC.* Montreal, Canada, Aug. 2001, pp. 156–60.

[13]   Antonini G., Ekman J., Orlandi A. "Integral order selection rules for a full wave PEEC solver." *Proceedings of the 15th Zurich Symposium on EMC.* Zurich, Switzerland, Feb. 2003, pp. 431–36.

[14]   Antonini G., Orlandi A. "Lightning-induced effects on lossy MTL terminated on arbitrary loads: A wavelet approach." *IEEE Trans. Electromagn. Compat.* 2000, vol. 42(2), pp. 181–89.

[15]   Yutthagowith P., Ametani A., Nagaoka N., Baba Y. "Lightning induced voltage over lossy ground by a hybrid electromagnetic-circuit model method with Cooray-Rubinstein Formula." *IEEE Trans. Electromagn. Compat.* 2009, vol. 51(4), pp. 975–85.

[16]   Antonini G., Cristina S., Orlandi A. "PEEC modeling of lightning protection systems and coupling to coaxial cables." *IEEE Trans. Electromagn. Compat.* 1998, vol. 40(4-2), pp. 481–491.

[17]   Yutthagowith P., Ametani A., Nagaoka N., Baba Y. "Application of the partial element equivalent circuit method to tower surge response calculations." *IEEJ Trans. EEE.* 2011, vol. 6(4), pp. 314–30.

[18] Restle P., Ruehli A., Walker S. G. "Dealing with inductance in high-speed chip design." *Proceedings of the IEEE Int. Design Automation Conference.* New Orleans, LA, USA, Jun. 1999, pp. 904–09.

[19] Yutthagowith P., Ametani A., Nagaoka N., Baba Y. "Influence of a current lead wire and a voltage reference wire to a transient voltage on a vertical conductor." *IEEJ Trans. EEE.*, 2010, vol. 5(1), pp. 1–7.

[20] Yutthagowith P., Ametani A., Nagaoka N., Baba Y. "Influence of a measuring system to a transient voltage on a vertical Conductor." *IEEJ Trans. EEE.* 2010, vol. 5(2), pp. 221–28.

[21] Yutthagowith P., Ametani A., Nagaoka N., Baba Y. "Application of the partial element equivalent circuit method to analysis of transient potential rises in grounding systems." *IEEE Trans. Electromagn. Compat.* 2011, vol. 53(3), pp. 726–36.

[22] Antonini G. "Fast multipole method for time domain PEEC analysis." *IEEE Trans. Mobile Computing.* 2003, vol. 2(4), pp. 275–87.

[23] Ruehli A., Miekkala U., Heeb H. "Stability of discretized partial element equivalent EFIE circuit models." *IEEE Trans. Antennas. Propagat.* 1995, vol. 43(6), pp. 553–59.

[24] Ekman J. Electromagnetic modeling using the partial element equivalent circuit method. Ph.D. Thesis, The LuLea University of Technology, Sweden, 2003.

[25] Nitsch J., Gronwald F., Wollenberg G. *Radiating Non-uniform Transmission Line Systems and the Partial Element Equivalent Circuit Method.* John Wiley and Sons, Ltd, 2009.

[26] Kochetov S., Wollenberg G. "Stable and effective full-wave PEEC models by full-spectrum convolution macromodeling." *IEEE Trans. Electromagn. Compat.* 2007, vol. 49(1), pp. 25–34.

[27] Ruehli A. E. "Inductance calculations in a complex integrated circuit environment." *IBM Journal of Research and Development.* 1972, vol. 16(5), pp. 470–81.

[28] Brennan P. A., Ruehli A. E. "Efficient capacitance calculations for three-dimensional multiconductor systems." *IEEE Trans. Microwave Theory. Tech.* 1973, vol. 21(2), pp. 76–82.

[29] Ametani A. "Application of the fast Fourier transform to electrical transient phenomena." *JEEE.* 1972, vol. 4(2), pp. 277–87.

[30] Bellen A., Heeb H., Miekkala U., Ruehli A. E. "Stable time domain solutions for EMC problems using PEEC circuit models." *Proceeding of the IEEE Int. Symposium on EMC*, Chicago, IL, USA, Aug. 1994, pp. 371–76.

[31] Feldmann P., Freund R. W. "Efficient linear circuit analysis by Pade approximation via the Lanczos process." *IEEE Tranaction on Computer-Aided Design.* 1995, vol. 14(5), pp. 639–49.

[32] Odabasioglu A., Celik M., Pileggi T. "PRIMA: Passive reduce-order interconnect macromodeling algorithm." *IEEE Transaction on Computer-Aided Design.* 1998, vol. 17(8), pp. 645–54.

[33]   Bellen A., Guglielmi N., Ruehli A. "Methods for linear systems of circuit delay differential equations of neutral type." *IEEE Trans. Circuits Sys.* 1999, vol. 46(1), pp. 212–16.

[34]   Kochetov, S. V., Wollenberg, G. "Stable time domain solution of EFIE via full-wave PEEC modeling." *Proceedings of the IEEE Int. Symposium on Electromagnetic Compatibility and Electromagnetic Ecology*, Saint Petersburg, Russia, Jun. 2005, pp. 371–76.

[35]   Liniger W., Ruehli A. E. "Time domain integration methods for electric field integral equations." *Proceedings of the 1995 Zurich Symposium on EMC*, Zurich, Switzerland, Aug. 1995, pp. 209–14.

[36]   Garrett J., Ruehli E., Paul C. "Accuracy and stability improvements of integral equation models using the partial element equivalent circuit PEEC approach." *IEEE Trans. Antennas. Propagat.* 1998, vol. 46(12), pp. 1824–31.

[37]   Uman M. A., McLain D. K., Krider E. P. "The electromagnetic radiation from a finite antenna." *Amer. J. Phys.* 1975, vol. 106(D13), pp. 33–38.

[38]   Yutthagowith P., Ametani A., Rachidi F., Nagaoka N., Baba Y. "Application of a partial element equivalent circuit method to lightning surge analyses." *Electric Power System Research Journal.* 2013, vol. 94(4), pp. 30–37.

[39]   Ishii M., Kawamura T., Kouno T., Ohasaki E., Murotani K., Higushi T. "Multistory transmission tower model for lightning surge analysis." *IEEE Trans. Power Del.* 1991, vol. 6(3), pp. 1327–35.

[40]   Motoyama H. "Development of a new flashover model for lightning surge analysis." *IEEE Trans. Power Del.* 1996, vol. 11, pp. 972–79.

[41]   Mozumi T., Baba Y., Ishii M., Nagaoka N., Ametani A. "Numerical electromagnetic field analysis of archorn voltages during a back-flashover on 500-kV twin-circuit line." *IEEE Trans. Power Del.* 2003, vol. 18(1), pp. 207–13.

[42]   Stojkovic Z., Savic M. S., Nahman J. M., Salamon D., Bukorovic B. "Sensitivity analysis of experimentally determined grounding grid impulse characteristics." *IEEE Trans. Power Del.* Oct. 1998, vol. 13(4), pp. 1136–42.

[43]   Ala G., Buccheri P. L., Romano P., Viola F. "Finite difference time domain simulation of earth electrodes soil ionization under lightning surge condition." *IET Sci. Meas. Technol.* May 2008, vol. 2(3), pp. 134–45.

[44]   Geri A. "Behaviour of grounding systems excited by high impulse currents: The model and its validation." *IEEE Trans. Power Del.* 1999, vol. 14(3), pp. 1008–17.

Chapter 8

Lightning surges in renewable energy system components

K. Yamamoto*

8.1 Lightning surges in a wind turbine

In recent years, the steady increase of the number of wind turbines (WTs) has resulted in many accidents caused by natural phenomena such as lightning and typhoons. In particular, this section focuses on the extensive damage on WTs caused by lightning [1–7].

In order to benefit from better wind conditions, WTs are often constructed on hilly terrain or at the seashore where few other tall structures exist; therefore, WTs are often struck by lightning. For promoting wind power generation, lightning protection methodologies should be established for WTs.

To establish the lightning protection methodologies, a number of simulation methods have been proposed. The simulation results are compared, in general, with measured results, and the accuracy and effectiveness are verified. In this section, numerical analyses of surge phenomena on a WT are introduced.

8.1.1 Overvoltage caused by lightning surge propagation on a wind turbine

There are experimental and analytical studies for the lightning protection of WTs [8] by using a reduced-size model of a WT and the finite-difference time-domain (FDTD) method [9, 10]. This study focuses on the overvoltage observed at the wave front of lightning surges. Assuming lightning strikes at the tip of a blade and the rear portion of a nacelle, a fast-front current is injected into the points, and the experiments and the analyses are carried out in the cases of the upward and downward lightning. The voltage at the current-injected point and various voltage differences are measured with various wave-front durations of the injected current and various earthing resistances. For the analytical studies based on the measured results, the FDTD method is used.

*Chubu University, Japan

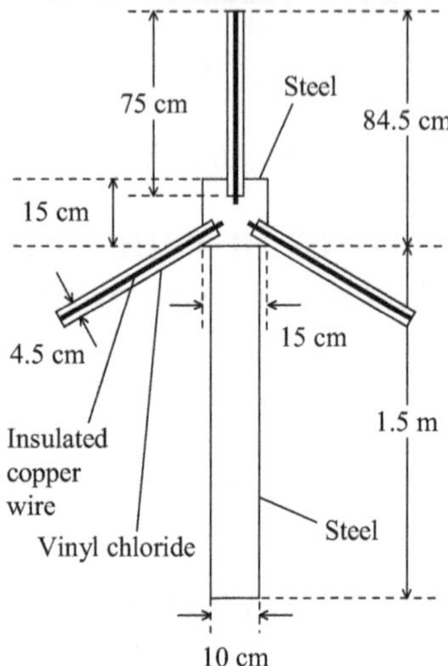

Figure 8.1   Reduced-size model of a wind turbine [8]

© Elsevier 2009. Reprinted with permission from K. Yamamoto, T. Noda, S. Yokoyama, A. Ametani, 'Experimental and analytical studies of lightning overvoltages in wind turbine generator systems', *Electric Power Systems Research*, vol. 79, no. 3, March 2009

8.1.1.1   Experimental conditions

A reduced-size model, as shown in Figure 8.1, was used in the experiments. It is a 3/100th-scale model of an actual WT with the blade length of 25 m and the tower height of 50 m. The material of the blades is vinyl chloride, and an insulated copper wire with a cross-sectional area of 2 mm^2 is traced on each blade to represent a lightning conductor. The actual tower is tapered; however, the tower of the scale model is of the tubular type with an outer diameter of 10 cm and a thickness of 3 mm. The nacelle is a metal cube with a side length of 15 cm.

The above-mentioned reduced-size model was set up as illustrated in Figure 8.2. Many aluminum plates that have a thickness of 2 mm were embedded in the earth so that the aluminum floor could be considered to be infinite in the time dimension of the measurements. The assumed point of a lightning strike is the tip of one of the blades, as shown in Figure 8.2(a), or the rear portion of the nacelle, as shown in Figure 8.2(b). In the case of the measurements for downward lightning, the return stroke propagates from the WT to the cloud. Here, the current is injected into the coaxial cable from the pulse generator and is led to the lightning-strike point. The current is injected into the reduced-size model of a WT through the resistance R_i. While the current propagates on the current lead wire of the coaxial cable, electromagnetic field exists in the coaxial cable and does not exist around the cable. After the current arrived at the lightning-strike point, the current I is injected into the lightning-strike point and the current $-I$ is reflected on the sheath of the coaxial cable. This current represents the return stroke propagating from the WT to

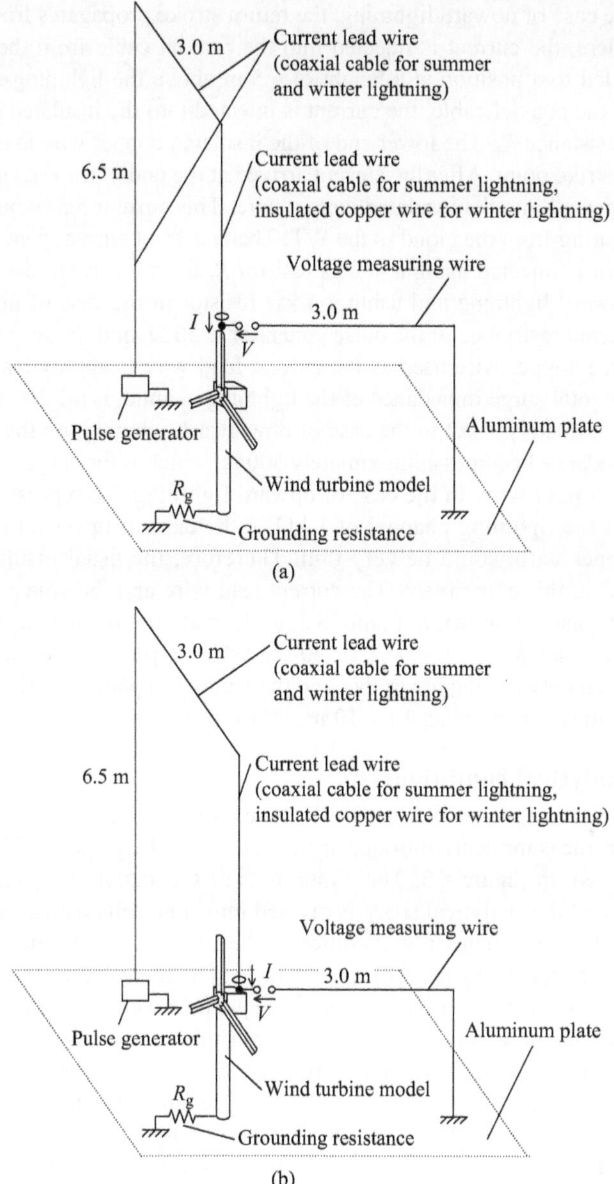

Figure 8.2   *Configurations of experimental set-ups: (a) cases of downward and upward lightning strikes on one of the blades; (b) cases of downward and upward lightning strikes at the rear portion of the nacelle [8]*

© Elsevier 2009. Reprinted with permission from K. Yamamoto, T. Noda, S. Yokoyama, A. Ametani, 'Experimental and analytical studies of lightning overvoltages in wind turbine generator systems', *Electric Power Systems Research*, vol. 79, no. 3, March 2009

a cloud. In the case of upward lightning, the return stroke propagates from the cloud to the WT. Here, the current is injected into the coaxial cable from the pulse generator and is led to a position at a height of 6.5 m above the lightning-strike point. At the end of the coaxial cable, the current is injected into the insulated copper wire through the resistance R_i. The lower end of the insulated copper wire is connected to the lightning-strike point. After the current arrived at the end of the coaxial cable, the current I is injected into the insulated copper wire. This current represents the return stroke propagating from the cloud to the WT. The fast-front current generated by the pulse generator is injected using a 480 Ω resistor R_i from a current lead wire in the case of downward lightning and using a 4 kΩ resistor in the case of upward lightning. The internal resistance of the pulse generator is 50 Ω, and the surge impedance of the insulated copper wire used as the current lead wire is approximately 500 Ω. Therefore, the total surge impedance of the lightning channel is $(480 + 50 + 500)$ Ω which is approximately 1 kΩ in the case of downward lightning; on the other hand, the surge impedance becomes approximately 500 Ω, which is the surge impedance of the insulated copper wire, in the case of upward lightning. To represent the surge impedance of the lightning channel of 1 kΩ in the case of upward lightning, the insulated copper wire would be very thin. Therefore, the usual insulated copper wire was used in this experiment. The current lead wire and the voltage measuring wire are orthogonal, as shown in Figure 8.2, to decrease the mutual electromagnetic induction. A resistor R_g (0 Ω, 9.4 Ω, or 20 Ω), which represents the earthing resistance, is inserted between the tower foot and the aluminum plate. The wave front τ of the injected current is varied as 4 ns, 10 ns, 20 ns, or 60 ns.

8.1.1.2   Analytical conditions

Using the electromagnetic field analysis based on the FDTD method we can reproduce the measurements introduced in section 8.1.1.2. The FDTD analytical spaces are shown in Figure 8.3. The dimensions of the analytical space are 6 m × 5 m × 7.5 m, and the analytical space is divided into cube cells with a side length of 2.5 cm. The absorbing boundary condition is the Liao's second-order approximation. A thin-wire model to model the current lead wire, voltage-measuring wire, and lightning conductor on a blade is used [11]. The nacelle is a cubic conductor with a side length of 15 cm, and the tower is a tube conductor with a stair-like surface. The lumped resistance representing the earthing resistance R_g is inserted between the tower foot and the aluminum plate. The current source paralleled with the resistance R_i is connected exactly above the lightning-strike point in the case of the downward lightning. In the case of the upward lightning, it is connected at a height of 6.5 m above the lightning-strike point.

8.1.1.3   Lightning surge propagation on a wind turbine

In Figures 8.4–8.7, the injected current I and the voltage V at the stroke points are shown. These values are normalized to 1 A for the peak value of the injected current.

The graphs in part (b) of Figures 8.4–8.7 show the voltages at the stroke points. Each solid line indicates the measured results and the broken line indicates the calculation results. The voltage waveforms in Figures 8.4 and 8.6 can be explained

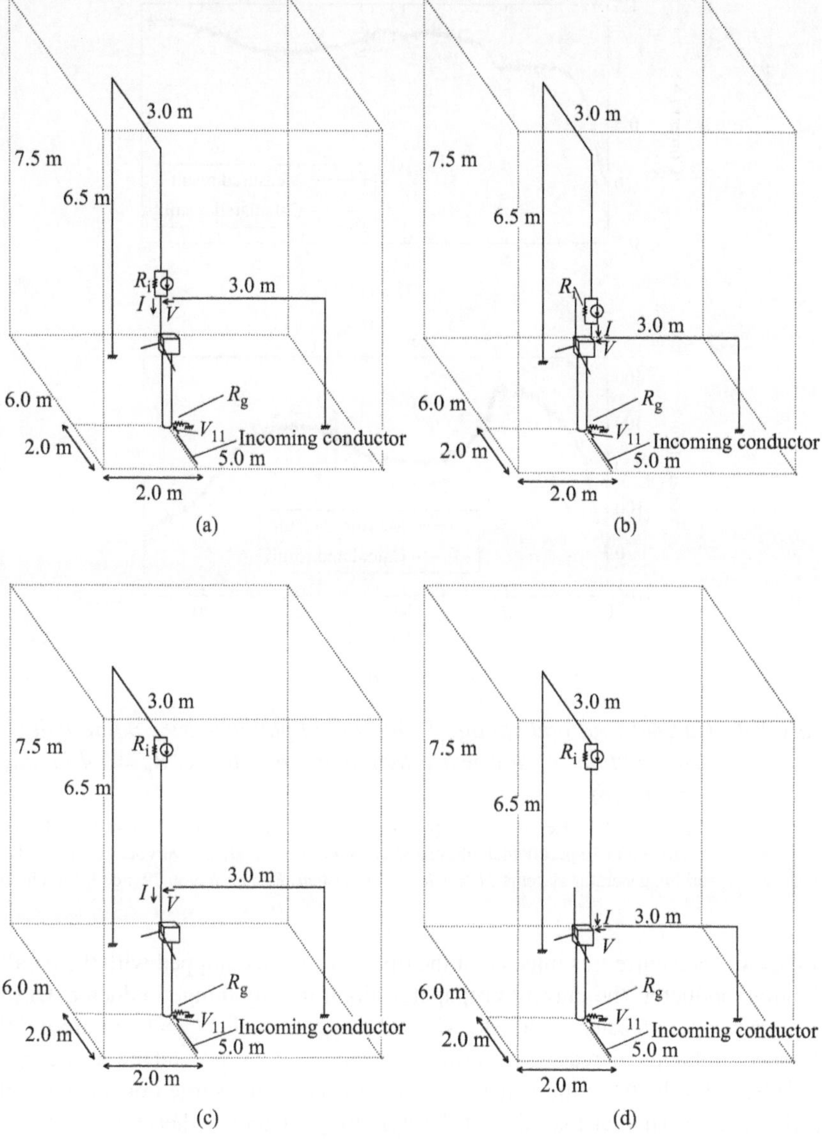

Figure 8.3   *Configurations of FDTD analyses: (a) case of a downward lightning strike on one of the blades; (b) case of a downward lightning strike at the rear portion of the nacelle; (c) case of a upward lightning strike on one of the blades; (d) case of a upward lightning strike at the rear portion of the nacelle [8]*

© Elsevier 2009. Reprinted with permission from K. Yamamoto, T. Noda, S. Yokoyama, A. Ametani, 'Experimental and analytical studies of lightning overvoltages in wind turbine generator systems', *Electric Power Systems Research*, vol. 79, no. 3, March 2009

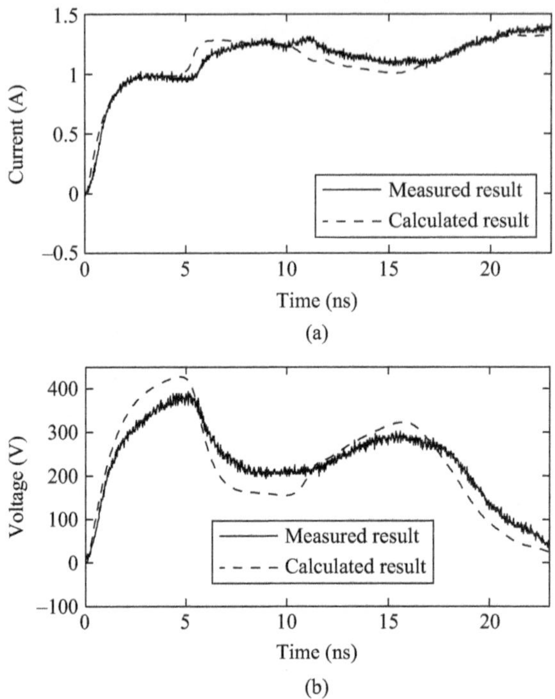

Figure 8.4   *Current I (a) and voltage V (b) at the lightning-strike point under the conditions of downward lightning, tip of blade, $R_g = 9.4\ \Omega$, and $\tau = 4\ ns$ [8]*

© Elsevier 2009. Reprinted with permission from K. Yamamoto, T. Noda, S. Yokoyama, A. Ametani, 'Experimental and analytical studies of lightning overvoltages in wind turbine generator systems', *Electric Power Systems Research*, vol. 79, no. 3, March 2009

as follows: when current is injected at the tip of the blade equipped with the parallel lightning conductor, the current propagates along the lightning conductor. During this process, the intensity of the electric and magnetic fields generated around the lightning conductor increases, resulting in an increase in the voltage at the tip of the blade. After the traveling wave reaches the nacelle, the surge impedance of the nacelle becomes smaller than that of the lightning conductor; therefore, a negative reflection is caused at the nacelle, and the negative traveling wave returns to the tip of the blade. The electric and magnetic fields around the blade start weakening; further, the voltage at the tip attains its peak value when the negative reflected wave reaches the tip again. Since the return propagation time of the blade is 5 ns, the peak of the voltage appears after approximately 5 ns. The surge impedance of the tower mainly influences the second peak of the voltage at approximately 17 ns, and the traveling waves on the lightning conductors that have an open end are superimposed on the voltage. Meanwhile, the voltage waveforms in Figures 8.5 and 8.7 can be explained as follows: as soon as current is injected into the rear portion

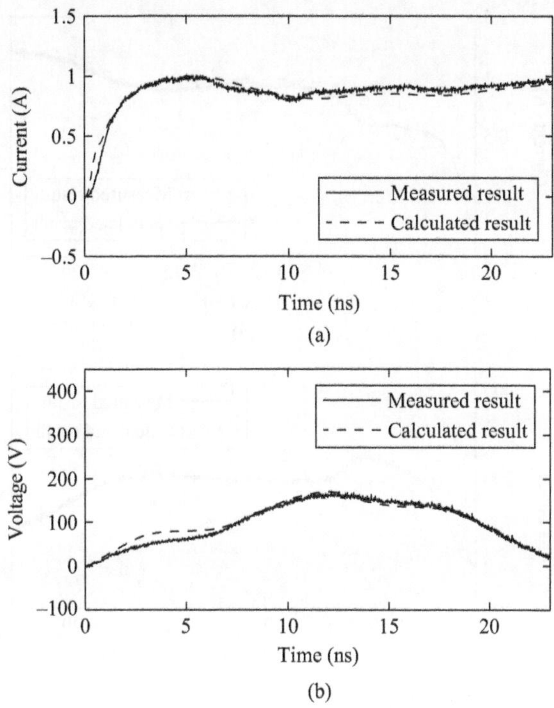

Figure 8.5   *Current I (a) and voltage V (b) at the lightning-strike point under the conditions of downward lightning, rear portion of a nacelle, $R_g = 9.4\ \Omega$, and $\tau = 4\ ns$ [8]*

© Elsevier 2009. Reprinted with permission from K. Yamamoto, T. Noda, S. Yokoyama, A. Ametani, 'Experimental and analytical studies of lightning overvoltages in wind turbine generator systems', *Electric Power Systems Research*, vol. 79, no. 3, March 2009

of the nacelle, the traveling wave disperses to the three blades and the tower. Because these surge impedances are connected in parallel, the voltage rise is small for 5 ns. After 5 ns, the positive traveling waves return to the nacelle from the tips of the blades, and the voltage at the rear portion of the nacelle starts increasing because of the current flow through the tower. The voltage rise exhibits a peak when the negative traveling wave at the tower foot reaches the nacelle (after approximately 11 ns).

In both upward and downward lightning, the shapes of the voltage curves are almost the same, as shown in Figures 8.4–8.7. However, the magnitudes in both these conditions are different because a part of the traveling wave from above is reflected to the upper current lead wire, while a part of it permeates through the WT in the case of upward lightning.

The results calculated using the FDTD for $R_g = 0\ \Omega$, 9.4 Ω, and 20 Ω are shown in Figure 8.8. The voltage waveforms obtained for $R_g = 0\ \Omega$, 9.4 Ω, and 20 Ω change after approximately 17 ns because the reflected wave at the tower foot

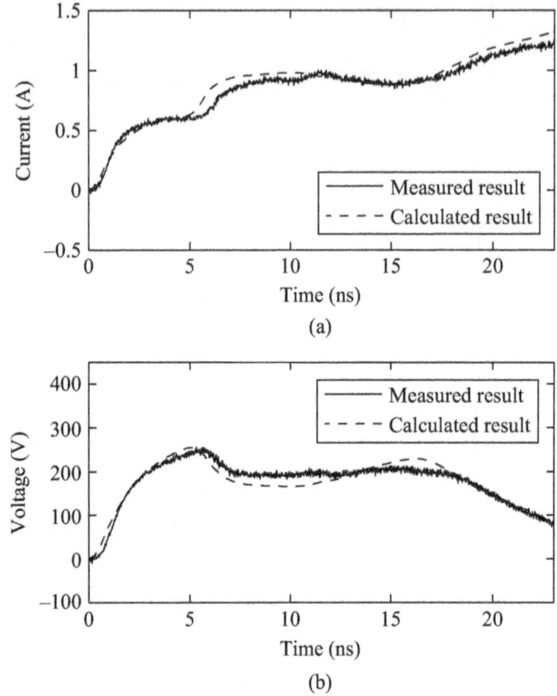

Figure 8.6   *Current I (a) and voltage V(b) at the lightning-strike point under the conditions of upward lightning, tip of blade, $R_g = 9.4\ \Omega$, and $\tau = 4\ ns$ [8]*

© Elsevier 2009. Reprinted with permission from K. Yamamoto, T. Noda, S. Yokoyama, A. Ametani, 'Experimental and analytical studies of lightning overvoltages in wind turbine generator systems', *Electric Power Systems Research*, vol. 79, no. 3, March 2009

is different in each condition. The calculation results for $\tau = 4$ ns, 10 ns, 20 ns, and 60 ns are shown in Figure 8.9. In the case of downward lightning at the top of the blade, the second peak of V becomes larger than its first peak when the wave front of the injected current becomes large.

8.1.1.4   Overvoltage caused by potential rise

V_{11} to V_{14} in Figure 8.10 represent the voltage difference between the tower foot and the incoming conductor directed from a distant point and earthed at the remote end; depending on the earthing resistance, this voltage difference is caused by the voltage increase at the tower foot. In particular, the voltage difference produces an overvoltage between a power line and a power converter or a transformer installed inside the tower on the earth level, or that between a communication line and a telecommunication device. The incoming conductor is an insulated copper wire or coaxial cable traced on the aluminum plate from a distant point and earthed at the

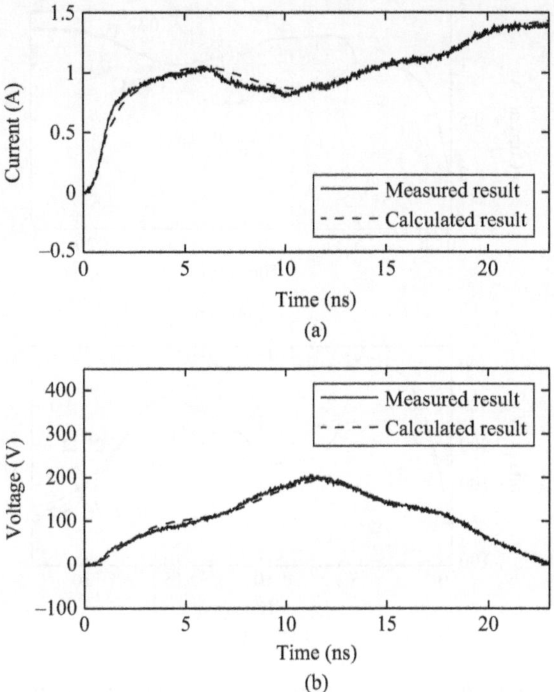

Figure 8.7   *Current I (a) and voltage V (b) at the lightning-strike point under the conditions of upward lightning, rear portion of a nacelle, $R_g = 9.4\ \Omega$, and $\tau = 4$ ns [8]*

© Elsevier 2009. Reprinted with permission from K. Yamamoto, T. Noda, S. Yokoyama, A. Ametani, 'Experimental and analytical studies of lightning overvoltages in wind turbine generator systems', *Electric Power Systems Research*, vol. 79, no. 3, March 2009

remote end. The length of the conductors is 4.5 m. The large earth capacitance of the traced conductor is comparable to that of the underground cables in actual WTs.

The experimental and calculation results are shown in Figure 8.11 under the conditions of downward lightning, top of the blade, $R_g = 9.4\ \Omega$, and $\tau = 4$ ns. In the FDTD analyses, it is difficult to model a coaxial cable because a side of the cube cells that is part of the FDTD space is not sufficiently larger than the radius of the coaxial cable. In this FDTD analysis, only V_{11} is calculated. V_{11} to V_{13} become nearly identical to the waveform shown in Figure 8.11. The same tendency is observed in the other conditions. The potential of the insulated copper wire and the metal sheath of the coaxial cable become nearly equal because they are connected to the aluminum plate at the remote end. Therefore, there is no large difference between V_{11} and V_{12}. The potential of the core of the coaxial cable becomes nearly equal to that of the metal sheath that covers the core. Therefore, the difference between V_{12} and V_{13} is not significant. In order to suppress the overvoltage between the incoming conductor and the tower foot, the metal sheath of the coaxial cable is

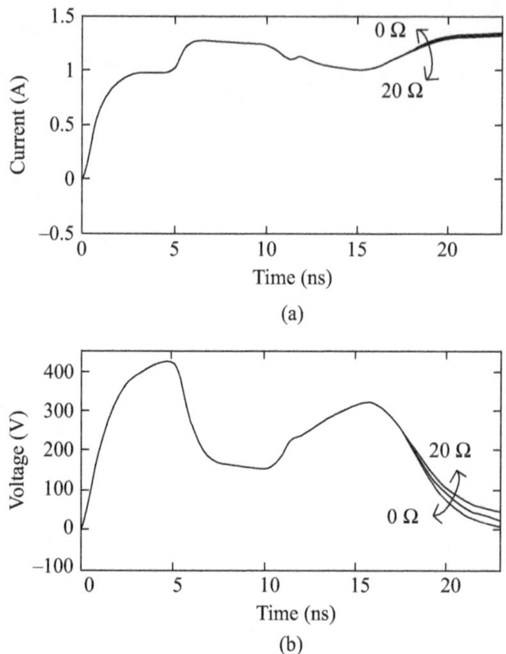

Figure 8.8   *Relations between V, I, and R_g under the conditions of downward lightning, top of the blade and $\tau = 4$ ns: (a) current I from FDTD analytical results; (b) voltage V from FDTD analytical results [8]*

© Elsevier 2009. Reprinted with permission from K. Yamamoto, T. Noda, S. Yokoyama, A. Ametani, 'Experimental and analytical studies of lightning overvoltages in wind turbine generator systems', *Electric Power Systems Research*, vol. 79, no. 3, March 2009

connected to the earth of the tower foot. The voltage difference V_{14} in such case is suppressed, 40–80%, as compared with V_{13}.

The relations between the maximum value of V_{11} to V_{13} and τ under the conditions $R_g = 0$ Ω, 9.4 Ω, and 20 Ω are shown in Figure 8.12. The relations between the maximum values of V_{11} to V_{13} and R_g under the conditions $\tau = 4$ ns, 10 ns, 20 ns, and 60 ns are shown in Figure 8.13. The maximum value of V_{11} to V_{13} decreases with an increase in τ and is proportional to R_g. Moreover, in the case of $R_g = 0$ Ω, the maximum value of V_{11} to V_{13} becomes 12 V/A, which is caused by the electromagnetic induction from the current following through the tower and aluminum plates.

The same tendency is observed in the relations between the maximum value of V_{14} and τ and that between the maximum value of V_{14} and R_g. In the case of $R_g = 0$ Ω, the maximum value of V_{14} becomes 5.9 V/A.

8.1.2   *Earthing characteristics of a wind turbine*

Damage caused to WTs due to lightning affects the safety and reliability of these systems. Most of the breakdowns and malfunctions of the electrical and control systems inside or in the vicinity of WTs are caused by earth potential rise due to

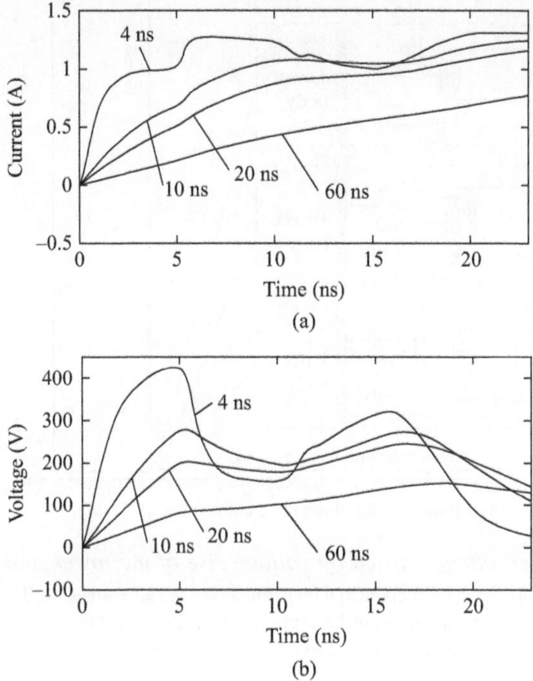

Figure 8.9 *Relations between V, I, and τ under the conditions of downward lightning, top of the blade, and $R_g = 9.4\ \Omega$: (a) current I from FDTD analytical results; (b) voltage V from FDTD analytical results [8]*

© Elsevier 2009. Reprinted with permission from K. Yamamoto, T. Noda, S. Yokoyama, A. Ametani, 'Experimental and analytical studies of lightning overvoltages in wind turbine generator systems', *Electric Power Systems Research*, vol. 79, no. 3, March 2009

lightning [4, 8, 12]. To solve the mechanism of generation of the potential rise in these multifarious earth systems, impulse tests at many kinds of WT sites are necessary [13–27].

In this section, the measurements of transient earthing characteristics carried out at WTs [13–26] are described. Some of the measured earthing characteristics have already been verified [17–26] using the FDTD method. From these compared results, it has been obvious that these earthing characteristics can be simulated by the FDTD method accurately.

8.1.2.1 Measurement sites

The measurements of earthing characteristics of WTs have been completed at the 11 sites shown in Figure 8.14. The relations between lightning current characteristic and overvoltage on electrical and electronics equipment at each site have been clarified. Sites 1–5 are in winter lightning area around the Sea of Japan and Sites 6–11 are in summer lightning area. At most of the sites, the lightning current waveforms had been recorded using Rogowski coils whose frequency band is about 0.1 Hz to 1 MHz [5–7].

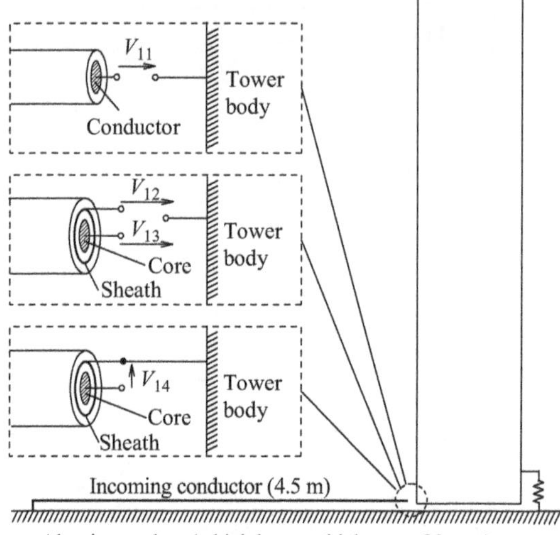

Figure 8.10   Overvoltage caused by voltage rise of the tower foot [8]

© Elsevier 2009. Reprinted with permission from K. Yamamoto, T. Noda, S. Yokoyama,
A. Ametani, 'Experimental and analytical studies of lightning overvoltages in wind
turbine generator systems', *Electric Power Systems Research*, vol. 79, no. 3, March 2009

Once an accident on electrical or electronics equipment occurs by a lightning strike,
we can make the threshold overvoltage or current on the equipment clear due to the
earthing characteristic measurements shown in this section.

8.1.2.2   Multifarious earthing systems of wind turbines

The earthing systems of WTs consist of the foundation, ring earth electrodes,
earthing meshes, foundation feet, earthing copper plates, counterpoises, and so on.
Insulating coating or bare wire is used for the counterpoise, which is sometimes
utilized for bonding between several earthing systems.

There are multifarious earthing systems of WTs as shown in Figure 8.15. The
shapes of the foundations and foundation feet are designed in consideration of
structural stability, not lightning protection for electrical and electronics equipment. However, the earthing characteristic of a WT depends on the shapes of these
foundations and foundation feet considerably. The author is afraid that there are
few earthing systems which are designed in consideration of both transient and
steady-state earthing characteristics efficiently. It should be important to establish
efficient design method to realize better earthing characteristic with cost effectiveness. It is introduced that the FDTD method is effective for such designs.

8.1.2.3   Experiments at actual wind turbines

Figure 8.16 shows one of the layouts of experimental components. In the experiment, a polyvinyl chloride wire of 5.5 mm^2 cross-sectional area was laid about
150 m from the impulse generator at a height of 1 m from the earth surface to an

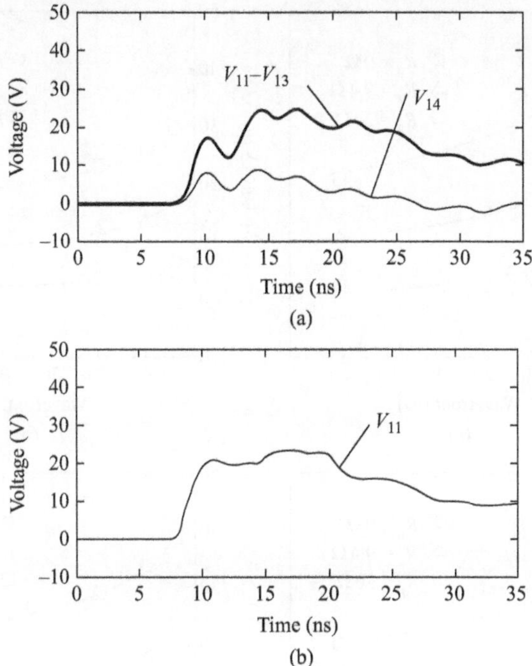

Figure 8.11   *Voltage differences between the incoming conductor and the tower foot under the conditions of downward lightning, top of the blade, $R_g = 9.4\ \Omega$, and $\tau = 4$ ns: (a) voltage V_{11}–V_{14} of measured results; (a) voltage V_{11} of FDTD analytical results [8]*

© Elsevier 2009. Reprinted with permission from K. Yamamoto, T. Noda, S. Yokoyama, A. Ametani, 'Experimental and analytical studies of lightning overvoltages in wind turbine generator systems', *Electric Power Systems Research*, vol. 79, no. 3, March 2009

area near the point of current injection (at the tower foot of the WT), and was connected to the foundation through a resistor of 500 Ω (resistance that is sufficiently greater than the earthing impedance of the WT). By injecting current into the foundation through the resister of 500 Ω, the impulse generator is regarded as a source of current and steep wave front current can be injected. The current with steep front of about 40 A in peak value and about 0.2 μs in wave front duration was injected as shown in Figure 8.17. Such steep front current contains wide frequency band components. Once a time-domain response such as a potential rise of an earthing system is measured, wide frequency characteristic can be obtained.

The measured potential rise is the potential difference between the voltage measuring wire earthed at a remote point and the foundation. Since the surge impedance of the voltage reference line is about 400 Ω, the sum of the earth resistance value of the earthing electrode at the remote end of the voltage measuring wire and the value of a matching resistor inserted between the voltage measuring wire and the earthing electrode at the remote end is adjusted to the value of 400 Ω. This enables various types of noise induced in the voltage reference line to be discharged to the earth smoothly.

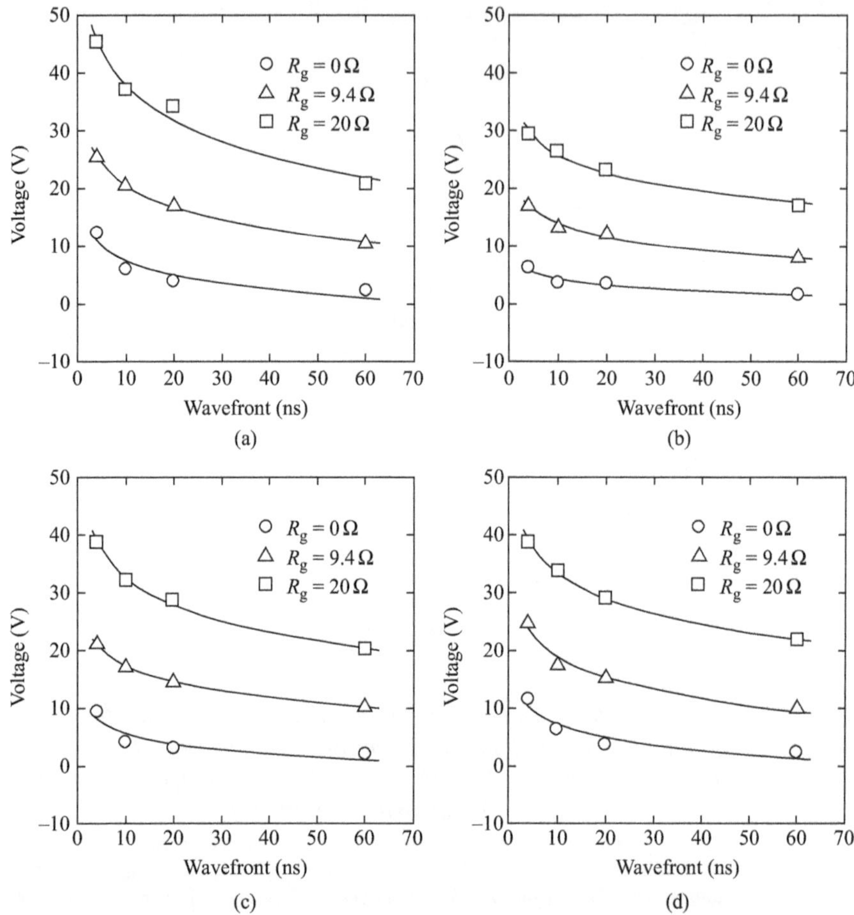

Figure 8.12   Relations between the maximum values of V_{11} to V_{13} and τ: (a) case of the downward lightning and stroke at the top of the blade; (b) case of the downward lightning and stroke at the rear portion of the nacelle; (c) case of the upward lightning and stroke at the top of the blade; (d) case of the upward lightning and stroke at the rear portion of the nacelle [8]

© Elsevier 2009. Reprinted with permission from K. Yamamoto, T. Noda, S. Yokoyama, A. Ametani, 'Experimental and analytical studies of lightning overvoltages in wind turbine generator systems', *Electric Power Systems Research*, vol. 79, no. 3, March 2009

As shown in Figure 8.16, the current lead and the voltage measuring wires are not orthogonal in some experiments because of restrictions of available space for experiments. If the angle is less than 90 degrees, inductive effects appear on the voltage measuring wire, and the measured result of the potential rise may not be correct. However, most of the results are verified using the FDTD method and we

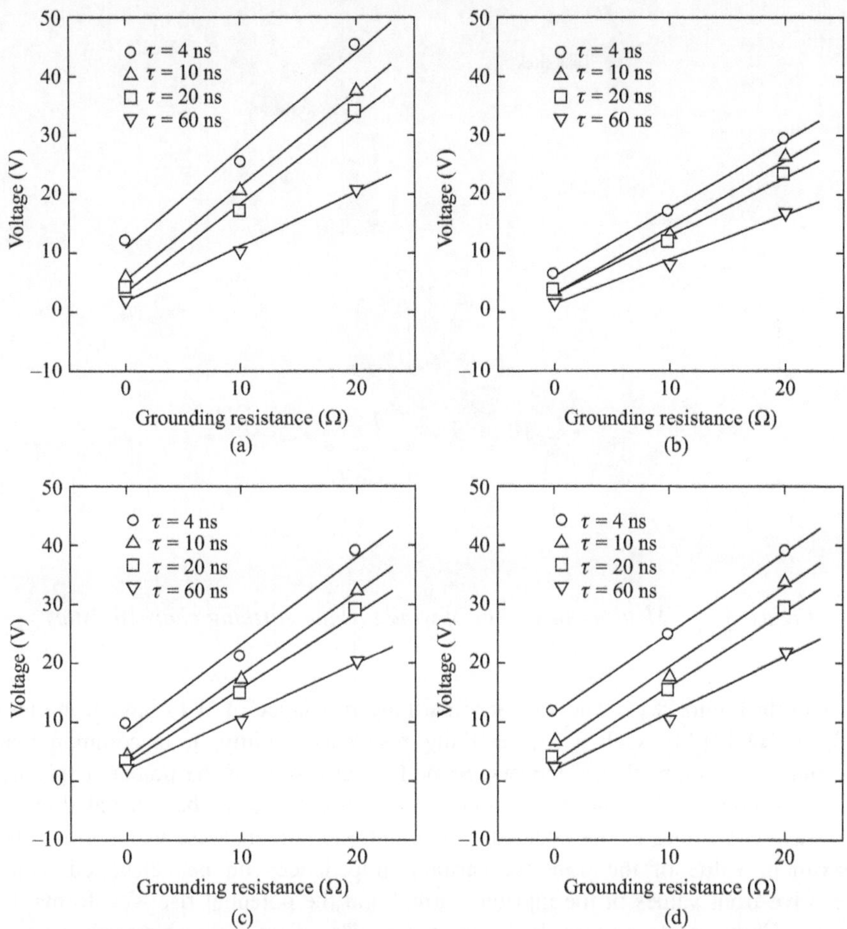

Figure 8.13   *Relations between the maximum values of V_{11} to V_{13} and R_g: (a) case of the downward lightning and stroke at the top of the blade; (b) case of the downward lightning and stroke at the rear portion of the nacelle; (c) case of the upward lightning and stroke at the top of the blade; (d) case of the upward lightning and stroke at the rear portion of the nacelle [8]*

© Elsevier 2009. Reprinted with permission from K. Yamamoto, T. Noda, S. Yokoyama, A. Ametani, 'Experimental and analytical studies of lightning overvoltages in wind turbine generator systems', *Electric Power Systems Research*, vol. 79, no. 3, March 2009

estimate the induction level. From those comparisons between measurements and simulations, it has been obvious that influence of up to 30% appears on the measured results of the potential rise.

When the steep wave front current such as that shown in Figure 8.17 is injected into earthing systems, the potential rise at all sites shows inductive potential, as can

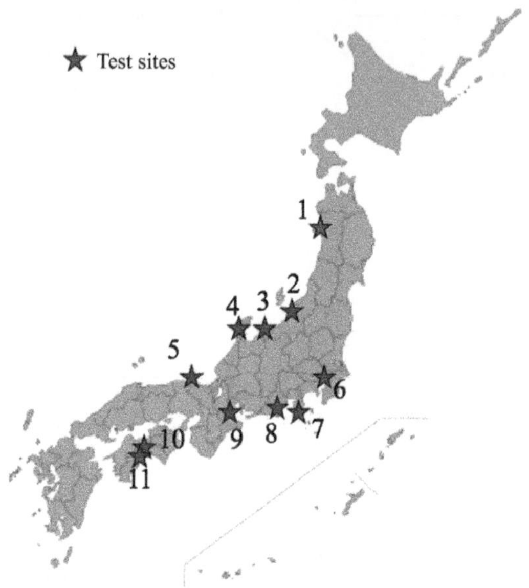

Figure 8.14   Measurement sites of wind turbine earthing characteristics

be seen in Figure 8.14. The values of earthing resistance of WTs have to be less than 10 Ω [2]. In such a low earthing resistance facility, it is common that the inductive potential rise appears. Some typical results of the potential rise are shown in Figure 8.18. The steady-state earthing resistance can be calculated using the wave tail values of the injected current and the potential rise waveforms. The maximum value of the transient earthing impedance can be calculated using the wave front values of the injected current and the potential rise waveforms. In terms of lightning protection, both the values of transient earthing impedance and the steady-state earthing resistance become important. Specially, the former and latter are important for WTs constructed at summer and winter lightning area, respectively.

It is also obvious that the potential rise depends on the stratified earth resistivity at each WT site. High-frequency components of the injected current in the wave front propagate near the earth surface. The depth which depends on the frequency f of the injected current, soil permeability μ, and resistivity ρ is calculated approximately as follows:

$$d = \sqrt{\frac{\rho}{\pi f \mu}} \tag{8.1}$$

If low-resistivity layer exists at near surface, surge can propagate easily. It means that the potential rise at the wave front becomes lower. The stratiform earth resistivity at each site is measured using the Wenner method. Please see the details shown in [18].

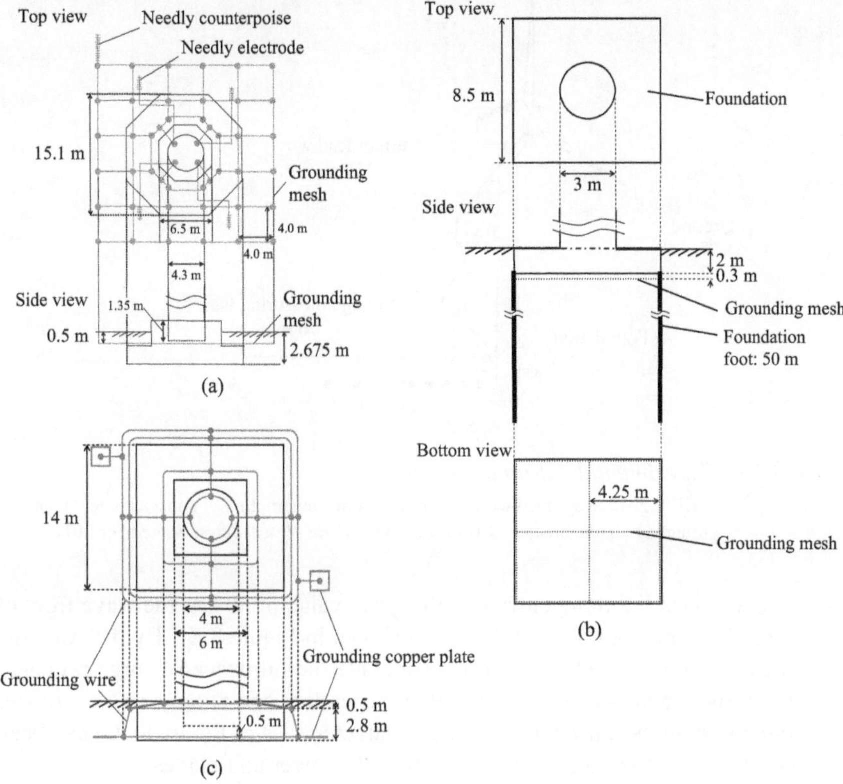

Figure 8.15　　*Multifarious earthing systems of wind turbines: (a) earthing system of Site 3; (b) earthing system of Site 6; (c) earthing system of Site 9 [27]*

© IEEE 2012. Reprinted with permission from Yamamoto, K., Yanagawa, S., 'Transient grounding characteristics of wind turbines', *IEEE Proceedings*, September 2012

8.1.2.4　Simulations using the FDTD method

In some sites shown in Figure 8.14, reproductions of measurements have been done using the FDTD method. Good agreements between measured and simulated results as shown in Figure 8.19 have been realized. Lately, the modeling methods of the WTs are introduced in several technical papers [17–26], and high accuracy simulations have become possible.

8.1.2.5　Frequency characteristics of wind turbines

Once the injected current and potential rise waveforms such as that shown in Figures 8.17–8.19 have been specified, the frequency characteristics of the earthing system as shown in Figure 8.20 can be calculated by using the Laplace transform. From the frequency characteristics of the absolute and phase values of the earthing impedance, the time responses to all the different lightning wave shapes can be calculated. As examples, potential rise responses to a step current with a peak value

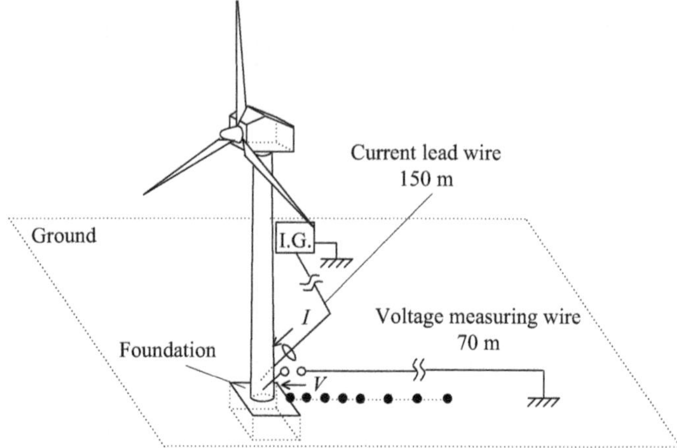

Figure 8.16   Experimental set-up at Site 6 [27]

© IEEE 2012. Reprinted with permission from Yamamoto, K., Yanagawa, S., 'Transient grounding characteristics of wind turbines', *IEEE Proceedings*, September 2012

of 1 A and a typical lightning current with a peak value of 30 kA, the wave front of 5.5 μs, and the stroke duration of 75 μs are shown in Figure 8.21. By the way, the above-mentioned measured results do not include the influence of surge propagations at the tower and blades; they are the independent and pure characteristics of the earthing system, because the frequency characteristics of Figure 8.20 have been calculated from the FDTD simulation without the tower and blades.

8.1.3   Example of lightning accidents and its investigations

In this section, lightning protection studies for a wind farm are introduced [26]. In Taikoyama Wind Farm constructed in the north part of Kyoto Prefecture and near the Sea of Japan, WTs have often been damaged by lightning in winter. To solve the problem and build lightning protection methodologies, a meeting of the committee had been convened from March 2003; many professionals of lightning protection and WTs had joined. As a result of the discussions at the meeting, a lightning protection tower was built near one of the WTs in December 2003. After the construction of the lightning protection tower, many lightning discharges have been caught by the tower; the efficiency of the tower has been recognized.

In the winter from 2011 to 2012, two typical accidents have occurred. In one of the accidents, a fire occurred inside the WT close to the lightning protection tower on January 28, 2012. Hereinafter, this accident is called "Accident 1." Accident 1 has been caused by a lightning to the WT itself; a back-to-back (BTB) converter system has remarkably burned out. Another accident has been the breakdown of the anemometers placed on the lightning protection tower on March 24, 2012. Hereinafter, this accident is called "Accident 2."

In this section, the causes and the protection methodologies for these accidents have been introduced. To confirm reliability of the protection methods,

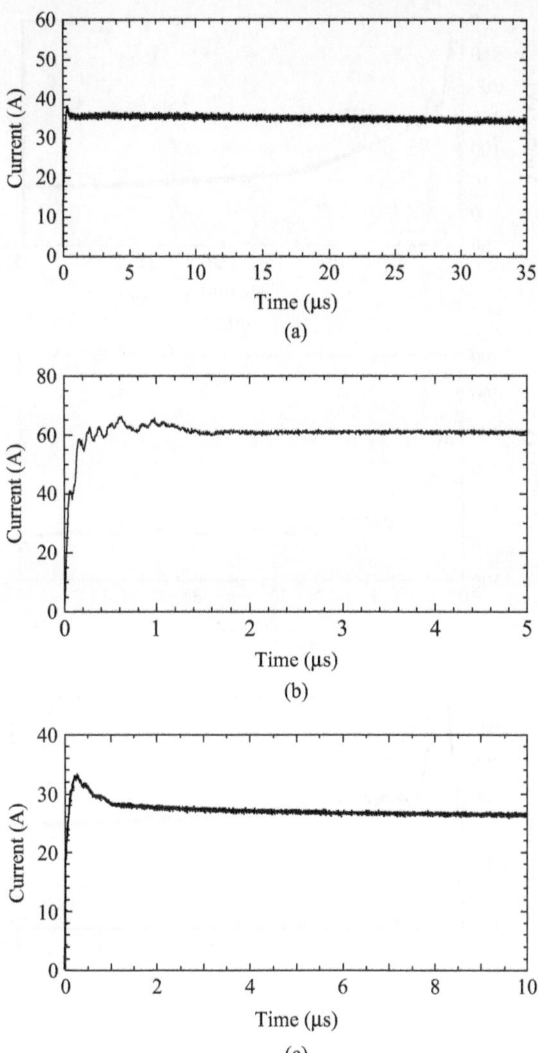

Figure 8.17   *Typical injected current waveforms: (a) injected current waveform at Site 3; (b) injected current waveform at Site 6; (c) injected current waveform at Site 9 [27]*

© IEEE 2012. Reprinted with permission from Yamamoto, K., Yanagawa, S., 'Transient grounding characteristics of wind turbines', *IEEE Proceedings*, September 2012

experiments of transient earthing characteristics and simulations using the FDTD method have been done. From comparison of these results, the degree of confidence in the simulation model can be testified, and the level of overvoltages at the WT and the lightning protection tower can be understood.

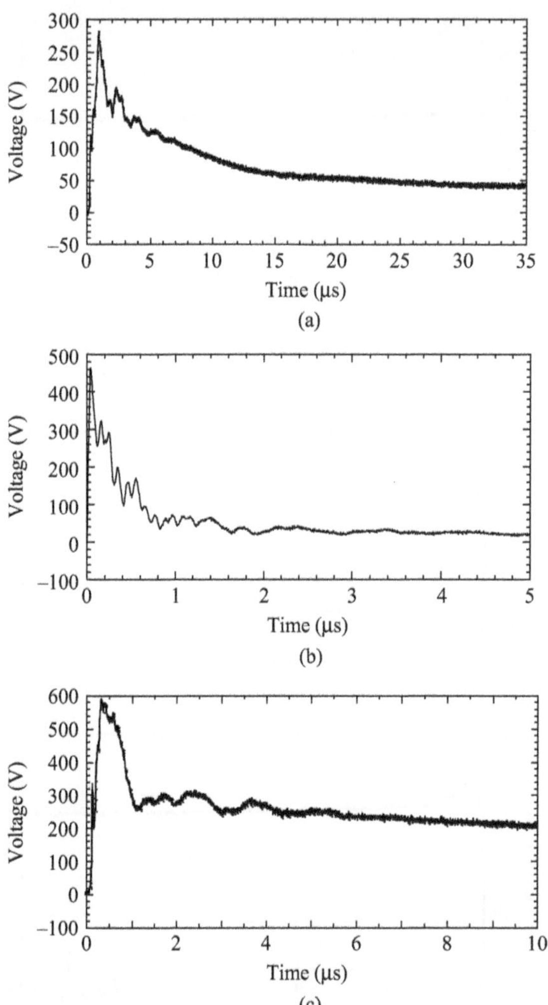

Figure 8.18   *Typical potential rise waveforms: (a) potential rise waveform at Site 3; (b) potential rise waveform at Site 6; (a) potential rise waveform at Site 9 [27]*

© IEEE 2012. Reprinted with permission from Yamamoto, K., Yanagawa, S., 'Transient grounding characteristics of wind turbines', *IEEE Proceedings*, September 2012

8.1.3.1   Situations of the accidents

Accident 1 has unfortunately occurred during anemometer installations; those anemometers were planned to be set up at the height of 49.7 m, 81 m, and 93.7 m as shown in Figure 8.22. Metal wires of control and power lines are connected to the three anemometers and lead to the inside of the nearby WT by way of the connection box around the foot of the lightning protection tower. The details of the

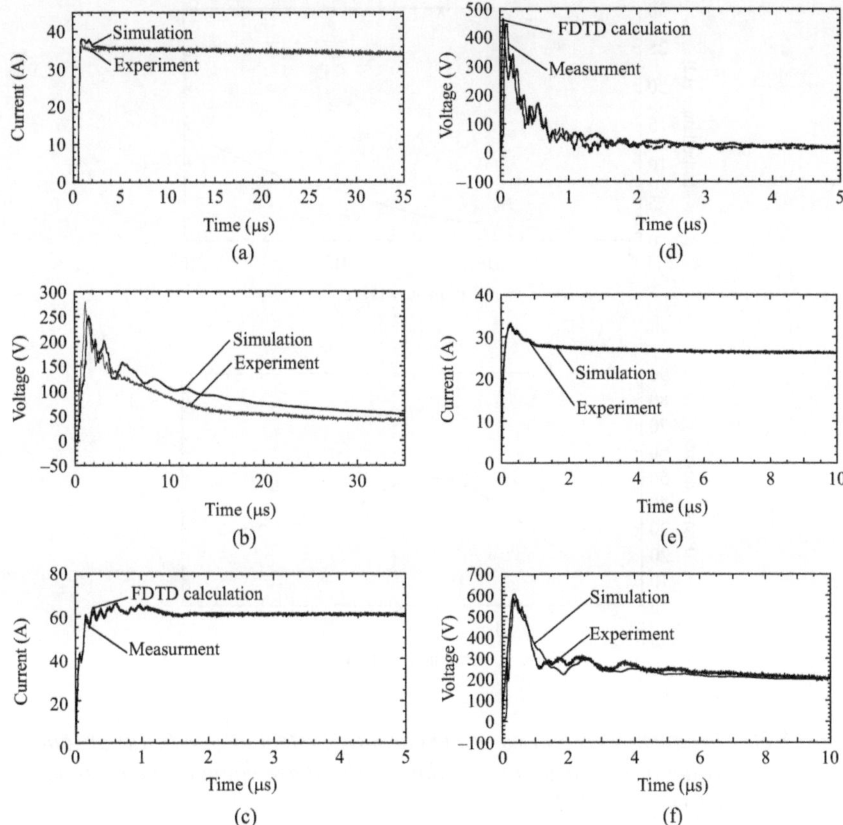

Figure 8.19   Comparisons between measurements and FDTD simulations:
(a) injected current waveform at Site 3; (b) potential rise waveform
at Site 3; (c) injected current waveform at Site 6; (d) potential
rise waveform at Site 6; (e) injected current waveform at Site 9;
(f) potential rise waveform at Site 9 [27]

© IEEE 2012. Reprinted with permission from Yamamoto, K., Yanagawa, S., 'Transient
grounding characteristics of wind turbines', *IEEE Proceedings*, September 2012

circuit of the anemometer system involved in Accident 1 are shown in Figure 8.23(a).
When lightning strike the WT, potential rise of the WT foundation occurred as shown
in Figure 8.24. It can be guessed that the potential rise caused discharges between the
metal wire lead from the lightning protection tower and earthed parts inside the WT
[12]; finally the discharges gave rise to the fire. The pictures of Accident 1 are shown
in Figure 8.25.

After Accident 1, the set-up of the anemometer system was finished. The circuit
of the system is shown in Figure 8.23(a). The system didn't have an enough lightning
protection system. Therefore, there was possibility of breakdowns of anemometers
and other equipment. Lightning to the lightning protection tower caused voltage

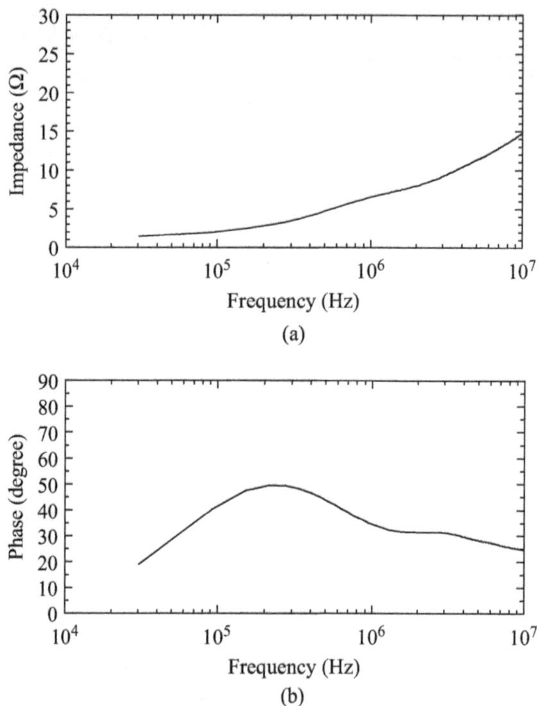

Figure 8.20   Frequency responses of the earthing system at Site 6: (a) absolute value of the earthing impedance; (b) phase value of the earthing impedance [18]

© 2010 IEEE. Reprinted with permission from Yamamoto, K., Yanagawa, S., Yamabuki, K., Sekioka, S., Yokoyama, S., 'Analytical surveys of transient and frequency-dependent grounding characteristics of a wind turbine generator system on the basis of field tests', *IEEE Transactions on Power Delivery*, vol. 25, no. 4, October 2010

difference between the structures of the lightning protection tower and power and signal lines leading to the anemometers. Breakdowns have occurred at the weakest point. The breakdown point is shown in Figure 8.26, where a spark track can be seen.

8.1.3.2   Protection methodologies

To prevent these accidents, the protection system shown in Figure 8.23(b) was proposed. In the protection system, anemometer system is gathered at the side of the lightning protection tower; there is no metal wire between the lightning protection tower and the WT. However, this system could not be established at site because of its higher cost exceeding the budgetary limits. Instead, the protection system using potential equalization method shown in Figure 8.23(c) was set up.

The concept of the protection system shown in Figure 8.23(b) is "isolation" between the lightning protection tower and the nearby WT. However, electro-optic

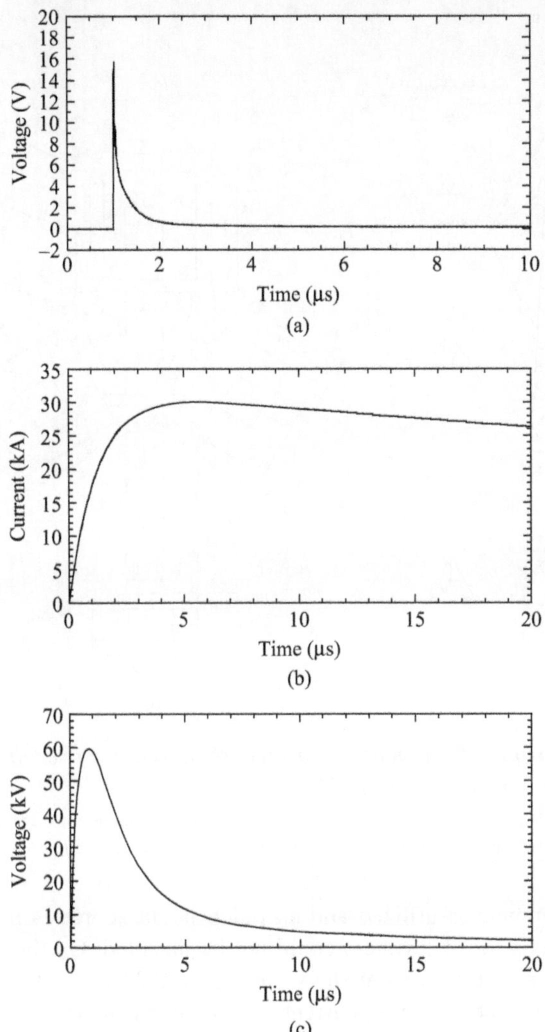

Figure 8.21   *Potential rise response to step and typical lightning currents at Site 6: (a) potential rise response to a step current; (b) typical lightning current waveform; (c) potential rise response to the typical lightning current [18]*

© 2010 IEEE. Reprinted with permission from Yamamoto, K., Yanagawa, S., Yamabuki, K., Sekioka, S., Yokoyama, S., 'Analytical surveys of transient and frequency-dependent grounding characteristics of a wind turbine generator system on the basis of field tests', *IEEE Transactions on Power Delivery*, vol. 25, no. 4, October 2010

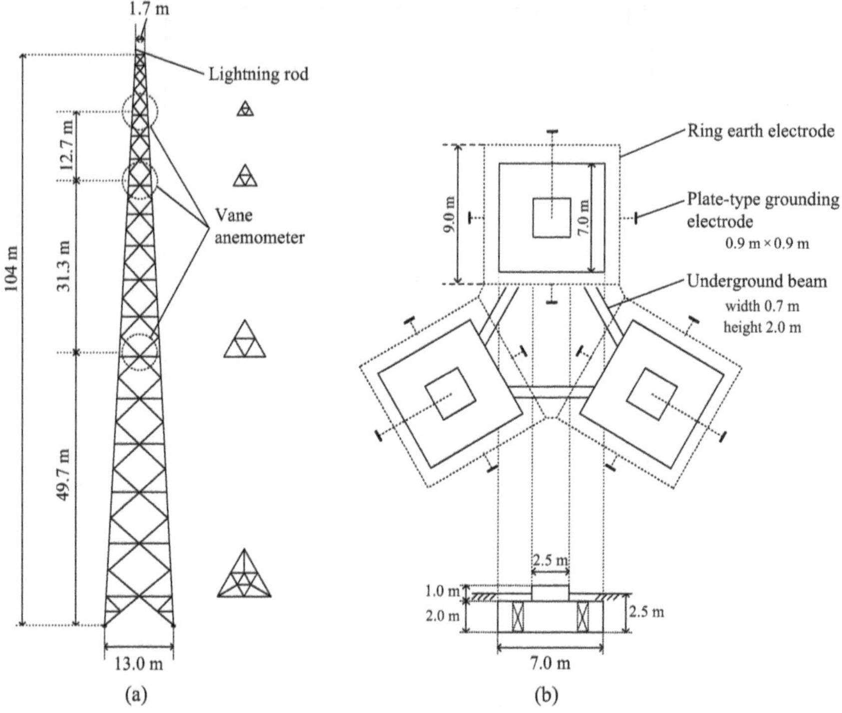

Figure 8.22   Details of the lightning protection tower: (a) overall view;
(b) earthing system [26]

© IEEE 2013. Reprinted with permission from Yamamoto, K., Sumi, S., 'Validations of lightning protections for accidents at a wind farm', *IEEE Proceedings*, October 2013

conversion equipment is utilized and an independent source is needed. Because this system costs a lot, the system could not be installed. On the other hand, the concept of the protection system shown in Figure 8.23(c) is "potential equalization." To decrease the lightning current into the metal wires such as control and power lines, earthing wires between the lightning protection tower and the nearby WT are set up.

8.1.3.3   Measurements and simulations of transient earthing characteristic

To confirm reliability of the protection methods shown in Figure 8.23(c), the author has a plan to make a simulation model using the FDTD method. For the validation of the FDTD model, the earthing potential rises and shunt currents are compared on the FDTD simulations and the measurements.

Figures 8.27 and 8.28 show the earthing wires between the foundation of WT and the lightning protection tower, and the details of the WT earthing system. The earthing resistivity has been measured by using the Wenner method. Figures 8.29 and 8.30 show the layer of earthing resistivity around the test site and the

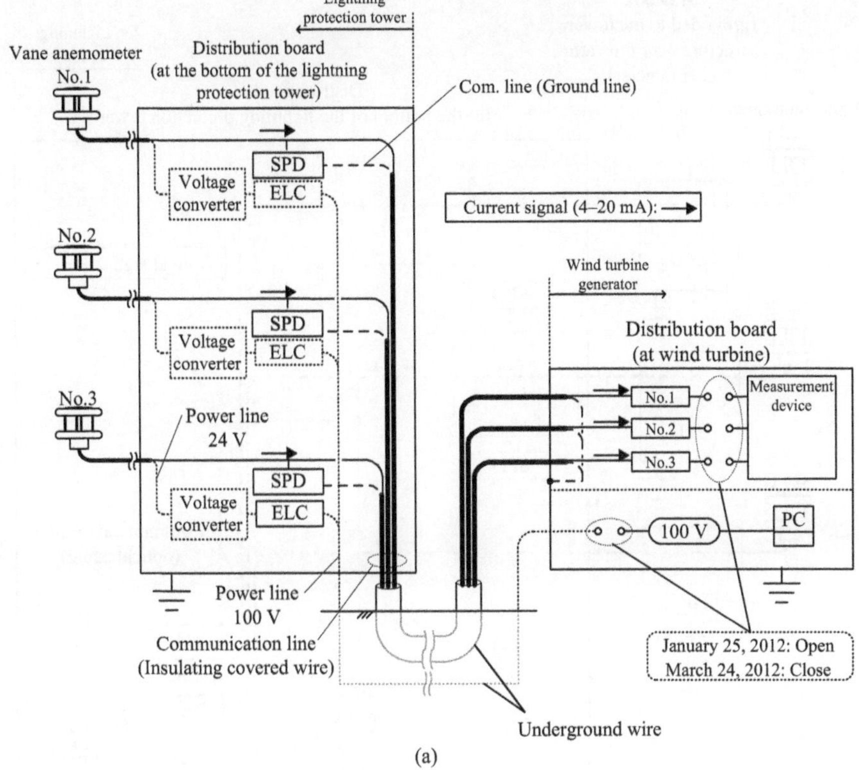

Figure 8.23   *Wire diagram of anemometer system: (a) wire diagram at accidents; (b) recommended lightning protection system; (c) adopted lightning protection system [26]*

© IEEE 2013. Reprinted with permission from Yamamoto, K., Sumi, S., 'Validations of lightning protections for accidents at a wind farm', *IEEE Proceedings*, October 2013

measuring place of the Wenner method. Based on the experimental set-up shown in Figure 8.31, the transient characteristic of the earthing system has been measured.

In the analysis, the experiment layout is expressed as faithfully as possible using analytical spaces as shown in Figure 8.32. The shape of the analytical space is rectangular, with 166 m in the x direction, 141 m in the y direction, and 124 m in the z direction, and grid spacing is all set to 0.5 m. Six faces that surround the analytic spaces have an open space simulated using the Liao's second-order absorbing boundary condition. The earth in the experiment is expressed by filling the space with the substance of resistivity $\rho = 17.8$ Ωm and relative permittivity 10 with 18 m high from the bottom of the analytic space, then, over the top of it, filling the space with the substance of resistivity $\rho = 427$ Ωm and relative permittivity 10 with 5 m thickness high.

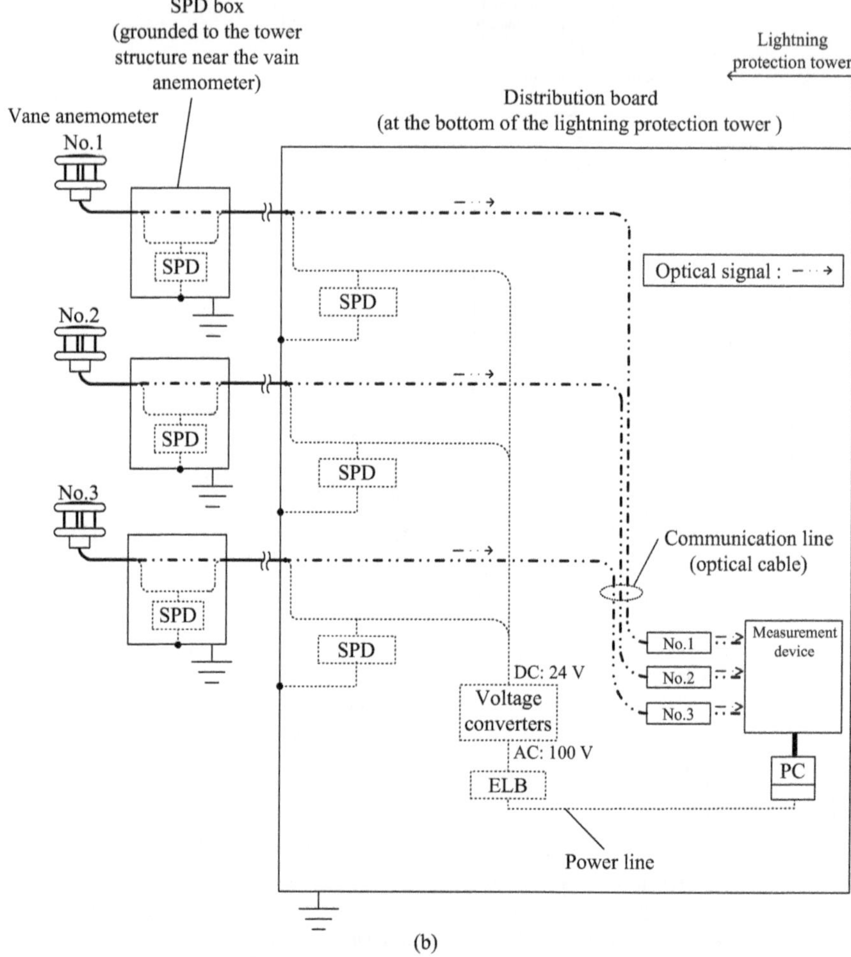

(b)

Figure 8.23　(Continued)

　　The polyvinyl chloride wires are used for the current lead wire and the voltage measuring wire; earthing wires are expressed by thin wire model with actual radius of those wires. The foundation and plate-type earthing electrodes are simulated as perfect conductors. The WT near the lightning protection tower is connected with other WTs by counterpoises. It is impossible to model all WTs in the wind farm in a same FDTD analytical space because of shortage of performance of personal computers. In this simulation model, the earthing system of other WTs is modeled as a perfect conductor of 100 m $\times$ 20 m $\times$ 30 m.

　　The injected current in the analysis is set to be the same as that of the experimental result shown in Figure 8.33. The current is injected to the top of the foundation via 500 Ω resistance. In order to identify transient earthing

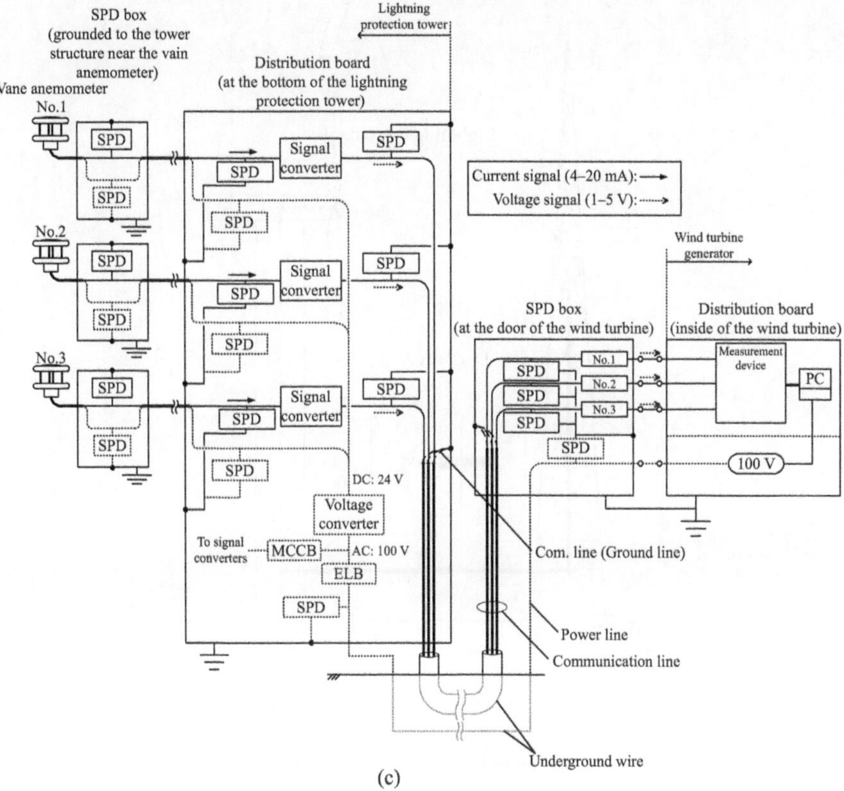

Figure 8.23   (Continued)

characteristics of the WT, current that flows into the foundation, the potential rise of the foundation, and the shunt current through the earthing wires between the WT and the lightning protection tower have been calculated.

The comparisons between experimental and simulation results are shown in Figures 8.33–8.35. The current shown in Figure 8.33 has been injected and the potential rise at the foundation of the WT in Figure 8.14 has been measured and calculated in the cases that the earthing system of the lightning protection tower is connected and disconnected with that of the nearby WT. Figure 8.35 shows the shunt currents into three connection earthing wires. The waveform of the total shunt current is shown in Figure 8.35(c); compared with the injected current, the maximum shunt current has become over 40% of the injected current. From the above-mentioned results, we can get good agreements between experimental and simulation results.

After the completion of the FDTD model, overvoltage and shunt current at surge protective devices (SPDs) in the anemometers are simulated. From these results, the lightning protection efficiencies can be confirmed.

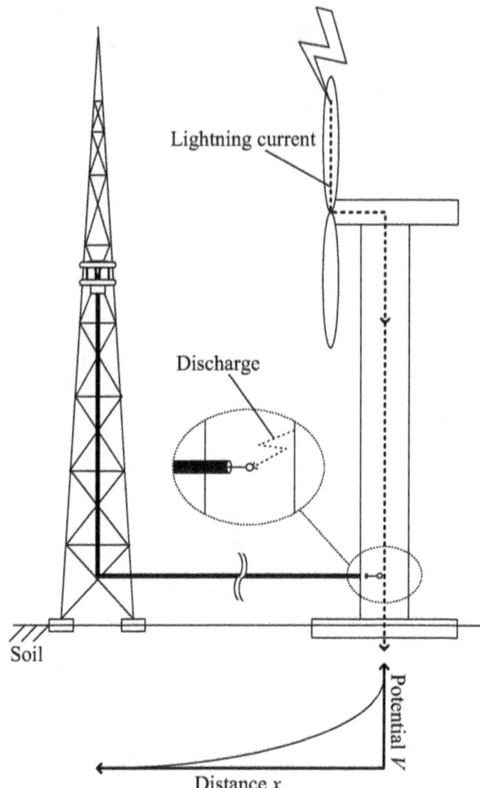

*Figure 8.24   Discharge caused by the potential rise of the wind turbine
foundation [26]*

© IEEE 2013. Reprinted with permission from Yamamoto, K., Sumi, S., 'Validations
of lightning protections for accidents at a wind farm', *IEEE Proceedings*, October 2013

8.2   Solar power generation system

In recent years, generation methods using renewable energy are rapidly spreading
as a countermeasure of global warming and energy depletion. Large-scale photovoltaic power plants (PVPPs) are often constructed in the same way as large-scale
wind farms.

In such large-scale PVPPs, PV modules are installed on the frame. PVPPs are
usually constructed in a large open space where a few tall structures exist to get
more solar energy. Therefore, such large-scale PVPPs are likely to be a target of
lightning. To protect them from lightning damages and establish appropriate
lightning protections using the lightning rod, surge protective device (SPD) and
such other devices are required. However, the mechanism of the lightning overvoltage generations has not been well discussed so far [28, 29].

In this section, examples of numerical simulations on PVPPs are introduced.

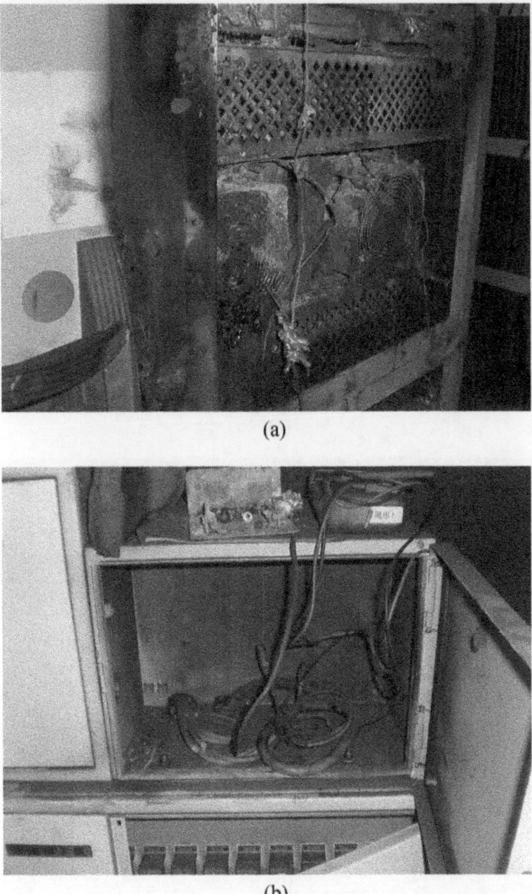

(a)

(b)

Figure 8.25 *Pictures of the accident: (a) burned-out board of BTB; (b) end terminals of communication and power line leaded from outside [26]*
© IEEE 2013. Reprinted with permission from Yamamoto, K., Sumi, S., 'Validations of lightning protections for accidents at a wind farm', *IEEE Proceedings*, October 2013

8.2.1 *Lightning surges in a MW-class solar power generation system*

In the large-scale PVPPs such as MW-class plants, there are many power conditioning systems (PCSs) that are highly capacious and expensive. These expensive facilities have to be protected from lightning overvoltage.

In this section, at first, the case that lightning overvoltage at the DC side of a PCS has been experimentally studied using the 1/10 scale model is introduced [30]. Additionally, the case that the measurements have been verified using the FDTD method is explained. Finally, its sensitive analysis depending on the wave front duration and the earthing resistivity has been introduced.

(a)

(b)

Figure 8.26   Damaged anemometer: (a) appearance of the anemometer; (b) trace of the breakdown discharge [26]

© IEEE 2013. Reprinted with permission from Yamamoto, K., Sumi, S., 'Validations of lightning protections for accidents at a wind farm', *IEEE Proceedings*, October 2013

8.2.1.1   Experiments using the 1/10th scale model

PV array and frame shown in Figure 8.36 are constructed in large open spaces where a few tall structures exist. Therefore, these are likely to be a target of lightning. To protect PV modules from direct lightning strikes, sometimes lightning conductors are installed at the top of the frame as shown in Figure 8.37. When lightning strikes this lightning conductor, lightning current branches off from the lightning conductor to the frame, then it flows into the earth through the earthing system of the PVPP. The lightning conductor is installed on the top of the flame; its interval is about 10 cm. It is hard to think that the conductor shows good protection efficiency for lightning protection because of such interval.

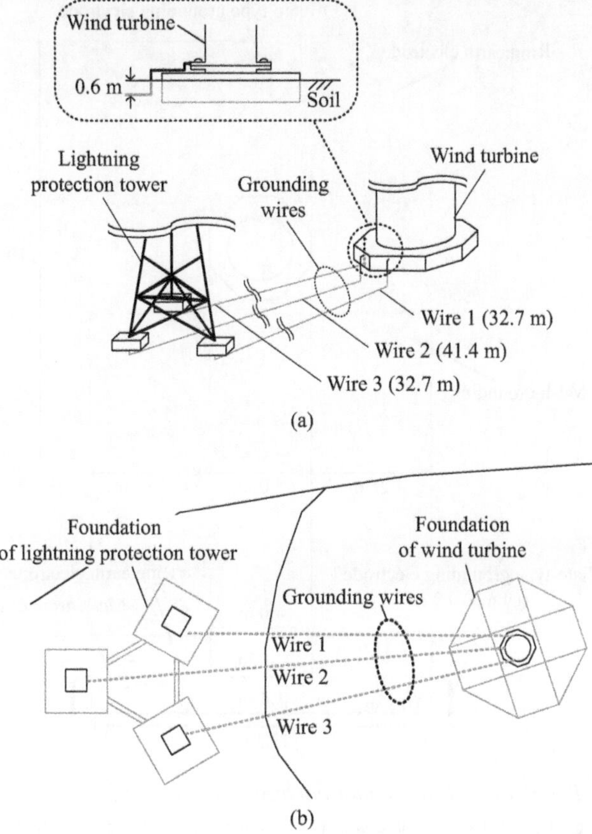

*Figure 8.27   Earthing wires between wind turbine foundation and lightning
protection tower: (a) Overhead view; (b) Top view [26]*

© IEEE 2013. Reprinted with permission from Yamamoto, K., Sumi, S., 'Validations
of lightning protections for accidents at a wind farm', *IEEE Proceedings*, October 2013

Lightning often strikes a PV module directly. In such a case, the PV module
struck by lightning is usually broken, and some PV modules installed near the
struck PV module may also be broken.

In large-scale PVPPs, there are many PCSs that are highly capacious and
expensive. A PCS is set up on each PV array as shown in Figure 8.36.

When lightning strikes the frame supporting PV arrays, surge voltage and
current are induced on DC power lines between PV arrays and a PCS and AC
power lines between a PCS and a substation. These induced voltage and current
depend on the lightning current path on the structure according to the shape of the
frame and the earthing system. As a result, overvoltage appears between power
lines and a PCS due to the induced voltage and current, and the potential rise of the
earthing system.

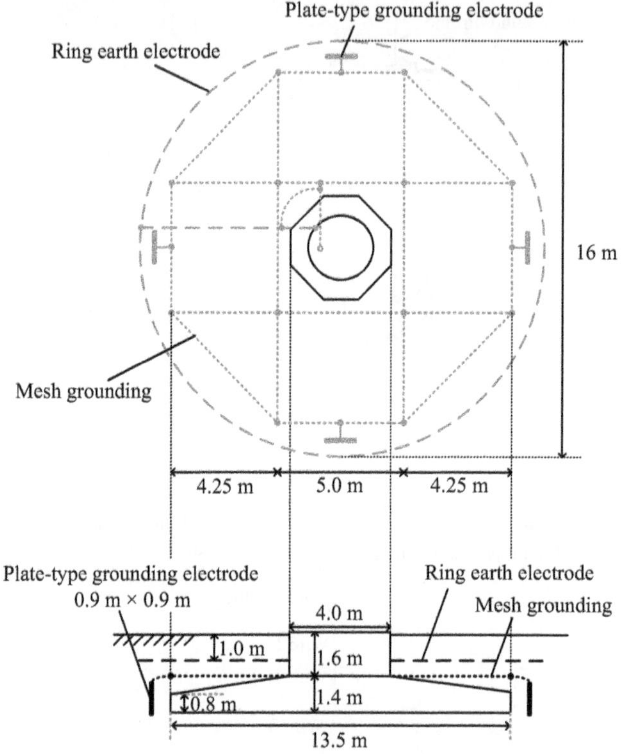

Figure 8.28   Details of wind turbine earthing system [26]

© IEEE 2013. Reprinted with permission from Yamamoto, K., Sumi, S., 'Validations of lightning protections for accidents at a wind farm', *IEEE Proceedings*, October 2013

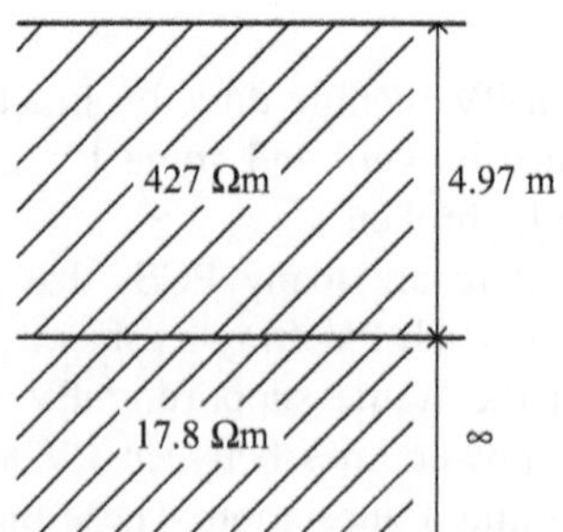

Figure 8.29   Earth resistivity [26]

© IEEE 2013. Reprinted with permission from Yamamoto, K., Sumi, S., 'Validations of lightning protections for accidents at a wind farm', *IEEE Proceedings*, October 2013

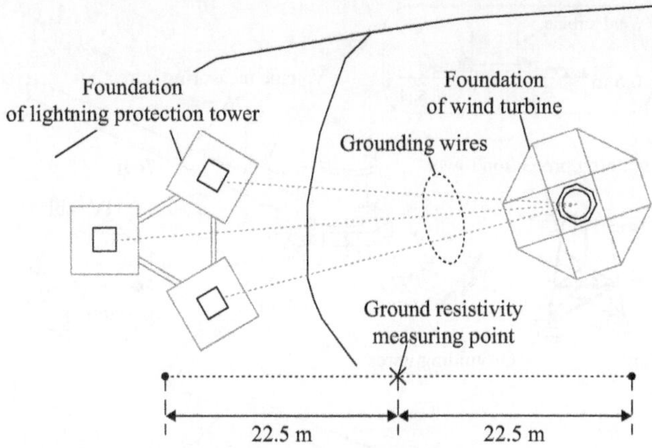

Figure 8.30   Measuring point of earth resistivity [26]

© IEEE 2013. Reprinted with permission from Yamamoto, K., Sumi, S., 'Validations of lightning protections for accidents at a wind farm', *IEEE Proceedings*, October 2013

In this section, at first, the overvoltage appearing at the DC side of the PCS has been experimentally discussed using the 1/10 scale model of the frame structure in the large-scale PVPPs shown in Figures 8.36 and 8.37. The models of PV modules are not included in this scale model. The effects of the PV module are ignored, and the overvoltage that is directly induced to the power line is just considered.

The experimental set-up is shown in Figure 8.38. In this experiment, 1/10 scale model of a frame supporting PV arrays in the large-scale PVPP is set up on the aluminum plate of 2 mm in thickness. Current injected points are Point 1 (corner of the frame) and Point 2 (center of the frame) shown in Figure 8.38. When current is injected to these assumed point of the lightning stoke, overvoltage at the DC side (V_1, V_2 [V]) of the PCS is measured with the injected current (I [A]) at the same time. The frame shown in Figure 8.38 is made of aluminum pipes whose radius is approximately 5 mm. The DC power line models are made of polyvinyl chloride wires whose radius is approximately 0.9 mm. The PCS is made of aluminum box that is a rectangular parallelepiped conductor, 15 cm on a side.

In general, PCS is installed in the metallic storage box as shown in Figure 8.39. Because this storage box is connected to the earth directly under it, the potential at the storage box is equal to the earth potential of the installation location. When the potential at the storage box fluctuates transiently by lightning strike to the frame of the supporting PV arrays in the large-scale PVPP, the potential of the PCS in the storage box also fluctuates with potential at its storage box. As a result, both potentials become approximately equal. By such a reason, in this experiment, potential at the PCS installed in the storage box and the potential at the storage box are dealt as same potential, and the overvoltage between the PCS and power line is defined as that between the storage box and the power lines.

Locations of the PCS are shown in Figure 8.40(a)–(c). These locations are typical examples of actual facilities. In Figure 8.40(a), the PCS is installed at the

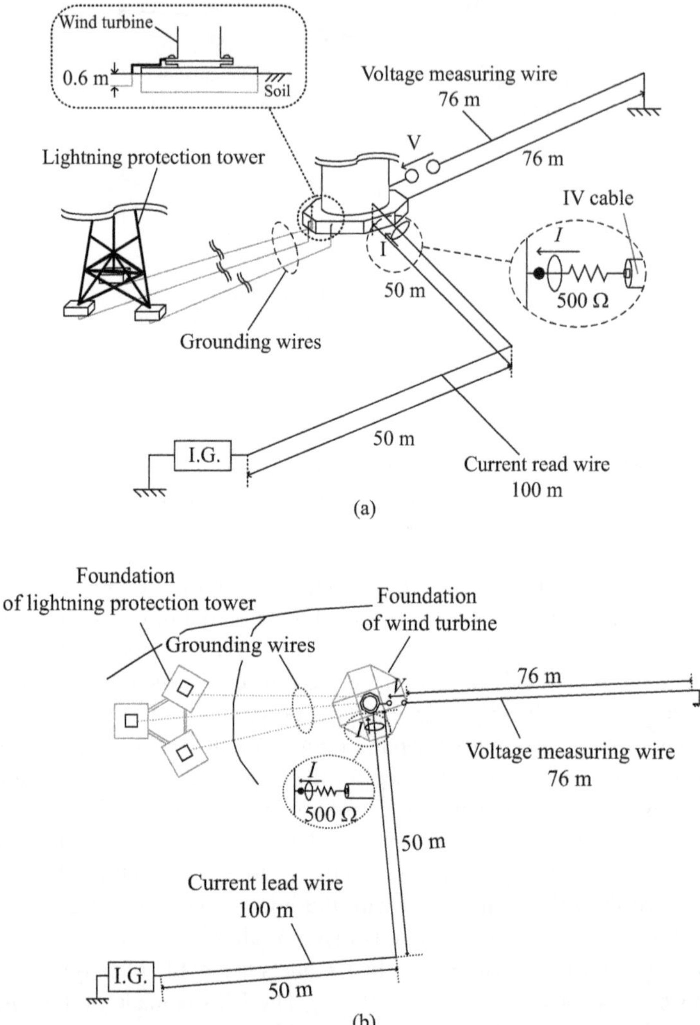

Figure 8.31   Experimental set-up: (a) Over view; (b) Top view [26]
© IEEE 2013. Reprinted with permission from Yamamoto, K., Sumi, S., 'Validations of lightning protections for accidents at a wind farm', *IEEE Proceedings*, October 2013

center of the frame, and the PCS gathers electrical power from bilateral PV arrays. In Figure 8.40(b), the PCS is installed at the side of the frame. But the PCS is still inside of the frame. The PCS is installed at the outside of the frame as shown in Figure 8.40(c). There is the characteristic that the length of the power line in Figure 8.40(b) and (c) is longer than that in Figure 8.40(a).

In the set-up shown in Figure 8.38, coaxial cable (3D2V) connected with the pulse generator (referred in the figure as P.G.) is laid up over the frame.

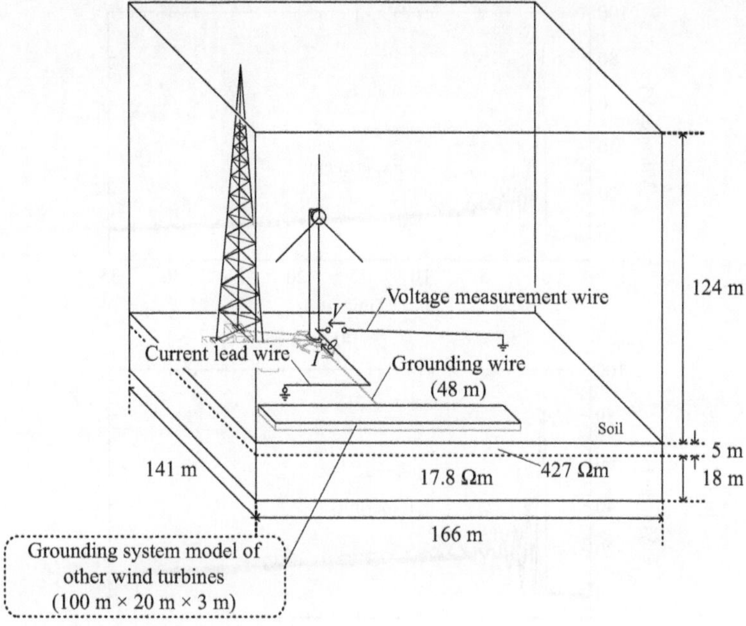

Figure 8.32   Analytical set-up [26]

© IEEE 2013. Reprinted with permission from Yamamoto, K., Sumi, S., 'Validations of lightning protections for accidents at a wind farm', *IEEE Proceedings*, October 2013

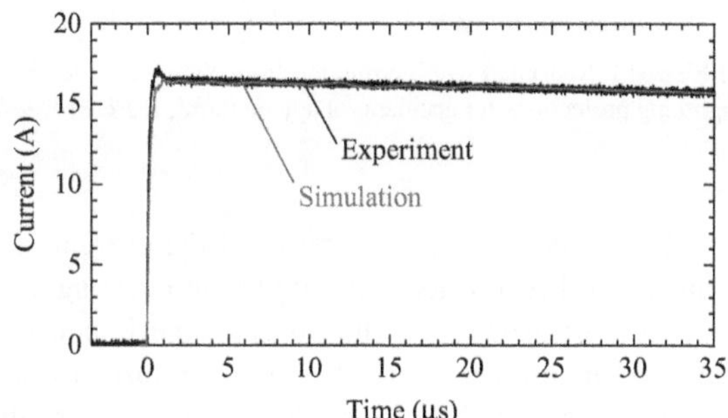

Figure 8.33   Injected current [26]

© IEEE 2013. Reprinted with permission from Yamamoto, K., Sumi, S., 'Validations of lightning protections for accidents at a wind farm', *IEEE Proceedings*, October 2013

The resistance of 500 Ω is inserted between the frame and the core wire of coaxial cable, and the steep current shown in Figure 8.41 (its peak value and wave front duration were approximately 1.2 A and 10 ns, respectively) is injected to the frame. The current is injected to the frame, and at the same time the different polarity current propagates upward from the injected point on the coaxial cable.

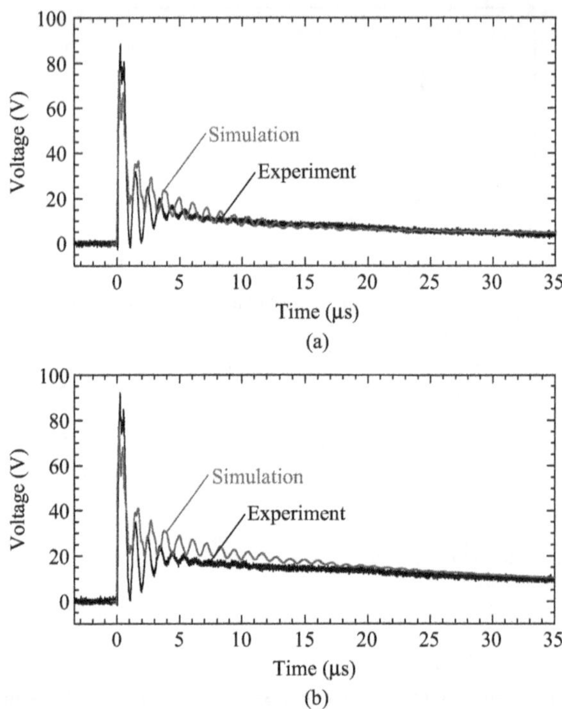

Figure 8.34   *Potential rise at the foundation of the wind turbine (foundations are disconnected): (a) Connected by three earthing wires; (b) Disconnected [26]*

© IEEE 2013. Reprinted with permission from Yamamoto, K., Sumi, S., 'Validations of lightning protections for accidents at a wind farm', *IEEE Proceedings*, October 2013

A part of the current reflects at the curved point of the current lead wire (5 m in height from the frame), and it returns to the injection point again. This feedback current influences on the measuring results. In this experiment, the time range of $10 \text{ m}/(3 \times 10^8 \text{ m/s}) \cong 30$ ns until the reflection wave returns to the injected point from the curved point is validated, because there is no such reflected current in actual lightning phenomenon. The current injection method using such a coaxial cable is explained in [31].

Under these conditions, the experiments of Cases 1.A–2.C shown in Table 8.1 have been conducted. Measuring data are the injected current I [A] and the potential difference between the PCS and DC power line V_1, V_2 [V].

The voltage waveforms of V_1 and V_2 in Cases 1.A–2.C are shown in Figures 8.42 and 8.43. In such experiments, the position of the voltage probe cable greatly influences the experimental results because of the electromagnetic induction from the current flowing on the current lead wire and the frames. To suppress such an electromagnetic induction, the voltage probes must be set perpendicularly

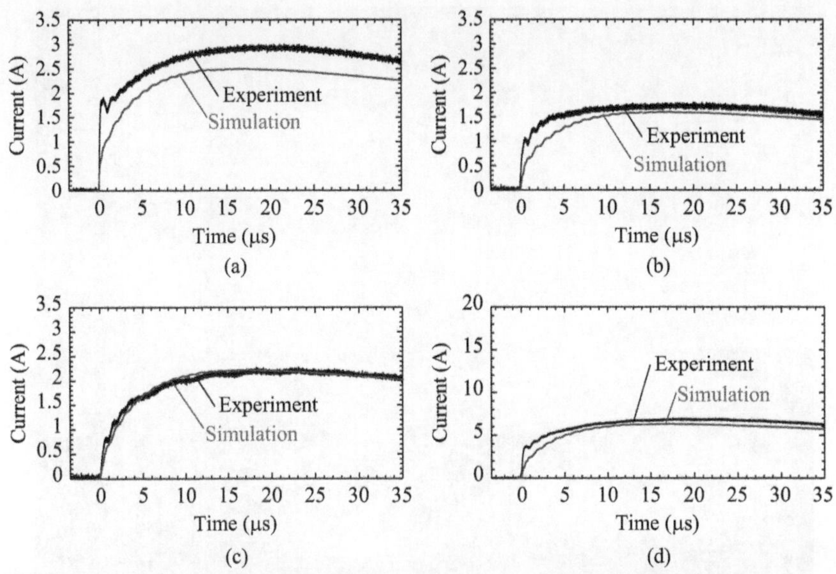

Figure 8.35   *Shunt current waveforms: (a) Wire 1; (b) Wire 2; (c) Wire 3; (d) Sum of the currents [26]*

© IEEE 2013. Reprinted with permission from Yamamoto, K., Sumi, S., 'Validations of lightning protections for accidents at a wind farm', *IEEE Proceedings*, October 2013

against all conductors where steep current flows. The distance between conductors and voltage probes must be larger [12, 32].

However, in these experiments, there is no best set-up that voltage probe cable is placed in vertically against all conductors, because the frame model contains diagonal bracings. Moreover, the distance between the voltage probe and conductors cannot be larger, because the voltage measuring points exist inside the frame. So, in this experiment, the voltage probe cable is extended to the backside of the PCS and frame vertically against the current lead wire and the top conductor consisting the frame. That is the positive direction of the x-axis from the measuring point at V_1, V_2 shown in Figure 8.38.

Next, the amplitude of fluctuations in voltages depending on the position of the voltage probe is explained. When the angle of the voltage probe cable is changed to 10–20 degrees to the y, z direction, voltage fluctuation of 3–5 V occurs. Even if voltage probes are set up vertically against all conductors so that steep current flows on, small induction still remains on it. The induced voltage cannot be removed completely. I thought that it is because of these reasons the voltage fluctuation of 3–5 V occurs.

Because these experimental results are expected to show measurement errors, analyses using the FDTD method have been made in section 8.2.1.2 and the experimental results using the scale model are verified. Furthermore, in section 8.2.1.3, analytical studies using an actual-scale FDTD model have been conducted.

Figure 8.36   Solar array of a large-scale photovoltaic power plant [30]

© IEEJ 2012. Reprinted with permission from K. Yamamoto, J. Takami, N. Okabe, 'Overvoltages on DC side of power conditioning system caused by lightning strike to structure anchoring photovoltaic panels', *IEEJ Trans*, vol. 132, no. 11, pp. 903–913, November 2012

8.2.1.2   Simulations using the FDTD method of the 1/10 scale model experiments

The analytical space is shown in Figure 8.44. The scale experiments described in section 8.2.1.1 are verified using this model. The dimension of the analytical space is 1.5 m × 5.0 m × 1.6 m and it is divided into cubic cells with a side length of 0.01 m. The absorbing boundary condition is the Liao's second-order condition.

The frame model is installed on the aluminum plate; its resistivity and thickness are 2.65×10^{-8} Ωm and 2 mm, respectively. This frame model is constructed by using thin wire models [11]. Thin wire models also model the power lines and current lead wire, and the PCS is made of a rectangular parallelepiped conductor and thin wire models. Shape and arrangement of all models shown in Figures 8.38 and 8.40 are expressed as faithfully as possible.

The current source and parallel resistance of 500 Ω are arranged above the current injected point shown in Figure 8.44, and connected between the frame and a lightning channel modeled by a thin wire.

As the injected current waveform is approximately equal to that in Figure 8.41, it is omitted in this chapter. In addition, 1 pF capacitor is inserted into the measuring point of potential difference because the voltage probe used in the experiments has the input capacitance.

Figure 8.37   *Lightning conductor on a structure of a solar array [30]*
© IEEJ 2012. Reprinted with permission from K. Yamamoto, J. Takami, N. Okabe,
'Overvoltages on DC side of power conditioning system caused by lightning strike to
structure anchoring photovoltaic panels', *IEEJ Trans*, vol. 132, no. 11, pp. 903–913,
November 2012

Under such analytical conditions, the calculations of Cases 1.A–2.C shown in Table 8.1 have been conducted. Output data are the injected current I [A] and the potential difference between the PCS and DC power line V_1, V_2 [V].

The potential difference is calculated by multiplying the electric field at the calculated point by the distance between the PCS and DC power line. The distance is one cell size in this model. Therefore, the effect of the induced voltage appeared on the voltage probe in the scale model experiments is not contained.

Calculation results of Cases 1.A–2.C are shown in Figures 8.45 and 8.46. Furthermore, maximum overvoltage value of each case is summarized in Table 8.2. As shown in Figures 8.45 and 8.46, the calculation and experimental results have been qualitatively matched, but a little difference has appeared. The main cause of the difference is thought to be the induced voltage to voltage probes (cf. section 8.2.1.1).

Approximately 10 ns periodic oscillation is shown in Figures 8.45 and 8.46. This periodic oscillation conforms to the resonant frequency of the conductors consisting the frame. The surge current flowing on the conductors reflects and passes at the connection point, and causes the induced voltage and current on the DC power lines. Finally, the induced voltage becomes the overvoltage at the DC side of the PCS. Moreover, it has been obvious that the effect of induced voltage occurred by the DC power line working as the dipole antenna [33].

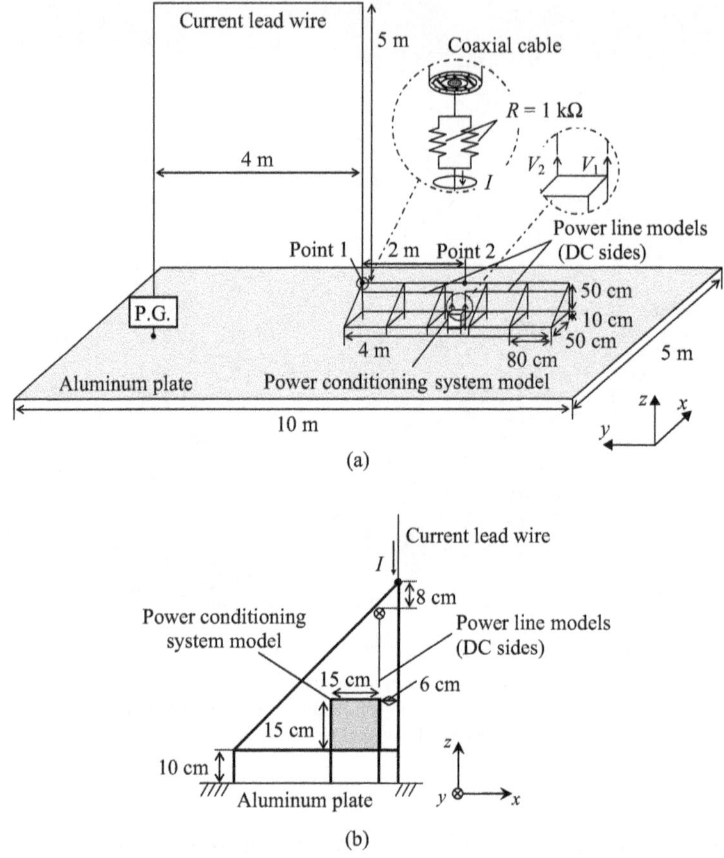

Figure 8.38 *Experimental set-up of the 1/10 scale model of frame supporting PV arrays in a large-scale photo-voltaic power plant: (a) Overall view; (b) Side view [30]*

© IEEJ 2012. Reprinted with permission from K. Yamamoto, J. Takami, N. Okabe, 'Overvoltages on DC side of power conditioning system caused by lightning strike to structure anchoring photovoltaic panels', *IEEJ Trans*, vol. 132, no. 11, pp. 903–913, November 2012

In the scale experiments and calculations, the steep current having 10 ns wave front duration is injected (this wave front duration is equivalent to 0.1 μs in the actual level). This represents very steep wave front duration of an actual subsequent stroke. The wave front duration becomes steeper, and the induced voltage on the DC power line becomes larger.

8.2.1.3 Simulations in case of an actual-scale model

In this section, at first, the case that the current is injected into the frame supporting a full-scale PV array has been studied. The frame is on the earth where the earth

Figure 8.39   Actual box of a power conditioning system [30]

© IEEJ 2012. Reprinted with permission from K. Yamamoto, J. Takami, N. Okabe, 'Overvoltages on DC side of power conditioning system caused by lightning strike to structure anchoring photovoltaic panels', *IEEJ Trans*, vol. 132, no. 11, pp. 903–913, November 2012

resistivity is 100 Ωm, and the wave front duration of the current is 1 μs. After that study, the overvoltage characteristics dependences on the wave front duration and earth resistivity have been studied.

An analytical model of an actual-scale PVPP is shown in Figure 8.47. The dimension of the analytical space is 15 m × 50 m × 18 m and it is divided into cubic cells with a side length of 0.1 m. The absorbing boundary condition is the Liao's second-order condition. The earth level is 7 m from the bottom of the analytical space; the earth resistivity is 100 Ωm, and the relative permittivity is 10. The frame supporting a full-scale PV array is earthed; its earthing resistance is approximately 3 Ω. Thin wire models of 5 cm radius compose the frame. The power line is also composed by thin wire models of 5 mm radius. The PCS is modeled as a rectangular parallelepiped conductor of 1.5 m on a side.

The injected current is shown in Figure 8.48. It is a step wave including wide frequency components, and its peak value and wave front duration were approximately 1 A and 1 μs, respectively. Analytical cases are shown in Table 8.1. Other analytical conditions are the same as described in section 8.2.1.2.

Calculation results are shown in Figures 8.49 and 8.50. Moreover, maximum overvoltage value is summarized in Table 8.3. Data in Table 8.3 show that the

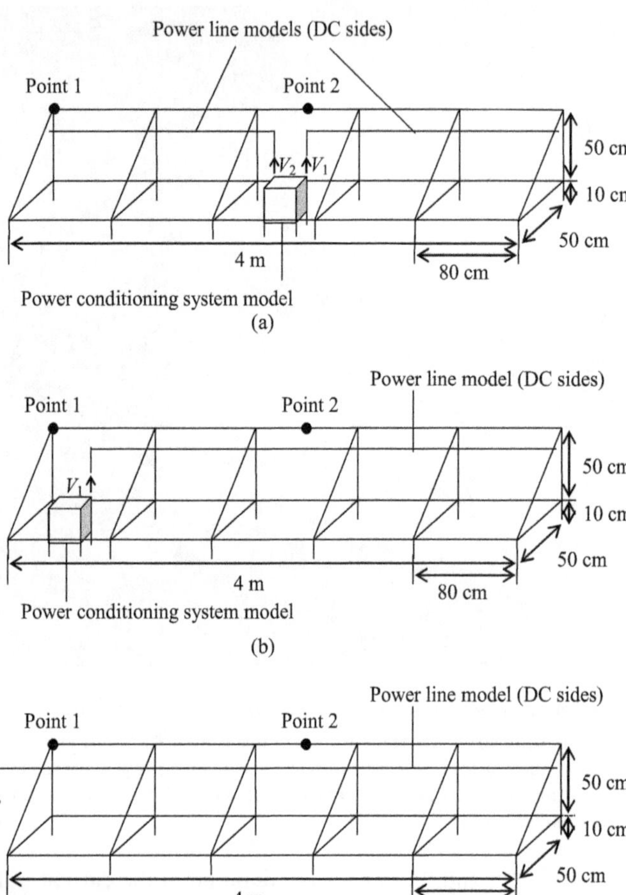

Figure 8.40   Positions of PCS: (a) Position A; (b) Position B; (c) Position C [30]
© IEEJ 2012. Reprinted with permission from K. Yamamoto, J. Takami, N. Okabe, 'Overvoltages on DC side of power conditioning system caused by lightning strike to structure anchoring photovoltaic panels', *IEEJ Trans*, vol. 132, no. 11, pp. 903–913, November 2012

120–230 kV overvoltage is applied to PCS DC side when the lightning of the wave front duration of 1 μs and peak current of 30 kA strikes the frame installed in the large-scale PV plant [34].

Next, features of overvoltage waveforms are discussed. In Figures 8.49 and 8.50, the oscillation appearing in the waveforms depends on the size of the scale of the frame [12, 32].

These oscillated waveforms are also overvoltage occurred by the surge current reflecting and passing on the frame, and by the DC power lines working as dipole

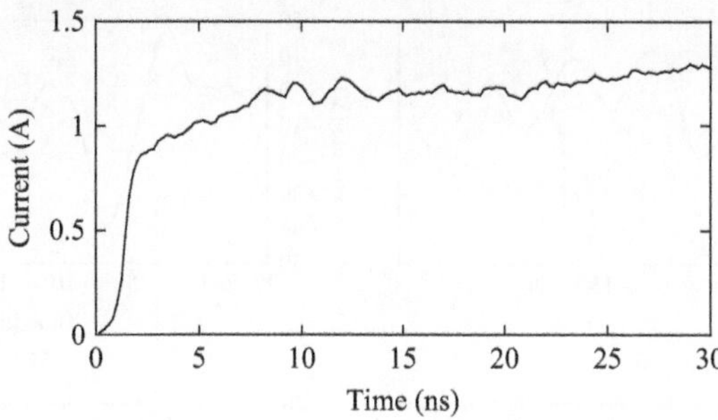

Figure 8.41   Injected current into the 1/10 scale model of frame supporting PV arrays in a large-scale photo-voltaic power plant [30]

© IEEJ 2012. Reprinted with permission from K. Yamamoto, J. Takami, N. Okabe, 'Overvoltages on DC side of power conditioning system caused by lightning strike to structure anchoring photovoltaic panels', *IEEJ Trans*, vol. 132, no. 11, pp. 903–913, November 2012

Table 8.1   Experimental cases and conditions [30]

Case number	Current injected point	Position of PCS
1.A	Point 1	A
1.B	Point 1	B
1.C	Point 1	C
2.A	Point 2	A
2.B	Point 2	B
2.C	Point 2	C

© IEEJ 2012. Reprinted with permission from K. Yamamoto, J. Takami, N. Okabe, 'Overvoltages on DC side of power conditioning system caused by lightning strike to structure anchoring photovoltaic panels', *IEEJ Trans*, vol. 132, no. 11, pp. 903–913, November 2012

antennas [33]. Especially, the surge current that flows on the conductor existing horizontally and close to the power line has a significant influence. All overvoltage waveforms are oscillating in a period of 0.13 μs or 0.27 μs cycle. Those oscillations are caused because the have half-wave length of each oscillation is equivalent to the approximate distances of 20 m or 40 m between the current injected point and the end of the frame. In Case 1.A, 0.13 μs periodic oscillations mainly occurred because the DC power line has 20 m length. In Cases 2.A, 2.B, and 2.C, because the length from current injected point to the end of the frame is also 20 m, 0.13 μs periodic oscillation mainly appeared. On the other hand, in Cases 1.B and 1.C, because the distance between the current injected point and the end of the frame is 40 m and the length of DC power line is 40 m, 0.27 μs periodic oscillation appeared.

The calculation results target the phenomena on the full-scale model, and its overvoltage has shown the oscillated and inductive waveform with the rise time of

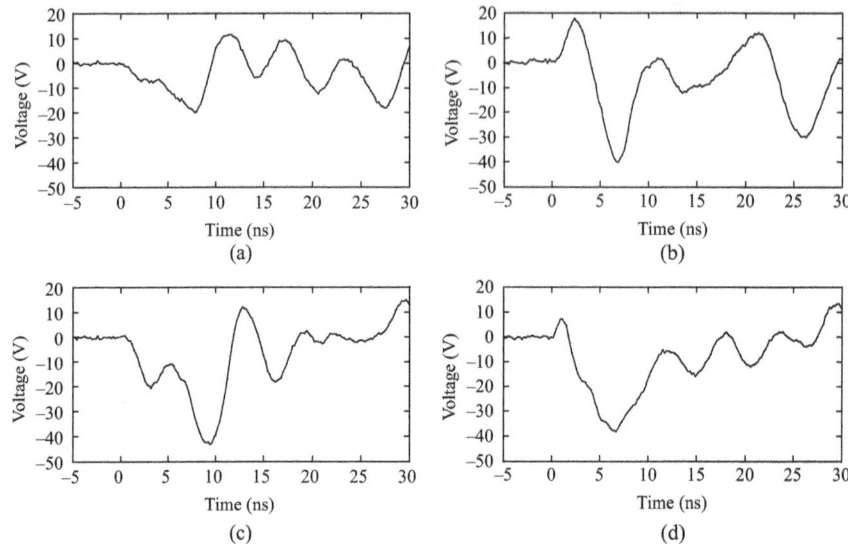

Figure 8.42   *Experimental results of V_1 and V_2 in Cases 1.A−1.B: (a) V_1 in Case 1.A; (b) V_2 in Case 1.A; (c) V_1 in Case 1.B; (d) V_1 in Case 1.C [30]*

© IEEJ 2012. Reprinted with permission from K. Yamamoto, J. Takami, N. Okabe, 'Overvoltages on DC side of power conditioning system caused by lightning strike to structure anchoring photovoltaic panels', *IEEJ Trans*, vol. 132, no. 11, pp. 903–913, November 2012

approximately 0.5 μs. When lightning strikes the large-scale PVPP, the potential at the frame supporting PV arrays rises depending on its earthing impedance. As the DC power lines are surrounded by the structures of the frame, the potential of the power line rises with that of the surrounding space. On the other hand, the potential of the PCS storage box installed inside the frame and earthed near the frame is mainly affected by the potential rise of the frame foot located closest to the PCS storage box. It has been thought that V_1 and V_2 have shown the negative polarity because the potential rise of the PCS is higher than that of the DC power lines around the both wave front and tail. In the study using the scale model, the earthing impedance of the frame is zero because the frame is directly connected to the aluminum plates spread all over the floor. However, even if the earthing impedance of the frame is zero, the transient potential rise appears due to electromagnetic fields formed around the frame [12, 35, 36]. The results, shown in Figures 8.45 and 8.46, in the scale model have indicated the negative polarity due to those reasons.

Sensitive analyses have been done using several different wave front durations of 0.2, 1.0, 5.5, and 10.0 μs. The analytical conditions in these sensitive analyses are the same as the previous calculations. Calculation results of Cases 1.A–2.C are shown in Figures 8.51 and 8.52. Maximum and steady-state values of overvoltage

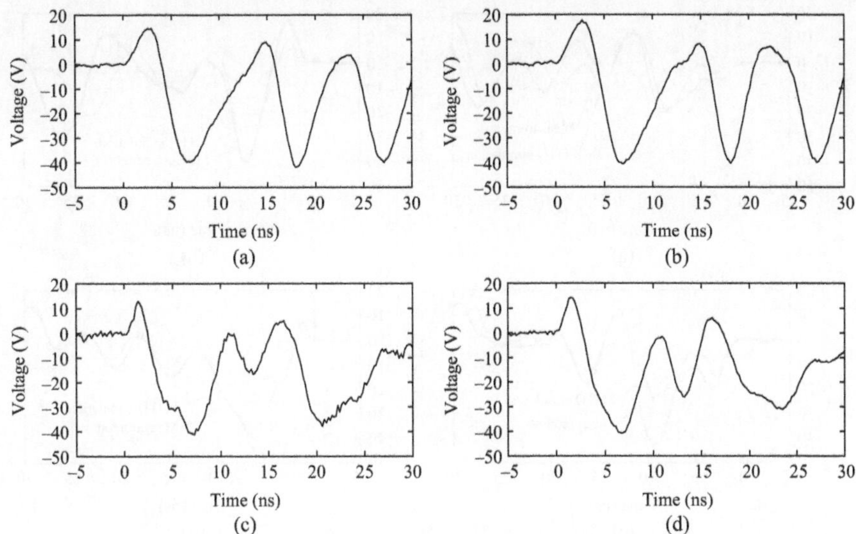

Figure 8.43   *Experimental results of V_1 and V_2 in Cases 2.A–2.C: (a) V_1 in Case 2.A; (b) V_2 in Case 2.A; (c) V_1 in Case 2.B; (d) V_1 in Case 2.C [30]*

© IEEJ 2012. Reprinted with permission from K. Yamamoto, J. Takami, N. Okabe, 'Overvoltages on DC side of power conditioning system caused by lightning strike to structure anchoring photovoltaic panels', *IEEJ Trans*, vol. 132, no. 11, pp. 903–913, November 2012

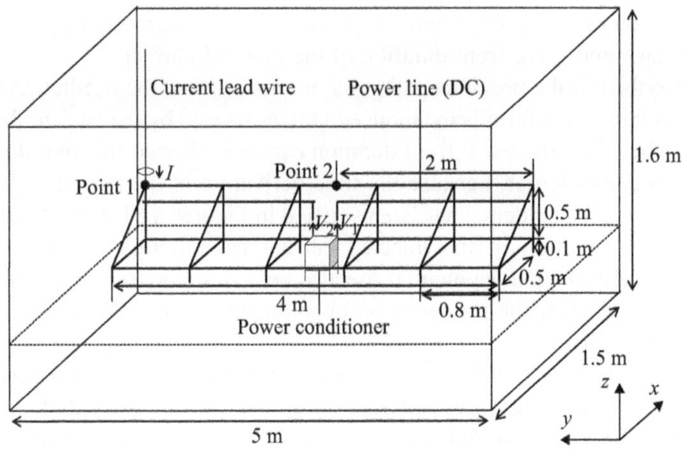

Figure 8.44   *Analytical space of the reduced-size model [30]*

© IEEJ 2012. Reprinted with permission from K. Yamamoto, J. Takami, N. Okabe, 'Overvoltages on DC side of power conditioning system caused by lightning strike to structure anchoring photovoltaic panels', *IEEJ Trans*, vol. 132, no. 11, pp. 903–913, November 2012

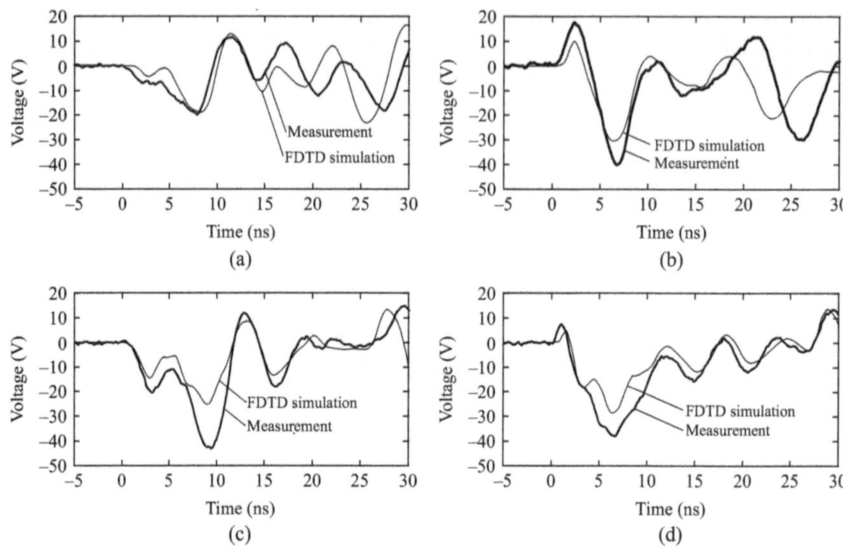

Figure 8.45   Analytical results of V_1 and V_2 in Cases 1.A−1.C: (a) V_1 in Case 1.A; (b) V_2 in Case 1.A; (c) V_1 in Case 1.B; (d) V_1 in Case 1.C [30]

© IEEJ 2012. Reprinted with permission from K. Yamamoto, J. Takami, N. Okabe, 'Overvoltages on DC side of power conditioning system caused by lightning strike to structure anchoring photovoltaic panels', *IEEJ Trans*, vol. 132, no. 11, pp. 903–913, November 2012

are summarized in Table 8.4. Figure 8.53 shows the relationship diagram between the overvoltage and wave front duration of the injected current.

As described in the previous paragraph in this section, the oscillation included in the overvoltage waveform is the induced voltage caused by the current flowing on the frame. Therefore, the wave front duration becomes shorter; the amplitude of the oscillation becomes larger, because the time derivative of the current (dI/dt [A/s]) becomes larger. This tendency has been verified in Figures 8.51–8.53. Furthermore, the wave front duration of the injected current gets shorter, and the inductive characteristic of the negative potential rise also becomes more sensible. On the other hand, it is obvious that the influence on overvoltage in the wave tail by the wave front duration is small.

Table 8.4 and Figure 8.53 represent the quantitative relationship between the maximum and steady-state overvoltage value and the wave front duration. For example, in V_2 in Case 1.A, 540 kV overvoltage may be applied to the DC side of the PCS when the lightning with the wave front duration of 0.2 µs and the peak current of 30 kA strikes the frame constructed on the earth with its resistivity of 100 Ωm [34]. Because this overvoltage level cannot be ignored on lightning protection designs, it is obvious that establishing the appropriate lightning protection for the PCS is important.

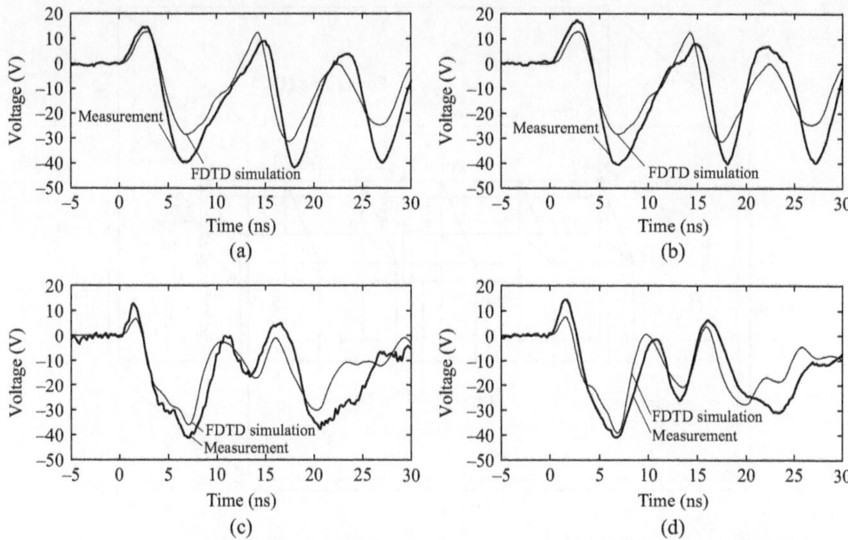

Figure 8.46   *Analytical results of V_1 and V_2 in Cases 2.A–2.C: (a) V_1 in Case 2.A;*
(b) V_2 in Case 2.A; (c) V_1 in Case 2.B; (d) V_1 in Case 2.C [30]

© IEEJ 2012. Reprinted with permission from K. Yamamoto, J. Takami, N. Okabe, 'Overvoltages on DC side of power conditioning system caused by lightning strike to structure anchoring photovoltaic panels', *IEEJ Trans*, vol. 132, no. 11, pp. 903–913, November 2012

Table 8.2   *Maximum values of overvoltage in the cases of the 1/10 scale model [30]*

Case number	Maximum voltage (V)	
	Measurement	**FDTD**
1.A	−40.1	−30.4
1.B	−43.3	−25.3
1.C	−38.2	−28.8
2.A	−41.8	−31.4
2.B	−41.2	−36.1
2.C	−40.8	−39.3

© IEEJ 2012. Reprinted with permission from K. Yamamoto, J. Takami, N. Okabe, 'Overvoltages on DC side of power conditioning system caused by lightning strike to structure anchoring photovoltaic panels', *IEEJ Trans*, vol. 132, no. 11, pp. 903–913, November 2012

Sensitive analyses using several different values of earth resistivity of 1 Ωm, 10 Ωm, 100 Ωm, 1,000 Ωm have also been done. Calculation results of Cases 1.A–2.C are shown in Figures 8.54 and 8.55. Maximum and steady-state values of overvoltage are summarized in Table 8.5. Figure 8.56 is the relationship diagram between the overvoltage and earth resistivity.

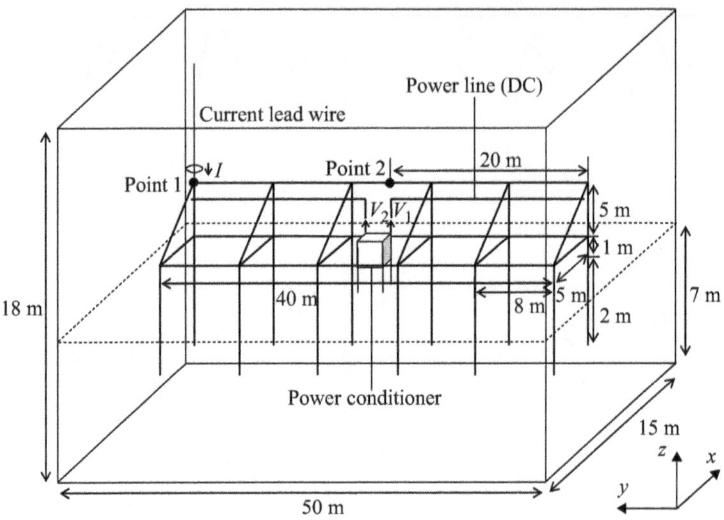

Figure 8.47   Analytical model of an actual-scale photovoltaic power plant [30]

© IEEJ 2012. Reprinted with permission from K. Yamamoto, J. Takami, N. Okabe, 'Overvoltages on DC side of power conditioning system caused by lightning strike to structure anchoring photovoltaic panels', *IEEJ Trans*, vol. 132, no. 11, pp. 903–913, November 2012

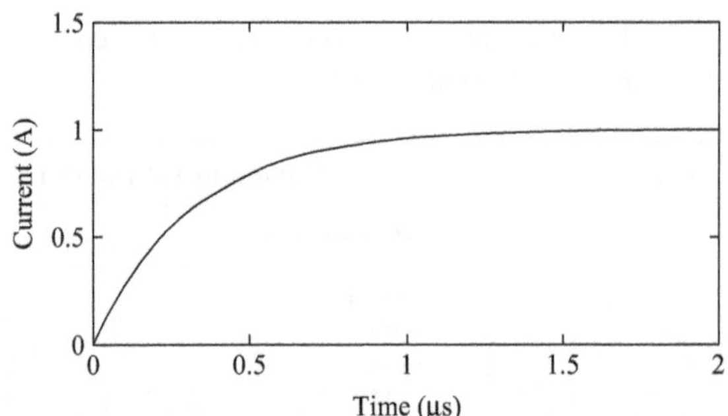

Figure 8.48   Injected current into the actual-scale model of frame supporting PV arrays in a large-scale photo-voltaic power plant [30]

© IEEJ 2012. Reprinted with permission from K. Yamamoto, J. Takami, N. Okabe, 'Overvoltages on DC side of power conditioning system caused by lightning strike to structure anchoring photovoltaic panels', *IEEJ Trans*, vol. 132, no. 11, pp. 903–913, November 2012

As mentioned in the previous paragraph in this section, the potential of the DC power lines rises with that of the surrounding space. On the other hand, the potential of the PCS storage box is mainly affected from the potential rise of the frame foot located near the PCS. As the earth resistivity is larger, the earthing

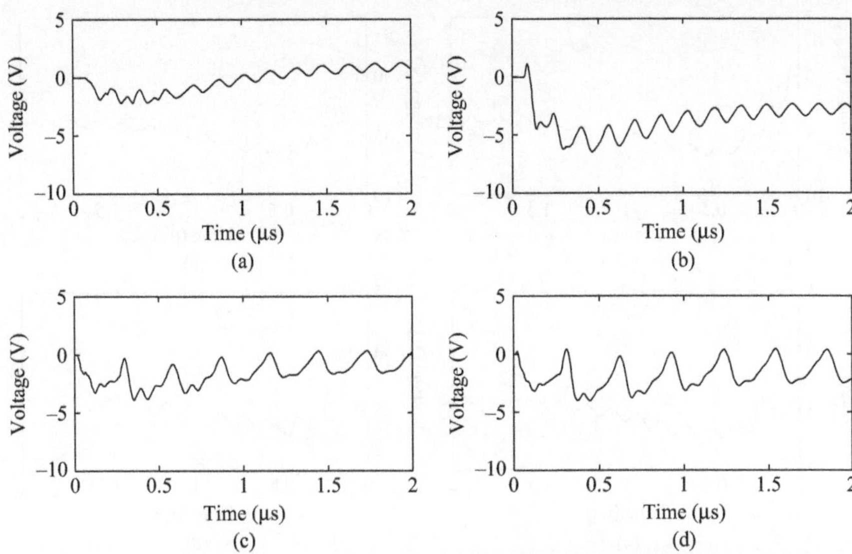

Figure 8.49   *Analytical results in Cases 1.A–1.C [30]*

© IEEJ 2012. Reprinted with permission from K. Yamamoto, J. Takami, N. Okabe, 'Overvoltages on DC side of power conditioning system caused by lightning strike to structure anchoring photovoltaic panels', *IEEJ Trans*, vol. 132, no. 11, pp. 903–913, November 2012

resistance of the frame is also larger. Therefore, the potential rise of the frame foot near the PCS is also larger. On the other hand, the potential of the DC power lines is also larger with the increase of the potential of the frame. Because the potential of the DC power lines is larger than that of the PCS storage box at the wave tail in Figures 8.54 and 8.55, the V_1 and V_2 become positive. At the wave front, the potential of the PCS storage box becomes larger than that of the DC power lines because the inductive earthing potential rise appears around the PCS.

Table 8.5 and Figure 8.56 represent the quantitative relationship between the overvoltage values at the wave front and tail and the earthing resistivity. For example, in V_1 in Case 1.A, 340 kV overvoltage may be applied to the DC side of the PCS when the lightning with the wave front duration of 1.0 μs and the peak current of 30 kA strikes the frame constructed on the earth with its resistivity of 1 kΩm. This overvoltage level cannot be ignored while preparing lightning protection designs, and the problem should be resolved immediately.

8.2.2   Overvoltage caused by a lightning strike to a solar power generation system

It is thought that lightning overvoltage in PV systems depends on the type of modules, the method of installing the module mount, positions of wirings and the earthing system, etc. However, there is no report investigating the lightning tolerance on actual equipment.

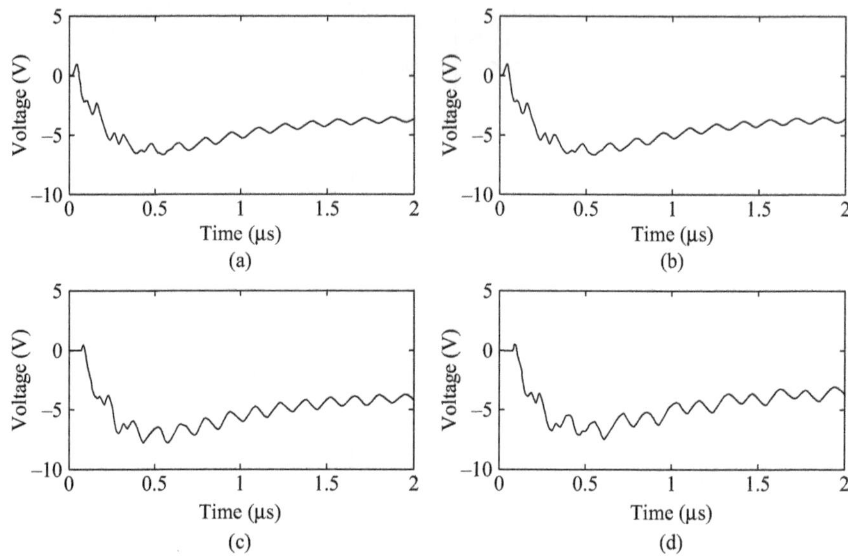

Figure 8.50   Analytical results in Cases 2.A–2.C [30]
© IEEJ 2012. Reprinted with permission from K. Yamamoto, J. Takami, N. Okabe, 'Overvoltages on DC side of power conditioning system caused by lightning strike to structure anchoring photovoltaic panels', *IEEJ Trans*, vol. 132, no. 11, pp. 903–913, November 2012

Table 8.3   Maximum values of overvoltage in the cases of the actual-scale model [30]

Case number	Maximum voltage (V)
1.A	−6.47
1.B	−3.88
1.C	−4.01
2.A	−6.66
2.B	−7.77
2.C	−7.44

© IEEJ 2012. Reprinted with permission from K. Yamamoto, J. Takami, N. Okabe, 'Overvoltages on DC side of power conditioning system caused by lightning strike to structure anchoring photovoltaic panels', *IEEJ Trans*, vol. 132, no. 11, pp. 903–913, November 2012

The overvoltage on the DC wirings in an actual small PV system has been measured and analyzed numerically when surge current is injected into the frame of the PV array. In this section, the results are described. Furthermore, the level of the overvoltage and the generation mechanism of the overvoltage are explained [37].

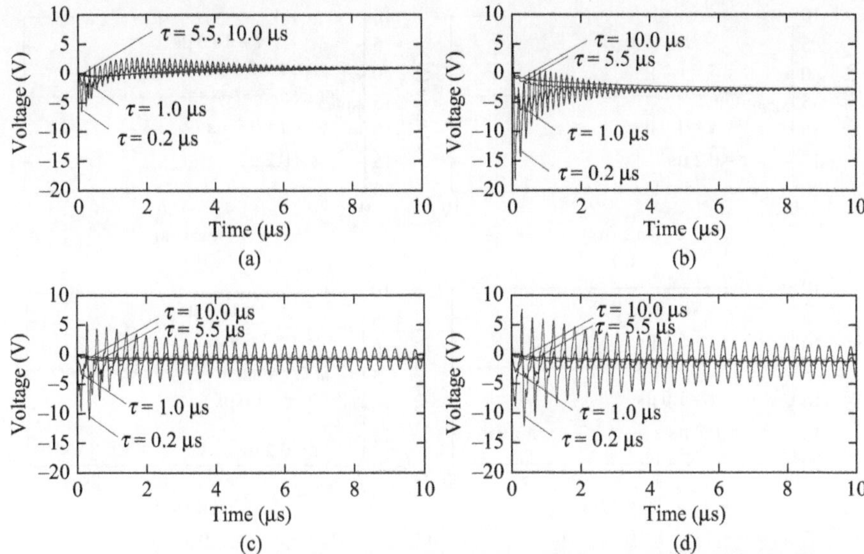

Figure 8.51   Analytical results in Cases 1.A–1.C for sensitive analyses using several different wave front durations [30]

© IEEJ 2012. Reprinted with permission from K. Yamamoto, J. Takami, N. Okabe, 'Overvoltages on DC side of power conditioning system caused by lightning strike to structure anchoring photovoltaic panels', *IEEJ Trans*, vol. 132, no. 11, pp. 903–913, November 2012

8.2.2.1   Investigations by experiments

Figure 8.57 shows the experimental set-up, Figure 8.58 shows the PV array outline, and Figure 8.59 shows the installation of PV arrays. An impulse generator (referred in the figure as I.G.) was set up at approximately 100 m away from the PV power generation system. A current injection wire with the cross-sectional area of 5.5 mm^2 was laid approximately 1 m above the earth.

As shown in Figure 8.60, the DC wires from the array are incoming into the steel junction box through the metal pipe shown in Figure 8.59(c), and into the distribution board through the vinyl pipe shown in Figure 8.59(d). For the wirings from the distribution board to the PCS located in the building near the PV system, the cross-linked polyethylene cables with a cross-sectional area of 8 mm^2 are used. The earthing system of the building in which the PCS is located and the distribution board are connected with an insulated wire with a cross-sectional area of 5.5 mm^2. The earthing systems of the array and the distribution board are also connected usually by the same insulated wire. This means that the earthing systems of the array, the distribution board, and the building are connected.

The PV system is composed of 12 units of arrays. One array is composed of 10 single-crystal modules (maximum output of 65 W, maximum output voltage of

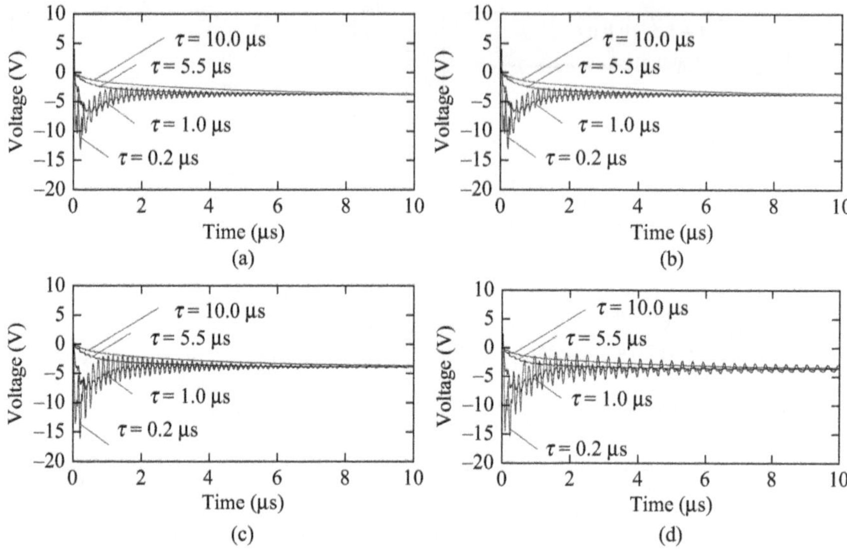

Figure 8.52   Analytical results in Cases 2.A–2.C for sensitive analyses using several different wave front durations [30]

© IEEJ 2012. Reprinted with permission from K. Yamamoto, J. Takami, N. Okabe, 'Overvoltages on DC side of power conditioning system caused by lightning strike to structure anchoring photovoltaic panels', *IEEJ Trans*, vol. 132, no. 11, pp. 903–913, November 2012

Table 8.4   Maximum and steady-state values of overvoltages [30]

Case number / Calculated voltage	Overvoltage (V)							
	Values at the wave front				Values at the wave tail			
	$\tau = 0.2\ \mu s$	$\tau = 1.0\ \mu s$	$\tau = 5.5\ \mu s$	$\tau = 10.0\ \mu s$	$\tau = 0.2\ \mu s$	$\tau = 1.0\ \mu s$	$\tau = 5.5\ \mu s$	$\tau = 10.0\ k\mu s$
1.A / V_1	−6.33	−2.18	−0.54	−0.31	1.05	1.05	1.05	1.05
1.A / V_2	−18.03	−6.47	−2.74	−2.74	−2.73	−2.74	−2.74	−2.74
1.B / V_1	−11.13	−3.88	−1.07	−0.81	−0.79	−0.79	−0.79	0.80
1.C / V_1	−12.31	−4.01	−1.44	−1.37	−1.33	−1.35	−1.34	1.34
2.A / V_1	−12.90	−6.66	−3.57	−3.58	−3.57	−3.57	−3.57	3.57
2.A / V_2	−12.90	−6.66	−3.57	−3.58	−3.57	−3.57	−3.57	−3.57
2.B / V_1	−16.02	−7.77	−3.88	−3.88	−3.88	−3.88	−3.87	−3.86
2.C / V_1	−14.81	−7.44	−3.43	−3.42	−3.42	−3.41	−3.41	−3.40

© IEEJ 2012. Reprinted with permission from K. Yamamoto, J. Takami, N. Okabe, 'Overvoltages on DC side of power conditioning system caused by lightning strike to structure anchoring photovoltaic panels', *IEEJ Trans*, vol. 132, no. 11, pp. 903–913, November 2012

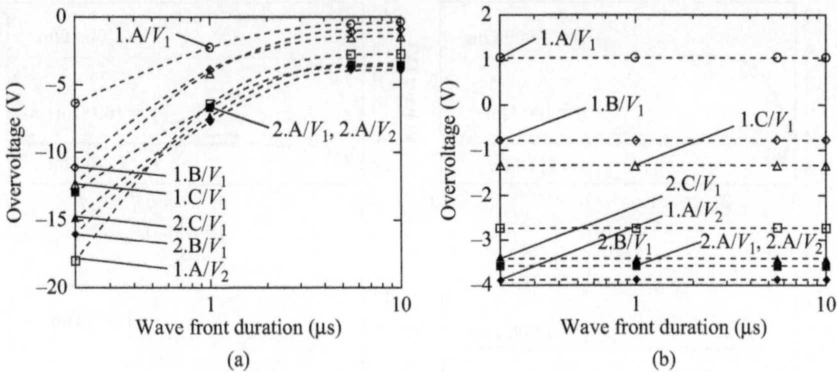

Figure 8.53   *Overvoltage characteristics depending on the wave front duration:*
(a) Values at the wave front; (b) Values at the wave tail [30]

© IEEJ 2012. Reprinted with permission from K. Yamamoto, J. Takami, N. Okabe,
'Overvoltages on DC side of power conditioning system caused by lightning strike to
structure anchoring photovoltaic panels', *IEEJ Trans*, vol. 132, no. 11, pp. 903–913,
November 2012

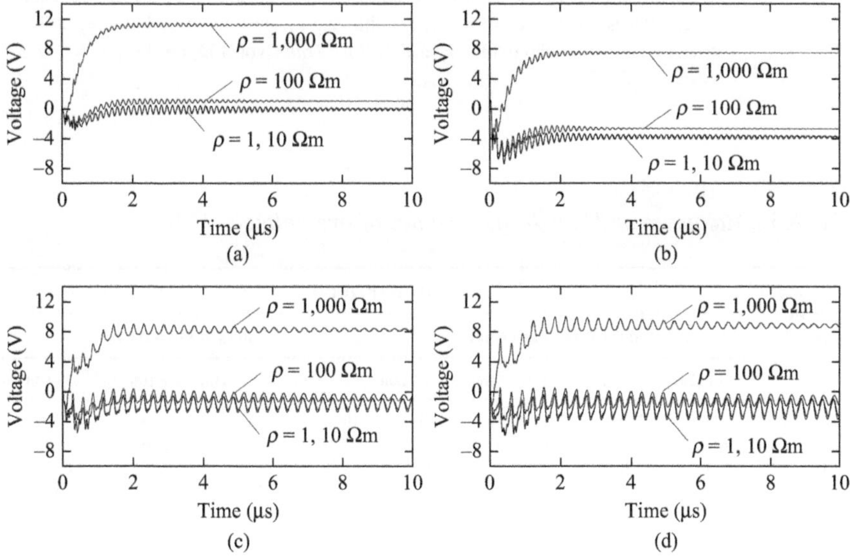

Figure 8.54   *Analytical results in Cases 1.A–1.C for sensitive analyses using*
several different values of earth resistivity [30]

© IEEJ 2012. Reprinted with permission from K. Yamamoto, J. Takami, N. Okabe,
'Overvoltages on DC side of power conditioning system caused by lightning strike to
structure anchoring photovoltaic panels', *IEEJ Trans*, vol. 132, no. 11, pp. 903–913,
November 2012

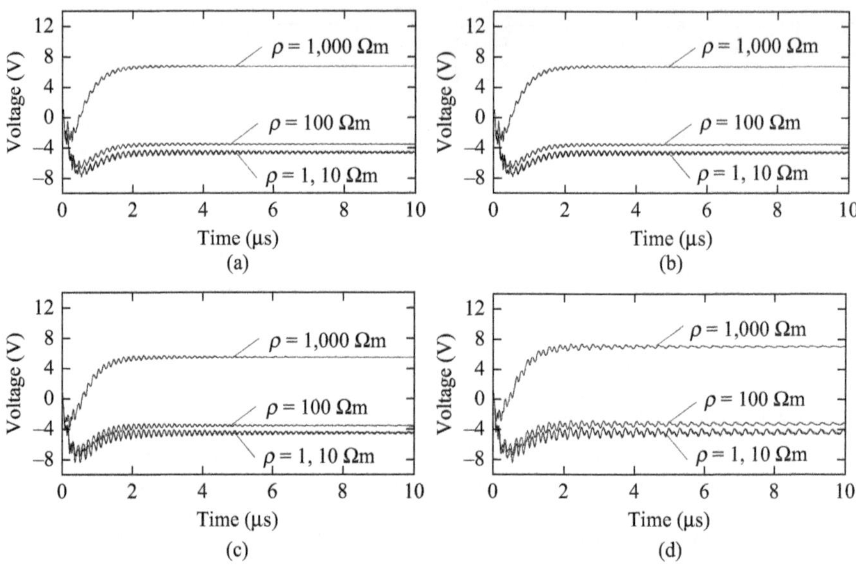

Figure 8.55   *Analytical results in Cases 2.A–2.C for sensitive analyses using several different values of earth resistivity [30]*

© IEEJ 2012. Reprinted with permission from K. Yamamoto, J. Takami, N. Okabe, 'Overvoltages on DC side of power conditioning system caused by lightning strike to structure anchoring photovoltaic panels', *IEEJ Trans*, vol. 132, no. 11, pp. 903–913, November 2012

Table 8.5   *Maximum and steady-state values of overvoltages [30]*

Case number / Calculated voltage	Overvoltage (V)							
	Values at the wave front				Values at the wave tail			
	$\rho=1\ \Omega\mathrm{m}$	$\rho=10\ \Omega\mathrm{m}$	$\rho=100\ \Omega\mathrm{m}$	$\rho=1\ \mathrm{k}\Omega\mathrm{m}$	$\rho=1\ \Omega\mathrm{m}$	$\rho=10\ \Omega\mathrm{m}$	$\rho=100\ \Omega\mathrm{m}$	$\rho=1\ \mathrm{k}\Omega\mathrm{m}$
1.A / V_1	−2.86	−2.75	−2.18	−1.41	−0.09	0.02	1.05	11.39
1.A / V_2	−7.49	−7.35	−6.47	−3.88	−3.87	−3.76	−2.74	7.58
1.B / V_1	−5.56	−5.36	−3.88	−2.29	−1.78	−1.69	−0.79	8.23
1.C / V_1	−6.04	−5.79	−4.01	−2.01	−2.40	−2.33	−1.35	8.86
2.A / V_1	−7.83	−7.61	−6.66	−3.15	−4.70	−4.60	−3.57	6.73
2.A / V_2	−7.83	−7.61	−6.66	−3.15	−4.70	−4.60	−3.57	6.73
2.B / V_1	−8.67	−8.63	−7.77	−4.06	−4.86	−4.77	−3.88	5.11
2.C / V_1	−8.46	−8.31	−7.44	−3.02	−4.55	−4.44	−3.41	6.78

© IEEJ 2012. Reprinted with permission from K. Yamamoto, J. Takami, N. Okabe, 'Overvoltages on DC side of power conditioning system caused by lightning strike to structure anchoring photovoltaic panels', *IEEJ Trans*, vol. 132, no. 11, pp. 903–913, November 2012

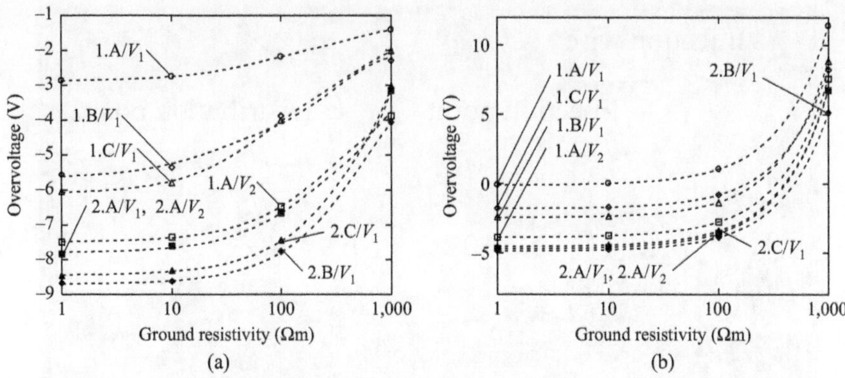

Figure 8.56   *Overvoltage characteristics depending on the earth resistivity:*
(a) Values at the wave front; (b) Values at the wave tail [30]

© IEEJ 2012. Reprinted with permission from K. Yamamoto, J. Takami, N. Okabe, 'Overvoltages on DC side of power conditioning system caused by lightning strike to structure anchoring photovoltaic panels', *IEEJ Trans*, vol. 132, no. 11, pp. 903–913, November 2012

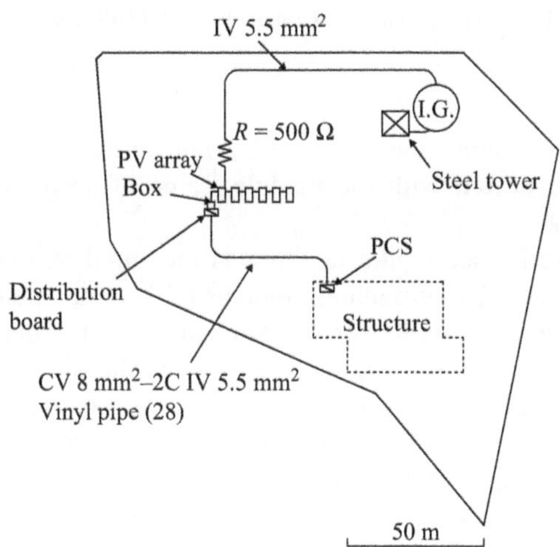

Figure 8.57   *Experimental set-up [37]*

© IEEE 2013. Reprinted with permission from Sakai, K. and Yamamoto, K., 'Lightning protection of photovoltaic power generation system: Influence of grounding systems on overvoltages appearing on DC wirings', *IEEE XPLORE 2013*, XII SIPDA, pp. 418–422, October 2013

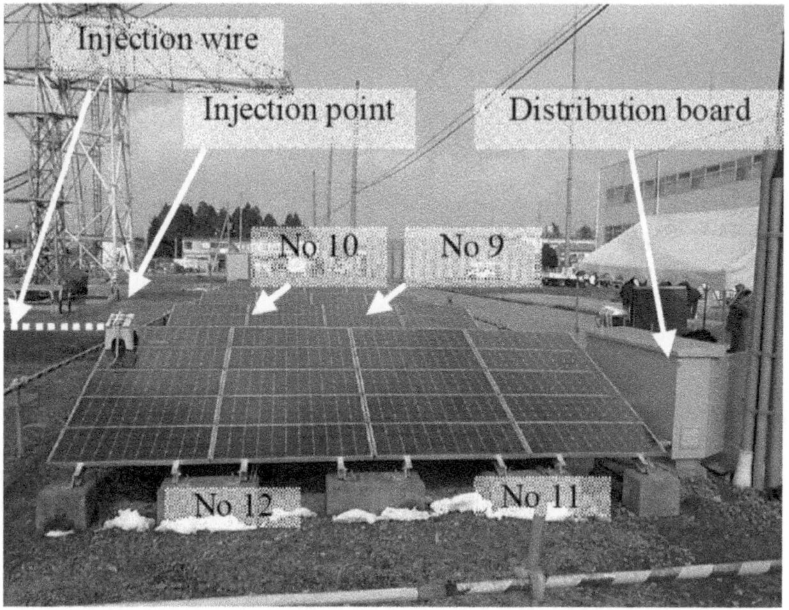

Figure 8.58   PV arrays outline [37]

© IEEE 2013. Reprinted with permission from Sakai, K. and Yamamoto, K., 'Lightning protection of photovoltaic power generation system: Influence of grounding systems on overvoltages appearing on DC wirings', *IEEE XPLORE 2013*, XII SIPDA, pp. 418–422, October 2013

21.3 V, and maximum output current of 3.05 A) in series. The injection wire of the surge current is connected with the module frame of array no. 12 through the resistance of 500 Ω.

The earthing resistance of the facilities is measured. A one-rod electrode is installed for two arrays. The installation mounts of the array are connected to each other with insulated wires with a cross-sectional area of 2 mm^2. The combined earthing resistance of the entire array is about 14 Ω. The earthing resistance of the entire PV system connected with the entire array mounts, the distribution board, and the building is about 0.5 Ω, which is almost equal to the earthing resistance of the building that has low earthing resistance. In this study, it was impossible to measure the earthing resistance of the distribution board individually because the underground connection wire was not able to be disconnected.

Two single-core CV cables with a cross-sectional area of 5.5 mm^2 protected by metal piping with a cross-sectional area of 22 mm^2 are utilized as the wires from the array to the positive and negative electrodes at the distribution board. The overvoltage between the positive and negative DC wirings of the array and the earth at the distribution board has been measured.

The experiments are conducted following two conditions. In the first condition, the earthing systems of the rod electrode of each array, the installation

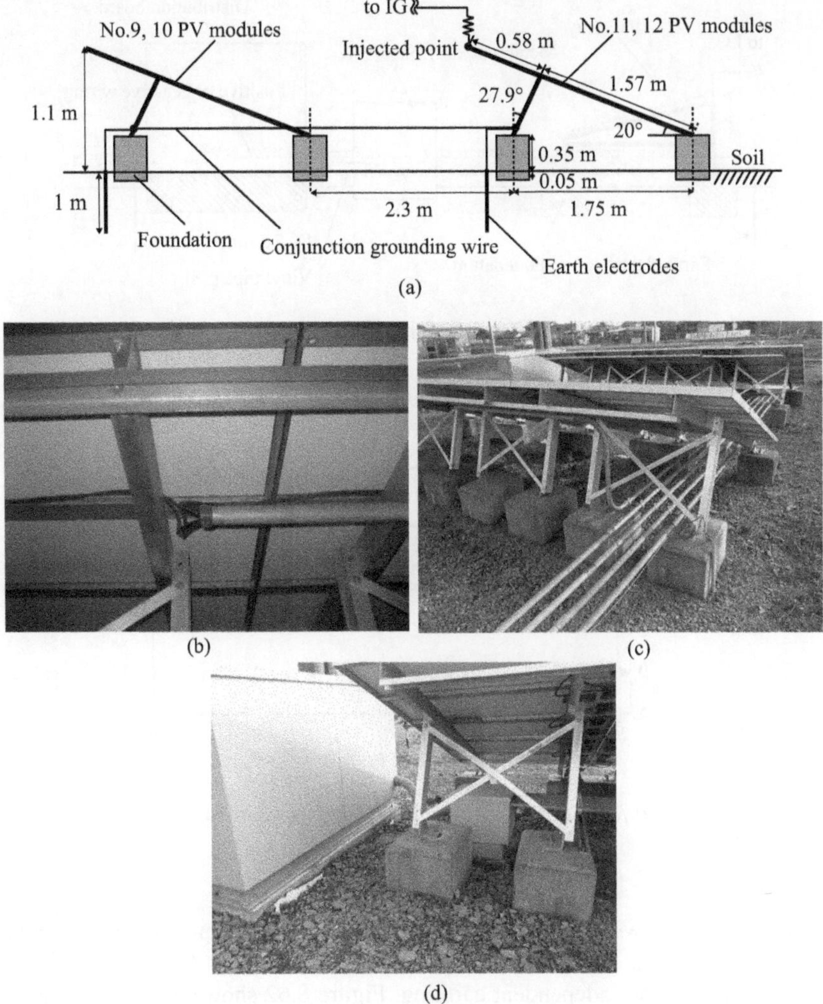

Figure 8.59   *Installation of PV arrays: (a) Earthing system of PV arrays;*
(b) Wiring layout; (c) Metal pipes under a PV array; (d) Layout of
PV modules in a PV array [37]

© IEEE 2013. Reprinted with permission from Sakai, K. and Yamamoto, K., 'Lightning
protection of photovoltaic power generation system: Influence of grounding systems on
overvoltages appearing on DC wirings', *IEEE XPLORE 2013*, XII SIPDA, pp. 418–422,
October 2013

mount, and the distribution board are connected. This condition is called
"integrated earthing." In the second condition, the earthing systems of each array
and the rod electrode are connected, but the earthing systems of the array and
distribution board are not connected. This condition is called "independent
earthing."

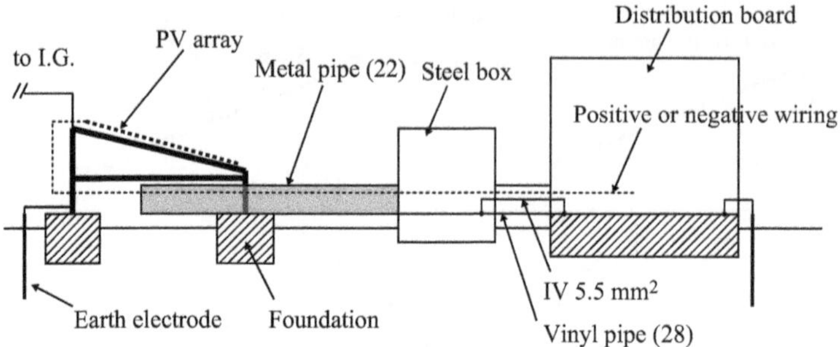

Figure 8.60   Arrangement of wirings

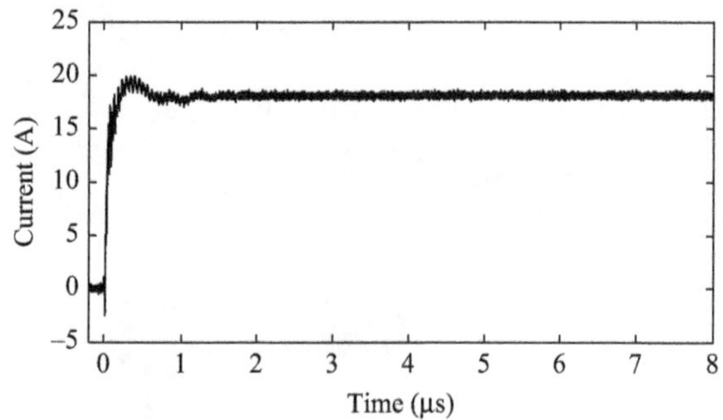

Figure 8.61   Injected current into the PV arrays shown in Figure 8.58

Figure 8.61 shows the injected current waveform, a steep-fronted wave when the injected current is used; its maximum current is 18.4 A at 0.35 μs.

In the case of the independent earthing, Figure 8.62 shows the overvoltages at the array no. 12. The overvoltage at other arrays became smaller than that at no. 12, and the shape of the waveform was similar. Therefore, these waveforms are not shown in this chapter. The overvoltages between the positive and negative DC wirings of the array and the earth at the distribution board became very similar because the positive wire was placed close to the negative wire. In this chapter, only the overvoltage between the negative wire and the earth is shown in Figure 8.62.

At array no. 12, a larger voltage between the positive/negative wirings and the earth appeared as compared to those at other arrays. The voltage reached at 335 V at 0.25 μs. Then, the voltage droped to 190 V and kept the constant value of about 210 V until 80 μs during the measurement period. The voltage between the positive and negative wirings became almost zero because the waveforms of the positive and negative wirings were very similar.

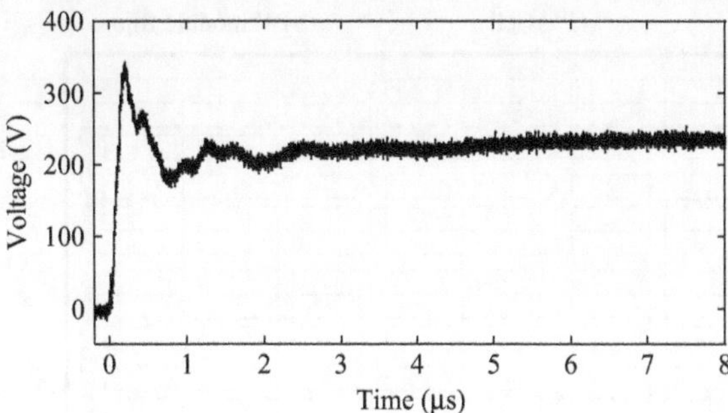

Figure 8.62   Overvoltage in the case of the independent earthing

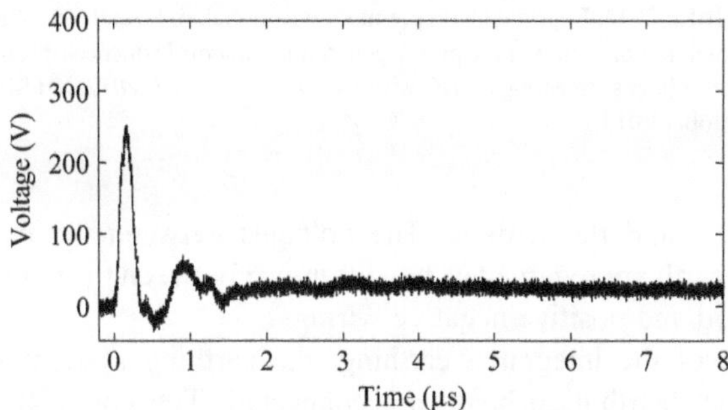

Figure 8.63   Overvoltage in the case of the integrated earthing

In the case of the integrated earthing, Figure 8.63 shows the overvoltage at the array no. 12. The waveforms of the overvoltage between the negative power line and the earth appeared in array no. 12 showed damped oscillation. The maximum overvoltage between the negative power line and the earth was 250 V at 0.2 μs, and then the voltage between the power lines and the earth kept the constant of 9 V. The voltage between the positive and negative wirings became almost zero because the waveforms of the positive and negative wirings were very similar.

8.2.2.2 Discussions of overvoltage

In the case of the independent earthing, the potential of the mounts and supporting structures of the arrays rises when the current is injected. The potential of the distribution board also rises depending on the potential rise around the mounts near the distribution board. However, both the earthing systems of the mounts and the distribution board are not connected; the potential of the mounts and the supporting structures is larger than that of the distribution board. The potential of the positive/ negative wirings rises due to the inductive and electrostatic coupling between the

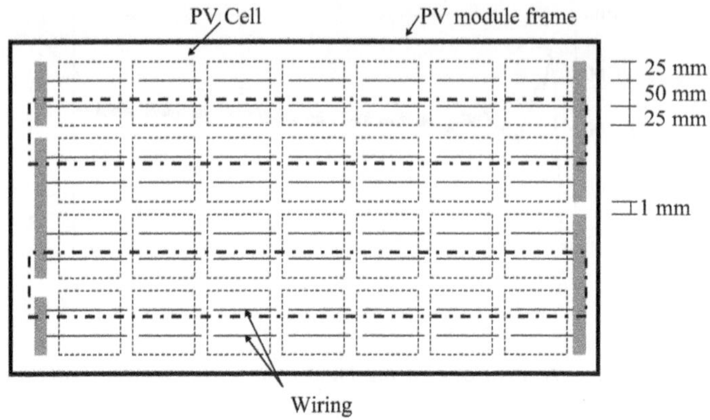

Figure 8.64   Inside wirings of a PV module [37]

© IEEE 2013. Reprinted with permission from Sakai, K. and Yamamoto, K., 'Lightning protection of photovoltaic power generation system: Influence of grounding systems on overvoltages appearing on DC wirings', *IEEE XPLORE 2013*, XII SIPDA, pp. 418–422, October 2013

support structures and the wirings. The voltages between the positive/negative wirings and the earth are regarded as the difference between the potential rise of the distribution board and positive/negative wirings.

In the case of the integrated earthing, the earthing systems of the support structure and the distribution board are connected. The potential of the support structures rises when the current is injected. The potential of the support structure and the distribution board becomes similar because both are connected. The potential of the positive wiring and negative wiring rises due to the inductive and electrostatic coupling between the support structure and the wirings. For such a reason, the positive and negative wiring voltages to earth at the wave tail became smaller than that in case of the independent earthing.

Figure 8.64 shows the inside wirings of a PV module. Figure 8.65 shows the picture of the PV module wirings. The area of the module is 0.51 m^2. The area encircled by wirings, which influences electromagnetic coupling, is 0.12 m^2. The injected current spreads over the support structures, and generates magnetic flux around it. Due to this magnetic flux, the voltage occurs in the wirings connected to the modules. However, the wirings between modules are usually processed at a factory. It is difficult to decrease the area encircled by this wiring.

8.2.2.3   Investigations by the FDTD simulations

The FDTD simulation model has been made to reproduce and verify the measured results in the previous sections. Figure 8.66 shows the analytical space. The dimension of the analytical space is 24.9 m × 8.275 m × 5.625 m and it is divided into cubic cells with a side length of 5 mm. The absorbing boundary condition is the Liao's second-order condition.

Figure 8.65   PV module wirings [37]

© IEEE 2013. Reprinted with permission from Sakai, K. and Yamamoto, K., 'Lightning protection of photovoltaic power generation system: Influence of grounding systems on overvoltages appearing on DC wirings', *IEEE XPLORE 2013*, XII SIPDA, pp. 418–422, October 2013

The model of the PV system is installed on the earth shown in Figure 8.67. The stratified earth structures of the resistivity are measured by the Wenner method. The current injection wire using a thin wire model is connected to the frame of the array no. 12. The same current in measurements is injected through the 500 Ω. Thin wire models are also used for the power lines, structures supporting modules, and wirings in the modules. The wirings in the metal pipes are modeled as thin wires in a metal pipe model shown in Figure 8.68.

The earthing resistance values are compared at the experimental and calculation set-ups. The comparisons are shown in Table 8.6. Under such analytical conditions, V_+ and V_-, which are the voltages between positive/negative power lines and the earth of the distribution board, are calculated.

Figure 8.69 shows the injected current waveforms, and Figures 8.70 and 8.71 show the comparisons of the overvoltage between the negative wiring and the earth at the distribution board in cases of the independent and integrated earthings, respectively.

In the case of the independent earthing, at array no. 12, the voltage between the negative wiring and the earth reached at 330 V at 0.25 μs. Then, the voltage kept the constant value about 200 V. The voltage between the positive and negative wirings became almost zero because the waveforms of the positive and negative wirings were very similar. These features are very similar with the measurements.

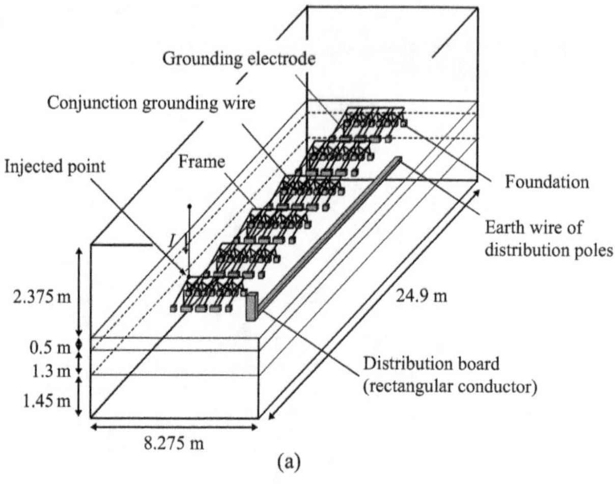

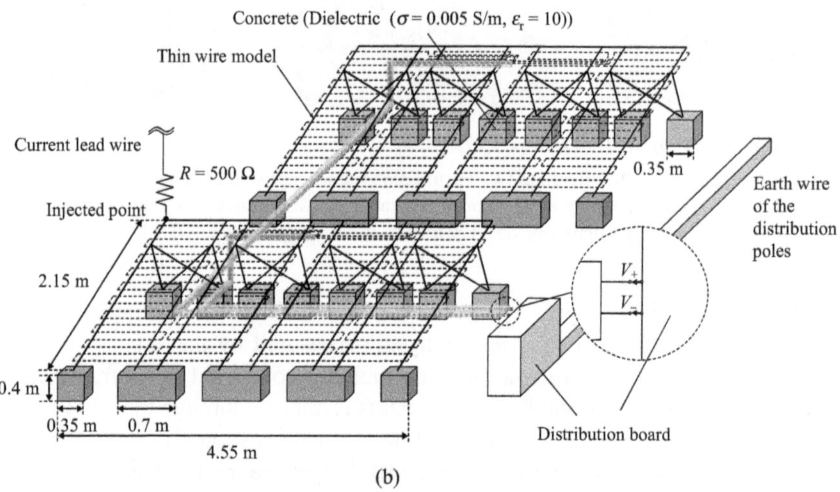

Figure 8.66   PV system model: (a) Outline; (b) Inside wiring

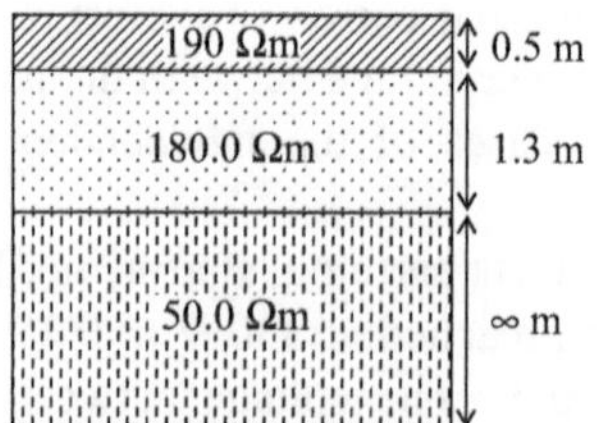

Figure 8.67   Earth resistivity at the experimental site

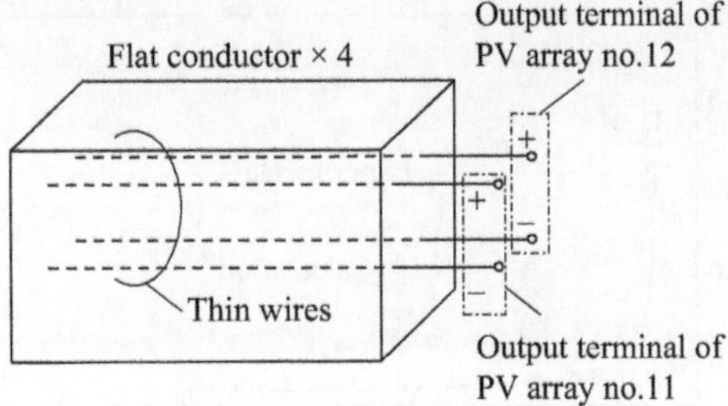

Figure 8.68   Metal pipe model

Table 8.6   Earthing resistance value

	Experimental [Ω]	Calculated [Ω]
PV array	14–30	11
Earth electrode	120–300	260
Distribution board	0.52	0.5
PV arrays and distribution board	0.51	0.32

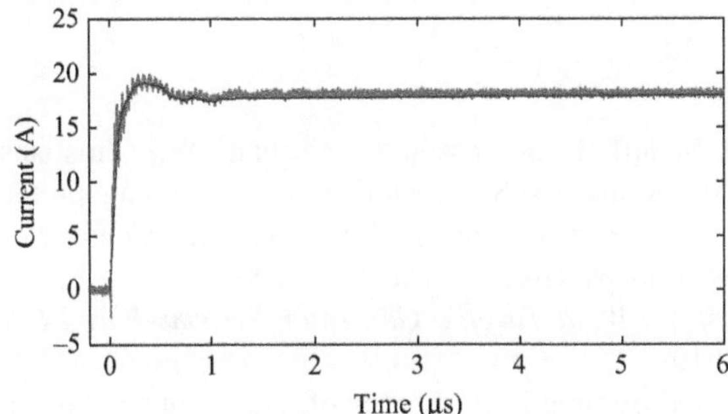

Figure 8.69   Injected current into the PV arrays

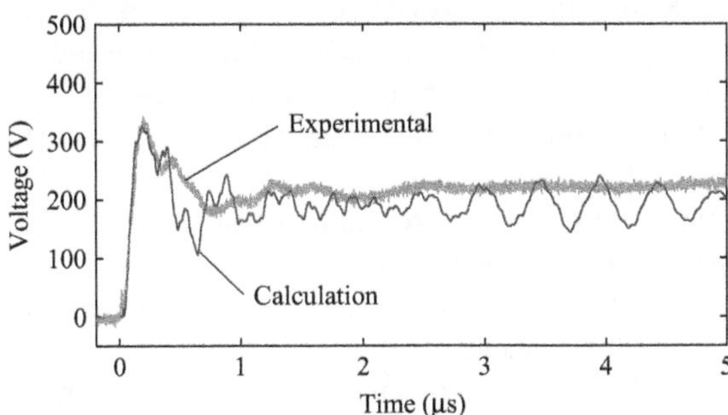

Figure 8.70   Comparison of overvoltage in the case of the independent earthing

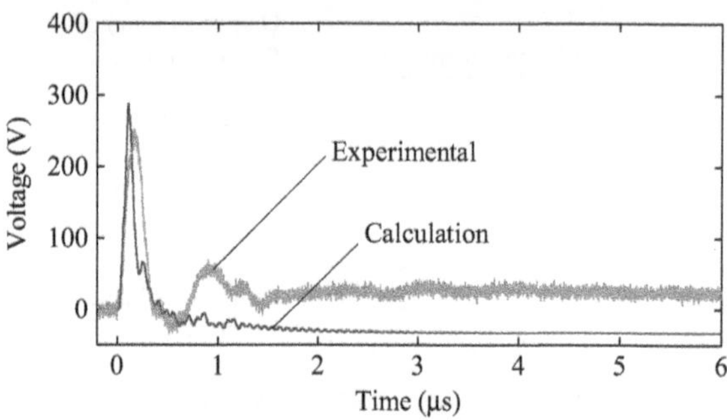

Figure 8.71   Comparison of overvoltage in the case of the integrated earthing

In the case of the integrated earthing, the waveforms of the overvoltage between the negative power line and the earth appeared in array no. 12 showed inductive. The maximum overvoltage was 290 V at 0.15 μs, and then the voltage between the power lines and the earth kept the constant of −20 V. The voltage between the positive and negative wirings became almost zero because the waveforms of the positive and negative wirings were very similar. For all the waveforms, the calculation results agree well with the measured ones.

References

[1]   Cotton I., Mcniff B., Soerenson T., Zischank W., Christiansen P., Hoppe-Kilpper M., Ramakers S. ... Muljadi E. 'Lightning protection for wind turbines'. *Proceedings of the 25th International Conference on Lightning Protection*; Rhodes, Greece, September 2000; pp. 848–853.

[2]   IEC 61400-24: *Wind Turbine Generator Systems-Part 24: Lightning Protection*. 2010.

[3]   Ribrant J. and Bertling L.M. 'Survey of failures in wind power systems with focus on Swedish wind power plants during 1997–2005'. *IEEE Transactions on Energy Conversion*; March 2007; vol. 22, no. 1, pp. 167–173.

[4]   Rachidi F., Rubinstein M., Montanya J., Bermudez J. -L., Rodriguez Sola R., Sola G. and Korovkin N. 'A review of current issues in lightning protection of new-generation wind-turbine blades'. *IEEE Transactions on Industrial Electronics*; June 2008; vol. 55, no. 6, pp. 2489–2496.

[5]   NEDO: 'Wind Turbine Failures and Troubles Investigating Committee annual report'. 2006 (in Japanese).

[6]   NEDO: 'Wind Turbine Failures and Troubles Investigating Committee annual report'. 2007 (in Japanese).

[7]   NEDO: 'Wind Turbine Failures and Troubles Investigating Committee annual report'. 2008 (in Japanese).

[8]   Yamamoto K., Noda T., Yokoyama S., Ametani A. 'Experimental and analytical studies of lightning overvoltages in wind turbine generator systems'. *Electric Power Systems Research*; March 2009; vol. 79, no. 3, pp. 436–442.

[9]   Yee K. S., Chen J. S., Chang A. H. 'Conformal finite-difference time-domain (FDTD) with overlapping grids'. *IEEE Transactions on Antennas and Propagation*; 1992; vol. 40, pp. 1068–1075.

[10] Kunz K. S., Luebbers, R. J. *The Finite Difference Time Domain Method for Electromagnetics*. Boca Raton, FL: CRC; 1993.

[11] Noda T., Yokoyama S. 'Thin wire representation in finite difference time domain surge simulation'. *IEEE Transactions on Power Delivery*; July 2002; vol. 17, pp. 840–847.

[12] Yamamoto K., Noda T., Yokoyama S., Ametani A. 'An experimental study of lightning overvoltages in wind turbine generation systems using a reduced-size model'. *Electrical Engineering in Japan*; March 2007; vol. 158, no. 4, pp. 22–30.

[13] Yamamoto K., Yanagawa S., Sekioka S., Yokoyama S. 'Transient grounding characteristics of an actual wind turbine generator system at a low resistivity site', *IEEJ Transactions on Electrical and Electronic Engineering*; January 2010; vol. 5, no. 1, pp. 21–26.

[14] Yamamoto K., Yanagawa S. 'Measurements of transient grounding characteristics at a wind turbine under construction'. *ISWL2011: 3rd International Symposium on Winter Lightning*; June 2011; no. 3–4, pp. 97–101.

[15] Yamamoto K., Yanagawa S., Ueda T. 'Field tests of grounding at an actual wind turbine generator system'. *International Conference on Lightning Protection (ICLP)*; Cagliari, Italy, September 2010; no. 1307.

[16] Yanagawa S., Natsuno D., Yamamoto K. 'A measurement of transient grounding characteristics of a wind turbine generator system and its considerations'. *Asia-Pacific International Conference on Lightning (APL)*; Chengdu, Chinea, November 2011; pp. 401–404.

[17] Niihara J., Ametani A., Yamamoto K. 'Transient grounding characteristics at a wind turbine with a counterpoise'. *Asia-Pacific Symposium on Electromagnetic Compatibility (APEMC)*; Singapore, May 2012.

[18] Yamamoto K., Yanagawa S., Yamabuki K., Sekioka S., Yokoyama S. 'Analytical surveys of transient and frequency dependent grounding characteristics of a wind turbine generator system on the basis of field tests'. *IEEE Transactions on Power Delivery*; October 2010; vol. 25, no. 4, pp. 3034–3043.

[19] Yamamoto K., Yanagawa S. 'Grounding characteristics of a wind turbine measured immediately after its undergrounding'. *Asia-Pacific Symposium on Electromagnetic Compatibility (APEMC)*; Singapore; May 2012.

[20] Yamamoto K., Yanagawa S., Ueda T. 'Verifications of transient grounding impedance measurements of a wind turbine generator system using the FDTD method'. *International Symposium on Lightning Protection (SIPDA)*; Brazil, October 2011; pp. 255–260.

[21] Yanagawa S., Natsuno D., Yamamoto K. 'An analytical surveys of transient grounding characteristics of a wind turbine generator system and its considerations'. *International Conference on Lightning Protection (ICLP)*; Vienna, Austria, September 2012.

[22] Yamamoto K. 'Measurements of transient grounding characteristics of a wind turbine generator system in mountainous area'. *2013 Asia-Pacific International Conference on Lightning*; June 2013; LPRE-437, pp. 431–434.

[23] Yoshikawa N., Ametani A., Yamamoto K., Yanagawa S. 'Improvement of wind turbine grounding system using various grounding wires'. *2013 Asia-Pacific Symposium on Electromagnetic Compatibility & Technical Exhibition on EMC, RF/Microwave Measurement & Instrumentation*; May 2013.

[24] Yamamoto K., Sumi S., Yanagawa S. 'Transient characteristic of an actual wind turbine grounding system'. *2013 Asia-Pacific Symposium on Electromagnetic Compatibility & Technical Exhibition on EMC, RF/Microwave Measurement & Instrumentation*; May 2013.

[25] Yamamoto K., Sumi S., Yanagawa S., Natsuno D. 'Transient grounding characteristic of a 600 kW class wind turbine'. *2012 CIGRE SC C4 Colloquium in Japan*; October 2012; no. III-4, pp. 79–84.

[26] Yamamoto K., Sumi S., 'Validations of lightning protections for accidents at a wind farm'. *XII SIPDA: International Symposium on Lightning Protection*; October 2013; pp. 412–417.

[27] Yamamoto K., Yanagawa S. 'Transient grounding characteristics of wind turbines'. *International Conference on Lightning Protection (ICLP)*; Vienna, Austria, September 2012; no. 74.

[28] Takahashi M., Kawasaki K., Matsunobu T. 'Lightning surge characteristic and protection methods of photovoltaic power generation system'. *IEEJ Transactions on Power and Energy*; October 1989; vol. 109, no. 10, pp. 443–450.

[29] Hernandez J. C., Vidal P. G., Jurado F. 'Lightning and surge protection in photovoltaic installations'. *IEEE Transactions on Power Delivery*; December 2008; vol. 23, no. 4, pp. 1961–1971.

[30] Yamamoto K., Takami J., Okabe N. 'Overvoltages on DC side of power conditioning system caused by lightning strike to structure anchoring photovoltaic panels'. *IEEJ Trans. PE*; November 2012; vol. 132, no. 11, pp. 903–913.

[31] Yamamoto K., Chikara T., Ametani A. 'A study of transient magnetic fields in a nacelle of a wind turbine generator system due to lightning'. *IEEJ Transactions on Power and Energy*; May 2009; vol. 129, no. 5, pp. 628–636.

[32] Matsuura S., Noda T., Asakawa S., Yokoyama S. 'An experimental study of surge characteristics of a distribution line using a reduced-scale model'. *IEEJ Transactions on Power and Energy*; January 2008; vol. 128, no. 1, pp. 226–234.

[33] Kubo Y., Uno H., Nagaoka N., Baba Y., Ametani A. 'Resonance characteristic of structure for PV panel and its influence on lightning surge'. *Proceedings of the Twenty-Second Annual Conference of Power & Energy Society. IEE of Japan*; August 2011; session 27, no. 254, pp. 5, 6 (in Japanese).

[34]   Betz H. D., Schumann U. *Lightning: Principles, Instruments and Applications – Review of Modern Lightning Research*, ISBN: 978-1-4020-9078-3, Springer, New York, USA; 2009.

[35]   Baba Y., Rakov V. A. 'On the mechanism of attenuation of current waves propagating along a vertical perfectly conducting wire above ground: Application to lightning'. *IEEE Transactions on Electromagnetic Compatibility*; August 2005; vol. 47, no. 3, pp. 521–532.

[36]   Baba Y., Rakov V. A., 'On the interpretation of grounding reflections observed in small-scale experiments simulating lightning strikes to tower'. *IEEE Transactions on Electromagnetic Compatibility*; August 2005; vol. 47, no. 3, pp. 533–542.

[37]   Sakai K., Yamamoto K. 'Lightning protection of photovoltaic power generation system'. *XII SIPDA: International Symposium on Lightning Protection*; October 2013; pp. 418–422.

Chapter 9

Surges on wind power plants and collection systems

*Y. Yasuda**

Definitions of the symbols and abbreviations used are included at the end of this chapter.

9.1 Introduction

Wind power generation has been considered as one of the most realistic options to solve the confused energy problem in the global future. According to the latest report by Global Wind Energy Council (GWEC), the global cumulative installed wind capacity has reached 300 GW, which is approximately a hundred times that of a nuclear reactor [1].

Lightning protection systems (LPSs) for wind power generation are becoming a more important public issue since installations of wind turbines (WT) have greatly increased worldwide. WTs are often struck by lightning because of their open-air locations, special shapes, and very high construction heights. As well as serious blade damage, breakdown accidents often occur in low-voltage and control circuits in many wind power plants (WPPs).[1] An earthing (grounding) system is one of the most important components required for appropriate LPSs in WTs and WPPs. According to the latest report by NEDO (New Energy and Industrial Technology Development Organization) [2], the most frequent failures, more than 50%, in WT equipment are those occurring in low-voltage, control, communication circuits. Such frequent incidents in low-voltage circuit may cause a deterioration of the utilization rate and consequently cause increases in the cost of power generation.

Japan, in particular, suffers from frequent and heavy lightning strikes, an example being the notorious *"winter lightning"* found in coastal areas of the Sea of Japan [3, 4]. Indeed, many WTs in Japan have been hit by lightning, and winter lightning poses a specific threat due to its intense power and electric current which are much higher than the world average [5–9]. Furthermore, due to its narrow

---

*Kansai University, Japan

[1]The terminology "wind farm" may be also used in many literatures, which might suggest an idyllic image where wind energy can be harvested from a natural resource. However, in modern technology, there exist huge groups of more than 100 turbines which are almost equivalent to a large thermal plant by themselves. In this chapter, therefore, the author would like to use the terminology "wind power plant (WPP)" rather than "wind farm."

landmass, WPPs in Japan tend to be constructed in mountainous areas with high resistivity soil. Thus, earthing is one of the most important issues in Japan for protecting WTs and WPPs from lightning.

It has been investigated and cleared that winter lightning may sometimes lead to burnout incidents of surge protection devices (SPDs) installed far from the WT that was actually struck by lightning (Figure 9.1). The phenomenon of surge invasion from a WT that is struck by winter lightning to the collection system[2] in the WPP is quite similar to the case of *"back-flow surge"*, in which the lightning surge flows out from a high structure such as a communication tower into a utility's distribution line [12]. High resistivity soil often makes the SPDs at tower earthing systems operate in reverse and allows a surge current to

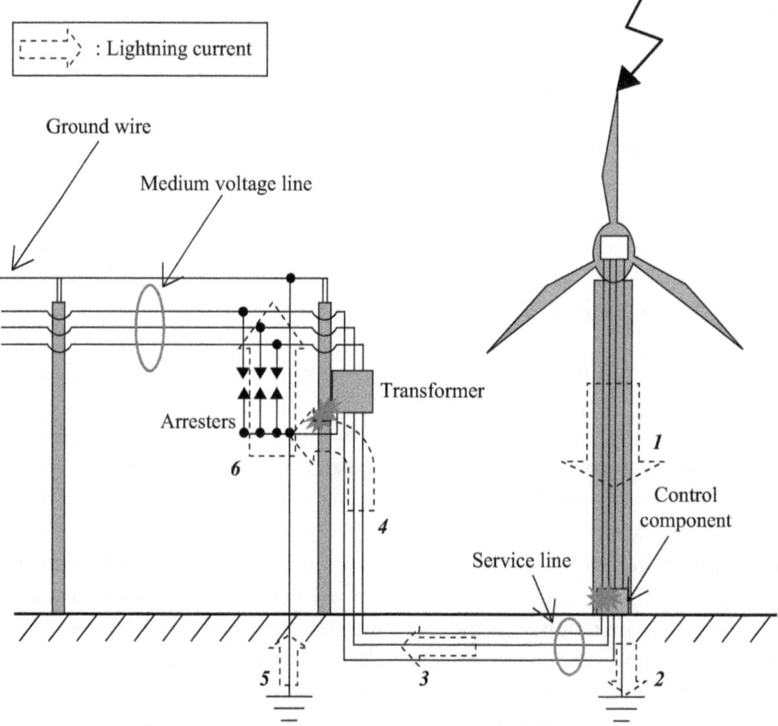

Figure 9.1   Concept of a "back-flow surge" from a WT [10]
© 2009 The Institute of Electrical Engineers of Japan. Reproduced by permission of The Institute of Electrical Engineers of Japan

[2]Although the collection system is often confused by a distribution system, the terminology is clearly defined by IEC 60050 International Electrotechnical Vocabulary; *electrical system that collects the power from a wind turbine generator system, and feeds it into a network step-up transformer or electrical loads* (IEV 415-04-06) [11]. In this chapter, the author will use the terminology "collection system" as the electrical network within a WPP whereas "distribution system" as that belongs to a utility or a system operator.

flow back to the network. It is reported that the back-flow surge can sometimes burn SPDs or break down low-voltage circuits even far from the point of the lightning-strike.

This chapter analyzes incidents of burnout to SPDs resulting from winter lightning at a WPP using ARENE [13] and PSCAD/EMTDC [14]. Calculations were performed to clarify the mechanism of how the back-flow surge propagates to other WTs from the directly struck WT. The calculations also clarified that burnout incidents could easily occur even in a WT that had not been struck by the lightning. It also became evident that burnout incidents can be reduced when interconnecting earthing wires are installed between WTs.

9.2   Winter lightning and back-flow surge

It is well known among many researchers that Japan has a unique and somewhat relentless environment that includes the notorious "winter lightning" in the coastal area of the Sea of Japan. Winter lightning is a special lightning that typically has upward flashes initiated by an upward leader from an earthed structure and has a duration of longer than several 10 milliseconds (ms). Therefore, the winter lightning sometimes has a heavy electric charge of more than 300 coulombs, which is the maximum value specified in IEC 61400-24:2010 [15]. In fact, a lot of incidents of WT blades have been reported due to the winter lightning in the coastal area of the Sea of Japan during the last decades.[3]

The phenomenon of surge invasion from a WT that is struck by winter lightning to the collection system in a WPP is quite similar to the case of the well-investigated phenomenon known "back-flow surge," where high-resistivity soil often makes SPDs in reverse and allows a surge current to flow back to the network. It is interesting to note that the back-flow surge sometimes burns SPDs or breaks down low-voltage circuits even on the WT or other equipment far from the WT of the lightning-strike.

Several breakdown and burnout incidents in low-voltage circuits and SPDs at WPPs are thought to be the result of the "back-flow surge." In practice, many of the incidents not only involved the WT that had actually been struck by lightning but also other WTs that had not been struck. The reason why WTs that had not been struck were nevertheless damaged has not been fully explained. The author has therefore concentrated on investigating surge analysis on a WPP [16, 17] and presents in this chapter a case study of surge analysis using a WPP model with multiple WTs connected to a power system. The introducing WPP models and analyses can be considered as a good example for a surge analysis on WPPs, which can clarify the influence of the electrical and geometrical configuration of a WPP when lightning surge is propagated through a collection system from a WT to other WTs.

---

[3]Incidents in blades are one of the most important issues on lightning protection in WT; however, it is beyond the scope of this book. For the detail information on the blade incidents and their counter-measures, see [8] and [9].

9.3   Earthing system of wind turbines and wind power plants

9.3.1   Earthing system of WTs

According to IEC 61400-24, a "Type A arrangement" (with vertical and/or horizontal electrode) and "Type B arrangement" (with ring earth electrode) are recommended for WT earthing. The type B arrangement is described as "this type of arrangement comprises a ring earth electrode external to the structure in contact with the soil for at least 80% of its total length or a foundation earth electrode. Additional vertical and horizontal earth electrodes may be used in combination with the ring electrode. The electrode should be buried to a depth of at least 0.5 m." This arrangement was originally defined as an earthing method for ordinary houses or buildings in IEC 62305-3:2006 (originally IEC 61024-1-2, which was abolished and the revised version was renumbered as the current standard). The concept of the earth electrode is to create equipotential bonding surrounding a house or a building to avoid values of step and touch voltages that conventionally are considered dangerous.

IEC 61400-24 states that, "For the ring earth electrode (or foundation earth electrode), the mean radius (r_e) of the area enclosed by the ring earth electrode (or foundation earth electrode) shall be not less than the value l_1:

$$r_e \geq l_1 \tag{9.1}$$

where l_1 is represented in Figure 2 (note: in this chapter, the figure is denoted as Figure 9.2) according to LPS levels I, II, III and IV."

IEC 61400-24 also continues with, "When the required value of l_1 is larger than the convenient value of r_e, additional horizontal or vertical (or inclined)

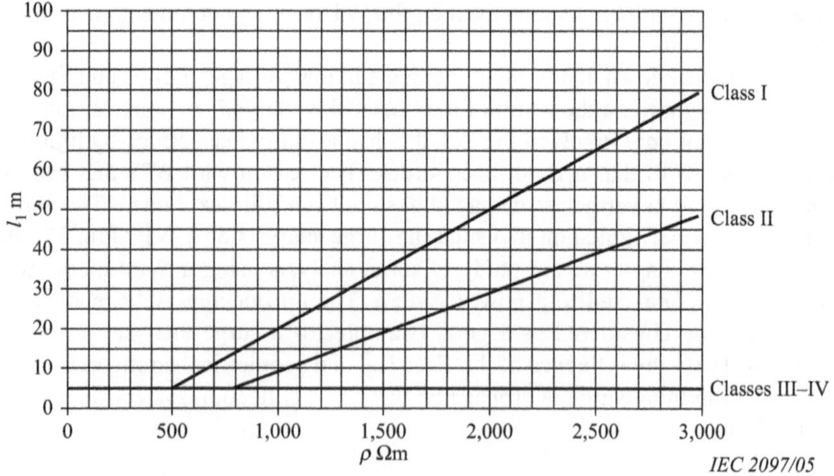

Figure 9.2   Minimum length l_1 defined in IEC 61400-24 [15]
© 2010 IEC Geneva, Switzerland. www.iec.ch. IEC 61400-24 ed.1.0

electrodes shall be added with individual lengths l_r (horizontal) and l_v (vertical) given by the following equations:

$$l_r = l_1 - r_e \tag{9.2}$$

and

$$l_v = (l_1 - r_e)/2 \tag{9.3}$$

The number of electrodes shall be not less than the number of the down-conductors, with a minimum of two."

9.3.2   Earthing system in WPPs

In IEC 61400-24, it is also mentioned on the earthing system in WPPs, where the following statement can be seen: "Each wind turbine shall have its own earthing system. The earthing systems of the individual wind turbines and the high voltage sub-station shall preferably be connected with horizontal earthing conductors, to form an overall wind farm earthing system. This is particularly beneficial in case good earthing resistance is difficult to obtain at each individual wind turbine position."

IEC 61400-24 also recommends, "The earthing system of a wind farm is very important for the protection of the electrical systems, because a low-resistance earthing system reduces the potential difference between the different structures of the wind farm and so reduces the interference injected into the electrical links." However, the recommendation should be carefully noted because the collection lines often has long distance more than several hundred meters between WTs and other equipment in a large WPP. A long distance wire/cable sometimes has inductive impedance that may rise an apparent resistance in high-frequency range such as lightning surge.

9.4   Wind power plant models for lightning surge analysis

9.4.1   WPP model

Figure 9.3 shows a WPP model made up of 10 wind power generators, identical in performance and condition. In this model, it is assumed that

- A high-voltage transformer (6.6/66 kV) interconnects the power grid system and an array of 10 wind power generators of the 1 MW class at 0.6 km intervals connected by over-ground wires with bared wires.
- Although a wind power generator consists of a gearbox, a synchronous or an induction generator, rectifier, three-phase inverter, and so on, for this simulation a stable synchronous generator is presumed for simplicity.
- Boost transformers for the generators (660 V/6.6 kV) are installed inside the WT towers, or in their vicinity. SPDs are attached to the primary and secondary terminals and connected to a common earth, as also shown in the detailed diagram in Figure 9.3.

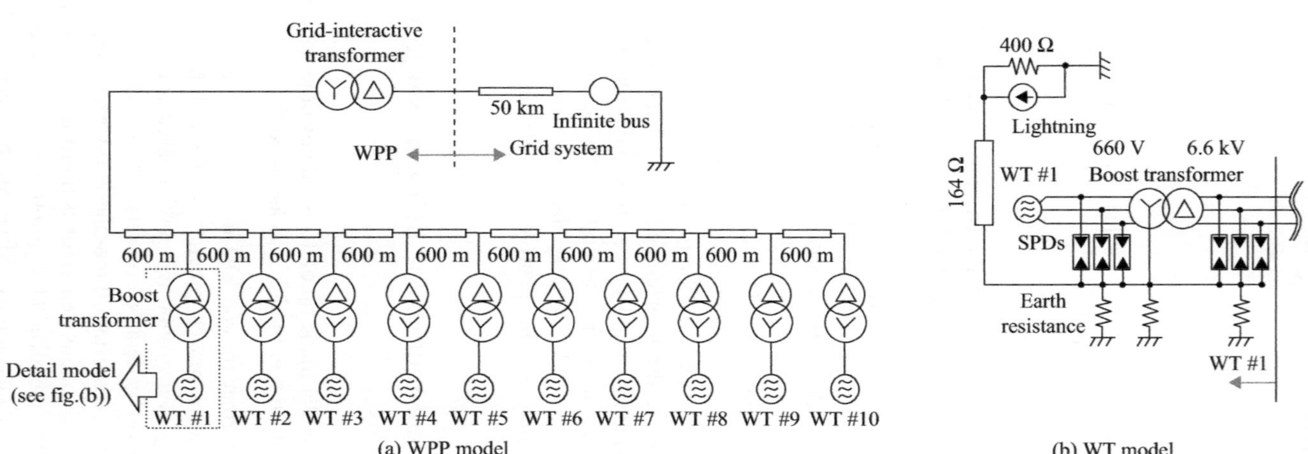

Figure 9.3   WPP model with 10 WTs: (a) WPP model; (b) WT model

Table 9.1   Analysis conditions for ARENE

WT (synchronous generator) model	
Rating power [MVA]	1.00
Direct-axis reactance X_d, X_d', X_d'' [p.u.]	2.00, 0.25, 0.20
Quadrature-axis reactance X_q, X_q', X_q'' [p.u.]	1.90, 0.50, 0.20
time constants T_{do}', T_{do}'', T_{qo}', T_{qo}'' [s]	6.0, 0.03, 0.50, 0.06

Transformer model (boost/grid-interactive)		
Connection method	**Y (neutral earthing)/Δ**	
Rating power [MVA]	1.0	10.0
Rating voltage [V]	600/6,600	6,600/66,000
Frequency [Hz]	60	60
% Impedance [%]	15.7	1.57
Mutual leakage inductance [mH]	18.2	18.2

Collection line model	
Positive-/zero-phase resistance [Ω/km][*]	0.00105/0.0210
Positive-/zero-phase inductance [mH/km][*]	0.83556/2.50067
Positive-/zero-phase capacitance [nF/km][*]	12.9445/6.4723
Frequency-dependent characteristics	Considered

[*]Values at 60 Hz.

- As shown in the right-side figure in Figure 9.3, the present WT model is connected to the earth resistance at three points, where each earth resistance is simulated as 20 or 50 Ω. Thus, the total value of the equipotentialized earth resistance of the WT is calculated as 6.67 or 16.6 Ω.
- Other details and constants used in the model are shown in Table 9.1.

In addition, since what we wished to observe in this simulation was the burnout of SPDs at WTs that had not themselves been struck, we assumed that blade burnout or explosive destruction and dielectric breakdown at the WT that had actually been struck was prevented by certain measures.

9.4.2   Model for winter lightning

A standard summer lightning event is generally assumed to have a crest peak of 30 kA, a crest width of 2 μs, and a wave tail of 70 μs. By contrast, since winter lightning has varying crest widths and crest peaks, a standardized model has not been established. Therefore, the present modeling of winter lightning is based on the model described in [11]. The parameters for crest peak and duration of wave tail are determined from a cumulative frequency distribution of lightning current wave shape. The assumptions concerning the lightning in the present analysis are as Table 9.2.

9.4.3   Model for surge protection device (SPD)

To provide protection from surge invasion, it was assumed that SPDs were installed in both the primary (LV side; WT side) and secondary (HV side; collection system

Table 9.2   Parameters for summer and winter lightning

	Summer lightning	**Winter lightning**
Crest peak	30 kA	51 kA (the 16% value of the cumulative frequency distribution of lightning)
Crest width	2 µs	2 µs
Duration of wave tail	70 µs	631 µs (the 16% value of the cumulative frequency distribution of lightning)
Polarity	Positive (normally negative but for comparison with winter lightning in this analysis)	Positive

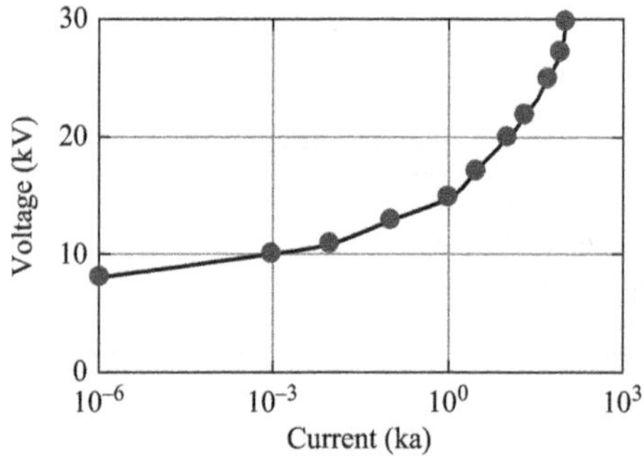

Figure 9.4   SPD model for surge analysis

side) terminals of the boost transformer near to each WT. The nominal discharge current of the SPD was assumed to be 2.5 kA and its characteristic curve, starting at $V_{1mA} = 8$ kV, is shown in Figure 9.4.

The burnout of an SPD depends on whether the heat produced by the current flowing through the SPD exceeds the thermal limit of the given SPD. To calculate the total heat E_{SPD} [kJ] produced by the SPD in the present analysis, it is necessary to specify the simultaneous power $P_{SPD}(t)$ [W] derived from the SPD's terminal voltage $V_{SPD}(t)$ [V] and the current flowing $I_{SPD}(t)$ [A] through the SPD. Then, the total electric energy W_{SPD} [Wh] can be calculated by integrating $P_{SPD}(t)$ from 0 to the time T [s] when $I_{SPD}(t)$ converges to 0 kA. The total thermal energy produced in the SPD E_{SPD} [kJ] is given by unit conversion from W_{SPD} [Wh]. This sequence is described by the following equations.

$$P_{SPD}(t) = V_{SPD}(t) \times I_{SPD}(t) \tag{9.4}$$

$$W_{SPD} = \frac{\int_0^T P_{SPD}(t)dt}{3{,}600} \tag{9.5}$$

$$E_{SPD}[kJ] = 3.6 \times W_{SPD}[Wh] \tag{9.6}$$

9.4.4   *Comparison analysis between ARENE and PSCAD/EMTDC*

In the analysis in this section, two different digital transient simulators were employed: ARENE (non-real-time PC version), that was once produced by EDF (Electricité de France), and PSCAD/EMTDC by Manitoba HVDC Research Centre. Both of the simulators have similar algorism of Schneider-Bergeron method as well as the conventional EMTP/ATP. It is also possible to describe the above-proposed model by the conventional EMTP and ATP [10].

The result by the two simulators is shown in Figure 9.5. The figures illustrate the graphs of impulse height of surges reaching at each WT. Figure 9.6 also gives clearly coincidence from the viewpoint of aspect of surge propagation calculated by two different simulators, ARENE and EMTDC. It shows the accuracy of the model description by ARENE and EMTDC for the lightning analysis on a WPP.

Figure 9.5 also shows an interesting tendency of the surge propagation in the WPP. It is clear that the surge from WT#1, which was struck by lightning, propagated to not only adjacent WT but also quite distant WTs far from lightning-strike one. Moreover, it is recognized that the further the WT from the lightning-strike one is, the longer the surge duration reached to the WT is. It suggests that a WT that was

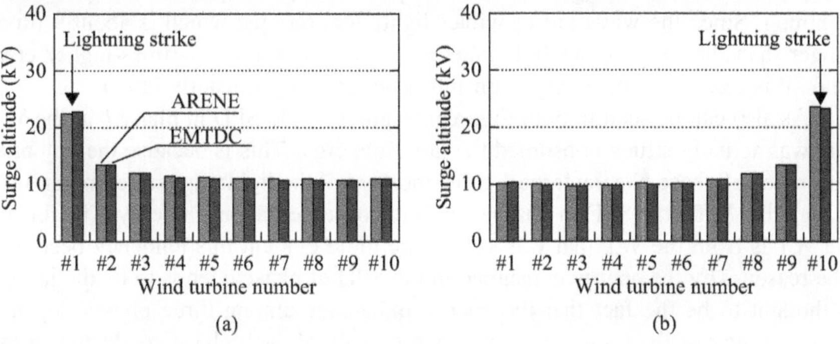

Figure 9.5   Comparison results of surge propagations ($R_g = 1.68\ \Omega$): (a) Case 1: Lightning on WT#1; (b) Case 2: Lightning on WT#10

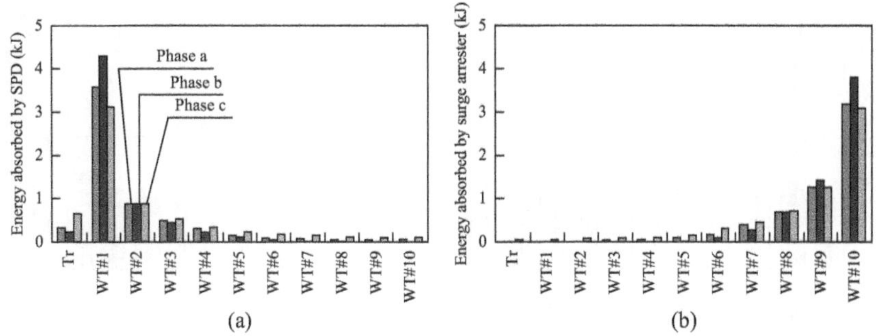

Figure 9.6   Energy consumed by SPDs in each WT (Summer lightning $R_g = 6.77\ \Omega$): (a) Case 1: Lightning on WT#1; (b) Case 2: Lightning on WT#10

not struck by lightning directly even at farthest point from the lightning-strike one has possibility to suffer from discharge incident due to a back-flow surge propagating along the collection line within the WPP. The mechanism of the back-flow surge propagation and SPD's incidents will be discussed in detail in the following section.

9.5 Mechanism of SPD's burnout incidents due to back-flow surge

9.5.1 Analysis of the surge propagations in WPP

In this section, an analysis is performed to compare the energy consumption of the SPDs during a summer and winter lightning. It is supposed that the lightning strikes either WT#1, which is the nearest WT to the grid, or WT#10, the furthest WT from the grid.

Figures 9.6 and 9.7 are the results of analyses done by ARENE (see section 9.4) in the cases of summer and winter lightning strikes, respectively. Each WT has an earth resistance of 6.67 Ω. In both figures, graphs (a) and (b) indicate the results in the cases of lightning strikes to WT#1 and WT#10.

Comparing Figures 9.6 and 9.7, it is clear that the energy consumed by each SPD is 10 to 30 times larger in the case by winter lightning than that by summer lightning. Since the wave tail of winter lighting is 631 μs, which is about 9 times longer than that of summer lightning, even more than the 1.7-times-higher crest peak, it is likely that the energy consumed becomes significantly larger.

As also can be seen in both Figures 9.6 and 9.7, the SPD in phase *b* at the WT that was actually struck consumed the largest energy. This is because the potential difference of phase *b* is the largest at the moment of the lightning strike. In contrast, at the other WTs, the SPD in phase *c* tends to consume the most energy. The further the WT is from the WT that was struck, the more evident this tendency becomes. The reason why the energy consumed in the SPD of phase *c* tends to be the largest is thought to be the fact that the energy imbalance among three phases depends upon a timing of lightning strike, i.e., phase angle of each phase conductor, at the struck WT. When the line voltage in phase *c* at the moment of the lightning strike is the highest among the three phases at the struck WT, the voltage between both

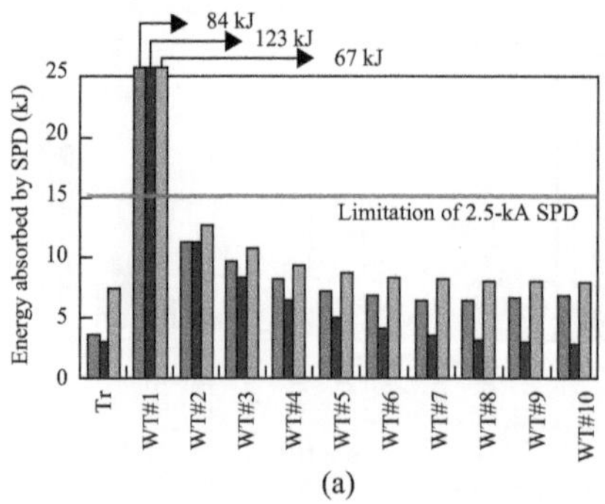

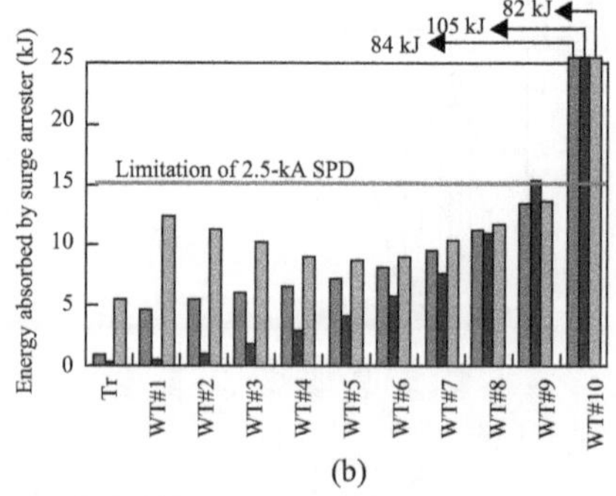

Figure 9.7   Energy consumed by SPDs in each WT (Winter lightning $R_g = 6.77\ \Omega$): (a) Case 1: Lightning on WT#1; (b) Case 2: Lightning on WT#10

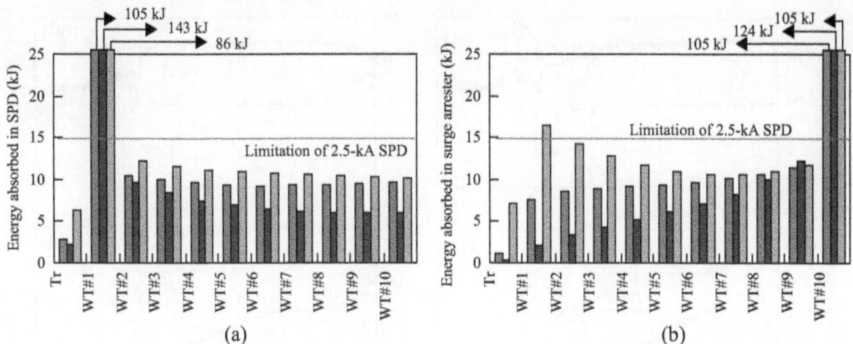

Figure 9.8   Energy consumed by SPDs in each WT (Winter lightning $R_g = 16.6\ \Omega$): (a) Case 1: Lightning on WT#1; (b) Case 2: Lightning on WT#10

terminals of the SPD tends to become the highest among the three phases. The further away the WT is, the stronger the tendency becomes.

Furthermore, to examine in more detail the results for winter lightning, another analysis was performed. Figure 9.8 is the result of the case where the earth resistance of the WTs was set at 16.7 Ω. In Case 1, the energy consumption of the SPDs surpassed their thermal limitation only in the WT#1, which had been directly struck, as shown in Figure 9.8(a).

On the other hand, in the case of a lightning strike to WT#10, the energy consumption in phase c of the SPDs is higher than the limit not only at WT#10 itself, but also at WT#1, the WT furthest away from the lightning strike. As can be seen in Figure 9.8(b), it is not always the case that the further a WT is from the WT that was struck, the lower the energy consumed by the SPD. From Figure 9.8(b), it is evident that, so far as the energy consumed by phase c is concerned, from WT#6 onward the further away the WT is, the higher the energy consumed. This tendency was also observed using other values of earth resistance. The longer the tail of the lightning strike, the clearer the tendency of this phenomenon becomes.

9.5.2   Detail analysis on surge waveforms

To clarify the reason of the reversal phenomenon described above, waveforms of voltage, current, and simultaneous power at each WT need to be drawn. Figure 9.9 is a set of graphs of the waveforms at WT#5 and WT#10 in the case of a lightning strike to WT#1 (Case 1). Correspondingly, the figure is also the set of waveforms at WT#1 and WT#5 when WT#10 has been directly struck (Case 2). Comparing the two graphs, it is evident that the summation of SPD's consumed energy of all phases at any WT is almost the same value of about 26 kJ. What is interesting is that the simultaneous power curve of phase c at WT#1 in Case 2 (see Figure 9.9 (a3)) exceptionally forms into a large arc and makes a large area. The consumed energy calculated from the integrated area is up to 15 kJ that is a thermal limitation of the present 2.5-kV class SPD. In this case, the current inrushing to WT#1 did not flow through the SPD in phase b but in phase c. The reason is considered to the difference of the potential at the moment of the lightning strike, where the initial phase θ of phase a in the present calculation is set at 0 degree.

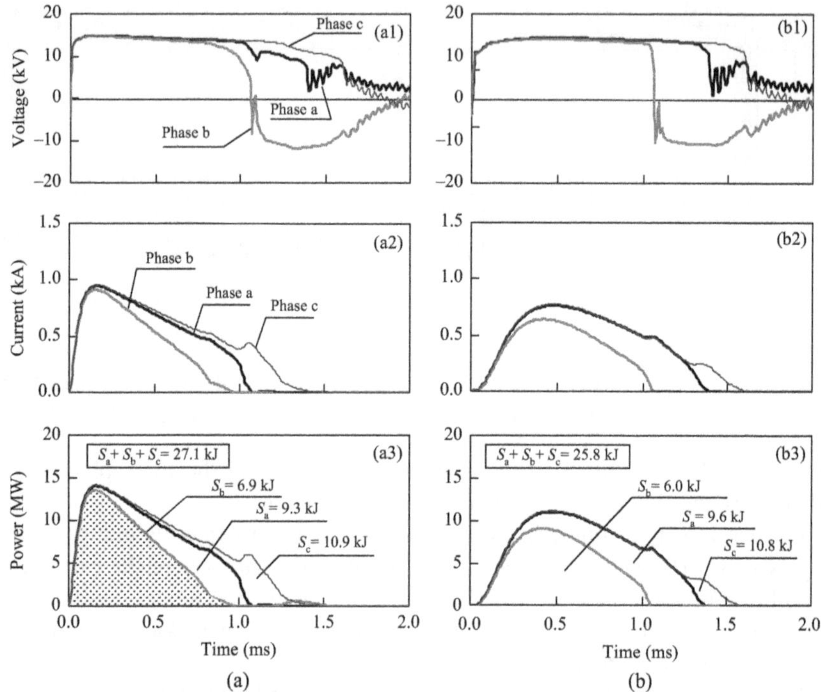

Figure 9.9   Waveforms of voltage, current, and simultaneous consumption power at SPDs in each WT (Winter lightning $R_g = 16.6$ Ω, Case 1): (a) WT#5; (b) WT#10 [16]

© 2008 IEEE. Reprinted, with permission, from Y. Yasuda, N. Uno, H. Kobayashi, T. Funabashi, 'Surge analysis on wind farm when winter lightning strikes', *IEEE Transactions on Energy Conversion*, vol. 23, no. 1, March 2008

Thus, from the above discussion, it becomes clear that during strong winter lightning strikes with a long tail there is a possibility of SPD burnout incidents not only at the WT that the lightning strikes and those nearby, but also at WTs far away from the one that was actually struck.

As a conclusion in this section, an analysis on burnout incidents of SPDs resulting from winter lightning at a WPP was performed. The results of calculations clarified the mechanism of how back-flow surge propagates to other WTs from the WT that has been directly struck by winter lightning. The calculations with various lightning point and earth resistance values demonstrated that burnout incidents could quite easily occur even in a WT far from the lightning-struck WT within the WPP. Consequently, it also became evident that the possibility of burnout incidents can become high in the case of high resistance at the WPP site and/or long-tail winter lightning.

9.6   Effect of overhead earthing wire to prevent back-flow surge

The aim of the analysis in the present section is to clarify the influence of earthing wire(s) of the collection line in a WPP. Reference [11] also noted that earthing wire(s) can reduce burnouts of SPDs in the case of a communication tower. This section tries to clarify if there is a similar effect from the installation of earthing wire(s) for WPP LPS. The model description and calculation has been done by PSCAD/EMTDC (see section 9.4).

The phenomenon of surge invasion to the WPP collection system from a WT that is struck by lightning is quite similar to the back-flow surge reported in [11]. In that report, the surge flowed from a communication tower into a utility's distribution line. High-resistivity soil often creates SPDs for tower earthing systems to operate in reverse and allows reflux of the surge current to the distribution line. It is reported that the back-flow surge can sometimes burn out SPDs or break down low-voltage circuits even on an electric pole far from the point of the lightning-strike.

Several breakdown and burnout incidents in low-voltage circuits and SPDs in WPPs are thought to be the result of the above back-flow surge. In practice, many of the incidents that have occurred not only involved the actual lightning-struck WT but also other WTs that had not been struck. The reason why WTs that had not been struck were nevertheless damaged has not been fully explained.

9.6.1   Model of a collection line in a WPP

Comparing the previous section, the simpler model for a WPP with two WTs is employed as illustrated in Figure 9.10. The main different point of a model of a collection system is the existence of earthing wire(s) as shown in Figure 9.10(b). The collection line in the present WPP model is assumed as an overhead line with three-phase conductors installed 10 m over the ground. Configuration details are shown in Figure 9.11 and a set of parameters is shown in Table 9.3.

The main aim of the present analysis is to confirm the effect of earthing wire(s). Various conditions, therefore, are simulated:

1. "Case 0": no earthing wires are installed above the overhead line.
2. "Case 1": one earthing wire is tensioned 1 m above the three phase conductors. Both terminals of the earthing wire are connected to the common earthing system of the WTs and the grid-interactive transformer.
3. "Case 2": two earthing wires horizontally separated at 0.7 m length above the conductors are installed.

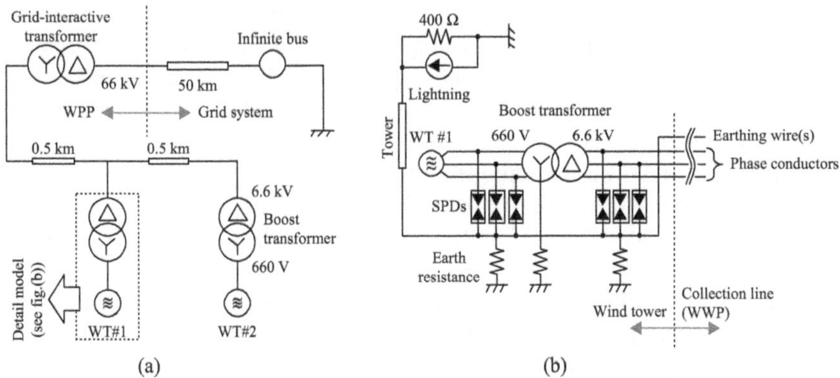

Figure 9.10   WPP and WT model: (a) WPP model with two WTs; (b) WT model

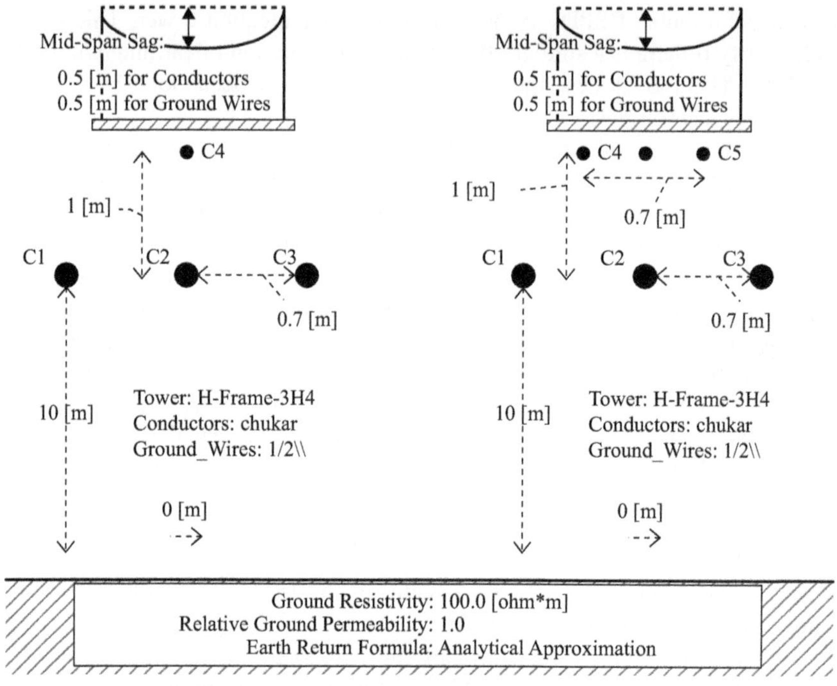

Figure 9.11   Collection line model in WPP by PSCAD description. (Left: one-earthing-wire model, Right: two-earthing-wires model) [16]

© 2008 IEEE. Reprinted, with permission, from Y. Yasuda, N. Uno, H. Kobayashi, T. Funabashi, 'Surge analysis on wind farm when winter lightning strikes', *IEEE Transactions on Energy Conversion*, vol. 23, no. 1, March 2008

In the following section, comparisons will be made between the energy consumption of the SPDs among the three cases with various numbers of earthing wires. It is assumed that the lightning strikes WT#1, which is the nearest WT to the grid.

9.6.2   Observation of waveforms around SPDs

Figure 9.12 sets out the results of EMTDC calculations in the cases of winter lightning strikes. Column (A) denotes the various waveforms measured around the SPD (phase a) installed at the high voltage terminal of the boost transformer of WT#1. In the WT#1 SPD, since the lightning surge invades to the common earthing system and operates the SPD in reverse, from earth to the line, the polarity of each waveform was inverted. Also, Columns (B) and (C) correspond to the waveforms around the SPD at WT#2 and the grid-interactive transformer, respectively. The phenomena of a back-flow surge in the WPP can easily be recognized.

On the other hand, the graphs in Row (1) show voltage waveforms between the terminals of the respective SPDs. Row (2) is for current waveforms flowing through the SPD, and Row (3) is for simultaneous power according to (9.6), in the same way. Every graph in Figure 9.12 has three curves due to the various conditions, i.e., Case 0, Case 1, and Case 2.

Table 9.3   Analysis conditions for EMTDC

WT (synchronous generator) model		
	Rating power [MVA]	1.0
	Rating voltage [kV]	0.66
Impedance	Resistance [W]	0.002
(R-L-C model)	Inductance [mH]	0.231
	Capacitance [mF]	0.001

Transformer model (boost/grid-interactive)		
Connection method	**Y/Δ**	
Rating power [MVA]	1.0	10.0
Rating voltage [kV]	0.66/6.6	6.6/66
Frequency [Hz]	60	
No load losses [p.u.]	0.0	
Copper losses [p.u.]	0.005	
Positive sequence leakage inductance [p.u.]	0.15	
Saturation	no	
Aircore reactance [p.u.]	0.2	
Magnetizing current [%]	1.0	

Collection Line Model in WPP		
Phase conductors	Height of all conductors [m]	10
	Configuration of conductors	Horizontal
	Spacing between phases [m]	0.7
	Conductor radius [mm^2]	20.3
	Sag for all conductors [m]	0.5
	Number of sub-conductors in a bundle	1
Earthing wire	Earthing wire radius [mm^2]	5.5
	Number of earthing wire(s)	0, 1, 2
	Height of earthing wire(s) [m]	11
	Spacing between earthing wires [m] (in case of two wires)	0.7
	Sag for all earthing wires [m]	0.5
	Resistivity of earth [Ωm]	100

Comparing Case 0 and Case 1, it can be clearly seen that the effect of the earthing wire to reduce the surge reaching the next WT and the grid transformer is quite significant. From the three graphs in Row (1), it is clear that the surge durations at every point in the WPP are reduced by half. The surge current passing through the SPD shown in Row (2) is also cut down by almost half or two-thirds. Consequently, the simultaneous power produced in the SPD becomes much lower, as shown in Row (3).

9.6.3   Evaluation of the possibility of the SPD's burning out

In Row (3) of Figure 9.12, the integral area surrounding the simultaneous power curve becomes equal to the thermal energy produced in the SPD. Summarizing the above integration, the bar graphs shown in Figure 9.13 are drawn to evaluate the

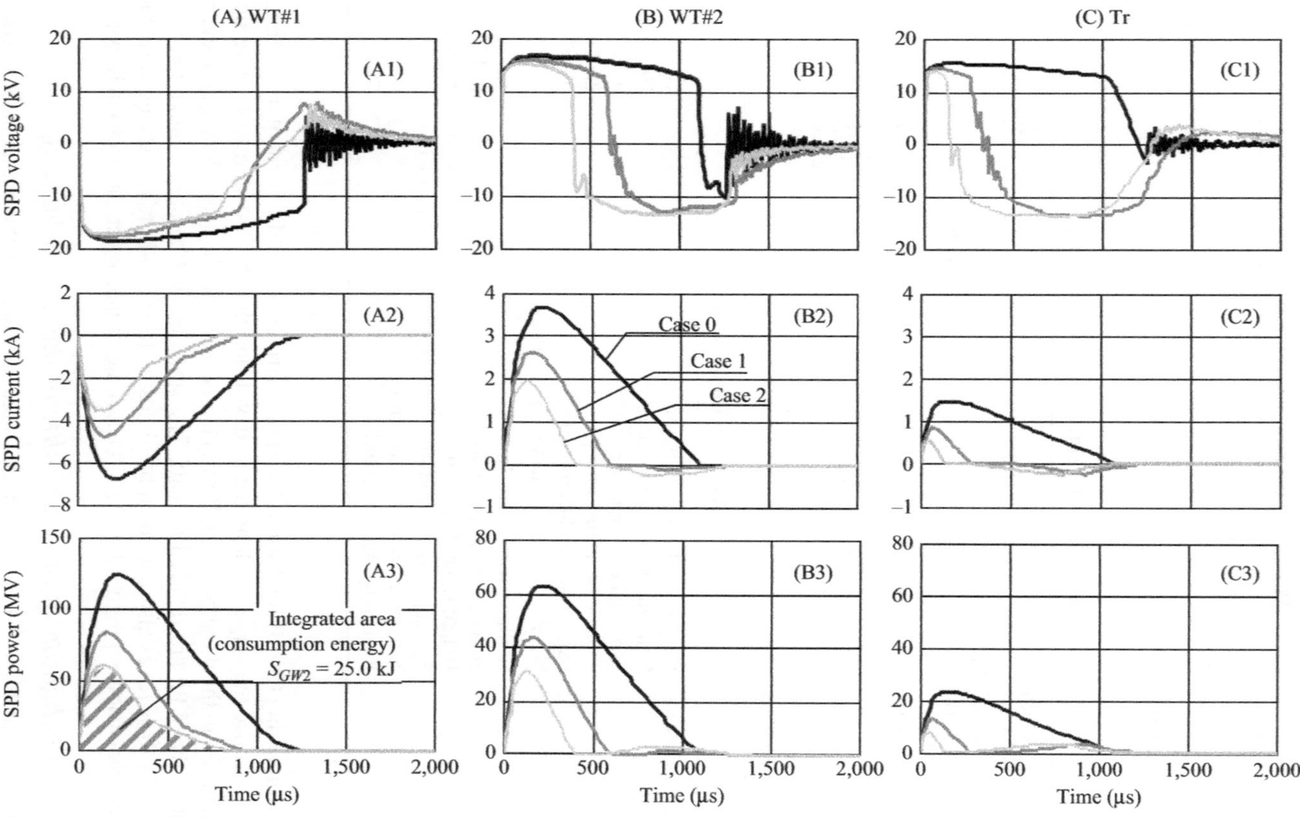

Figure 9.12   Calculated waveforms at SPDs at various points among the WPP in case of winter lightning (2/631 µs, 51 kA)

possibility of a burnout incident at the SPDs. Column (1) in Figure 9.13 illustrates the results of the integration area of simultaneous power curves, i.e., the energy consumption in each SPD. The graph of WT#1, which is directly struck by lightning, displays a tendency to produce huge thermal energy in the SPDs. This suggests that there is a definite possibility of burnout incidents under the conditions found with huge winter lightning strikes. Since the total earth resistance of the WT is assumed as 3.33 Ω in the present case, it becomes clear that a lower resistance or a higher rate for the SPD is needed to avoid burnouts.

The most important result presented here is shown in Row (B) in Figure 9.13. From this graph of the energy consumption in the SPD of WT#2, the successful effect of installing earthing wire(s) is evident. If an earthing wire was not installed (Case 0), a huge quantity of energy could surge in, even to the adjoining WT that was not directly struck by the lightning. By contrast, in the case of earthing wire employment (Case 1), the surge energy invading to WT#2 is cut down to less than half. Moreover, the results for Case 2 show that the surge energy can be suppressed by much less than 15 kJ, which is the thermal limitation of a 2.5-kA class SPD.

A comparison between the different numbers of earthing wires also gives an interesting result. From the results in Figure 9.13, a multiple earthing wire strategy provides a further margin of safety against lightning surge. As a similar tendency, to that in the present calculation for the back-flow surge, is noted in the case of a communication tower in [11], it becomes evident that a back-flow surge in a WPP can be reduced by the installation of earthing wire(s).

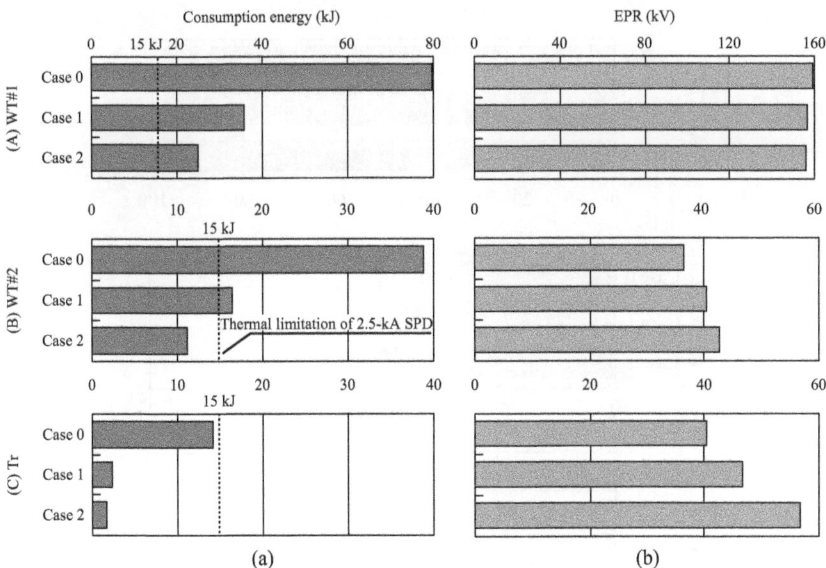

Figure 9.13   Energy consumption at SPDs and maximum EPR at the selected point in WPP in case of winter lightning (2/631 μs, 51 kA): (a) Consumption energy at SPD; (b) Maximum EPR

9.6.4   *Evaluation of potential rise of earthing system*

Finally, we need to also mention that a negative influence from an earthing wire(s) installation. Column (2) of Figure 9.13 shows a surprising result. The graph of an EPR at WT#2 demonstrates an upward trend according to increases in the number of earthing wires. The same tendency can be recognized in the result in the grid-interactive transformer (Tr). Even worse is that the additional installation of earthing wires does not contribute very much to a reduction of the EPR at WT#1's earthing system.

Another result of calculation in case of summer lightning (2/70 μs, 30 kA) is shown in Figure 9.14, where there is barely possibility of an SPD's burnout because the total energy of the back-flow surge due to summer lightning is much smaller than that of winter lightning. From Figure 9.14, it is also evident that the EPR at the equipment that are not struck by lightning tends to increase because of the installation of earthing wire(s). Comparing with the case of winter lightning, the altitude of the EPR due to summer lightning is relatively small. However, there still remains negative impact given by the earthing wire(s).

The reason of this negative impact seems to be because an earthing wire of 0.4 km has relatively strong inductive impedance against a high frequency domain of more than 1 MHz, such as found in a lightning surge. However, since the current flowing through the interconnecting earthing wire tends to be large, the EPR of the next WT or a grid-transformer displays an increasing tendency. This may give rise to a possibility of breakdowns of low-voltage circuits inside the WT. While it can be concluded that an earthing wire strategy is very effective against winter lightning, it may cause unexpected effects in the case of summer lightning.

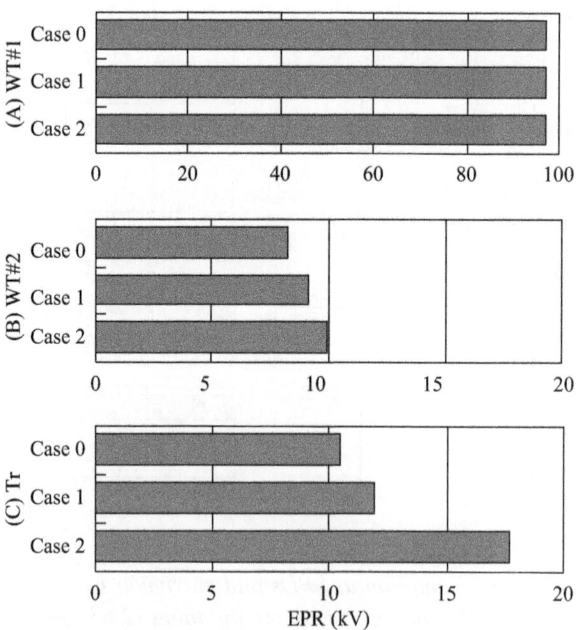

Figure 9.14   Maximum EPR at points around the WPP in case of summer lightning (2/70 μs, 30 kA)

9.7   Conclusions

In this chapter, lightning surge analyses on a WPP were performed using ARENE (section 9.5) and PSCAD/EMTDC (section 9.6). The analyses simulated a WPP using a simple model with WTs, boost transformers, SPDs, and a gird-interactive transformer connected with a collection system. From the results of several analyses the following conclusions were obtained: (1) A back-flow surge can propagate to other WTs from the WT that has been directly struck by winter lightning via the earthing system and the collection line in a WPP. (2) Burnout incidents of SPDs could quite easily occur even in a WT far from the lightning-struck WT. (3) Installing multiple earthing wires to the collection line in a WPP can reduce the burnout incidents due to winter lightning. (4) The earthing wire may not help, especially for summer lightning, to reduce the EPR (earth potential rise) of WTs and the grid-interactive transformer.

Consequently, the results of the present calculations suggest that an accurate earthing design and an LPS strategy must be implemented for WTs situated in WPPs. If a WPP is to be constructed in an area affected by heavy winter lightning, multiple earthing wires and higher rated SPDs should be installed to avoid burnouts of the SPDs and other equipment. If a given WPP also potentially suffers from summer lightning, the installation of earthing wire(s) is not recommended because the interconnection of earthing wires does not have a good effect on reducing the EPR. In both cases, trials to reduce earth resistance should be selectively done for the particular WT that would tend to suffer from lightning because of the prevailing wind direction or geographical condition.

The introduced WPP model in this chapter employed overhead earth wires in its collection lines, which is cheaper equipment than underground cables and can be seen in many existing WPPs. On the analysis on the WPPs with underground cables, the modeling and result of the analysis would be completely different. As bared cables may usually be used for the interconnection between WTs in a WPP, it is slightly difficult to model the horizontal bared cables in EMTP and other similar simulators as well as FDTD method. Here, the author would like only to suggest several foreseeing references [18, 19] and expect further investigations in future.

Symbols and abbreviations

$I_{SPD}(t)$ [A]	surge current that flow through an SPD
l_1 [m]	minimum length defined in IEC 62035-3
l_r [m]	horizontal length of buried electrode
l_v [m]	vertical length of buried electrode
$P_{SPD}(t)$ [kJ]	thermal energy that is produced in an SPD
R [Ω]	resistance to earth (in this chapter, simply "earth resistance")
R_g [Ω]	total earth resistance at a WT
T [s]	duration during a surge current flow through an SPD
$V_{SPD}(t)$ [V]	voltage that is imposed on an SPD
$W_{SPD}(t)$ [Wh]	electric energy that is consumed in an SPD
ρ [Ω m]	resistivity of soil

ARENE	(a commercial name of a digital transient simulator)
ATP	Alternative Transient Program
FDTD	Finite Deference Time Domain
EMTP	Electro-Magnetic Transient Program
EPR	Earth Potential Rise
GWEC	Global Wind Energy Council
GUI	Graphical User Interface
HV	high voltage
IEC	International Electro-technical Commission
LPS	Lightning Protection System
LV	low voltage
NEDO	New Energy and Development technology Organization, Japan
PSCAD/EMTDC	(a commercial name of a digital transient simulator)
SPD	surge protective device
WPP	wind power plant
WT	wind turbine

Acknowledgments

The author thanks the International Electrotechnical Commission (IEC) for permission to reproduce information from its International Standards IEC 61400-24 ed.1.0 (2010). All such extracts are copyright of IEC, Geneva, Switzerland. All rights reserved. Further information on the IEC is available from www.iec.ch. IEC has no responsibility for the placement and context in which the extracts and contents are reproduced by the author, nor is IEC in any way responsible for the other content or accuracy therein.

References

[1] GWEC. "Global installed wind power capacity in 2013—Regional Distribution," http://www.gwec.net/wp-content/uploads/2014/04/5_17-1_global-installed-wind-power-capacity_regional-distribution.jpg

[2] New Energy and Industrial Technology Development Organization (NEDO) Final Report on New Generation Wind Power Development Project, 2015 (to be issued).

[3] Yokoyama, S., Miyake, K., Suzuki, T., Kanao, S. Winter lightning on Japan Sea Coast—development of measuring system on progressing feature of lightning discharge, *IEEE Transaction on Power Delivery*, vol. 5, pp. 1418–1425, 1990.

[4] Motoyama, H., Shinjo, K., Matsumoto, Y., Itamoto, N. Observation and analysis of multiphase back flashover on the Okushishiku Test Transmission Line caused by winter lightning, *IEEE Transactions on Power Delivery*, vol. 13, no. 4, pp. 1391–1398, 1998.

[5]   Wada, A., Yokoyama, S., Numata, T., Hirose, T. Lightning observation on the Nikaho-Kogen Wind Farm, *Proc. of International Workshop on High Voltage Engineering (IWHV '04)*, vol. 1, pp. 51–55 (The papers of Joint Technical Meeting on Electrical Discharges, Switching and Protecting Engineering and High Voltage Engineering, IEE Japan, ED- 04-118, SP-04-29, HV-04-59), 2004.

[6]   Shindo, T., Suda, T. Lightning risk of wind turbine generator system, *IEEJ Transaction on Power and Energy*, vol. 129, no. 10, pp. 1219–1224, 2009.

[7]   Natsuno, D., Yokoyama, S., Shindo, T., Ishii, M., Shiraishi, H. Guideline for lightning protection of wind turbines in Japan, *Proc. of 30th International Conference on Lightning Protection (ICLP2010)*, No. SSA-1259, 2010.

[8]   Yasuda, Y., Yokoyama, S., Minowa, M., Satoh, T. Classification of lightning damage to wind turbine blades, *IEEJ Trans. on Electrical and Electronic Engineering*, vol. 7, no. 6, pp. 559–566, 2012.

[9]   Yasuda, Y., Yokoyama, S., Ideno, M. Verification of lightning damage classification to wind turbine blades, *Proc. of 30th International Conference on Lightning Protection*, No. 253, 2012.

[10]   Okamoto, H., Sekioka, S., Ebina, Y., Yamamoto, K., Yasuda, Y., Funabashi, T., Yokoyama, S. Energy absorption of distribution line arresters due to lightning back flow current and ground potential rise for lightning hit to wind turbine generator system, *IEEJ Trans. on Electric and Power*, vol. 129, no. 5, pp. 668–674, 2009.

[11]   IEC 60050: International Electrotechnical Vocabulary.

[12]   Nakada, K., Wakai, T., Taniguchi, H., Kawabata, T., Yokoyama, S., Yokota, T., Asakawa, A. Distribution arrester failures caused by lightning current flowing from customer's structure into distribution lines, *IEEE Transactions on Power Delivery*, vol. 14, no. 4, pp. 1527–1532, 1999.

[13]   Devaux, O., Lemerle, P., Delsol, O., Martino, S. ARENE: A new simulator reduces the cost of equipment tests, *Proc. of 15th International Conference on Electricity Distribution (CIRED '99)*, No. 4/p. 4, 1999, http://www.cired.net/publications/cired1999/papers/4/4_p4.pdf.

[14]   Manitoba HVDC Research Centre web page: PSCAD, https://hvdc.ca/pscad/.

[15]   IEC 61400-24: Wind Turbine—Part 24: Lightning Protection, 2010.

[16]   Yasuda, Y., Funabashi, T. Analysis on back-flow surge in wind farms, *Proc. of the International Conference on Power Systems Transients (IPST '07)*, 2007.

[17]   Yasuda, Y., Uno, N., Kobayashi, H., Funabashi, T. Surge analysis on wind farm when winter lightning strikes, *IEEE Transactions on Energy Conversion*, vol. 23, no. 1, pp. 267–262, 2008.

[18]   Nagao, M., Baba, Y., Nagaoka, N., Ametani, A. FDTD electromagnetic analysis of a wind turbine generator tower struck by lightning, *IEEJ Transaction of Electric and Power*, vol. 129, no. 10, pp. 1181–1186, 2009.

[19]   Sekioka, S., Hatziargyriou, N. Lorenzou, M. A study of grounding system design of wind turbine generator considering effective length of grounding conductors (II), *The papers of Joint Technical Meeting, IEE Japan*, FTE-14-027, HV-14-032, 2014.

Chapter 10

Protective devices: fault locator and high-speed switchgear

*T. Funabashi**

10.1 Introduction

A time domain model of a fault locator for transmission line faults is described in this chapter. This model uses the MODELS language in ATP-EMTP. Also, various fault types are modeled, using a constant or nonlinear fault resistance. It has been found that an impedance relay type method is influenced by the fault arc non-linearity but a current diversion ratio method is not. Errors caused by current transformer (CT) saturation have been examined, and the influence of the CT saturation has been found to be negligible.

In the case of transmission line faults, the power system and customer voltages drop instantaneously even if the customer has a nonutility generation system. Therefore, a device that separates the power system and the customer's facility rapidly after the voltage sag occurs has been developed. Operational characteristics of the new device are explained by using EMTP simulations.

10.2 Fault locator

Many fault locator algorithms were developed to be operated on digital relay data [1–7]. These methods were derived from frequency domain equations and were established by a phasor simulation. The algorithms were based on the assumption that a fault arc had a constant impedance and did not show a nonlinear effect. However, in practice, a fault arc in air is known to show nonlinearity due to its physical characteristic. Also an input device may cause error in the current/voltage detection, and this results in an error on a fault locating algorithm.

To study the influence of nonlinearity on the accuracy of a digital fault locator, a time domain simulation must be performed considering the characteristic of the fault arc and the input device. Time domain models of the protective relay were developed by using the electromagnetic transients program EMTP [8–10].

*Professor, Nagoya University, Japan

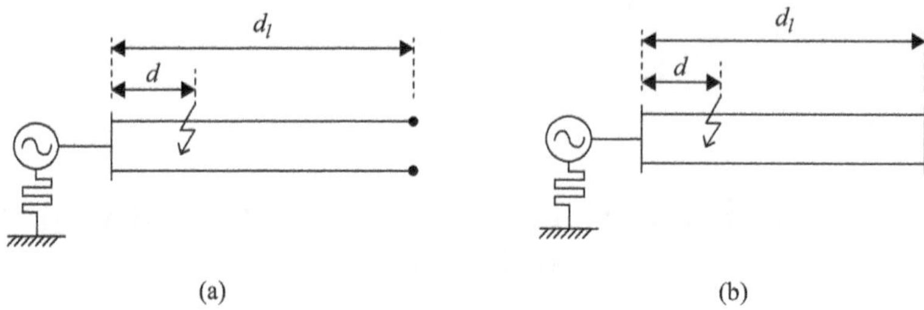

*Figure 10.1   Fault location: (a) Radial circuit (impedance relay type method);
(b) Double circuit (current diversion ratio method)*

These use a subroutine called transient analysis of control system (TACS) [11, 12] for representing the blocks of a relay simulation.

In this chapter, a time domain model of a fault locator is described using the MODELS [13, 14] language in the ATP-EMTP [11, 12]. Various fault types are modeled considering constant and nonlinear fault resistances, and also considering CT saturation and an input device error.

10.2.1   Fault locator algorithm

Digital fault locator algorithms are divided into two categories: one that uses data from one terminal of a transmission line [1–4], and the other that uses data from both terminals [5–7]. The former is superior in the economical viewpoint because it requires no data transfer for long distance. The latter is superior in the accuracy viewpoint of fault location, but requires a data transfer system.

This chapter deals with a one-terminal type fault locator. The reason is simplicity of input data and formula. The main purpose of a simulation is looking at a transition of located distance in time domain. The power systems to which the fault locator is applied are the 66-, 77-, and 154-kV high-resistance grounded systems widely used in Japan. The focus is on a phase-to-ground fault because this type of fault occupies about 90 percent of transmission line faults. One application is for a radial circuit and another is for a double circuit as illustrated in Figure 10.1. Corresponding formulas of fault location in a frequency domain are summarized below:

(a)   Impedance relay type method:

$$d = \frac{I_m\left(V \cdot I_{pol}^*\right)}{I_m\left(V_u \cdot I_{pol}^*\right)} \tag{10.1}$$

where

$$V_u = Z_s \cdot I + \sum Z_m I_{other}$$

(b)   Current diversion ratio method:

$$d = \frac{2 \cdot Re\left(I_{0m} \cdot V_{pol}^*\right)}{Re\left\{(I_0 + I_{0m}) \cdot V_{pol}^*\right\}} \cdot d_l \tag{10.2}$$

where

> d: distance from location point to fault point
> d_l: transmission line length
> $Re(A)$: real part of A (A: complex number)
> $I_m(A)$: imaginary part of A
> A^*: conjugate of A
> V: faulted-phase voltage at locator terminal
> V_u: line drop voltage per unit length
> V_{pol}: polarized voltage, e.g., $V_{pol} = V_{bc}\angle 90°$ in case of phase-a locating
> Z_s: self-impedance per unit length
> Z_m: mutual impedance per unit length
> I: faulted-phase current at locator terminal
> I_{other}: unfaulted phase current at locator terminal
> $I_{pol} = 3 \cdot I_0 = I_a + I_b + I_c$
> I_0: zero sequence current of faulted line
> I_{0m}: zero sequence current of the other line

Briefly speaking, the impedance relay type method makes fault location by dividing the fault phase voltage by the line voltage drop per unit length [1]. The current diversion ratio method [4], paying attention to that a fault current is a sum of currents on the faulted line and the other line, calculates the distance on the assumption that the faulted line's voltage drop equals to the other line's one. Thus, the latter is applicable only to a loop network.

10.2.2   Fault locator model description using MODELS

The procedure of a simulation at a time is illustrated in Figure 10.2. A time domain model of a fault locator is represented using the MODELS language in the ATP-EMTP. The fault locator model shown in Figure 10.2 basically consists of five basic parts, which are applied in sequence.

For both the impedance relay type and the current diversion ratio methods, two fault resistance values 0.1 mΩ and 20 Ω are adopted. The line length d_l is 20 km and the fault point is 5 km from the locating terminal. Parameters of the model systems are

> Source impedance $= 0.19981 + j2.0052$ (Ω)
> Substation neutral current in short circuit case $= 200$ (A)
> Self-impedance $= 0.210.79 + j0.6571$ (Ω/km)
> Mutual impedance within the circuit $= 0.1257 + j0.2700$ (Ω/km)

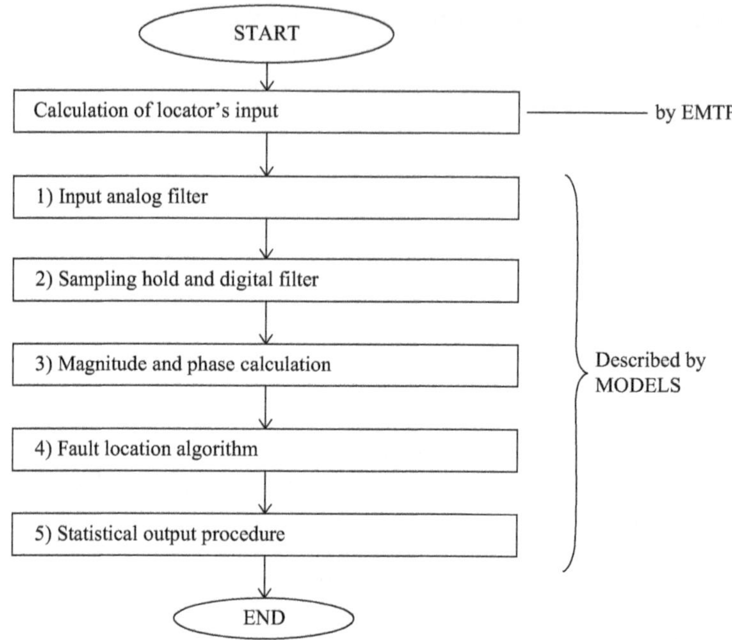

Figure 10.2   The procedure of a simulation

Mutual impedance between circuits $= 0.1258 + j0.2290$ (Ω/km)
Self-capacitive admittance $= 2.5320$ (μS/km)
Mutual capacitive admittance within the circuit $= -0.10.5110$ (μS/km)
Mutual capacitive admittance between circuits $= -0.1996$ (μS/km)

The calculated results for the radial circuit are shown in Figures 10.3 and 10.4. The oscillation which appears in the terminal current and voltage immediately after the fault occurrence is caused by the harmonics due to the line capacitance. But the harmonics can be eliminated by a filter, and cause no significant influence on the locating distance. Fault locating is delayed by about 30 ms from the fault initiation. The delay is caused by a digital filter, Fourier transformation, and output procedure, as in (1), (3), and (5), respectively, in Figure 10.2.

When the fault resistance R_f is 0.1 mΩ, the error is nearly zero. When R_f is 20 Ω, the error is about 0.4 km. These errors due to neglecting the capacitance of the line are acceptable considering the total length 20 km. The fault resistance value does not have a large influence on fault location in the presence of capacitance to the ground. Calculated results by the current diversion ratio method are shown in Figure 10.5. The terminal current and voltage are nearly the same as those in Figures 10.3 and 10.4. The location error is nearly zero. Thus the fault resistance has no influence on the location error when the resistance is held constant.

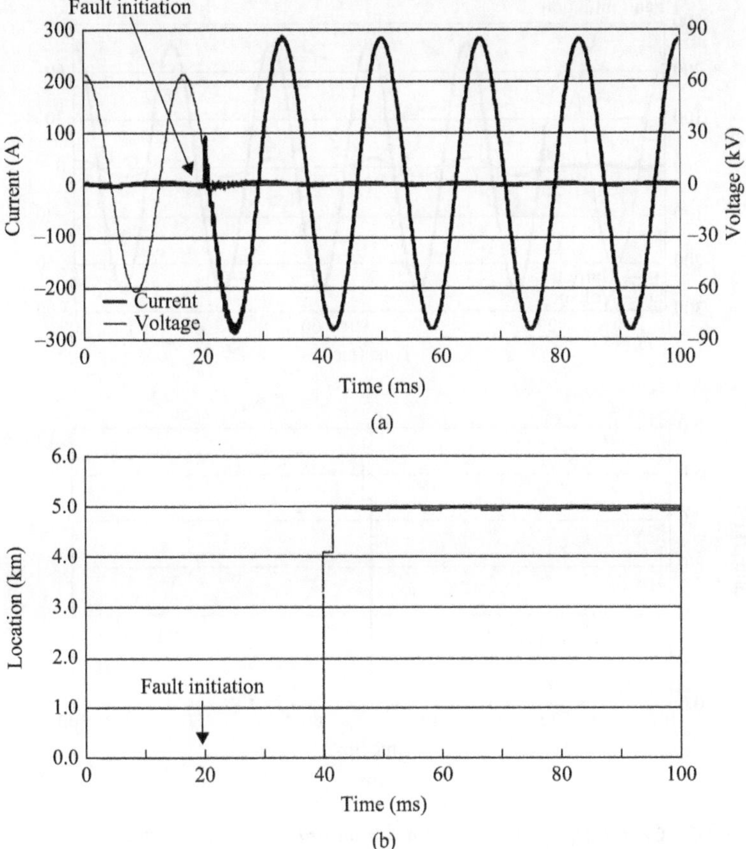

*Figure 10.3   Calculated results for the impedance relay type method: (a) Current/
voltage (faulted phase); (b) Locating distance*

10.2.3   Study on influence of fault arc characteristics

Here, the case where a nonlinear arc resistance model is added to the fundamental
network is examined. Many fault arc models have been proposed [15–18]. This
chapter adopts nonlinear arc resistance models and represents them by using the
MODELS language.

Model-1 (square voltage method)

Model-1 is a simple nonlinear resistance with the arc voltage represented as a
square wave [15, 16]:

$$R = \frac{V_a}{|I|} \tag{10.3}$$

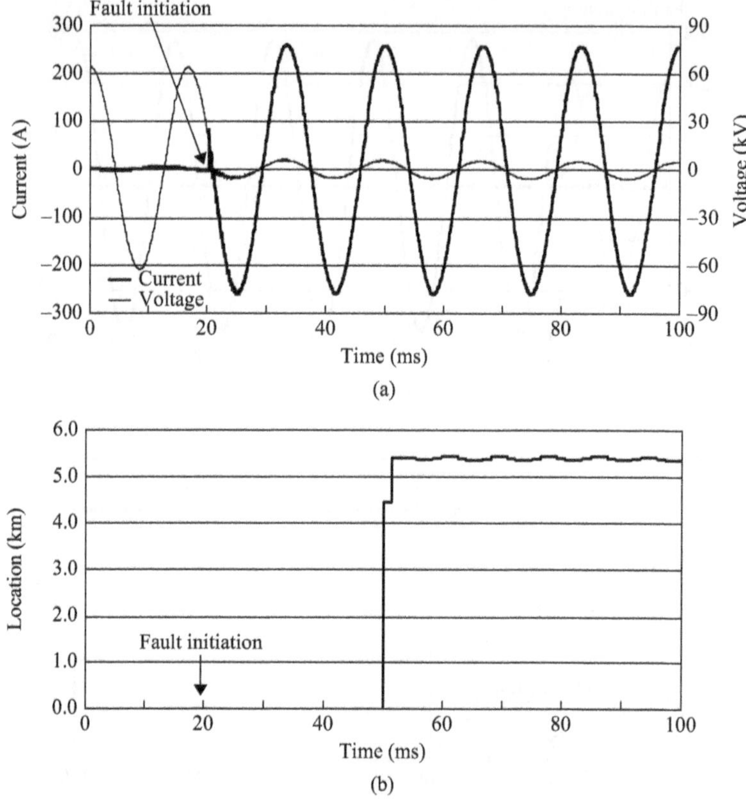

Figure 10.4   *Calculated results for the impedance relay type method: (a) Current/ voltage (faulted phase); (b) Locating distance*

where

R: arc resistance

I: arc current

V_a: amplitude of the square wave voltage approximation (10% of phase voltage RMS value = 6.28702 kV)

Model-2 (Kizilcay's time-dependent model)

Model-2 describes the dynamic behavior of a fault through air based on measurement with various system voltages and types of insulation [17, 18]:

$$R = \frac{l}{g} \tag{10.4}$$

$$\frac{dg}{dt} = \frac{l}{\tau}(G - g) \tag{10.5}$$

$$G = \frac{|i|}{(u + R_{arc} \cdot |i|) \cdot len} \tag{10.6}$$

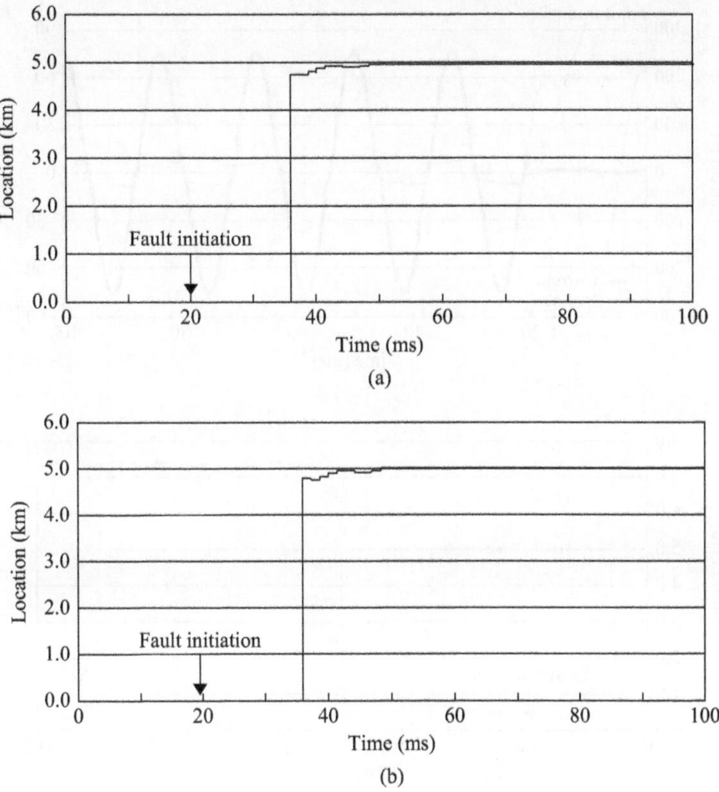

Figure 10.5   Calculated results for the current diversion ratio method: (a) Fault resistance = 0.1 mΩ; (b) Fault resistance = 20 Ω

where

R: arc resistance
g: time-varying arc conductance
G: stationary arc conductance
i: arc current
τ: arc time constant (0.10 ms)
u: constant voltage parameter per arc length (110.3 V/cm)
R_{arc}: resistive component per arc length (0.55 mΩ/cm)
len: time-dependent arc length (58 cm)

In this model the arc length is assumed to be constant (58 cm for the 77-kV system) because of the short arc length. Parameters τ, u, and R_{arc} are derived from the measurements given in [18]:

$$\tau = 0.10 \text{ (ms)}$$

$$u = 0.83 \times 1{,}000/58 = 110.3 \text{ (V/cm)}$$

$$R_{arc} = 32/58 = 0.55 \text{ (m}\Omega\text{/cm)}$$

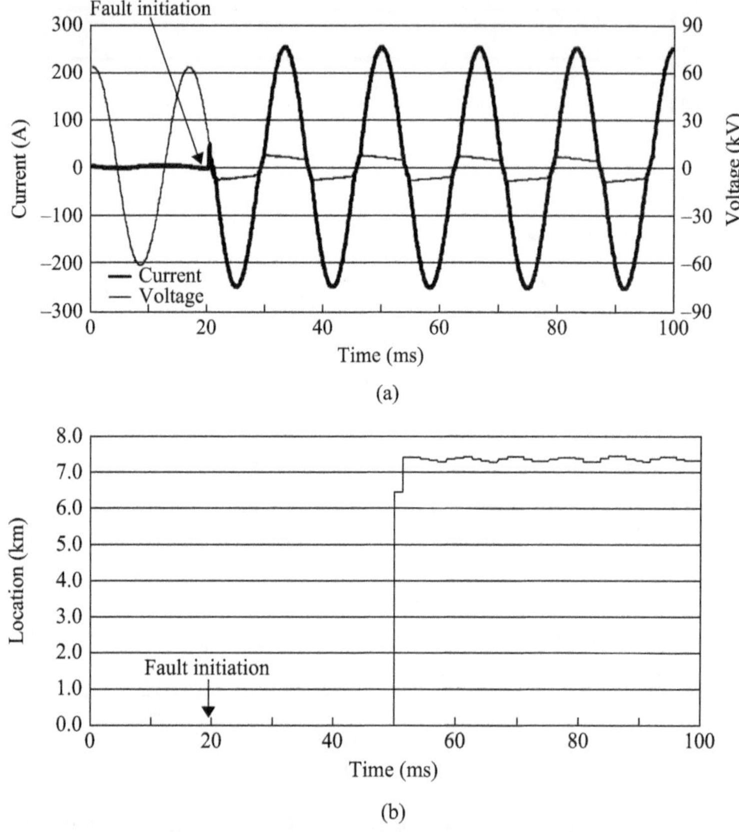

Figure 10.6   Calculated results for the impedance relay type method: (a) Current/ voltage (faulted phase); (b) Locating distance

In this case of steel towers, the value of the arc-voltage gradient lies between 12 and 15 V/cm [19]. The value of "u" lies in this range.

Calculated results are shown in Figures 10.6 to 10.8. The terminal current and voltage are nearly the same in both cases. It is clear that the impedance relay type method produces a large error, while the current diversion ratio method is not influenced by the arc resistance nonlinearity.

The effect of model parameters are investigated.

For Model-1:

V_a (kV) 1, 2, 5, 10, 20, 30% of phase voltage RMS value

For Model-2:

τ (ms)	0.2, 0.4, 0.6, 0.8
u (V/cm)	14.3 $\pm$ 10%, $\pm$ 20%
R_{arc} (mΩ/cm)	0.55 $\pm$ 10%, $\pm$ 20%
len (cm)	58 + 10%, + 20%

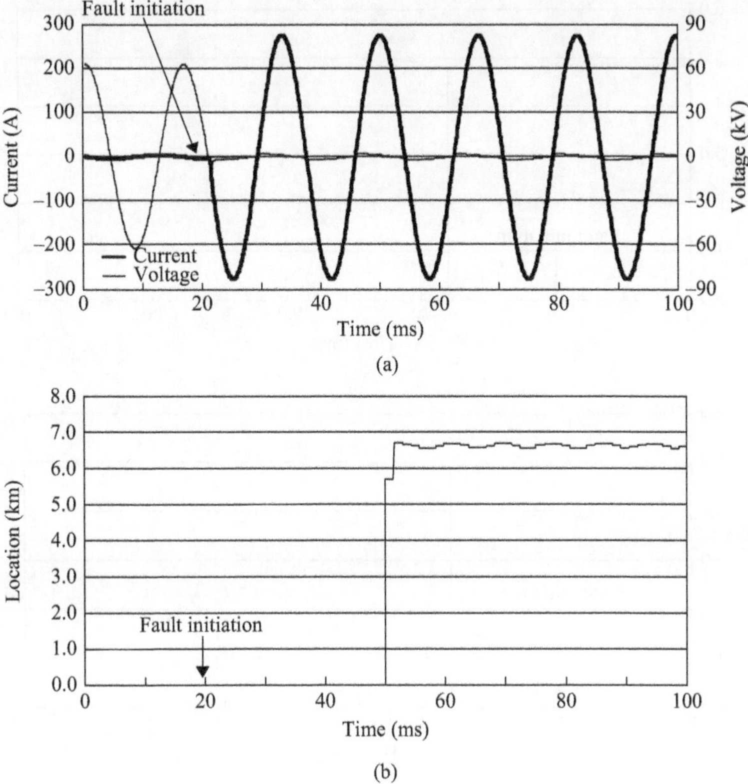

Figure 10.7   Calculated results for the impedance relay type method: (a) Current/ voltage (faulted phase); (b) Locating distance

The calculated results are shown in Figures 10.9 and 10.10. As the location errors, produced by the current diversion ratio method, are nearly zero, the graphs of those results are omitted. From the figures it is observed that the locating error is proportional to the parameter values except R_{arc}. A highly nonlinear characteristic of the fault arc produces larger fault location errors.

10.2.4   Study on influence of errors in input devices

Simulations typically omit errors in input values introduced by measuring devices such as CT, CCVT, etc. For the linear characteristics of CT and CCVT, a compensation method described in [20] is effective. But the nonlinear (saturable) behavior of a CT is a cause of a location error, and cannot be compensated by using a linear compensation method. Simulations were carried out considering a CT input error, and it was investigated how the CT's saturable current influences the fault location. Many CT/CCVT models have been proposed and the influence of their errors on protective relays or the compensation method have been studied [21–23]. In this chapter, two types of CT saturation have been simulated, one with hysteresis

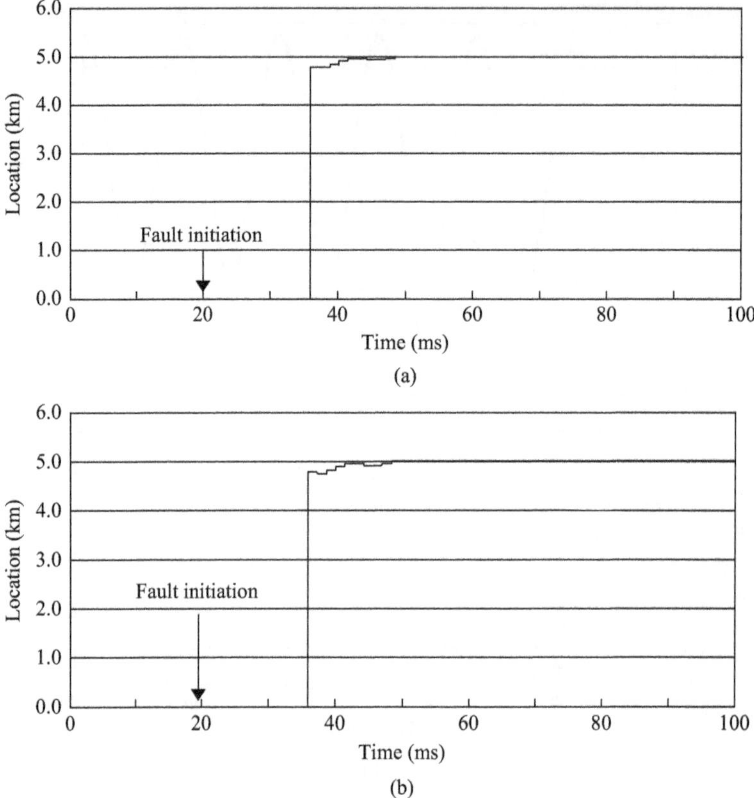

Figure 10.8   *Calculated results for the current diversion ratio method: (a) Arc resistance model-1; (b) Arc resistance model-2*

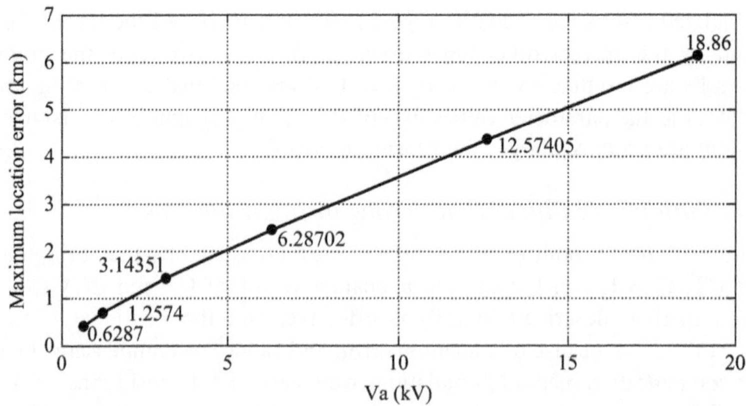

Figure 10.9   *Sensitivity analysis of Model-1 parameters for the impedance relay type method*

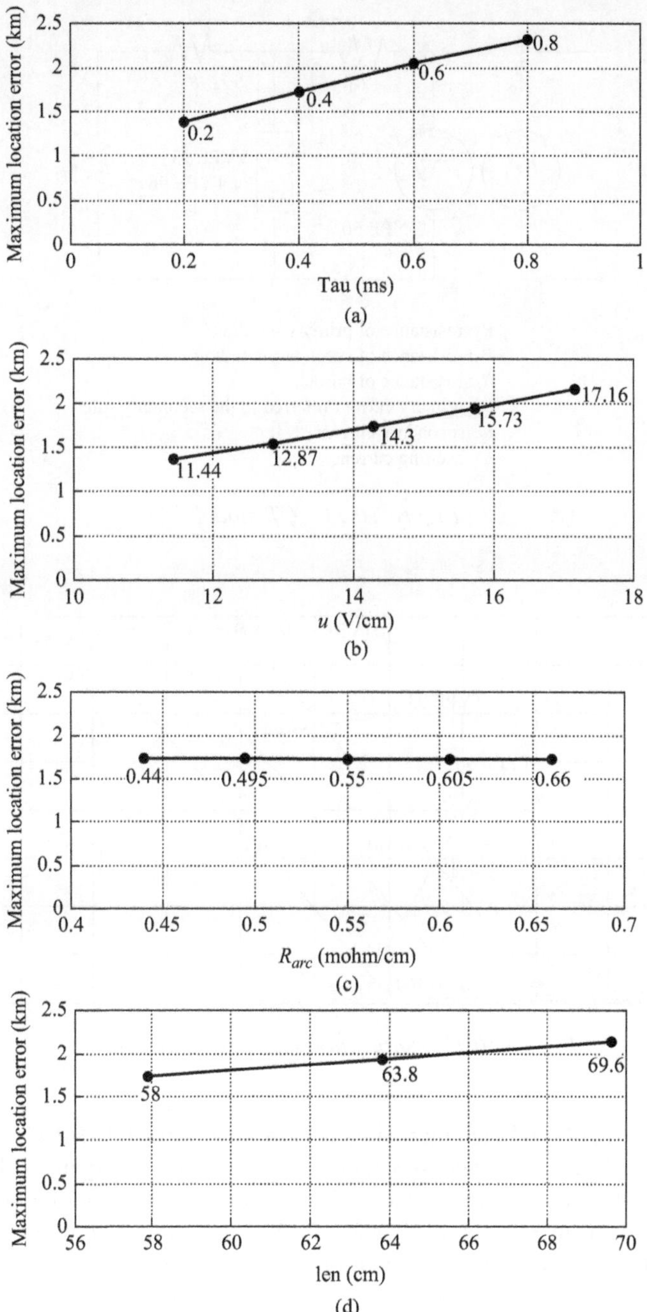

Figure 10.10   *Sensitivity analysis of Model-2 parameters for the impedance relay type method: (a) Time constant τ is varied; (b) u; (c) R_{arc}; (d) Arc length*

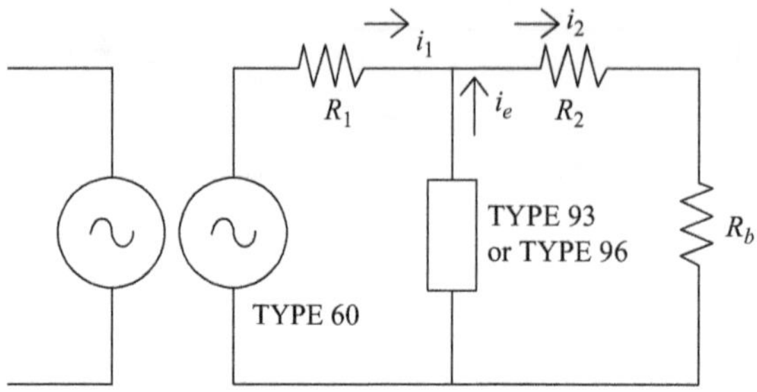

R_1: resistance of primary winding
R_2: resistance of secondary winding
R_b: resistance of burden
i_1: primary current referred to the secondary side
i_2: secondary current
i_e: exciting current

Figure 10.11   CT model

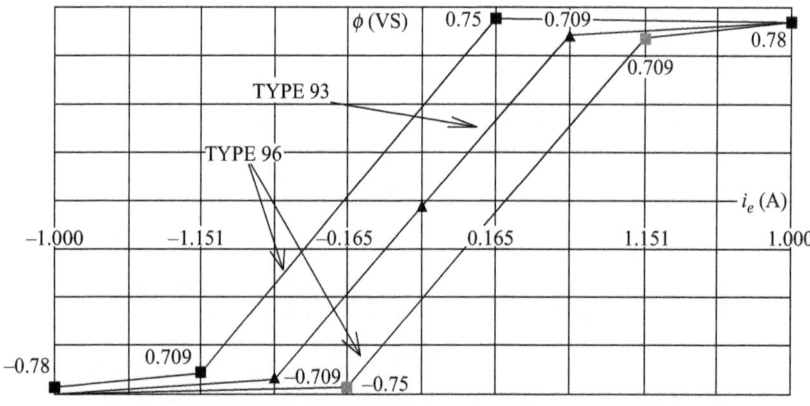

Figure 10.12   Saturation characteristics of CT

(ATP type-96) [12] and the other without hysteresis (type-93). The CT model has been used in Figure 10.11 and its saturation characteristic in Figure 10.12.

A phase-to-phase fault is located using the impedance relay type method, and results are given in Figures 10.13 and 10.14. These results show no significant locating errors, because most of the current measurement errors caused by the CT saturation disappear by the time when the locator determines the fault distance. The time constant of the circuit is about 8.4 ms. This means that the CT saturation error does not influence the correctness of the fault location. Although the operation of protection relays is greatly affected by CT saturation, its influence can be ignored in the application of digital fault location.

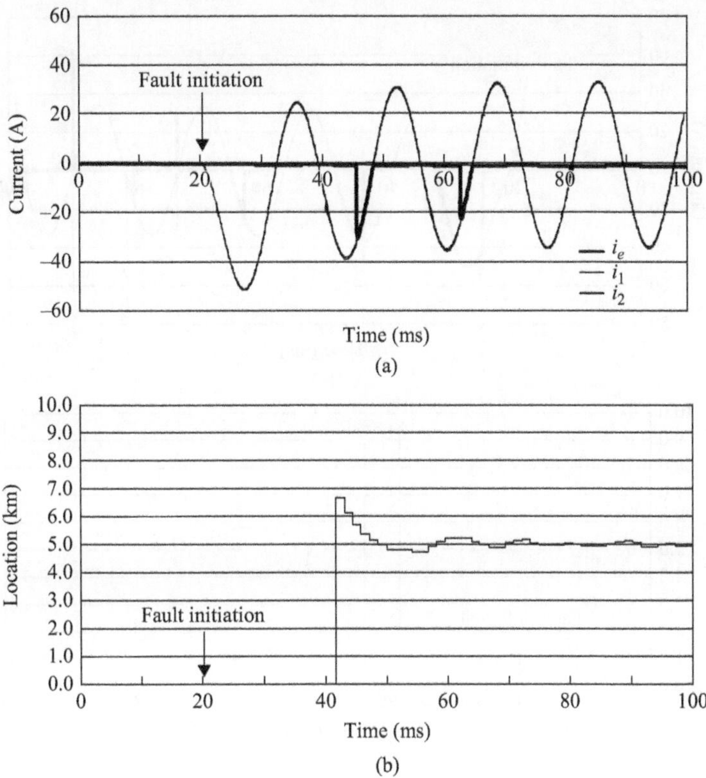

Figure 10.13   Calculated results for the impedance relay type method with CT model (without hysteresis): (a) CT circuit's currents (b-phase); (b) Locating distance

10.3   High-speed switchgear

A parallel operation of utility power system and nonutility generation system has been increasing. For example, many combined heat and power (CHP) systems are operated at huge industrial plants of factories. CHP will be increased not only at industrial plants but also at nonindustry facilities such as hospitals and hotels in aspect of effective use of electrical and heat energy. However, in the case of transmission line faults, the power system and customer voltages drop instantaneously even if the customer has the CHP generating system. This voltage sag affects customer's various devices. Especially, equipment with electronic and digital circuits is affected by the voltage sag, even if the duration of voltage sag is short such as 0.1 second. Even when a circuit breaker operates in 0.2 second and separates the system, load equipment with electronic and digital circuits can be damaged easily. Therefore, a device that would separate customer's facilities from the power system rapidly is required for some load equipment [24].

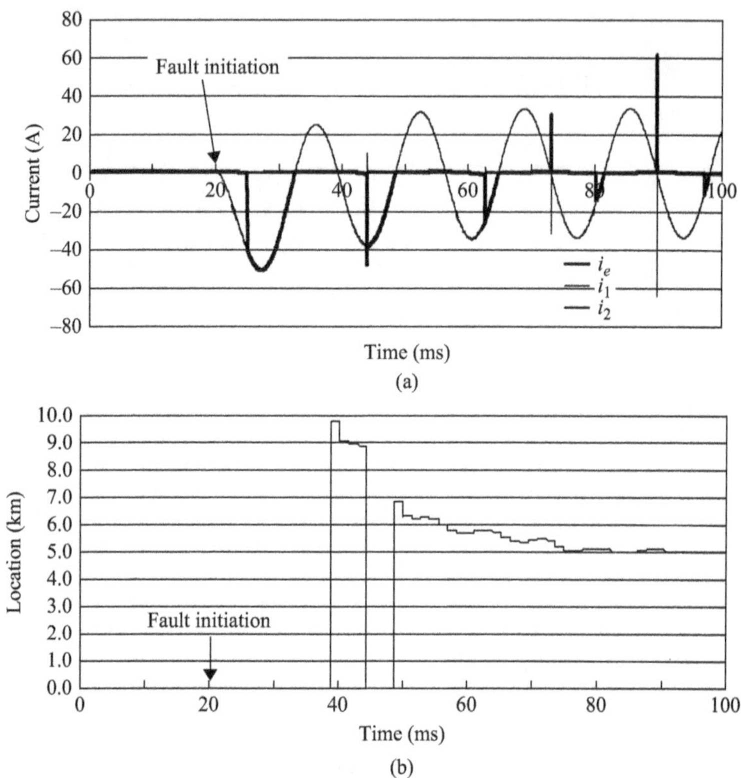

Figure 10.14   *Calculated results for the impedance relay type method with CT model (with hysteresis): (a) CT circuit's currents (b-phase); (b) Locating distance*

In this background, a high-speed switchgear that could separate a power system and a CHP by using thyristors were developed [25] in purpose of protecting most important loads from the voltage sag.

At first, the influence of the voltage sag and then the principle of the device (high-speed switchgear) are explained in this section. Second, the test results at a factory are presented and compared with simulation results by EMTP. The influence of the voltage sag magnitude is studied. Finally, waveforms recorded by the device in the real field is presented.

Examples of the influence of the voltage sag on loads are shown in reference [24]. Most devices are affected by the voltage sag, even if the time duration is within 0.1 second. It can be said that they are insufficient in an aspect of ride through capability to the voltage sag.

A high-speed switchgear was developed in 1993 [26]. The device consists of a thyristor switch, disconnecting switches, and the control circuit to detect voltage sag and synchronizing switches. The device is installed in series with the circuit breaker, which is interconnecting the bus of the customer's own power generation

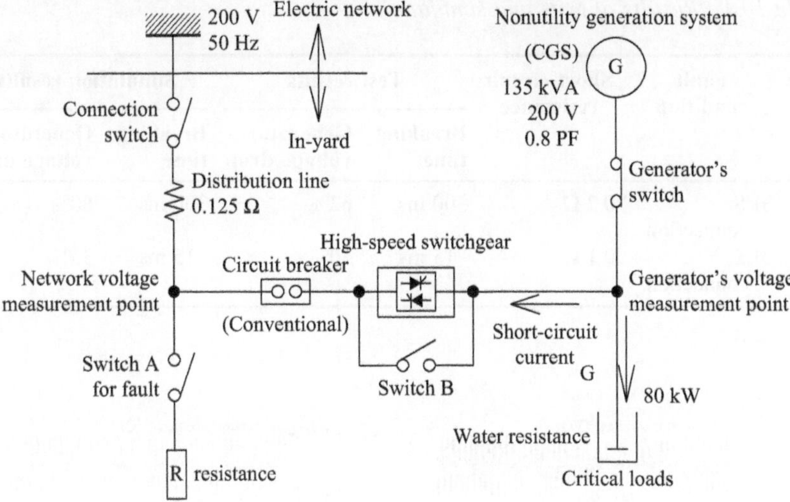

Figure 10.15   A test circuit

system and that supplied from a utility. It executes high-speed tripping, within one cycle of commercial frequency, when the voltage on the customer's bus drops.

10.3.1   Modeling methods

A test circuit is illustrated in Figure 10.15. A utility power system and a CHP are connected via a circuit breaker and a high-speed switchgear in series. The capacity of the critical load is set to be 75 percent of the generator-rated power. Instantaneous voltage sag is generated by short circuit in the utility power system side. The test is simulated under two conditions to compare the performance of a conventional circuit breaker with that of the high-speed switchgear.

Case-1: The high-speed switchgear is short-circuited by switch-B. The generator voltage, the network voltage, and the short circuit current are measured when the generator and the network are separated by the circuit breaker.

Case-2: Switch-B is opened. Performance of the high-speed switchgear and the power system is measured when the device works. The voltage drop detection level of the device is set to be 15 percent of the rated voltage.

Test and simulation results are given in Table 10.1 and the waveforms are shown in Figure 10.16.

10.3.2   Comparative study with measurement

In Case-1 of Figure 10.16(a), the circuit breaker separates the power system after 300 ms. The generator voltage drop (62 percent) is too large.

Case-2 of Figure 10.16(a) shows a waveform when the high-speed switchgear is operated 15 ms after the fault occurrence, and separates the power system from the CHP system. When the instantaneous voltage sag occurs, the generator voltage

Table 10.1   Results of tests and simulations

Case	Fault condition	Short-circuit resistance	Test results		Simulation results	
			Breaking time	**Generator voltage drop**	**Breaking time**	**Generator voltage drop**
1	3LS, connection	0.2 Ω	300 ms	62%	300 ms	60%
2	3LS, Y connection	0.1 Ω	15 ms	30%	15 ms	35%

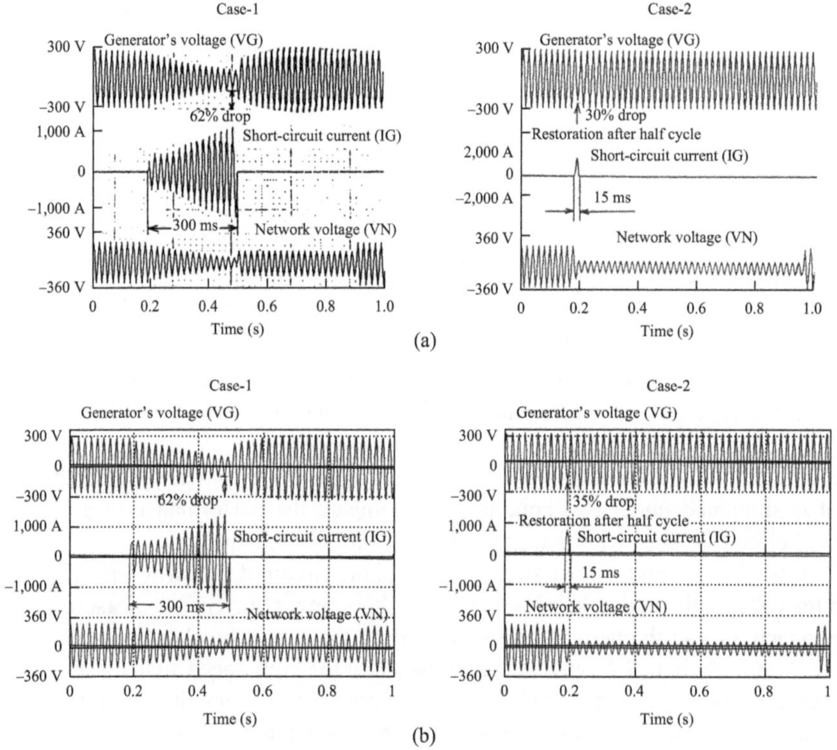

*Figure 10.16   Waveforms of test results and simulation results: (a) Test results;
(b) Simulation results*

becomes lower (30 percent drop) for half a cycle of the power system commercial frequency. By fast separation of the power systems by the device, the generator voltage reaches a steady state in the next half cycle. Compared with Case-1, it is confirmed that the device prevents an extreme voltage sag of the CHP, and makes a stable separation of critical loads possible.

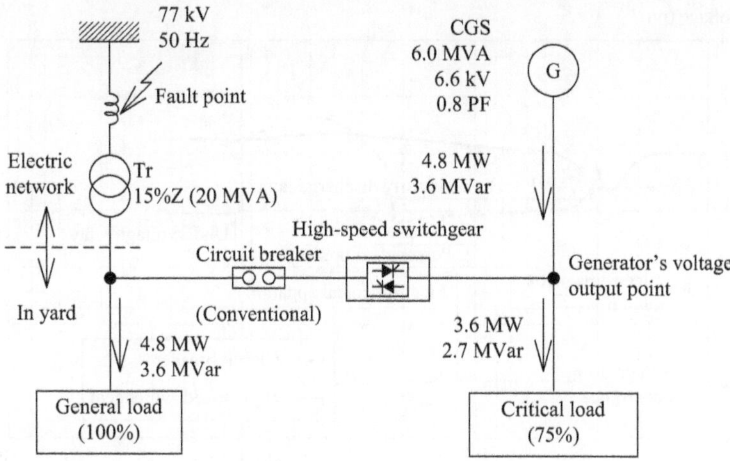

Figure 10.17   Simulation circuit

In the simulation using EMTP, the circuit breaker and the high-speed switch-gear are represented by an ordinary switch model, and a generator is represented by Park's model [11]. Opening time of the switch is determined as the same value as the test results. The generator voltage drop and the waveforms show a good agreement with the test results. Thus, the validity of an EMTP simulation for the device has been confirmed.

10.3.3   Influence of voltage sag magnitude

It was observed that most devices are affected by the voltage sag within 0.1 second. It has been said that the number of instantaneous voltage sags within 0.2 second is more than 80 percent of the total voltage sags. The instantaneous voltage sags whose magnitude is less than 20 percent occupy about 55 percent of the total voltage sags, and those less than 50 percent occupy about 80 percent. Taking the values into considerations, it can be said that a power-consuming device is not capable of withstanding the instantaneous voltage sag.

Consumption of critical loads and general loads is set to be of 75 percent and 100 percent of the generator-rated power, respectively. In Figure 10.17, a power system and a CHP are interconnected with each other through a circuit breaker and the device. In the simulation it is assumed that each device is under individual operation without information exchange. This is to study the influence of the interrupting time of each device. Occurrence time of an instantaneous voltage sag is about 20 ms after the simulation starts. The interrupting time of the circuit breaker is 0.2 second after the occurrence of the voltage sag. The interrupting time of the device is within 6.5 to 20 ms. This time is affected by the condition of the voltage sag and the phase angle of the current interruption. In this simulation the time is set to 20 ms, which is the worst case scenario. The circuit breaker and the device are represented by an ordinary switch model in EMTP. The generator is simulated by

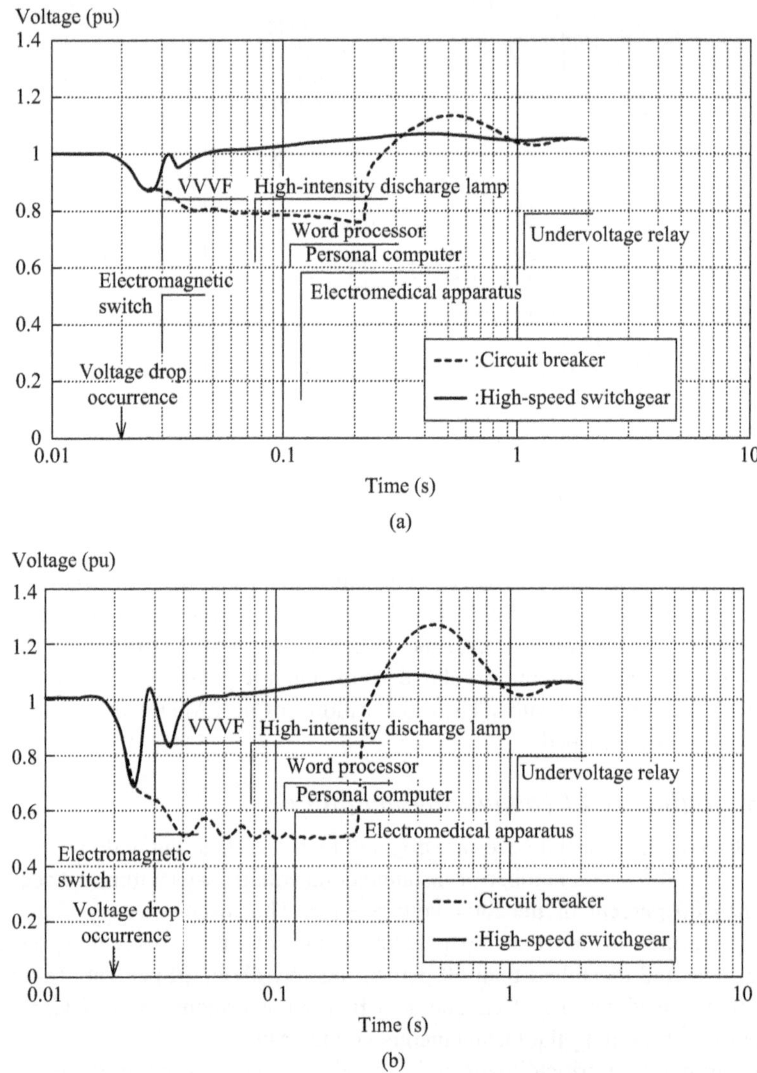

Figure 10.18   Simulation results: (a) Voltage sag: 20% (target value);
(b) Voltage sag: 50% (target value)

Park's model. The generator control circuit is simulated using TACS function of EMTP [11]. The magnitude of the voltage sag is set to 20 and 50 percent by changing the fault point impedance.

For the conditions of the instantaneous voltage sag of 20 and 50 percent, simulations were performed with the circuit breaker connecting buses and the device. Figure 10.18 shows the generator voltage simulated and the influence of the instantaneous voltage sag for various loads mentioned above. For each condition, maximum voltage sag in the faulted period and the voltage rise are given in

Table 10.2   Maximum voltage drop during the fault and voltage rise after clearing the fault

Voltage drop (target value)	High-speed switchgear		Circuit breaker	
	Max. voltage drop	Voltage rise after breaking	Max. voltage drop	Voltage rise after breaking
20%	12.0%	7.5%	23.5%	14.1%
50%	32.2%	9.8%	50.9%	26.8%

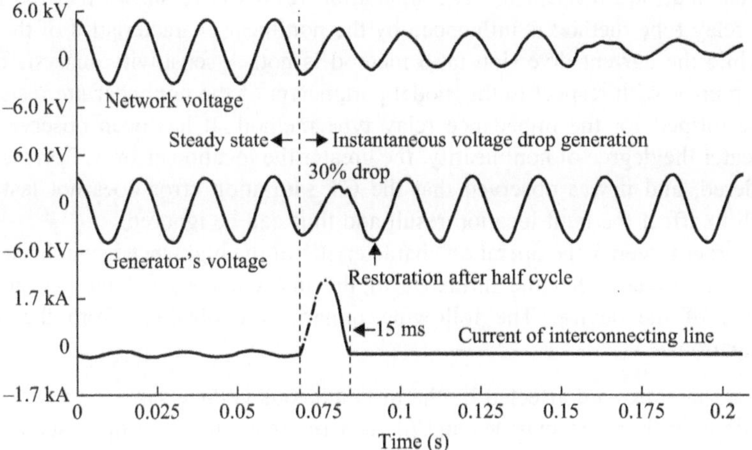

Figure 10.19   A recorded instantaneous voltage sag

Table 10.2. In the simulated waveform of Figure 10.18, the voltage amplitude is oscillating by the influence of a transient direct current component during the subtransient period ($T_d'' = 35$ ms).

When the system is interrupted by the circuit breaker under the condition of the voltage sag of 20 percent, a high intensity discharge lamp is influenced by the voltage sag. If interrupted by the device, it is not affected. Under the condition of the voltage sag of 50 percent, most load devices are influenced by the voltage sag. As the voltage rise after the interruption is greater than 20 percent, there is a possibility that an overvoltage protective relay operates depending on the time delay setting. In the case of interruption by the device, load equipment is not affected by the instantaneous voltage sag, and the voltage rise after interruption is lower.

A voltage sag waveform is shown in Figure 10.19. It was observed in August, 1995, by the device installed in a customer site with 3.3 kV rated voltage. The device breaks the short circuit current at 15 ms after the instantaneous voltage sag generated. This breaking makes the power system separated from the customer's facility at high speed so that the generator can continue to operate steadily and restrain the voltage sag (approximately 30 percent) of the important loads within half a cycle of the commercial frequency. However, the network voltage cannot recover even when 0.1 ms or longer time is passed. No trouble was observed in the critical loads when the high-speed switchgear was adopted.

10.4   Conclusions

This chapter has investigated the influence of the fault arc nonlinearity and the input device error on the accuracy of digital fault locators. A time domain simulation model of the digital fault locator was represented using the MODELS language in the ATP-EMTP. Various types of faults were simulated with constant and nonlinear fault arc resistances. The simulation results have shown that the impedance relay type method is influenced by the nonlinear characteristics of the fault arc, while the current diversion ratio method is not. A sensitivity analysis of the locating error with respect to the model parameters of the nonlinear arc resistance was performed for the impedance relay type method. It has been observed that the greater the degree of nonlinearity, the greater the location error. CT errors were considered, and it was observed that the CT saturation error does not last long enough to affect the fault location result and thus can be ignored.

In order to verify the operation characteristic of the high-speed switchgear, this chapter has investigated the influence of the instantaneous voltage sag and the principle of the device. The following remarks are obtained from the above investigation.

1. Load facilities are affected by the instantaneous voltage sag.
2. Adopting thyristor switches and the scheme that can sense the instantaneous voltage sag, this device can separate a CHP from a power system.
3. A comparison of measured results with simulated results shows that the ability of the device can be verified through the EMTP simulation.
4. Test and simulation results show that most critical (sensitive) device could be affected in the case of the circuit breaker operation. As a countermeasure against the instantaneous voltage sag, the device can protect most critical loads and is running well in the field.

References

[1] Takagi, T., Yamakoshi, Y., Yamaura, M., Kondou, R., Matsushima, T. 'Development of a new type fault locator using the one-terminal voltage and current data'. *IEEE Transaction on Power Apparatus and Systems*, vol. PAS-101, no. 8, August 1982, pp. 2892–2898.

[2] Eriksson, L., Saha, M. M., Rockfeller, G. D. 'An acculate locator with compensation for apparent reactance in the fault resistance resulting from remote-end infeed'. *IEEE Transaction on Power Apparatus and Systems*, vol. PAS-104, no. 2, February 1985, pp. 424–436.

[3]   Sachdev, M., Agarwl, R. 'A technique for estimating transmission line fault locations from digital impedance relay measurements'. *IEEE Transactions on Power Delivery*, vol. 3, no. 1, January 1988, pp. 121–129.

[4]   Suzuki, K., Ohura, Y., Kurosawa, Y., Takemura, T., Kumano, S., Yoshikawa, M., Funabashi, T., Iwatani, F. 'Existing condition and experiences of distance relays and fault locators using digital technology'. CIGRE Symposium Boumemouth S34-89, 1989.

[5]   Girgis, A. A., Fallon, C. M. 'Fault location techniques for radial and loop transmission systems using digital fault recorded data'. *IEEE Transactions on Power Delivery*, vol. 7, no. 4, October 1992, pp. 1936–1945.

[6]   Johns, A. T., Moore, P. J., Whittard, R. 'New technique for the accurate location of earth faults on transmission systems'. *IEE Proc.-Gener. Transm. Distrib.*, vol. 142, no. 2, March 1995, pp. 119–127.

[7]   Novosel, D., Hart, D. G., Udren, E., Saha, M. M. 'Fault location using digital relay data'. *IEEE Computer Applications in Power*, July 1995, pp. 45–50.

[8]   Peterson, J. N., Wall, R. W. 'Interactive relay controlled power system modelling'. *IEEE Transactions on Power Delivery*, vol. 6, no. 1, January 1991, pp. 96–102.

[9]   Wilson, R. E., Nordstrom, J. M. 'EMTP transient modeling of a distance relay and a comparison with EMTP laboratory testing'. *IEEE Transactions on Power Delivery*, vol. 8, no. 3, July 1993, pp. 984–992.

[10]   Chaudhary, A. K. S., Tam, K., Phadke, A. G. 'Protection system representation in the electromagnetic transients program'. *IEEE Transactions on Power Delivery*, vol. 9, no. 2, April 1994, pp. 700–711.

[11]   Dommel, H. W. *EMTP theory book*. Bonneville Power Administration, 1986.

[12]   Scott-Meyer, W. *ATP rule book*. Bonneville Power Administration, 1993.

[13]   Dube, L., Bonfani, I. 'MODELS: A new simulation tool in the EMTP'. *European Transactions on Electrical Power Engineering (ETEP)*, vol. 2, no. 1, January/February 1992, pp. 45–50.

[14]   Dube, L. *User's guide to MODELS in ATP*. Bonneville Power Administration, 1996.

[15]   Djuric, M. B., Terzija, V. V. 'A new approach to the arcing faults detection for fast autoreclosure in transmission systems'. *IEEE Transactions on Power Delivery*, vol. 10, no. 4, October 1995, pp. 1793–1798.

[16]   Sousa, J., Santos, D., Correia de Barros, M. T. 'Fault arc modeling in EMTP'. IPST'95-International Conference on Power Systems Transients, Lisbon, 3–7 September 1995, pp. 475–480.

[17]   Kizilcay, M., Pniok, T. 'Digital simulation of fault arcs in power systems'. *European Transactions on Electrical Power Engineering (ETEP)*, vol. 1, no. 1, January/February 1991, pp. 55–60.

[18]   Kizilcay, M., Koch, K. H. 'Numerical fault arc simulation based on power arc tests'. *European Transactions on Electrical Power Engineering (ETEP)*, vol. 4, no. 3, May/June 1994, pp. 177–185.

[19]   Cornick, K. J., Ko, Y. M., Pek, B. 'Power system transients caused by arcing faults'. *IEE Proc.*, vol. 128, pt. C, no. 1, January 1981, pp. 18–27.

[20]   Harada, E., Yashiro, M., Mizutori, T., Funabashi, T. 'Error compensation scheme for analog detector'. *Japanese patent* #1708068, November 11, 1992.

[21]   Working Group of the Relay Input Sources Subcommittee of the Power System Relay Committee. 'Transient response of current transformers'. *IEEE Transactions on Power Apparatus and Systems*, vol. PAS-96, no. 6, November/December 1977, pp. 1809–1814.

[22]   Transient Response of CCVTs Working Group of the Relay Input Sources Subcommittee of the Power System Relay Committee. 'Transient response of coupling capacitor voltage transformers IEEE committee report'. *IEEE Transactions on Power Apparatus and Systems*, vol. PAS-100, no. 12, December 1981, pp. 4811–4814.

[23]   Horowitz, S. H., Phadke, A. G. *Power system relaying.* 2nd Edition, Research Studies Press, Taunton, Somerset, England, 1992, pp. 59–61, 70–76.

[24]   Electric Technology Research Association Report *Instantaneous Voltage Sag.* Committee Report, 1990, vol. 46, no. 3.

[25]   Matsushita, K., Matsuura, Y., Funabashi, T., Takeuchi, N., Kai, T., Oobe, M. 'Study for the performance of high speed switchgear for protection of in-house generation system'. *IEEE Industry and Application Society I&CPS'96* (Industrial and Commercial Power Systems Technical Conference), May 6–9, 1996, New Orleans, Louisiana, USA, pp. 91–95.

[26]   Matsushita, K., Kataoka, Y., Ono, M. 'High speed switch gear for protection of generator'. *Proceeding of the Fourth Annual Conference of Power and Energy Society*, IEE of Japan (Regular Session II, Short Papers), July 28–30, 1993, Hokkaido University, Sapporo, Japan, pp. 484–485.

Chapter 11

Overvoltage protection and insulation coordination

*T. Ohno**

This chapter discusses the overvoltage protection and the insulation coordination with a focus on the transient analysis required. The first section gives definitions of temporary overvoltages, slow-front overvoltages, fast-front overvoltages, and very-fast-front overvoltages. The section also explains causes and characteristics of each overvoltage. It is necessary to understand them in order to perform the insulation coordination study.

The next section introduces the insulation coordination study based on the IEC 60071 series. According to the study procedure defined in these international standards, representative overvoltages first need to be determined for temporary, slow-front, and fast-front overvoltages. The determination of representative overvoltages requires extensive transient analyses, and it is the crucial part in the insulation coordination study.

The application of surge arresters is indispensable for the overvoltage protection. The selection of surge arresters is introduced in section 11.3, according to the IEC 60099 series. Specifications of surge arresters and their selection procedure are explained. Finally, an example of the transient analysis for the overvoltage protection and the insulation coordination is introduced in section 11.4.

11.1 Classification of overvoltages

Overvoltages are characterized by their origins, voltage shapes, and durations. Besides continuous (power-frequency) voltages, IEC 60071-1 classifies overvoltages into four categories—temporary overvoltages, slow-front overvoltages, fast-front overvoltages, and very-fast-front overvoltages [1]. Voltage shapes of these overvoltages are defined as in Table 11.1. This section explains these four categories of overvoltages in relation to the overvoltage protection and the insulation coordination.

*Manager, Power System Analysis Group, Tokyo Electric Power Co., Japan

Table 11.1   Voltage shapes of overvoltages

	Temporary	**Slow-front**	**Fast-front**	**Very-fast-front**
Frequency	10–500 Hz	–	–	0.3–100 MHz 30–300 kHz *1
Duration	0.02–3,600 s	–	–	–
Time to peak	–	20–5,000 µs	0.1–20 µs	$\leq$0.1 µs
Time to half value	–	$\leq$20 ms	$\leq$300 µs	–

*1: The very-fast-front overvoltage is superimposed on the base overvoltage. The frequency of the very-fast-front overvoltage is 0.3–100 MHz, while the frequency of the base overvoltage is 30–300 kHz.

11.1.1   Temporary overvoltage

Voltage levels of temporary overvoltages are normally lowest among four categories of overvoltages, but durations are normally longest. The long duration is caused by low-frequency components contained in temporary overvoltages, and the low frequency also leads to wider or deeper propagation of temporary overvoltages.

Difficulties can arise when performing the transient analysis of temporary overvoltages since it requires the modeling of a wider area of the power system (big model) [2]. Determining the extent of the modeled area is not a trivial question. An adaption of model reduction technique is also considered.

One can encounter other difficulties due to the fact that severe temporary overvoltages are often caused under specific network conditions [3, 4]. Finding the specific network condition can require a number of simulation runs with a big model.

Since the severe temporary overvoltage occurs only under specific network conditions, the probability of its occurrence is significantly low compared with other overvoltages. If such a condition exists, however, the consequence of the severe temporary overvoltage should not be ignored. Hence, the temporary overvoltage analysis is generally conducted with the deterministic method.

Temporary overvoltages are generally caused by the following phenomena:

- Single-line-to-ground fault
- Resonance
- Load rejection/system islanding

When a single-line-to-ground fault occurs, the voltages of healthy phases rise until the fault is cleared. IEEE Std. 142 defines the effectively grounded system as a system "grounded through a sufficiently low impedance such that for all system conditions the ratio of zero-sequence reactance to positive sequence reactance (X_0/X_1) is positive and not greater than 3, and the ratio of zero-sequence resistance to positive-sequence reactance (R_0/X_1) is positive and not greater than 1" [5]. In such a system, the voltages of healthy phases are normally limited to 80% of the line voltage.

The transient analysis to find the overvoltage in healthy phases is sometimes performed when a studied system is not effectively grounded. The overvoltage can also be checked when the ground fault overvoltage analysis is conducted to study slow-front overvoltages.

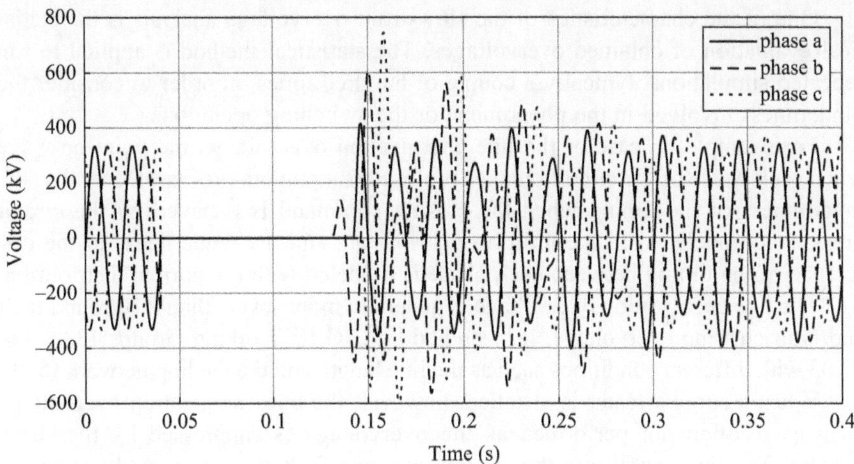

Figure 11.1   Example of the overvoltage caused by the load rejection

The resonance includes the series resonance, the parallel resonance, the open-phase resonance, and the ferroresonance. The transient analysis on the series, parallel and the open-phase resonance, in other words, linear resonance, is performed in all phases—feasibility study, implementation study, and forensic analysis—of the overvoltage protection design and the insulation coordination study [2–4]. The overvoltage caused by the ferroresonance, in other words, nonlinear resonance, is often difficult to predict. The transient analysis on the ferroresonance is normally conducted in the forensic analysis.

Extremely severe resonance conditions can be found in the resonance overvoltage analysis. Especially, the black-start restoration can present a major difficulty due to the low load condition. Depending on the severity of the resonance, it may not be realistic to suppress the overvoltage only by surge arresters. In such a case, the overvoltage should be avoided by operational countermeasures or other equipment such as shunt reactors.

Load rejection and system islanding can cause an oscillatory overvoltage composed of two-frequency components, one of which is always the nominal frequency component [2–4]. The superimposition of two-frequency components leads to a peculiar overvoltage waveform as shown in Figure 11.1. The overvoltage caused by the load rejection or the system islanding is suppressed by carefully selecting the compensation rate of the transmission line charging capacity or applying surge arresters.

11.1.2   Slow-front overvoltage

Major causes of slow-front overvoltages are the following phenomena or switching operations:

- Line energization and reenergization
- Ground fault and fault clearing
- Interruption of inductive and capacitive currents

One of the characteristics of the slow-front overvoltage analysis is the statistical evaluation of obtained overvoltages. The statistical method is applied to run repeated simulations, typically a couple of hundred times, in order to consider the randomness involved in the phenomena or the switching operations.

For example, in case of the line energization overvoltage, the point-on-wave when each phase of the circuit breaker is closed is a random process. Generally, the randomness of the timing when the closing command is received by the circuit breaker is modeled with the uniform distribution, and the randomness of the difference of the timing between each phase is modeled with the normal distribution.

The line energization overvoltage is generally more severe than the ground fault and fault clearing overvoltage. It is studied by CIGRE Working Groups 13.02 and 13.05 with different conditions such as the line length and the feeding network [6–9].

When a surge arrester is installed, however, the line energization overvoltage analysis is often not performed as the overvoltage is suppressed by the surge arrester. The same applies to the ground fault and fault current overvoltage analysis. The phase-to-earth overvoltage is limited to the protective level of the surge arrester, and the phase-to-phase overvoltage is limited to twice the protective level of the surge arrester. In many cases, the maximum phase-to-phase overvoltage is much lower than this value.

When the installation of surge arresters is not enough to suppress the slow-front overvoltage, the application of the point-on-wave switching or the pre-insertion resistor to a circuit breaker is considered.

When a line is switched off from the grid, for example, for a fault clearing or line maintenance, the residual voltage can remain on the line. The sign of the residual voltage is determined by the point-on-wave when the line is switched off. In case of the line maintenance, the line is normally grounded and discharged to ensure the safety of the maintenance work. In other cases, the line may not be discharged before reenergization, which can result in a severe reenergization overvoltage. In the most severe case, the reenergization occurs at the point-on-wave opposite to the one when the line is switched off. When an inductive voltage transformer is connected to the line, however, it is not necessary to study the reenergization overvoltage as the line is discharged through the voltage transformer even before the autoreclose.

The interruption of capacitive currents is studied for large shunt capacitors and long EHV/HV cables. The leading current interruption capability of circuit breakers is defined in IEC 62271-100 [10]. The transient analysis is performed in order to confirm that the overvoltage is lower than the test voltage of capacitor bank current switching tests defined in IEC 62271-100. Care should be taken since if a restrike occurs in a circuit breaker, an extremely severe overvoltage can be built up by repeated restrikes [11].

11.1.3   *Fast-front overvoltage*

Fast-front overvoltages are caused by lightning strikes. Direct strokes, back flashovers, and overvoltages induced by lightning strikes to the ground are considered. Induced overvoltages are generally studied only for LV circuits.

A modeled area for the fast-front overvoltage analysis can be much smaller compared with the one for the temporary overvoltage analysis. Only the targeted transmission line and the substation are normally included in the model, but they need to be modeled in detail. For example, the length of each section of the conductor is considered in the substation. An example of the simulation model and the fast-front overvoltage analysis is shown in section 11.4.

The magnitude of the fast-front overvoltage is highly dependent on the magnitude of the lightning strike current. For the direct lightning strike, the magnitude of the lightning strike current is determined considering the shielding failure with the ground wire shielding design applied. An improvement of the ground wire shielding design can result in the reduction of the assumed lightning strike current.

For back flashovers, the magnitude of the fast-front overvoltage is highly dependent also on the tower footing resistance (grounding resistance). The effect of the grounding resistance is shown in the example of the transient analysis in section 11.4.2.

Surge arresters are a crucial part of the insulation coordination of the fast-front overvoltage. It is often not realistic to suppress the fast-front overvoltage below withstand voltages without surge arresters. The selection and the arrangement of surge arresters play an important role for the overvoltage protection and the insulation coordination. The selection of surge arresters is discussed in section 11.3.

A switching of a short line with or without a restrike can also lead to the fast-front overvoltage. The magnitude of the overvoltage is lower than the one caused by lightning strikes. However, it can occur very close to equipment and can cause a severe inter-turn overvoltage in nearby transformers and other equipment with windings, such as shunt reactors. Special care should be taken when surge arresters cannot be installed close to such equipment.

11.1.4   Very-fast-front overvoltage

Very-fast-front overvoltages are caused by restrikes when opening a disconnector in GIS. A restrike can also occur in a circuit breaker, depending on the voltage between both sides of the circuit breaker, resulting in the very-fast-front overvoltage.

If the very-fast-front overvoltage is caused near a transformer or other equipment with windings, it can lead to a severe inter-turn overvoltage in the equipment. The transient analysis of the inter-turn overvoltage requires a detailed model inside the equipment. An example of such a detailed model can be found in [12].

One severe example is a switching of a shunt reactor. When a current into the shunt reactor is interrupted by a circuit breaker, a transient voltage expressed by $L\,di/dt$ appears on the shunt reactor side of the circuit breaker, where L is the inductance of the shunt reactor. If a restrike occurs in the circuit breaker, the very-fast-front overvoltage caused by it can lead to a severe inter-turn overvoltage in the nearby shunt reactor.

As a countermeasure to this example, a point-on-wave switching can be applied to circuit breakers of shunt reactors. By opening the circuit breaker when a current into the shunt reactor is zero, the transient voltage expressed by $L\,di/dt$ becomes zero. This prevents the occurrence of the restrike and, as a result, protects the shunt rector.

11.2   Insulation coordination study

11.2.1   Study flow

The transient analysis is an important part of the insulation coordination study. The insulation coordination is defined in IEC 60071-1 as [1]:

> *selection of the dielectric strength of equipment in relation to the operating voltages and overvoltages which can appear on the system for which the equipment is intended and taking into account the service environment and the characteristics of the available preventing and protective devices*

As the definition indicates, *the overvoltages which can appear on the system* need to be known for the insulation coordination, and they are generally obtained as a result of the transient analysis.

The procedure of the insulation coordination study is explained in IEC 60071-1 and IEC 60071-2 [1, 13]. According to the procedure as shown in Figure 11.2, the representative voltages and overvoltages are first determined. Here, the representative overvoltages include all four categories of the overvoltages discussed in section 11.1—temporary overvoltages, slow-front overvoltages, fast-front overvoltages, and very-fast-front overvoltages.

11.2.2   Determination of the representative overvoltages

The determination of the representative overvoltages is the process most relevant to the transient analysis in the insulation coordination study. Sometimes, the representative overvoltages can be determined without the transient analysis, for example, from past experiences or from the protective level of the surge arrester. In the other cases, the transient analysis needs to be performed in order to determine the representative overvoltages.

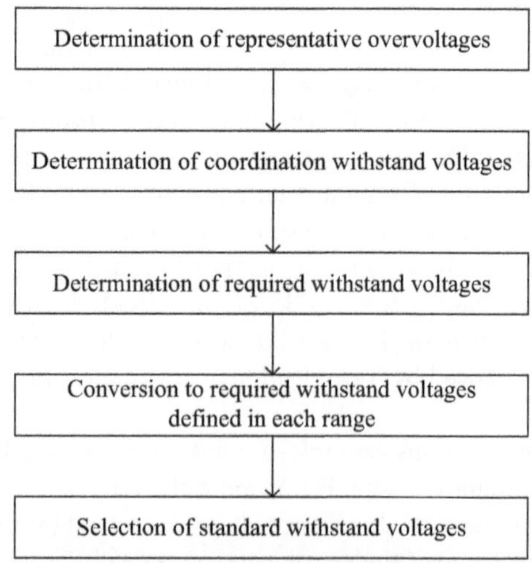

Figure 11.2   Simplified study flow diagram of the insulation coordination

Table 11.2　Standard voltage shapes

	Standard short-duration power-frequency voltage	**Standard switching impulse**	**Standard lightning impulse**
Frequency	48–62 Hz	–	–
Duration	60 s	–	–
Time to peak	–	250 μs	1.2 μs
Time to half value	–	2,500 μs	50 μs

The representative overvoltages are supposed to have standard voltage shapes defined by IEC 60071-1 as in Table 11.2 [1]. The overvoltages found in the transient analysis need to be converted to these standard voltage shapes in order to determine the representative overvoltages. The standard voltage shape for the very-fast-front overvoltage is not shown in the table as it is to be specified by the relevant apparatus committees.

The following explains major considerations when converting the overvoltage found in the transient analysis to the representative overvoltages.

11.2.2.1　Temporary overvoltage

The duration of the representative temporary overvoltage is always 60 s. The amplitude of the representative temporary overvoltage is determined by the amplitude and the duration of the maximum temporary overvoltage found in the transient analysis with the deterministic method. Using the amplitude/duration power-frequency withstand characteristic of the insulation considered, the amplitude of the maximum temporary overvoltage is converted to the amplitude of the representative temporary overvoltage, which poses the same stress in the insulation considered.

For example, when the amplitude/duration power-frequency withstand characteristic is expressed by the following equation:

$$V^n \times t = k \tag{11.1}$$

where

V: amplitude of the power-frequency withstand voltage [kV]
t: duration of the power-frequency withstand voltage [s]
k, n: constant

The amplitude of the representative temporary overvoltage is found as

$$V_{re} = \sqrt[n]{\frac{k}{t_{re}}} \tag{11.2}$$

where

V_{re}: amplitude of the representative temporary overvoltage [kV]
t_{re}: duration of the representative temporary overvoltage [s] ($t_{re} = 60$)

If the amplitude/duration power-frequency withstand characteristic of the insulation considered is not available, the amplitude of the representative temporary overvoltage is considered to be equal to the amplitude of the maximum temporary overvoltage.

11.2.2.2 Slow-front overvoltage

The time to peak and the time to half value of the representative slow-front overvoltage are 250 µs and 2,500 µs, respectively, as shown in Table 11.2. The amplitude of the representative slow-front overvoltage is normally considered equal to the 2% value found in the transient analysis. If the voltage shape of the slow-front overvoltage found in the transient analysis has a much longer time to peak than 250 µs, the amplitude of the representative slow-front overvoltage can be lowered from the 2% value considering its lower stress on the insulation.

When the statistical evaluation is performed against a performance criterion, the probability distribution of the representative slow-front overvoltage is characterized by the maximum value, the 2% value, and the standard deviation found in the transient analysis.

11.2.2.3 Fast-front overvoltage

The time to peak and the time to half value of the representative fast-front overvoltage are 1.2 µs and 50 µs, respectively, as shown in Table 11.2. The amplitude of the representative fast-front overvoltage is normally considered equal to the maximum value found in the transient analysis. Generally, the deterministic method is applied to the representative fast-front overvoltage. However, the statistical nature of the magnitude of the lightning strike current is considered when setting up the lightning strike current model for the transient analysis.

11.2.2.4 Very-fast-front overvoltage

The very-fast-front overvoltage analysis is important for the design of equipment, especially for the equipment with windings. It is also carried out in the forensic analysis but is normally not performed in the insulation coordination study in order to select the dielectric strength of equipment. Instead, a failure of equipment due to the very-fast-front overvoltage is avoided by the equipment design, the appropriate arrangement of surge arresters, and other mitigation measures such as the point-on-wave switching as discussed in section 11.1.4.

11.2.3 Steps following the determination of the representative overvoltages

According to the procedure of the insulation coordination study explained in IEC 60071-2, the following steps are necessary following the determination of representative overvoltages:

- Determination of coordination withstand voltages
- Determination of required withstand voltages
- Conversion to required withstand voltages defined in each range
- Selection of standard withstand voltages

In the deterministic method, the coordination withstand voltages are determined by multiplying the representative overvoltage by the coordination factor. The coordination factor is normally considered equal to one, expect for the case when the surge arrester limits the maximum slow-front overvoltage and affects the probability distribution. The coordination factor in such a case is given dependent on the proportion of the switching impulse protective level of the surge arrester to the 2% value of the phase-to-earth overvoltage [13].

In the statistical method, the coordination withstand voltages are determined through the evaluation of the risk of failure against the performance criterion, that is, the acceptable risk of failure.

In the next step, the required withstand voltages are determined by multiplying the coordination withstand voltages by the safety factor and the atmospheric correction factor. The recommended safety factors are 1.15 for internal insulation and 1.05 for external insulation when they are not specified by the relevant apparatus committees. The atmospheric correction factor is considered only for the external insulation and is given by

$$K_a = e^{m\left(\frac{H}{8,150}\right)} \tag{11.3}$$

where

K_a: atmospheric correction factor [kV]
m: exponent given in IEC 60071-2
H: altitude above sea level [m]

In range I (1 kV $< U_m \leq$ 245 kV, U_m: highest voltage for equipment), only the standard rated short-duration power-frequency withstand voltage and standard rated lightning impulse withstand voltage (LIWV) are defined [1]. As the standard rated switching impulse withstand voltage (SIWV) is not defined, the required SIWV determined above needs to be converted to the required short-duration power-frequency withstand voltage and the required LIWV.

In contrast, only the standard rated SIWV and the standard rated LIWV are defined in range II ($U_m >$ 245 kV). As the standard rated short-duration power-frequency withstand voltage is not defined, the required short-duration power-frequency withstand voltage is converted to the required SIWV. The conversion may not be necessary when the short-duration power-frequency withstand voltage is given for the equipment considered.

In the last step, the standard withstand voltages are finally selected. The minimum standard withstand voltages which exceed the required withstand voltages are normally selected.

The procedure of the insulation coordination study shown in Figure 11.2 is not a one-way study flow which does not require any review during the process. Rather, various conditions, such as assumed network conditions and applied surge arresters, are reviewed during the process, depending on the obtained results. In such a case, the transient analysis will be carried out again to find the representative overvoltages under the updated study conditions.

The procedure of the insulation coordination study is also explained in IEEE Std. 1313.2 [14]. The procedure includes fewer steps compared with that in IEC 60071-2, but the necessary transient analysis for the insulation coordination study is the same in both standards.

11.3   Selection of surge arresters

The selection of surge arresters is an integral part of the insulation coordination study. As discussed earlier, surge arresters are an indispensable element for the suppression of the fast-front overvoltage below withstand voltages. In addition, the overvoltage protection from severe temporary overvoltages also depends on the application of surge arresters.

This section discusses the selection of gapless metal-oxide surge arresters defined in IEC 60099-4 [15]. Recently, this type of surge arrester is most frequently applied to different types of substations—both air-insulated and gas-insulated, from LV to UHV. IEC 60099-5 gives recommendations for the selection and application of surge arresters of this type [16]. The second edition of IEC 60099-5 was released in 2013. This new edition also covers metal-oxide surge arresters with external series gap for overhead lines, which is recently seeing the increased number of applications worldwide.

In addition, it is expected that the third edition of IEC 60099-4 will soon be released. However, the update of these international standards will not have a significant impact on the transient analysis necessary for the selection of surge arresters.

The selection of surge arresters has both electrical and mechanical aspects. This book explains only about the electrical aspect, which is related to the transient analysis. The following list shows the electrical specification required for the selection of surge arresters:

(1)   Continuous operating voltage (COV)
(2)   Rated voltage
(3)   Nominal discharge current
(4)   Protective levels
(5)   Energy absorption capability (line discharge class)
(6)   Rated short-circuit current

11.3.1   *Continuous operating voltage*

The continuous operating voltage (COV) is, "The maximum permissible value of a sinusoidal power frequency voltage" by definition. COV is specified by the r.m.s. value, and can be continuously applied to the arrester terminals.

COV is basically decided according to the voltage control policy at the substation. In most power systems, COV should be the phase-to-earth voltage that is higher than or equal to the nominal system voltage plus 5–10%.

11.3.2   Rated voltage

The rated voltage is by definition, "The maximum power frequency voltage that is applied in the operating duty test for ten seconds." Considering the duration (10 s), the rated voltage is the phase-to-earth voltage that is considered in relation to the temporary overvoltage that can occur in the power system.

As discussed in section 11.1.1, the temporary overvoltage caused by the single-line-to-ground fault and the load rejection are normally considered for the selection of the rated voltage. An extremely severe resonance overvoltage should be disregarded, assuming that severe conditions are avoided by operational countermeasures or other equipment such as shunt reactors.

If an overvoltage longer than 10 s, such as the one caused by the Ferranti phenomenon, needs to be considered, the rated voltage is selected based on the TOV capability curve provided by the manufacturer.

11.3.3   Nominal discharge current

Nominal discharge current is the peak value of lightning (8/20) current impulse and is the main parameter for the selection of surge arresters. The choice of the nominal discharge current affects the protective levels and the energy absorption capability, which will be discussed later in this section. That is, when the nominal discharge current of a surge arrester has to be changed, it is highly likely that the entire process of the insulation coordination design/study has to be repeated.

Five standard nominal discharge currents—1.5 kA, 2.5 kA, 5 kA, 10 kA, 20 kA—are defined by IEC 60099-4 [15]. Among these values, only 10 kA or 20 kA is selected for the application in the HV system (rated voltage of the surge arrester > 132 kV). According to Table 5 in IEC 60099-4, line discharge classes 1–3 can be selected for the nominal discharge current 10 kA, while line discharge classes 4 and 5 can be selected for the nominal discharge current 20 kA. For the system with the highest voltage up to 420 kV, the nominal discharge current 10 kA with the line discharge class 3 is normally sufficient. However, severe overvoltages found in the transient analysis can require the selection of the nominal discharge current 20 kA with the line discharge class 5.

11.3.4   Protective levels

Two protective levels are normally specified, the lightning impulse protective level (LIPL) and the switching impulse protective level (SIPL). LIPL is the maximum residual voltage for the lightning current impulse. The peak value of the current impulse is the nominal discharge current. An example of the relationship between LIPL and the nominal discharge current is shown in Figure 11.3.

SIPL is the maximum residual voltage for the switching current impulse. The peak value of the current impulse is specified in Table 4 of IEC 60099-4 for different line discharge classes. Another protective level, the steep current impulse protective level is also included in the specification, but it is normally not necessary to consider this value for the selection of surge arresters.

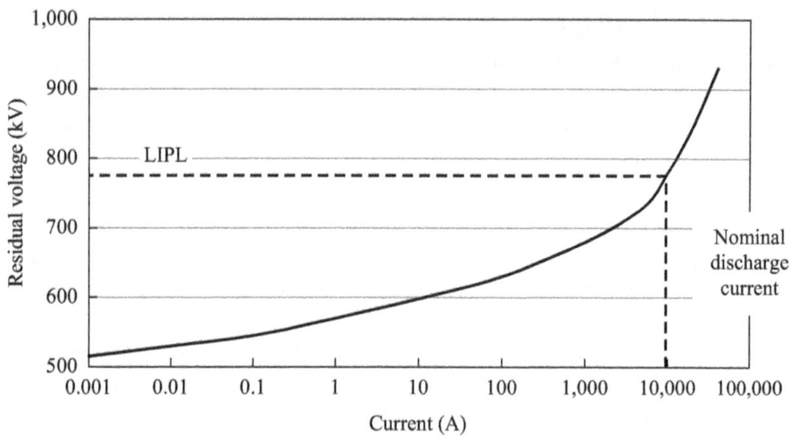

Figure 11.3   LIPL and nominal discharge current

Protective levels can be significantly different for different surge arresters even though their nominal discharge currents are equal. The selection of protective levels can have a large impact on the magnitude of the overvoltage that can propagate into a substation.

11.3.5   Energy absorption capability

The energy absorption capability is the capability of the surge arrester to discharge the energy through itself maintaining its thermal stability. It is normally expressed in the form of the specific energy, which is the energy divided by the rated voltage. The specific energy depends on the line discharge class and the switching impulse residual voltage.

The energy absorption capability required for the surge arrester can be determined through the transient analysis. The required energy absorption capability can be converted to the specific energy by dividing it by the rated voltage. If the obtained specific energy is higher than the specification, a higher line discharge class needs to be selected.

Surge arresters are installed mainly to suppress fast-front overvoltages. Although they are also intended to deal with slow-front overvoltages and temporary overvoltages in some severe cases, they are, in most applications, not necessary in order to suppress these overvoltages below withstand voltages. However, once surge arresters are applied, they must endure all kinds of overvoltages that arise in the network and must have the required energy absorption capability considering all kinds of overvoltages.

Severe temporary overvoltages can require very high energy absorption capability due to their long duration. In Japan, it is considered that the overvoltage caused by the load rejection requires the highest energy absorption capability among other overvoltages, and it decides the standard energy absorption capability of surge arresters.

11.3.6   Rated short-circuit current

The rated short-circuit current expresses the ability of the surge arrester to prevent violent shattering of its housing in case of its internal failure. It is the maximum short-circuit current with a duration of 200 ms with which the surge arrester can still satisfy this safety requirement. It has to be higher than the short-circuit current level where the surge arrester is installed.

11.3.7   Study flow

Summarizing the above explanations on the electrical specifications, the selection of surge arrester is performed in the procedure shown in Figure 11.4. The network analysis and the transient analysis required for the selection of surge arresters are shown on the left side of the figure.

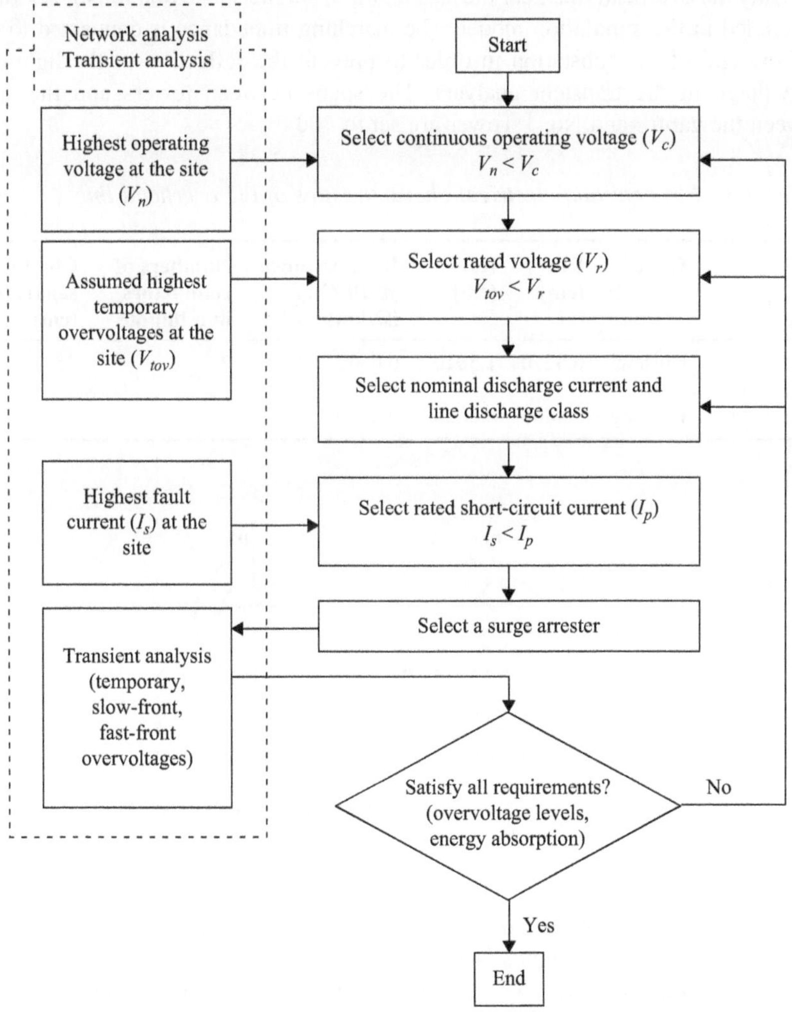

Figure 11.4   Simplified study flow diagram for the selection of surge arresters

11.4   Example of the transient analysis

This section shows an example of the fast-front overvoltage analysis for the insulation coordination. The propagation of a lightning overvoltage into a substation is studied, and the overvoltage protection with the surge arrester is discussed.

11.4.1   Model setup

11.4.1.1   Overhead line model

It is assumed that a single-circuit 400 kV overhead line is feeding a 400 kV substation. The physical and electrical characteristics of the overhead line are summarized in Table 11.3 and Figure 11.5. The overhead line is modeled by J. Marti model in order to consider its frequency-dependent characteristics.

Only the overhead line near the substation, from the substation to No. 5 Tower, is included in the simulation model. The matching impedance is connected to the opposite end of the substation in order to prevent the reflection of the lightning overvoltage in the transient analysis. The spans between towers and the span between the gantry and No. 1 Tower are set to 400 m.

Table 11.3   Physical and electrical characteristics of the overhead line

	Code	Rin (cm)	Rout (cm)	DC resistance @ 20°C (Ω/km)	Numbers of conductors in a bundle	Conductor separation (cm)
Phase conductor	Curlew	0.5270	1.5810	0.0542	2	45
Ground wire	Dorking	0.4805	0.8008	0.2770	–	–

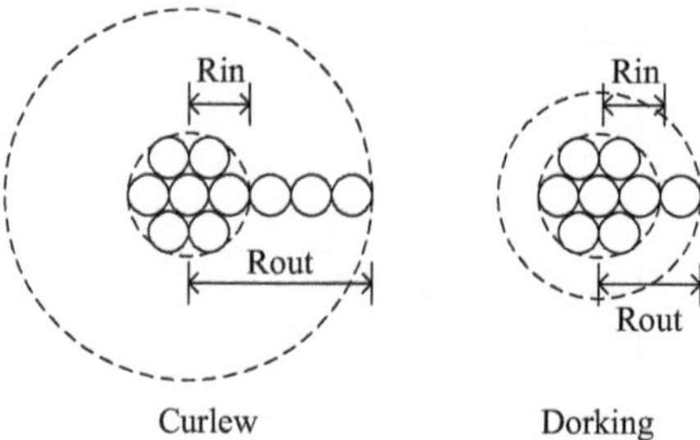

Figure 11.5   Cross section diagram of the overhead line

11.4.1.2   Tower

The tower model is created using the standard tower models and their surge impedances as shown in Figure 11.6 [17–19].

All towers are assumed to be suspension towers as shown in Figure 11.7. One tower is divided into three sections so that it becomes possible to express the tower

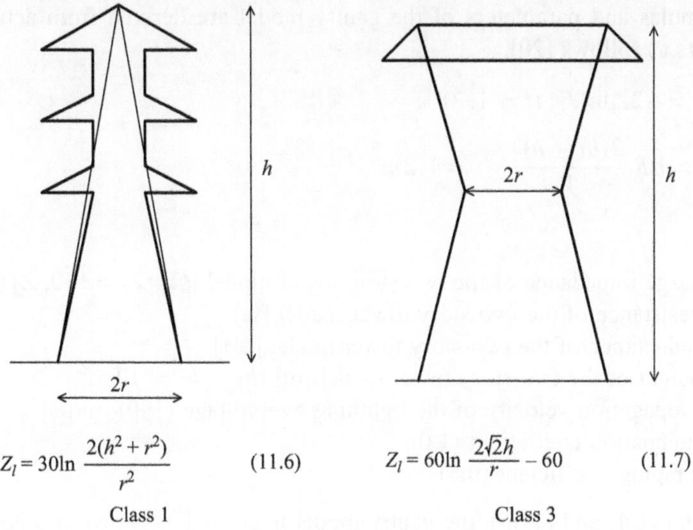

$$Z_l = 30\ln\frac{2(h^2 + r^2)}{r^2} \qquad (11.6)$$

Class 1

$$Z_l = 60\ln\frac{2\sqrt{2}h}{r} - 60 \qquad (11.7)$$

Class 3

Figure 11.6   Standard tower models

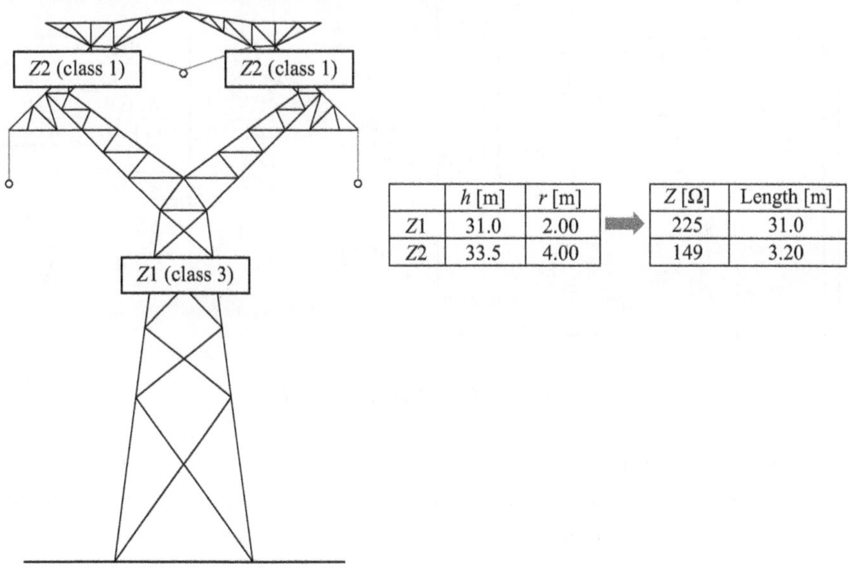

	h [m]	r [m]		Z [Ω]	Length [m]
Z1	31.0	2.00		225	31.0
Z2	33.5	4.00		149	3.20

Figure 11.7   Tower model

by the combination of standard tower models. Grounding resistance of each tower is set to 10 Ω.

11.4.1.3   Gantry

Gantry tower is modeled with the multistory tower model, which is the standard model in Japan. A two-story tower model is normally used to model a gantry.

The formulas and parameters of the gantry model are derived from actual measurements as follows [20].

$$R_i = -2Z_i \ln\sqrt{\gamma} \quad (i = 1, 2) \tag{11.4}$$

$$L_i = \alpha R_i \frac{2(h_1 + h_2)}{V} \quad (i = 1, 2) \tag{11.5}$$

where

Z_i: surge impedance of the two-story tower model [Ω] ($Z_1 = 130$, $Z_2 = 90$)
R_i: resistance of the two-story tower model [Ω]
L_i: inductance of the two-story tower model [μH]
h_i: height of the two-story tower model [m] ($h_1 = h_2 = 10.5$)
V: propagation velocity of the lightning overvoltage (300) [m/μs]
α: attenuation coefficient (1.0)
γ: damping coefficient (0.8)

From (11.4) and (11.5), the gantry model is created as shown in Figure 11.8. Grounding resistance of the gantry is set to 1 Ω.

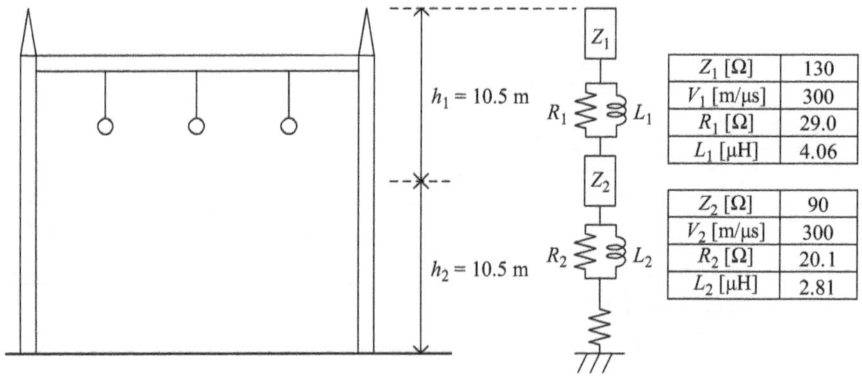

Z_1 [Ω]	130
V_1 [m/μs]	300
R_1 [Ω]	29.0
L_1 [μH]	4.06

Z_2 [Ω]	90
V_2 [m/μs]	300
R_2 [Ω]	20.1
L_2 [μH]	2.81

Figure 11.8   Gantry model

11.4.1.4   Lightning strike

The lightning strike current is modeled as the ramp wave with the following parameters as shown in Figure 11.9 [21]. The standard lightning impulse is assumed.

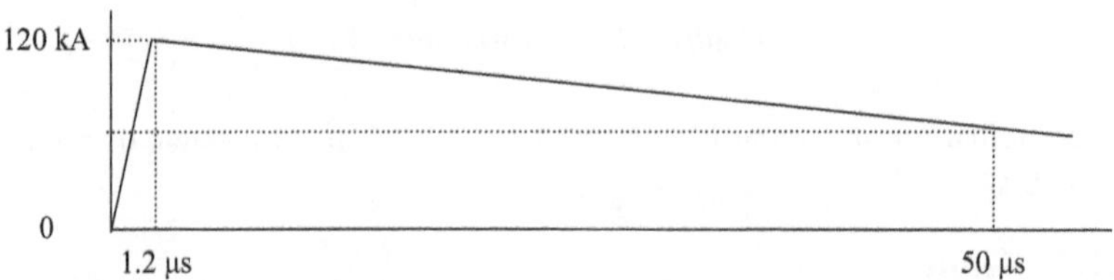

Figure 11.9   Lightning strike

The lightning path impedance is set to the value, 400 Ω, which is typically adopted in Japan [22, 23]. Only the lightning strike to the top of No. 1 Tower and the resulting back flashover are considered in this example. The direct lightning strike to a phase conductor is not studied as the back flashover, in general, causes more severe overvoltages.

Stroke current:	120 kA
Time to peak:	1.2 μs
Time to half value:	50 μs
Lightning path impedance:	400 Ω
Stroke point:	Top of No. 1 Tower

11.4.1.5 Back flashover

There are several methods to model the back flashover process [19]. This example applies the flashover model with the leader development proposed by Nagaoka in 1991 [24]. This model is based on the leader model proposed by Shindo and Suzuki but reduces required computational efforts by applying an approximate expression to the time-dependent leader development [25].

The applied model consists of two switches, the leader development switch (SW1) and the flashover switch (SW2), as shown in Figure 11.10. The former controls the inception of the leader progression, which depends on the gap voltage. The latter controls the inception of the back flashover, which depends on the gap current or the zero-crossing of the gap voltage.

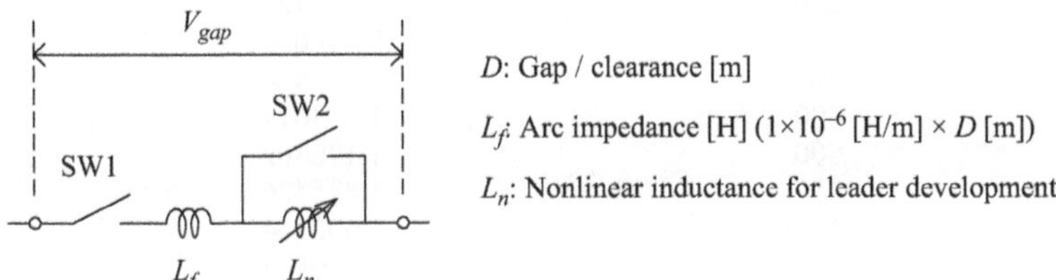

Figure 11.10   Flashover model with nonlinear inductance

Parameters of the back flashover model are obtained by the following formulas:

$$L_0 = \frac{1.23D - 0.432}{1000} \text{ [H]} \tag{11.8}$$

$$I_0 = 10^{-0.0343D - 0.00025R_b + 2.85} \text{ [A]} \tag{11.9}$$

$$n = -0.0743 - 0.000734R_b + 2.18 \tag{11.10}$$

$$L_n(i) = \frac{L_0}{(1 + i/I_0)^n} \text{ [H]} \tag{11.11}$$

$$\Phi_n(i) = \frac{L_0 I_0 \left\{ 1 - \frac{1}{(1+i/I_0)^{n-1}} \right\}}{n-1} - L_f i \ \ [\text{Wb-T}] \tag{11.12}$$

The standard value 200 Ω is applied to the equivalent impedance, R_b.
The two switches are closed when the following conditions are satisfied:

SW1 closed when $\qquad\qquad V_{gap} > 500D + 200 \ [\text{kV}]$ $\qquad\qquad$ (11.13)

SW2 closed when $\qquad\qquad I_{gap} > I_{FO}$ or $V_{gap} = 0$ $\qquad\qquad$ (11.14)

where $I_{FO} = I_0 \left\{ \left(L_0 \times 10^6 / D \right)^{1/n} - 1 \right\}$: flashover current

The nonlinear inductance for the leader development L_n is modeled with its current (i)–flux (Φ_n) characteristic. The characteristic is a function of clearance D. Assuming the clearance of 4.0 m, the characteristic is derived from (11.8)–(11.12) as shown in Table 11.4.

Table 11.4   Current-flux characteristic of nonlinear inductance L_n

i [A]	Φ_n [Wb-T]
1	0.004476
10	0.044010
50	0.204936
100	0.377681
150	0.525539
200	0.653723
250	0.766059
300	0.865428
400	1.033611
500	1.170862
750	1.425526
1,000	1.602264
2,000	1.980574
4,000	2.262186
8,000	2.444232
15,000	2.534149
25,000	2.559027

11.4.1.6   Substation

This example assumes a simple 400 kV air-insulated substation with a breaker-and-a-half bus configuration. Only a portion of the substation drawn with a bold line in the figure is used in the simulation (Figure 11.11). Although actual substations are generally more complicated, such simplifications are often made in practice so that the results can be translated into different substations.

This example first studies the fast-front overvoltage at CB1 and CB2. In this case (Case 1), both CB1 and CB2 are opened in order to produce severe overvoltages

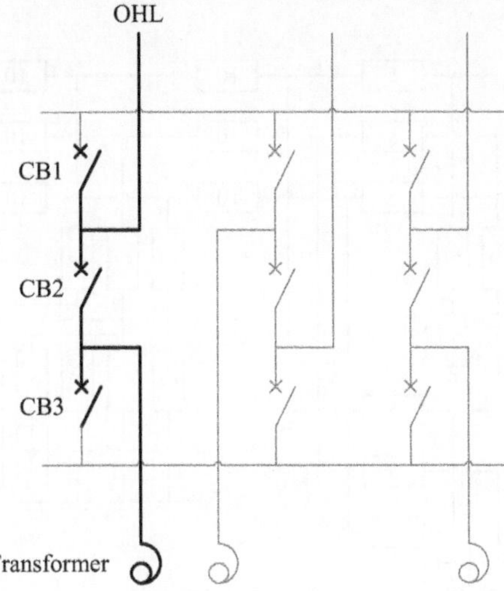

Figure 11.11   Single line diagram of the substation

at these open terminals. The simulation model of the substation can be expressed as in Figure 11.12. The number in each box shows the length of each bus section in meters.

The second case (Case 2) studies the fast-front overvoltage at the transformer. In this case, CB2 is closed so that the overvoltage can propagate from the overhead line to the transformer. CB3 is opened in order to assume a severe condition. The simulation model of the substation in this case is shown in Figure 11.13. The number in each box shows the length of each bus section in meters.

The following typical values are used to model the 400 kV air-insulated bus in the substation.

Surge impedance:	350 Ω
Propagation velocity:	300 m/μs
Attenuation:	Not considered

Other equipment in the substation, such as the 400 kV transformer, circuit breakers, and instrument transformers, are modeled only as floating capacitances. The circuit breaker model includes the capacitance between contacts when opened. Values of these capacitances are obtained from [21].

11.4.1.7 Surge arrester

This study models surge arresters at the line entrance and near the transformer. The assumed voltage (V)–current (i) characteristics of surge arresters are shown in Table 11.5. This is a surge arrester with nominal discharge current 10 kA and line discharge class 3, which is typically applied to the 400 kV system. The

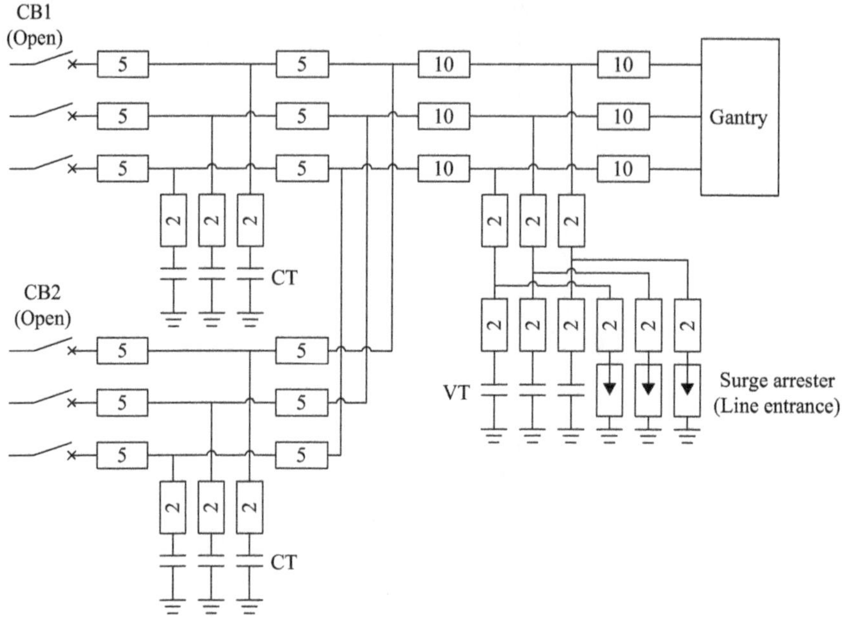

Figure 11.12   Node map of the substation—Case 1

lightning impulse protective level should be around 780 kV according to the *V–i* characteristic.

11.4.2   *Results of the analysis*

This section shows the results of the fast-front overvoltage analysis with the simulation model explained in the last section. The fast-front overvoltage at CB1 and CB2 is studied in the first case (Case 1).

Figure 11.14 shows the overvoltage at No. 1 Tower and CB1 without the surge arrester at the line entrance. The overvoltage at CB1 is higher than the overvoltage at No. 1 Tower because of the positive reflection of the overvoltage at the opened circuit breaker. The overvoltage at CB1 is greater than 1,425 kV, which is the highest lightning impulse withstand voltage (LIWV) for equipment whose highest voltage is 420 kV. This result shows the difficulty of the overvoltage protection from the lightning strike without the surge arrester (SA).

Figure 11.15 compares the overvoltage at CB1 with and without the surge arrester at the line entrance. The overvoltage is suppressed to 1,000 kV by the application of the surge arrester. As the surge arrester clips the voltage, its application introduces another point of reflection of the overvoltage. The over-voltage of higher frequency whose period is approximately 1 μs can be observed with the surge arrester in Figure 11.15.

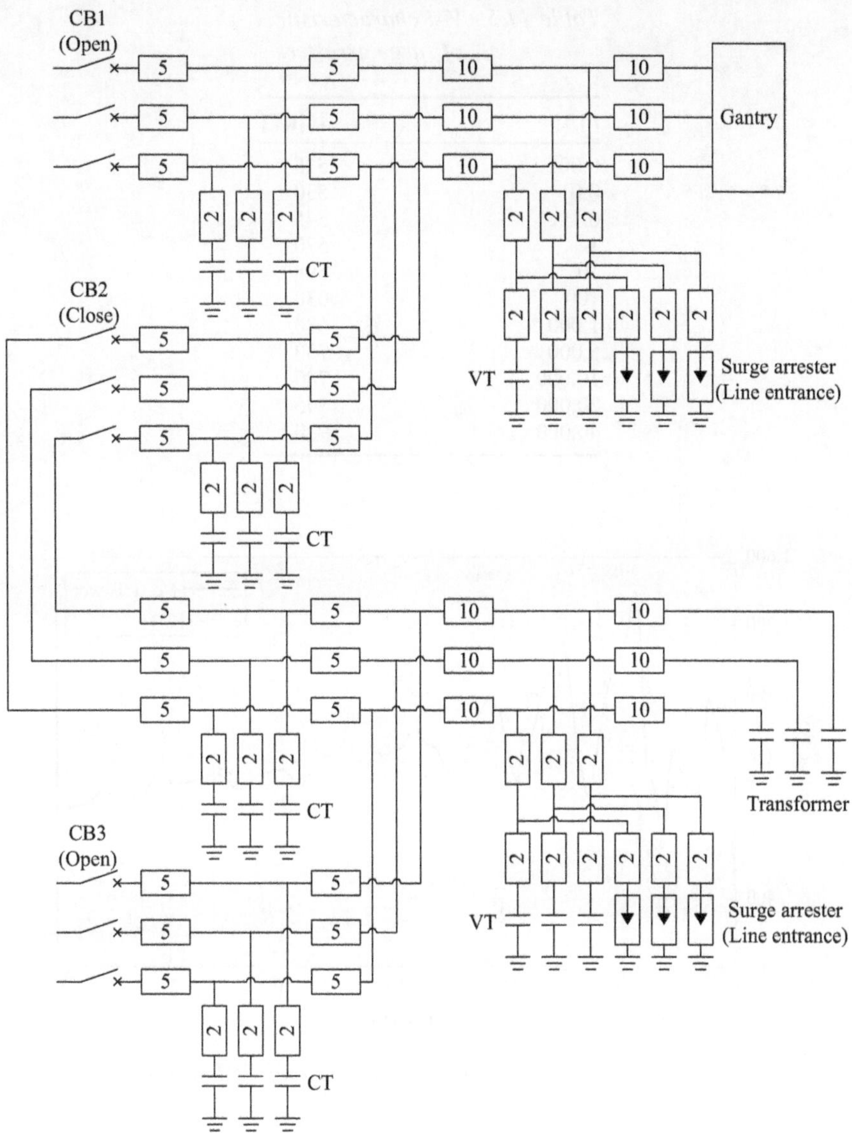

Figure 11.13   Node map of the substation—Case 2

The observed time to peak of the overvoltage is approximately 4 μs, which is much shorter than 8 μs. It is recommended to have a larger safety margin, considering these two factors, that is, the existence of the high frequency component and the short time to peak.

Figure 11.16 shows the overvoltage at the surge arrester and CB1. The surge arrester limits the overvoltage approximately to its lightning impulse protective

*Table 11.5   V–i characteristic
of surge arresters*

i [A]	V [kV]
0.001	515
0.01	530
0.1	545
1	570
10	598
100	630
1,000	680
5,000	730
10,000	780
20,000	845
40,000	930

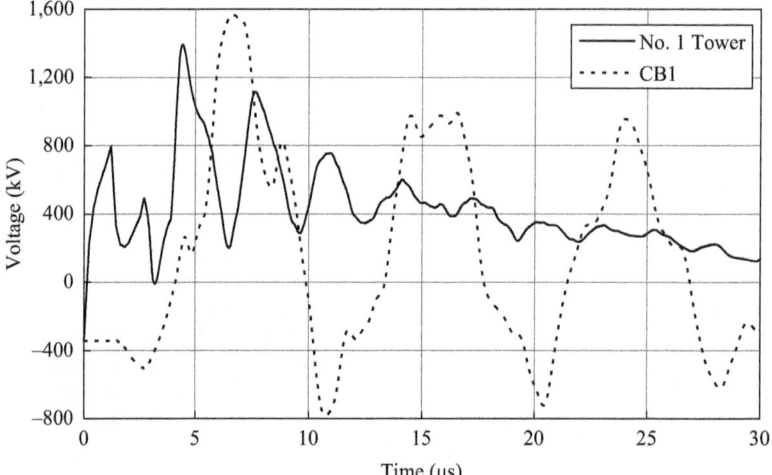

Figure 11.14   Overvoltage at No. 1 Tower and CB1 without SA

level (LIPL) at its location. The separation between the surge arrester and CB1 is 24 m, including the lead wire to the surge arrester. The overvoltage at CB1 becomes higher than the overvoltage at the surge arrester due to the positive reflection.

Figures 11.17 and 11.18, respectively, show the discharge current and the absorbed energy of the surge arrester at the line entrance. The discharge current is lower than the nominal discharge current 10 kA. As is often the case with the fast-front overvoltage analysis, the absorbed energy by the surge arrester is much lower

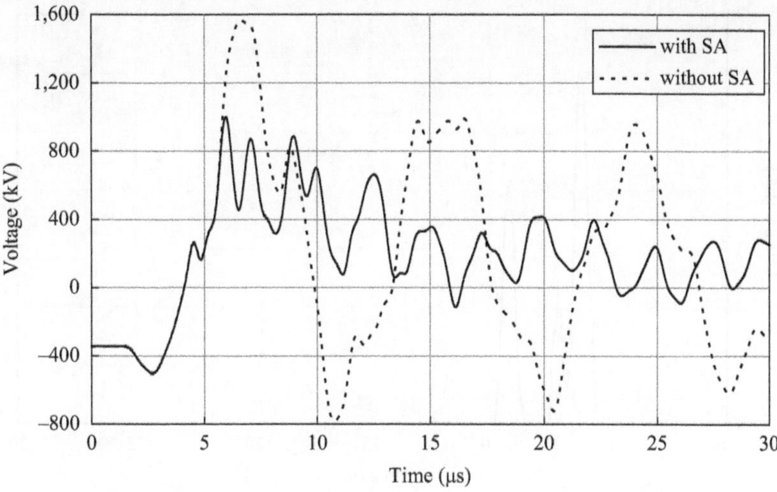

Figure 11.15   *Overvoltage at CB1 with and without SA*

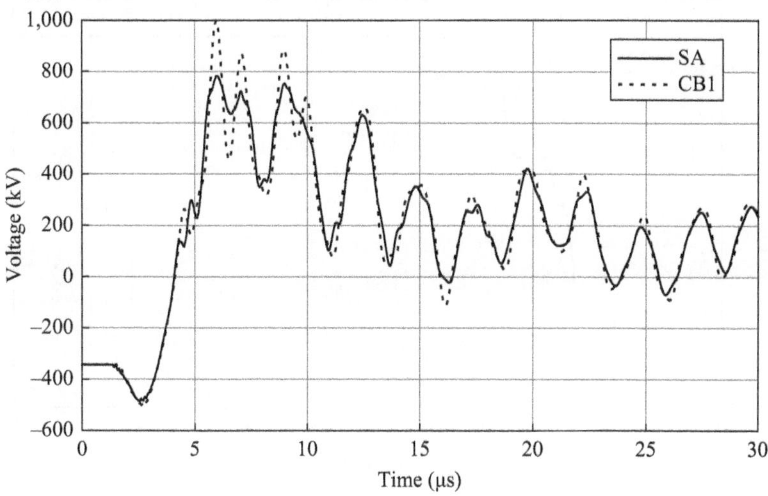

Figure 11.16   *Overvoltage at the surge arrester and CB1*

than its capability, considering the typical energy absorption capability (specific energy) around 3 kJ/kV for line discharge class 3.

The grounding resistance at No. 1 Tower has a significant impact on the overvoltage. Figure 11.19 compares the overvoltage at CB1 with the grounding resistance 10 Ω and 15 Ω. The surge arrester at the line entrance is removed in order to clearly

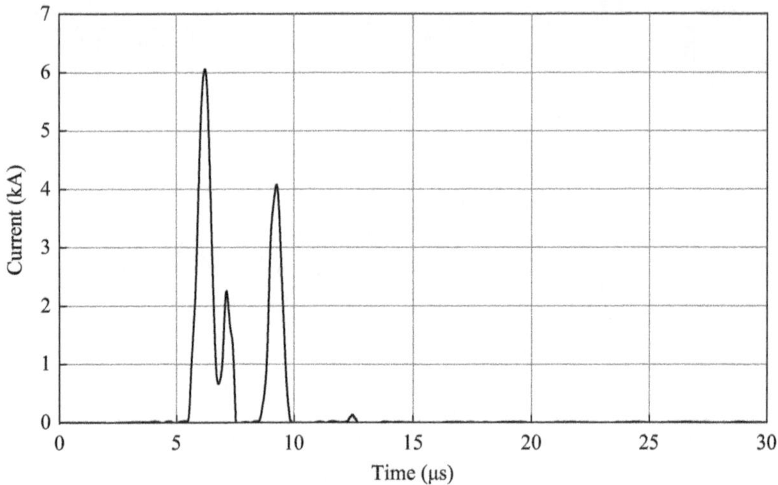

Figure 11.17   Discharge current of the surge arrester

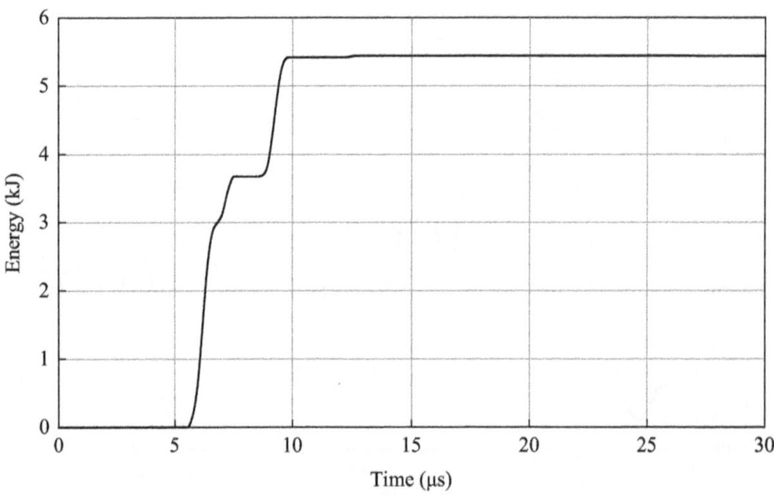

Figure 11.18   Energy absorption by the surge arrester

observe the effect of the grounding resistance. The magnitude of the overvoltage is increased by about 200 kV because of this larger grounding resistance.

The next case (Case 2) studies the fast-front overvoltage at the 400 kV transformer. Figure 11.20 shows the overvoltage with and without the surge arrester near the transformer. Since the results of Case 1 show the necessity of the surge arrester at the line entrance, it is considered to exist regardless of the existence surge arrester near the transformer.

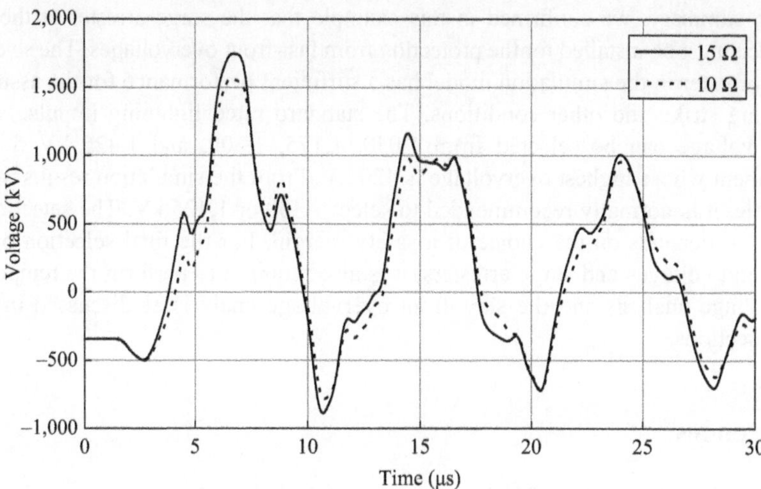

Figure 11.19   *Overvoltage at CB1 with grounding resistance 10 and 15 Ω without SA*

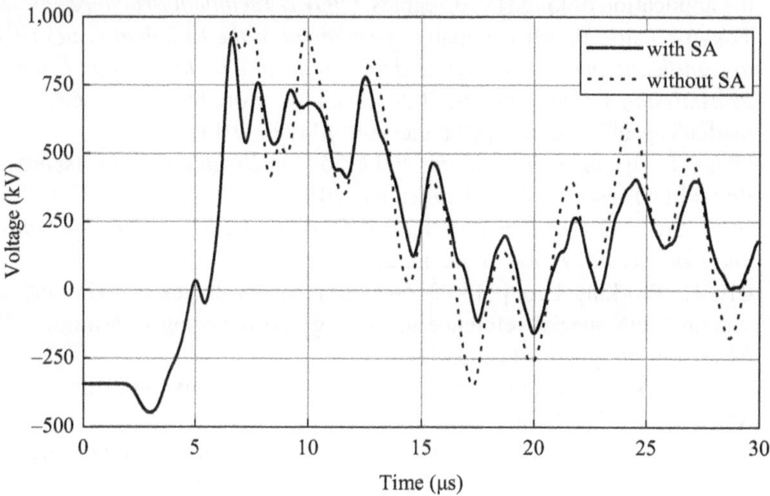

Figure 11.20   *Overvoltage at the transformer with and without SA near the transformer*

As the overvoltage is suppressed by the surge arrester at the line entrance, the surge arrester near the transformer does not have a major effect on the magnitude of the overvoltage. It is not necessary to install a surge arrester near the transformer if we only consider the lightning overvoltage from the overhead line. However, it is a common practice to install surge arresters near transformers, which are often most important equipment in a substation, in order to protect them from overvoltages that originate in a substation.

In summary, we confirmed in this example that the surge arrester at the line entrance must be installed for the protection from fast-front overvoltages. The selected surge arrester in the simulation model has a sufficient performance for the assumed lightning strike and other conditions. The standard rated lightning impulse withstand voltage can be selected from 1,050, 1,175, 1,300, and 1,425 kV for the equipment whose highest overvoltage is 420 kV. From the simulation results in this example, it is normally recommended to select 1,300 or 1,425 kV. The selection of 1,175 kV depends on the choice of a safety margin. For the final selection of the withstand voltages and surge arresters, it is also required to perform the temporary overvoltage analysis and the slow-front overvoltage analysis as discussed in previous sections.

References

[1]   IEC 60071-1 ed8.0. *Insulation co-ordination—Part 1: Definitions, principles and rules*; 2006.

[2]   CIGRE WG C4.502. Power system technical performance issues related to the application of long HVAC cables. *CIGRE Technical Brochure 556*; 2013.

[3]   Tokyo Electric Power Company. *Assessment of the technical issues relating to significant amounts of EHV underground cable in the all-island electricity transmission system* [online]; 2009. Available from http://www.eirgrid.com/media/Tepco%20Report.pdf (Accessed May 6, 2014).

[4]   Ohno, T. *Dynamic study on the 400 kV 60 km Kyndbyværket—Asnæsværket line*. PhD Thesis, Aalborg University; 2012.

[5]   IEEE Std. 142. *IEEE Recommended Practice for Grounding of Industrial and Commercial Power Systems*; 2007.

[6]   CIGRE Working Group 13.02. Switching overvoltages in EHV and UHV systems with special reference to closing and reclosing transmission lines. *Electra N°30*, pp. 70–122; 1973.

[7]   CIGRE Working Group 13.05. The calculation of switching surges. *Electra N°19*, pp. 67–78; 1971.

[8]   CIGRE Working Group 13.05. The calculation of switching surges—II. Network representation for energization and re-energization studies on lines fed by an inductive source. *Electra N°32*, pp. 17–42; 1974.

[9]   CIGRE Working Group 13.05. The calculation of switching surges—III. Transmission line representation for energization and re-energization studies with complex feeding networks. *Electra N°62*, pp. 45–78; 1979.

[10] IEC 62271-100 ed2.0. *High-voltage switchgear and controlgear—Part 100: Alternating current circuit-breakers*; 2008.

[11] Ametani, A., Ohno, T. *Cable system transients: Theory, modeling and simulation*, Wiley-IEEE Press; 2014.

[12] Greenwood, A. *Electrical transients in power systems second edition (section 11.1)*, Wiley-Interscience; 1991.

[13] IEC 60071-2 ed3.0. *Insulation co-ordination—Part 2: Application guide*; 1996.

[14] IEEE Std. 1313.2. *IEEE guide for the application of insulation coordination*; 1996.

[15] IEC 60099-4 ed2.0. *Surge arresters—Part 4: Metal-oxide surge arresters without gaps for a.c. systems*; 2004.

[16] IEC 60099-5 ed2.0. *Surge arresters—Part 5: Selection and application recommendations*; 2013.

[17] Sargent, M. A., Darveniza, M. "Tower surge impedance." *IEEE Trans. on PAS-88*, pp. 680–687, May 1969.

[18] Wagner, C. F., Hileman, A. R. "A new approach to the calculation of the lightning performance of transmission line III—a simplified method: Stroke to tower." *AIEE Trans. on PAS-79*, pp. 589–603, October 1960.

[19] EPRI. *AC Transmission line reference book—200 kV and above*, 3rd Edition. Chapter 6; 2005.

[20] Yamada, T., Mochizuki, A., Sawada, J., Zaima, E., Kawamura, T., Ametani, A., et al. "Experimental evaluation of a UHV tower model for lightning surge analysis." *IEEE Trans. on Power Delivery*, **vol. 10**, no. 10, January 1995.

[21] Fast Front Transients Task Force of the IEEE Modeling and Analysis of System Transients Working Group. "Modeling guidelines for fast front transients." *IEEE Trans. on Power Delivery*, **vol. 11**, no. 1, January 1996.

[22] Bewley, L. V. *Traveling waves on transmission systems*, Wiley; 1963.

[23] Ametani, A., Kawamura, T. "A method of a lightning surge analysis recommended in Japan using EMTP." *IEEE Trans. on Power Delivery*, **vol. 20**, no. 2, pp. 867–875, April 2005.

[24] Nagaoka N. "A flashover model using a nonlinear inductance." *IEEJ*, **vol. 111-B**, no. 5, pp. 529–534, 1991.

[25] Shindo, T., Tsuzuki, T. "A new calculation method breakdown voltage-time characteristics of long gaps." *IEEE Trans. on PAS*, **vol. 104**, no. 6, pp. 1556–1563, June 1985.

Chapter 12
FACTS: voltage-sourced converter

K. Temma[*]

FACTS (Flexible AC Transmission Systems) was advocated by Dr. Narain G. Hingorani in 1998 [1]. This system is a power electronics device applicable to a power grid. Since FACTS can control power and reactive power positively compared to general transmission devices, it gives flexibility to the power grid in resolving the problems of the power grid. It is one of the effective countermeasures for the stability problems and voltage problems of the power grid. FACTS technology has also received a lot of attention for the countermeasures for enhancement of renewable energy in the power grid [2–4]. When the renewable energy is connected to the power grid, local problems such as a voltage fluctuation by wind power may become obvious. FACTS is the effective device for the countermeasure for renewable energy also.

12.1 Category

STATCOM (static synchronous compensator), VSC-BTB (voltage-sourced converter back-to-back), SSSC (static synchronous series compensator), TCSC (thyristor-controlled series capacitor), and so on are categorized as representative FACTS based on VSC technology in Table 12.1 [5]. Many FACTS consist of a VSC.

STATCOM (also called SVG: static var generator) consists of a VSC with a converter transformer and is connected to a power grid in parallel. SSSC consists of a VSC with a converter transformer as in STATCOM and is connected to the power grid in series. In other words, STATCOM is the parallel-connected VSC device for the power grid and SSSC is the series-connected VSC device.

The VSC is applied to both STATCOM and SSSC. They have a DC capacitor on the DC side and can generate AC voltage by cutting out the constant DC voltage. They can supply both leading reactive power and lagging reactive power to the power grid by controlling the AC voltage output of the VSC.

[*]Mitsubishi Electric Co., Japan

Table 12.1   Category of FACTS and brief circuit configuration

FACTS	Circuit configuration
STATCOM	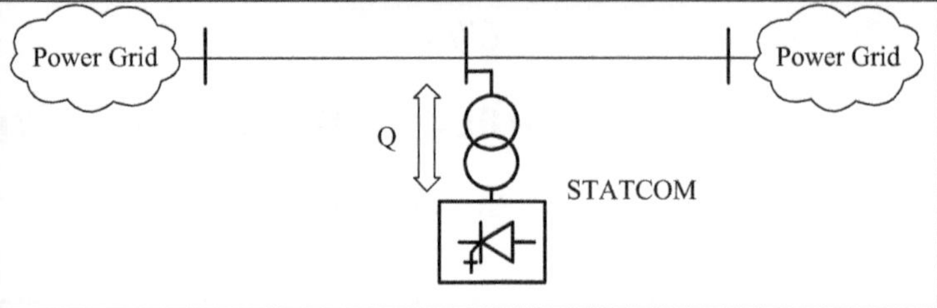 Power Grid — Power Grid; Q; STATCOM
SSSC	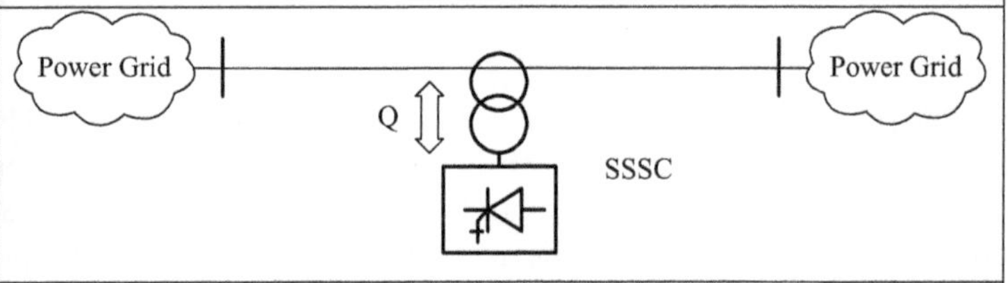 Power Grid — Power Grid; Q; SSSC
VSC-BTB	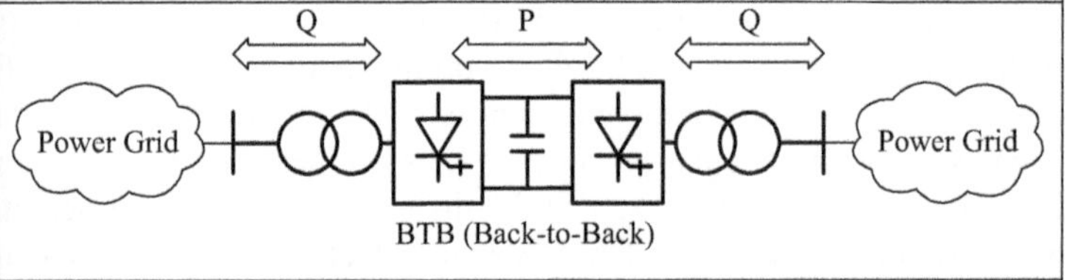 Q — P — Q; Power Grid — Power Grid; BTB (Back-to-Back)
UPFC	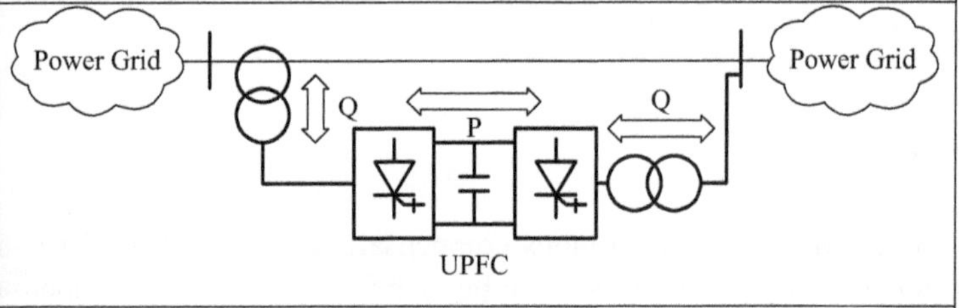 Power Grid — Power Grid; Q; P; Q; UPFC
TCSC	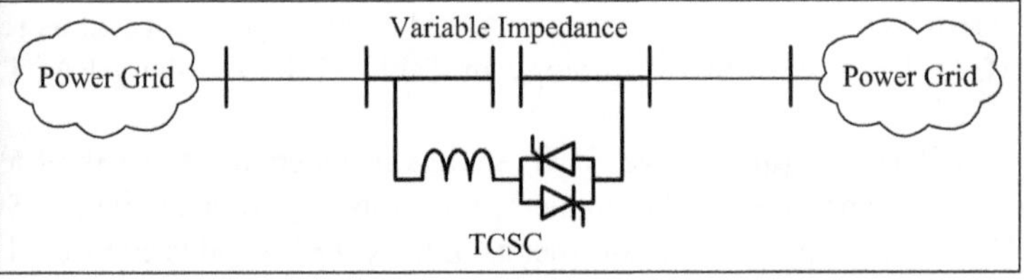 Power Grid — Power Grid; Variable Impedance; TCSC

VSC-BTB consists of two STATCOMs connected to each DC side (back-to-back) and can supply not only the reactive power but the power through the DC circuit. It can also control the power and the reactive power independently.

Unified power flow controller (UPFC) consists of STATCOM and SSSC connected to each DC side and can supply the power and the reactive power. IPFC (integrated power flow controller) is an applicable type of UPFC. TCSC is the device applied to a thyristor in these devices. It consists of a variable reactor with a thyristor switch and a capacitor connected in parallel.

12.2   Control system and simulation modeling

Main circuit configurations of many FACTS are based on VSC technology. This section takes STATCOM as an example and describes a control of the STATCOM. A typical control of FACTS has two levels: an upper-level control and a lower-level control (Figure 12.1). The upper-level control is the control to the power grid. The lower-level control is the control to operate the VSC stably. Figure 12.2 and Table 12.2 show the brief control block diagram of STATCOM.

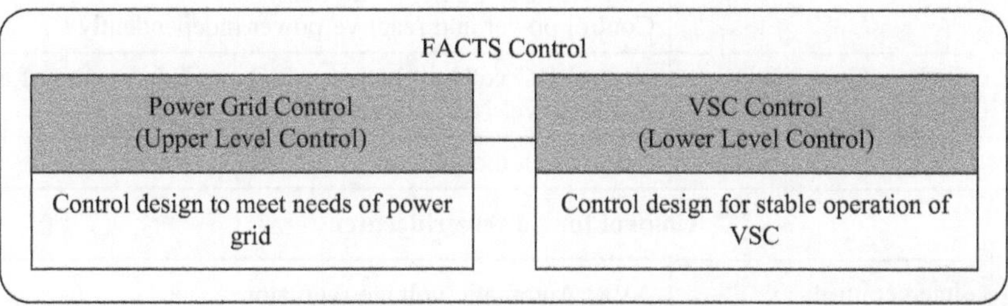

Figure 12.1   Level of FACTS control

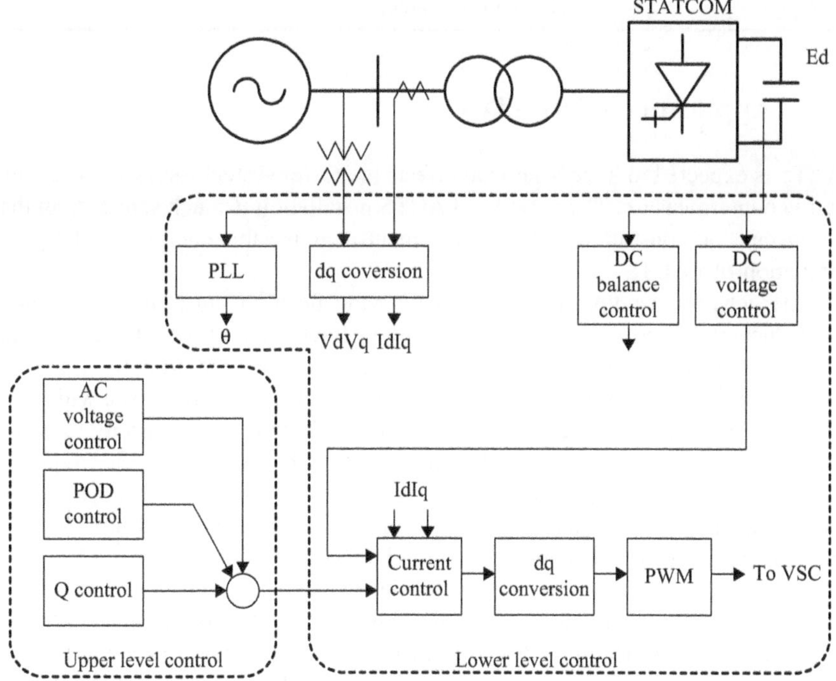

Figure 12.2   Control diagram of STATCOM

Table 12.2   STATCOM control (VSC control and power grid control)

VSC control	
Current control	Control the converter current
DC voltage control	Control the DC voltage of the converter
PLL	Phase locked loop Detect the phase of AC system
dq conversion	Detect AC voltage and AC current of VSC Convert three phases to dq axis Control power and reactive power independently
DC voltage balance control	Control DC voltages between P-C and C-N in case of a three-level NPC converter
PWM	Pulse width modulation
Control for power gridcenter	
AC voltage control	AVR: Automatic voltage regulator Control a bus voltage of a power grid
POD control	Power oscillation damping control Detect and stabilize power system oscillation
Q control	Control the reactive power output of VSC with reactive power reference

12.3   Application of STATCOM

FACTS is expected to have beneficial effects on power system issues and is a quite useful countermeasure. The period of FACTS installation is much shorter than that of transmission construction. An enhancement of renewable energy facilitates the installation of FACTS.

The generation output of a large-scale wind farm has an influence on wind speed fluctuation and it raises the power output fluctuation of the wind power and the voltage fluctuation of the grid connection point. When the proportion of wind power capacity to power system capacity is increased, a frequency issue will occur in the power grid. However, since the proportion of wind power capacity to power system capacity is not so large in the early stage of the installation, the frequency issue concerned with the power does not tend to occur. On the other hand, a voltage fluctuation may occur by the power flow fluctuation from the wind farm in local area around the substation connected to the large-scale wind farm.

It is difficult for an existing voltage regulator such as a load tap changer, a shunt reactor, or a shunt capacitor to suppress this voltage fluctuation frequently. One of the fundamental countermeasures for it is to increase a short circuit capacity at the grid connection point, but it doesn't seem realistic. FACTS such as STATCOM can operate with fast response and suppress the voltage fluctuation continuously.

Table 12.3   STATCOM application for power system issues

Power system issue	STATCOM control	Objective
Voltage fluctuation	AVR	Countermeasure for voltage fluctuation
Small-signal stability	Q bias, AVR, POD	Improvement in small-signal stability Improvement in power oscillation Improvement in transmission capacity
Transient stability	AVR, POD, Transient stability control	Improvement in transient stability Prevention of loss of synchronism of generator
Voltage stability	AVR, Voltage stability control	Improvement in voltage instability Improvement in voltage collapse
Overvoltage suppression	AVR, Overvoltage suppression	Suppression of grid overvoltage (Ferranti effect)

Therefore, the installation of STATCOM has increased in Europe. Thus, STATCOM has become an effective solution for the voltage fluctuation caused by large-scale wind farms. STATCOM can also resolve a lot of other power system issues, as identified in Table 12.3 [6]. STATCOMs are in practical use for various power system stability problems; they are applied as countermeasures [7–16]. The control methods of STATCOM corresponding to each power system issue are described as follows.

12.3.1   Voltage fluctuation

There are two types of control methods to control the voltage of a grid, as identified in Table 12.4. One method is "Vref-type AVR" which is given a reference of the grid voltage and controls a desirable voltage. The other is "ΔV-type AVR" which detects the variation in a voltage fluctuation and compensates it. Many of STATCOMs for a grid have an AVR (Automatic Voltage Regulator) control.

1. Vref-type AVR
    (a) Keep the voltage of a grid constant.
    (b) Output a reactive power to control a desirable voltage for a voltage reference.
    (c) Slope reactance of STATCOM changes the sensitivity of a reactive power output.

2. ΔV-type AVR
    (a) Compensate the variation in a voltage fluctuation.
    (b) Output zero Mvar of a reactive power in steady-state operation.
    (c) Output a reactive power for the variation in a voltage fluctuation.
    (d) Decide a period to suppress a fluctuation by the time constant of a reset filter.
    (e) Slope reactance of STATCOM changes the sensitivity of a reactive power output.

Table 12.4   Voltage control method

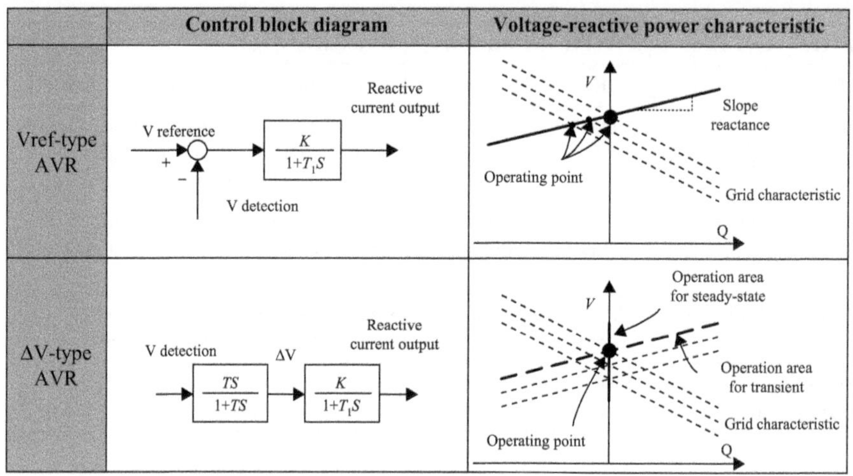

12.3.2   Small-signal stability

There is a small-signal stability constraint under heavy power flow on a long-distance transmission line, and the desirable power cannot be transmitted [7–9]. STATCOM can improve small-signal stability in this condition. Controls of AVR, POD (power oscillation damping), and Q bias, as shown in Table 12.5, are applied to the control system of STATCOM in order to improve the small-signal stability. The combination of AVR, POD, and Q bias can enhance the effect on the small-signal stability.

1.   Q bias control
     (a)   Output a reactive power according to a reference of a reactive power or current.
     (b)   Improve small-signal stability and transmission capacity by providing a reactive power constantly under heavy power flow.

2.   AVR
     (a)   Maintain synchronism of a grid with a reactive power output.
     (b)   Detect a voltage drop with heavy load and compensate a voltage deviation.
     (c)   Vref-type AVR and ΔV-type AVR can be applicable.

3.   POD
     (a)   Detect a power system oscillation inherent in a characteristic of a grid.
     (b)   Improve damping power of a grid with a reactive power.

4.   Application to small-signal stability

     An example of STATCOM application to improve small-signal stability is 450 MVA STATCOM installed at Toshin substation of Chubu Electric in Japan. Here a newly installed, large-capacity thermal power plant is connected to the main power grid through 500 and 275 kV transmission lines. When one circuit of the transmission lines that consists of long-distance double circuits is opened, small-signal stability

Table 12.5   Control of small-signal stability improvement

	Control block diagram	**Controlled object [17]**
Q bias	Q reference — Reactive current reference of STATCOM; Q bias control	• increase in rotor angle through a nonoscillatory mode due to lack of synchronizing torque
AVR	V reference (+/−) — AVR — Reactive current reference of STATCOM; V detection	• increase in rotor angle through a nonoscillatory mode due to lack of synchronizing torque • rotor oscillations of increasing amplitude due to lack of sufficient damping torque
POD	Power flow — POD — Reactive current reference of STATCOM	• rotor oscillations of increasing amplitude due to lack of sufficient damping torque

problems may occur because of loss of synchronism. The 450 MVA STATCOM improves small-signal stability of this grid. Figure 12.3(a) shows the characteristic of voltage-reactive power of the STATCOM. As one circuit of a transmission line is opened, small-signal instability occurs without STATCOM but STATCOM can improve small-signal stability, as identified in Figure 12.3(b).

(a)   Voltage-reactive power of characteristics of STATCOM.
(b)   Small-signal stability improvement in opening one circuit of transmission line.

12.3.3   Voltage stability

A heavy load on transmission lines may cause voltage instability and voltage collapse [10–12]. A long-term planning stop or a retirement of generators will decrease an upthrust of power flow from a load side and become a heavy load, and voltage stability problems may occur. A P–V curve is used as an index of voltage stability. STATCOM can enhance the P–V curve in P-axis direction as identified in Figure 12.4. When one circuit is opened in double circuits of a transmission line, the P–V curve will change rapidly and voltage instability will occur. In such a case, an effect to avoid the voltage instability or voltage collapse can be expected by fast response of STATCOM. When STATCOM has a coordination control with a load tap changer or shunt devices (shunt reactors and shunt capacitors), STATCOM can be more effective for voltage stability.

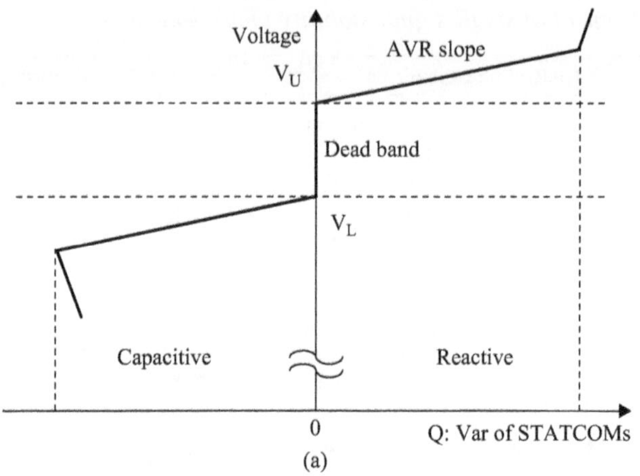

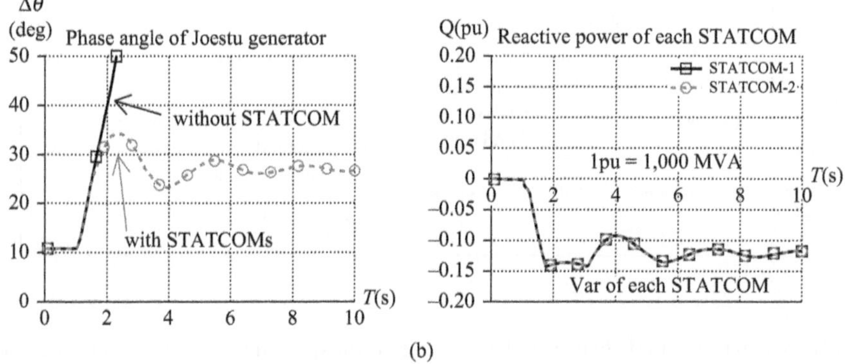

Figure 12.3   450 MVA STATCOM – Effect on small-signal stability and overvoltage:
(a) Voltage-reactive power of characteristics of STATCOM;
(b) Small-signal stability improvement in opening one circuit of
transmission line

© IEEE 2010. Reprinted with permission from Fujii, T., Temma, K., Morishima, N.,
Akedani, T., Shimonosono, T., Harada, H., '450MVA GCT-STATCOM for stability
improvement and over-voltage suppression', *IEEE Proceedings*, June 2010

1. AVR
    (a) Compensate voltage drop.
    (b) Keep a grid voltage constant, and enhance *P–V* curve and improve voltage stability.
    (c) Compensate voltage drop rapidly by one circuit outage on a heavy load and improve voltage stability.

2. Coordination control with load tap changer of transformer
    (a) Need to allow a leading reactive power output margin for prevention of voltage collapse by one circuit outage on heavy load.
    (b) When AVR keeps a grid voltage constant, STATCOM outputs the reactive power and it may not be the enough margin of the reactive power to

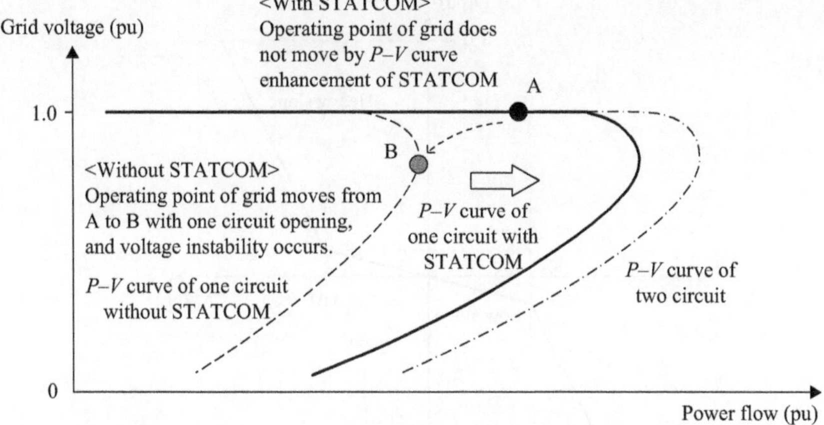

Figure 12.4   Effect on voltage stability improvement by STATCOM

avoid voltage collapse; thus, as STATCOM has a fast response AVR, it operates in advance of a load tap changer of a transformer, which is an existing voltage regulator, and therefore it is possible to allow a reactive power output margin.

(c) It is essential that a load tap changer operates for normal voltage fluctuation and that STATCOM operates for a voltage stability problem; the coordination control for both operations should be installed.

(d) AVR control for voltage stability consists of AVR of STATCOM and the coordination control of the load tap changer.

3. Coordination control with shunt devices (shunt reactors and/or shunt capacitors)

(a) As STATCOM has a fast response AVR, it operates in advance of shunt devices (shunt reactors and/or shunt capacitors), which are existing voltage regulators, and therefore it is possible to allow a reactive power output margin of STATCOM.

(b) When a coordination control of shunt devices is applied to STATCOM system, the control can shift shunt devices from STATCOM output and allow the margin of STATCOM output.

4. Application to voltage stability

An example of STATCOM application to improve voltage stability is 80 MVA STATCOM at Kanzaki substation of the Kansai electric power in Japan. The power supply and reactive power supply to an urban area will decrease by retirement of a thermal power plant in the area, and a transmission line from the main power grid to the urban area will become heavily loaded. As a result, voltage stability problem may occur by one circuit outage of the transmission line. This STATCOM is applied as countermeasure for the voltage stability problem. It has a coordination control of the load tap changer of an existing transformer. The load tap changer operates in case of a normal voltage fluctuation. The STATCOM operates in case

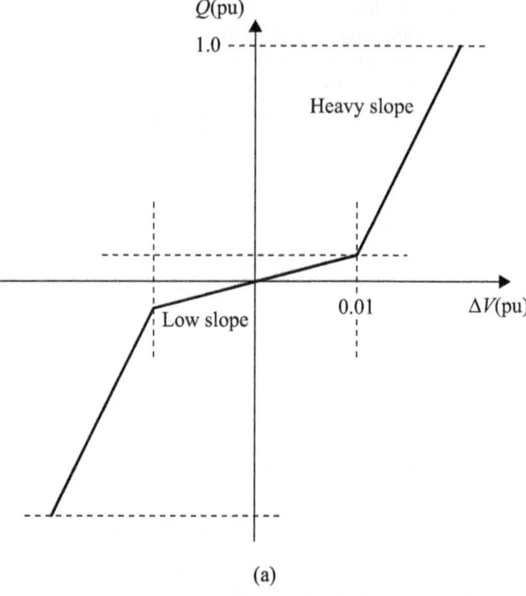

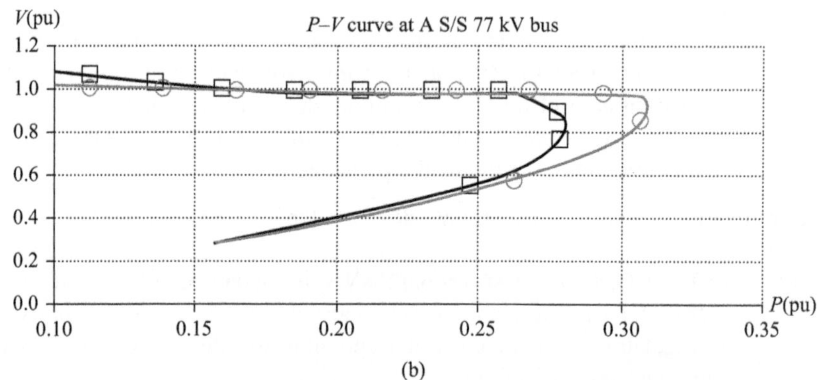

Figure 12.5   *STATCOM for voltage stability: (a) V–Q (voltage-reactive power)
characteristic of STATCOM; (b) Effect on P–V curve by STATCOM*
© IEEJ 2004. Reprinted with permission from 'Improvement of voltage stability
with STATCOM', *Proceedings of Technical Meeting on Power Engineering of
IEEJ, PE-04-1*, pp. 1–6, 2004

of one circuit outage and it supplies reactive power in order to avoid voltage
instability and voltage collapse. Figure 12.5 shows voltage-reactive power $(V\text{–}Q)$
characteristic of the STATCOM and the effect on power grid by STATCOM
installation. STATCOM enhances the $P\text{–}V$ curve in P-axis direction.

(a) $V\text{–}Q$ (voltage-reactive power) characteristic of STATCOM.
(b)   Effect on $P\text{–}V$ curve by STATCOM.

12.3.4   Transient stability

STATCOM is a new countermeasure for transient stability problem [13, 14]. The fast response of STATCOM can avoid first swing instability or nth swing instability of generators. A control of transient stability improvement of STATCOM is called power system stabilizer (PSS) or power oscillation damping (POD). The control perceives the state of power oscillation of the grid by detecting power flow of transmission lines. STATCOM supplies the reactive power to the grid to control the acceleration and the deceleration of generators according to the control, and can improve transient stability.

1. AVR
    (a) Compensate voltage drop due to large disturbance and maintain synchronism.
    (b) Avoid first swing instability of generators.
    (c) Compensate voltage fluctuation of power oscillation after large disturbance.

2. PSS/POD
    (a) Detect power flow fluctuation.
    (b) Control damping corresponding to power oscillation.
    (c) Improve damping power of generators and avoid loss of synchronism.
    (d) AVR output and PSS/POD output are references of reactive current of STATCOM.

3. Coordination control
    (a) Coordinate within two controls, AVR and PSS/POD, as identified in Figure 12.6.
    (b) Actualize optimal transient stability improvement.
    (c) Depend on power system configuration and state.
    (d) Need to be customized according to power system characteristic.

4. Application to transient stability

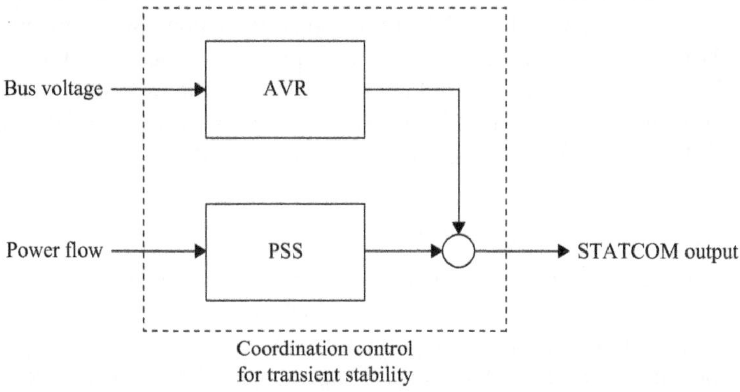

Figure 12.6   Example of control of transient stability improvement

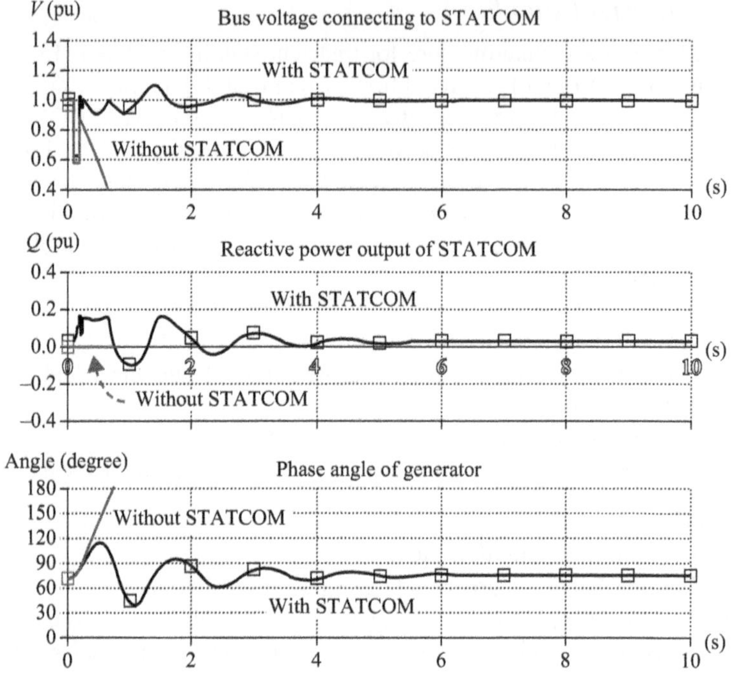

Figure 12.7   Effect on transient stability improvement by STATCOM
© IEEJ 2011. Reprinted with permission from 'Development of STATCOM for transient stability improvement', *Proceedings of Technical Meeting on Power Engineering of IEEJ, PE-11-187*, pp. 27–30, 2011

An example of STATCOM application to improve transient stability is 130-MVA STATCOM installed at Inuyama substation of the Kansai electric power in Japan. The objectives of the STATCOM are prevention of first swing instability and loss of synchronism. When AVR, PSS, and coordination control are applied to the STATCOM, it can improve transient stability. The capacity for transient stability is 130 MVA. As identified in Figure 12.7, first swing instability occurs by a large disturbance without STATCOM. However, first swing instability can be avoided and transient stability improved with STATCOM.

12.3.5   Overvoltage suppression

Countermeasure for overvoltage suppression is basically a method similar to countermeasure for voltage fluctuation, and STATCOM absorbs reactive power rapidly by AVR and suppresses overvoltage [7–9]. Either Vref-type AVR or ΔV-type AVR can be applied to STATCOM for overvoltage suppression. Furthermore, current forcing control can be applied to very fast overvoltage suppression. Figure 12.8 is an example of AVR in combination with current forcing control.

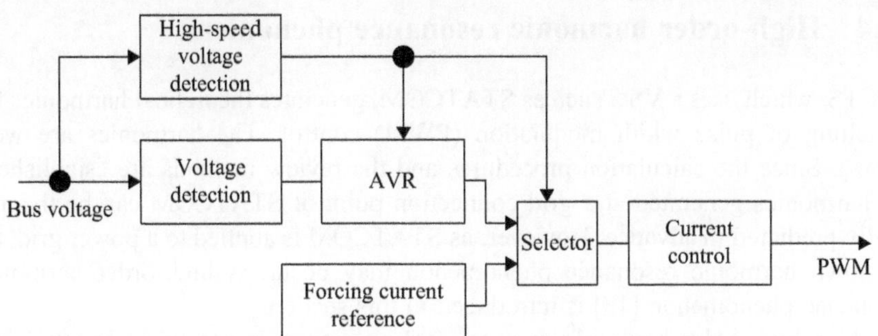

Figure 12.8   Control block diagram of overvoltage suppression

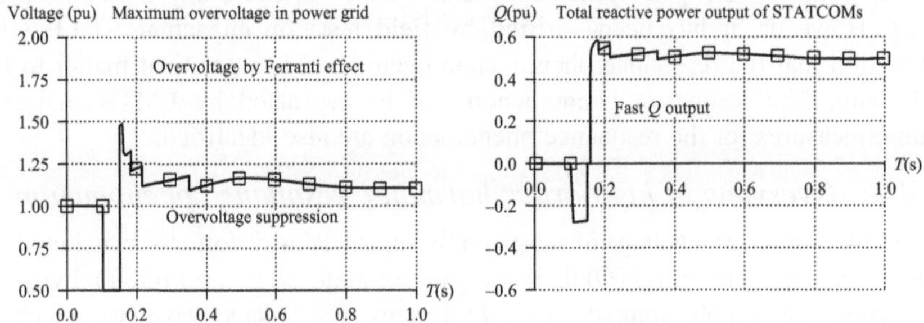

Figure 12.9   Effect on overvoltage suppression by 450 MVA STATCOM

© IEEE 2010. Reprinted with permission from Fujii, T., Temma, K., Morishima, N., Akedani, T., Shimonosono, T., Harada, H., '450MVA GCT-STATCOM for stability improvement and over-voltage suppression', *IEEE Proceedings*, June 2010

1. AVR
    (a) Detect overvoltage and suppress overvoltage.
    (b) Decide the output corresponding to the volume of voltage drop or voltage fluctuation.

2. Current forcing control
    (a) Fast response and fast overvoltage suppression compared to AVR.
    (b) Detect overvoltage and output required reactive current.

3. Application to overvoltage suppression

An example of STATCOM application to suppress overvoltage is 450 MVA STATCOM at Toshin substation as previously mentioned. One of the objectives of the STATCOM is overvoltage suppression. Many shunt capacitors are installed in the grid where STATCOM is installed in order to compensate the reactive power losses of long-distance transmission lines on heavy load. It is possible that Ferranti phenomenon occurs by large disturbance because of isolated power grid that consists of generators, long-distance transmission lines, and many shunt capacitors. The overvoltage suppression control is applied to the STATCOM and can suppress overvoltage, as identified in Figure 12.9.

12.4   High-order harmonic resonance phenomena

FACTS, which uses a VSC such as STATCOM, generates theoretical harmonics by switching of pulse width modulation (PWM) control. The harmonics are well known. Since the calculation procedures and the review methods are established, the harmonics generated at a grid connection point of STATCOM can be theoretically predicted in advance. However, as STATCOM is applied to a power grid, an unknown harmonic resonance phenomenon may occur. A high-order harmonic resonance phenomenon [18] is introduced in this section.

This high-order harmonic resonance phenomenon is identified by studying STATCOM installation. Furthermore, the mechanism of the harmonic resonance phenomenon is clarified. It has been generally estimated that this high-order harmonics would be damped immediately and would not cause any harmonic problems. However, it has been verified by field tests on an actual STATCOM installation that the resonance phenomenon occurs and is a practical matter to be dealt with. The resonance phenomenon can be estimated by EMTP analysis. Countermeasures for the resonance phenomenon are also identified.

12.4.1   Overview of high-order harmonic resonance phenomenon

STATCOM produces a voltage of substantially sinusoidal waveform by switching DC voltage with gate turn off (GTO) thyristor, gate commutated turn-off (GCT) thyristor, or insulated gate bipolar transistor (IGBT) devices. The voltage waveform which is substantially sinusoidal waveform without lower harmonics can be generated by waveform synthesis with a PWM control, either with or without a multiple converter transformer. However, the higher frequency harmonics are still generated in both cases.

12.4.1.1   Harmonics of a voltage-sourced converter

Harmonics of three-phase VSCs using PWM control are given by (12.1) and (12.2).

$$V_{f0} = \frac{\sqrt{3}}{2}\alpha E_d$$

$$V_{fh} = \frac{\sqrt{3}}{2}\frac{4E_d}{n\pi}J_k\left(\frac{\alpha n\pi}{2}\right)$$

(12.1)

where:

V_{f0}: amplitude of the fundamental-frequency ω_0 component
V_{fh}: amplitude of the fundamental-frequency $n\omega_s \pm k\omega_0$ component:

$$k = 3(2m-1) \pm 1, \quad m = 1,2,\cdots \quad (n = 1,3,5,\cdots)$$

$$k = \begin{cases} 6m+1, & m = 0,1,\cdots \\ 6m-1, & m = 1,2,\cdots \end{cases} \quad (n = 2,4,6,\cdots)$$

(12.2)

J_k: k-th order Bessel function
α: amplitude modulation ratio of VSC
E_d: DC voltage of VSC
ω_0: angular frequency of control signal
ω_s: angular frequency of carrier signal

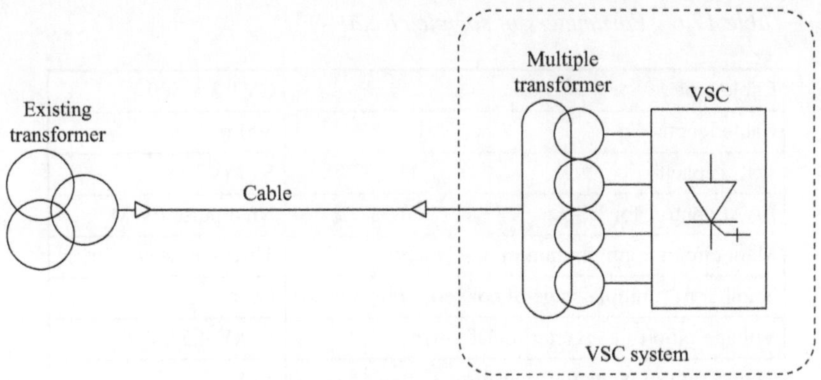

Figure 12.10   Voltage-sourced converter system and power transformer of grid

© IEEE 2005. Reprinted with permission from Temma, K., Ishiguro, F.,
Toki, N., Iyoda, I., Paserba, J.J., 'Clarification and measurements of high frequency
harmonic resonance by a voltage sourced converter', *IEEE Transactions on Power
Delivery*, vol. 20, no. 1, January 2005

Various harmonic orders are included in the total harmonic voltage of STATCOM.
A constant dead time "T_d" between an on-pulse and an off-pulse has an influence
on the various harmonic orders, but T_d is necessary to prevent failure of switching
devices. Therefore, a VSC will generate a lot of small harmonics in the range of a
lower to higher order.

When the VSC is connected to a grid that has resonant and anti-resonant
characteristics, the harmonics from the VSC may be magnified and have a negative
impact on the grid.

Figure 12.10 illustrates a sample circuit for high-order harmonic resonance
phenomenon. A VSC system such as STATCOM that consists of a VSC and the
grid connection point of the VSC system is the tertiary side of an existing power
transformer of grid through a cable. The length of the cable is about 500 m.
Table 12.6 lists the parameters of the sample circuit. Harmonic voltages of the VSC
system are reduced by shifting each phase of the switching pulses of each four-
stage converter through the multiple converter transformer.

12.4.1.2 EMTP analysis of high-order harmonic resonance

Electromagnetic transient program (EMTP) analysis is performed. The VSC is
connected to the grid as illustrated in Figure 12.10 in the analysis. Here, the lagging
reactive power output of the VSC is set to 37.5 Mvar. Very high order harmonics
are included in the voltage (V_h) on the primary side of the multiple converter
transformer, as identified by the EMTP results shown in Figure 12.11. A fast
Fourier transform (FFT) of EMTP is performed in order to analyze the voltage V_h
in detail with the high-order harmonics on the primary side of the converter
transformer. The FFT results of high-order harmonics are listed in Table 12.7. The
harmonics around the 130th order is the most significant one.

When the VSC system is disconnected from the existing transformer of
Figure 12.10 and the VSC, only the converter transformer and the cable remain;

Table 12.6   Parameters of sample circuit

Cable type	CVT 3 × 500
Cable length	500 m
VSC capacity	53 MVA
PWM control for VSC	Nine-pulse PWM
Main circuit configuration of VSC	Three-phase converter
Number of multiple stage of converter transformer	Four stage
Voltage ratio of converter transformer	66 kV/4.6 kV
Connection of converter transformer	Y-Δ
%L – Inductance of converter transformer	20%

© IEEE 2005. Reprinted with permission from Temma, K., Ishiguro, F., Toki, N., Iyoda, I., Paserba, J.J., 'Clarification and measurements of high frequency harmonic resonance by a voltage sourced converter', *IEEE Transactions on Power Delivery*, vol. 20, no. 1, January 2005

harmonic voltages generated from the VSC on the primary side of the converter transformer are analyzed (V_{ho}). Table 12.8 lists the result of the FFT analysis.

As listed in Table 12.7, the ratios V_h/V_0 of each harmonic voltage V_h on the primary side of the multiple converter transformer range from 0.34% to 4.16% of fundamental voltage with the connection to the grid. As listed in Table 12.8, the ratios V_{ho}/V_0 of each harmonic voltage V_{ho} on the primary side of the multiple converter transformer range from 0.2% to 3.74% of fundamental voltage without connection to the grid. The results indicate that the ratios of both sets of results are approximately the same. However, when the magnified ratios of the two conditions (column labeled V_h/V_{ho} in Table 12.8) are examined, it is verified that the harmonic voltages of the 131st to 139th order are magnified with the connection of the VSC to the grid. The magnified ratio V_h/V_{ho} of 135th harmonic voltage is the largest found here and is nearly 24 times.

12.4.1.3   Frequency–impedance (f–Z) characteristics of grid

Based on the observations of FFT results as described previously, it is assumed that a series and a parallel resonant circuit exist in this grid since the amplitude of the harmonic voltages near the 130th order is magnified in the EMTP simulation results. Furthermore, frequency-domain analysis is performed in order to investigate this phenomenon. This is accomplished in two stages. The VSC is replaced with a current source at the first stage. Then frequency versus impedance (f–Z) characteristics of the grid can be calculated as viewed from the secondary side of the converter transformer by varying the frequency of injected current over a range of interests. At the second stage, the VSC and the converter transformer are replaced by a current source. The f–Z characteristics as viewed from the primary side of the converter transformer can be calculated in the same way. Figure 12.12 illustrates the concept.

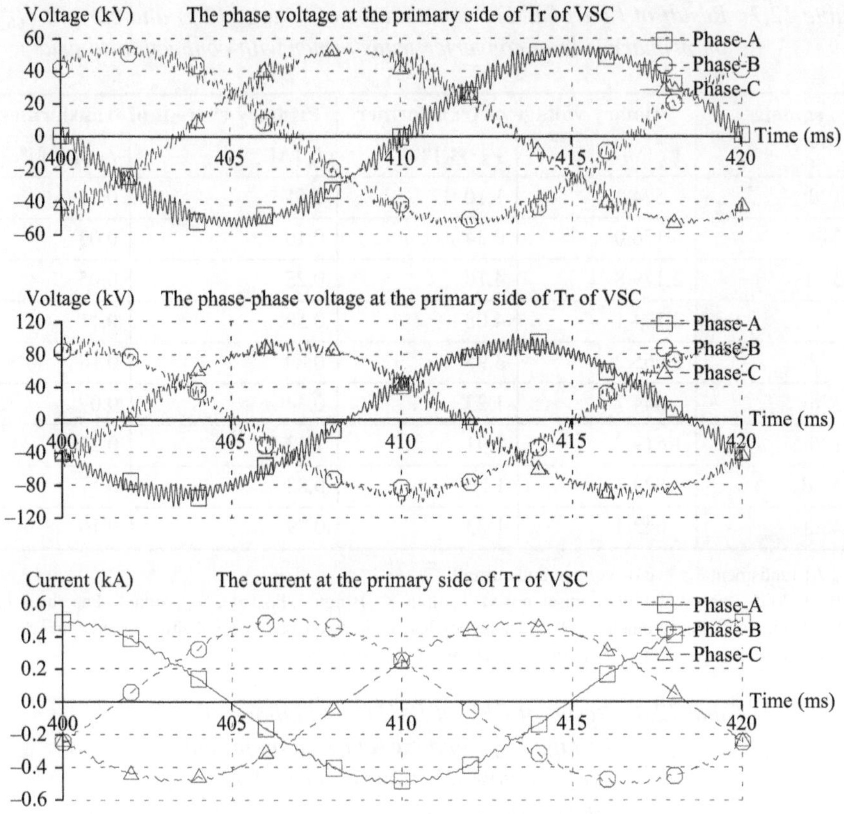

Figure 12.11   EMTP result of voltage and current on primary side of converter transformer

© IEEE 2005. Reprinted with permission from Temma, K., Ishiguro, F., Toki, N., Iyoda, I., Paserba, J.J., 'Clarification and measurements of high frequency harmonic resonance by a voltage sourced converter', *IEEE Transactions on Power Delivery*, vol. 20, no. 1, January 2005

Figure 12.13 shows the results of f–Z viewed from the primary side and Figure 12.14 shows the results of f–Z viewed from the secondary side. From these figures we can make the following observations:

1. f–Z characteristics from the primary side: A parallel resonant exists in this circuit and the parallel resonance point is 130.1th. The resonance impedance is 267.3 kΩ.
2. f–Z characteristics from the secondary side: A series resonant exists in this circuit and the series resonance point is 135.8th. The resonance impedance is 23.8 Ω.

It is verified that a series resonant circuit and a parallel resonant circuit exist within a very narrow frequency band near the 130th harmonic in this frequency domain analysis.

Table 12.7   Result of FFT of EMTP simulations of voltage (V_h) and current (I_h) on primary side of converter transformer with connection to grid

Harmonic order	Primary voltage of transformer		Primary current of transformer	
	V_h [V]	V_h/V_0 [%][a]	I_h [A]	I_h/I_0 [%][a]
119th	574.3	1.10	0.55	0.11
121st	176.0	0.34	0.16	0.03
131st	2,179.8	4.16	0.25	0.05
133rd	2,134.1	4.08	0.58	0.11
135th	2,065.2	3.95	0.89	0.18
137th	644.7	1.23	0.34	0.07
139th	1,519.7	2.91	1.13	0.23
143rd	829.7	1.59	0.83	0.17
145th	642.4	1.23	0.79	0.16

[a]V_0, I_0: fundamental wave of voltage and current.

© IEEE 2005. Reprinted with permission from Temma, K., Ishiguro, F., Toki, N., Iyoda, I., Paserba, J.J., 'Clarification and measurements of high frequency harmonic resonance by a voltage sourced converter', *IEEE Transactions on Power Delivery*, vol. 20, no. 1, January 2005

Table 12.8   Result of FFT of EMTP simulations of voltage (V_h) and current (I_h) on primary side of converter transformer without connection to grid

Harmonic order	Primary voltage of transformer		$\dfrac{V_h}{V_{ho}}$
	V_{ho} [%]	V_{ho}/V_0 [%][a]	
119th	1,653.0	3.74	0.35
121st	502.7	1.14	0.35
131st	1,651.0	3.73	1.32
133rd	867.4	1.96	2.46
135th	87.6	0.20	23.58
137th	136.9	0.31	4.71
139th	1,083.0	2.45	1.40
143rd	1,133.0	2.56	0.73
145th	1,243.0	2.81	0.52

[a]V_0, fundamental wave of voltage.

© IEEE 2005. Reprinted with permission from Temma, K., Ishiguro, F., Toki, N., Iyoda, I., Paserba, J.J., 'Clarification and measurements of high frequency harmonic resonance by a voltage sourced converter', *IEEE Transactions on Power Delivery*, vol. 20, no. 1, January 2005

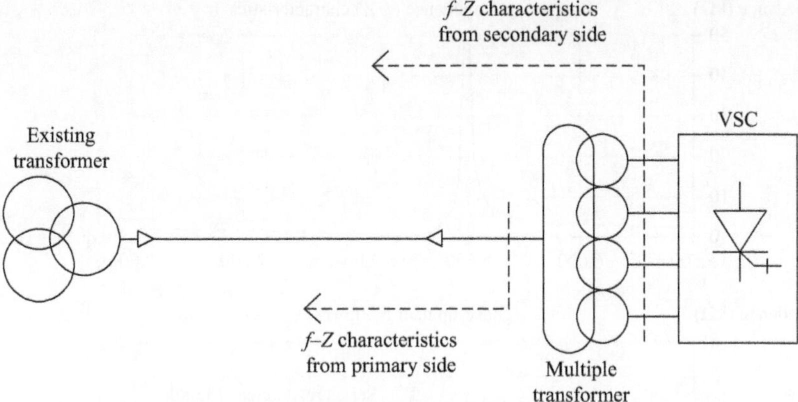

Figure 12.12   *Frequency–impedance characteristics for power system*

© IEEE 2005. Reprinted with permission from Temma, K., Ishiguro, F., Toki, N., Iyoda, I., Paserba, J.J., 'Clarification and measurements of high frequency harmonic resonance by a voltage sourced converter', *IEEE Transactions on Power Delivery*, vol. 20, no. 1, January 2005

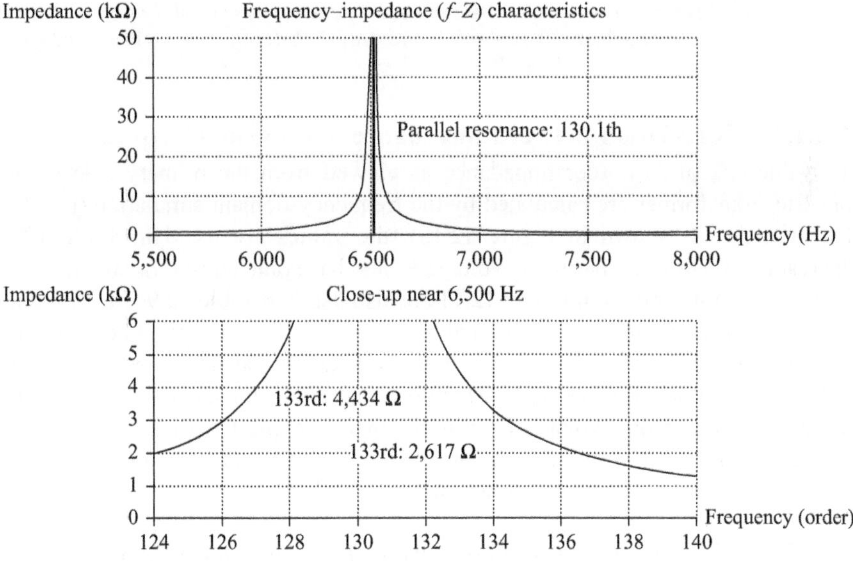

Figure 12.13   *Frequency-impedance (f–Z) characteristics as viewed from the primary side of the converter transformer by EMTP results*

© IEEE 2005. Reprinted with permission from Temma, K., Ishiguro, F., Toki, N., Iyoda, I., Paserba, J.J., 'Clarification and measurements of high frequency harmonic resonance by a voltage sourced converter', *IEEE Transactions on Power Delivery*, vol. 20, no. 1, January 2005

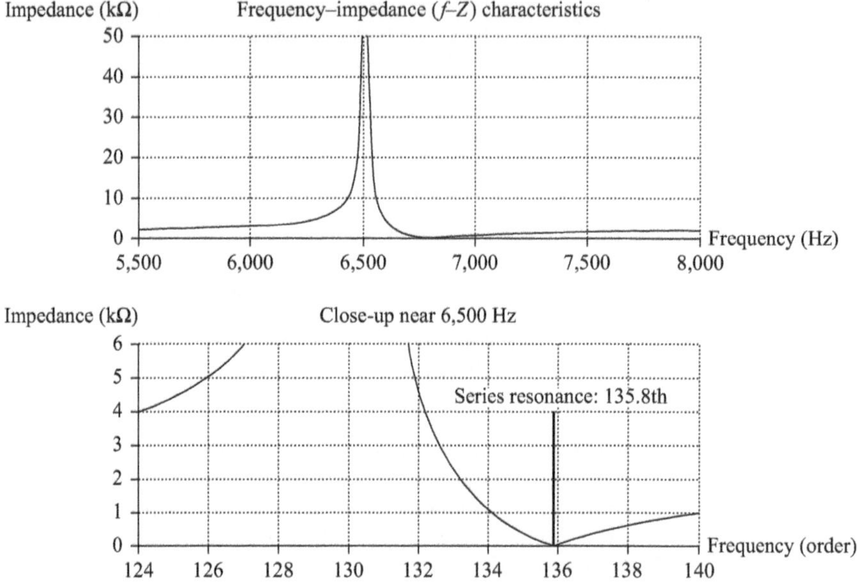

Figure 12.14   Frequency-impedance (f–Z) characteristics as viewed from the secondary side of the converter transformer by EMTP results

© IEEE 2005. Reprinted with permission from Temma, K., Ishiguro, F., Toki, N., Iyoda, I., Paserba, J.J., 'Clarification and measurements of high frequency harmonic resonance by a voltage sourced converter', *IEEE Transactions on Power Delivery*, vol. 20, no. 1, January 2005

12.4.1.4   Correlation between impedance and harmonic voltage

The values Z_h of Nth-order impedance as viewed from the primary side of the converter transformer are calculated by the frequency-domain simulation (i.e., f–Z characteristics as shown in Figure 12.13); the values are listed in Table 12.9. The results of Nth-order harmonic voltage V_h and harmonic current I_h calculated by the time-domain analysis of Figure 12.11 are also listed in Table 12.9. Each product of Z_h (from the frequency-domain analysis) times I_h (from the time-domain analysis) is approximately equal to V_h (from the time domain analysis), which is listed in the last column of Table 12.9. Therefore, harmonic impedance Z_h, harmonic current I_h, and harmonic voltage V_h from the two different domain analyses have a correlation with each other and it is verified that both time-domain and frequency-domain simulation results are appropriate.

12.4.2   *Principle of high-order harmonic resonance phenomenon*

It is presumed that the circuit in the vicinity of the VSC causes the high-order harmonic resonance voltage as shown in Figure 12.10. A basic equivalent circuit that consists of the VSC system, cables, and tertiary winding of the existing transformer in the grid is applied to the analysis presented here.

The basic equivalent circuit that defines the phenomenon consists of the leakage inductance L_1 of the converter transformer, the capacitance C of the cable,

Table 12.9   Correlation between Nth order harmonic voltage and Nth order impedance

Harmonic order	V_h [V] Time-domain	I_h [A] Time-domain	Z_h [Ω] Frequency-domain	$Z_h \times I_h$ [V]
119th	574.3	0.55	1,067	585.9
121st	176.0	0.16	1,319	211.0
131st	2,179.8	0.25	15,008	3,752.0
133rd	2,134.1	0.58	4,434	2,571.7
135th	2,065.2	0.89	2,617	2,329.1
137th	644.7	0.34	1,863	633.4
139th	1,519.7	1.13	1,450	1,638.5
143rd	829.7	0.83	1,009	837.5
145th	642.4	0.79	877	692.8

© IEEE 2005. Reprinted with permission from Temma, K., Ishiguro, F., Toki, N., Iyoda, I., Paserba, J.J., 'Clarification and measurements of high frequency harmonic resonance by a voltage sourced converter', *IEEE Transactions on Power Delivery*, vol. 20, no. 1, January 2005

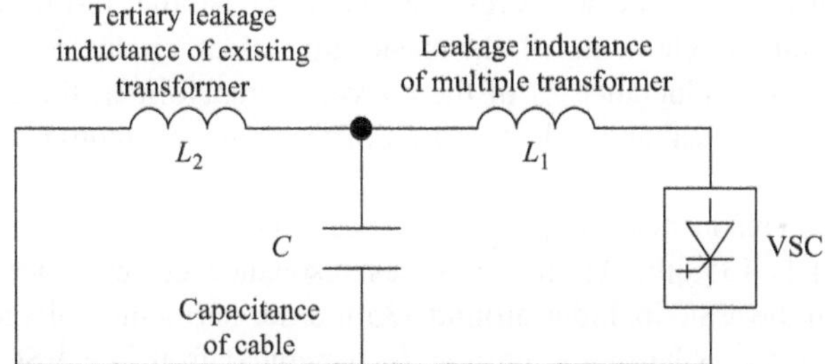

Figure 12.15   Basic equivalent circuit close to voltage-sourced converter system
© IEEE 2005. Reprinted with permission from Temma, K., Ishiguro, F., Toki, N., Iyoda, I., Paserba, J.J., 'Clarification and measurements of high frequency harmonic resonance by a voltage sourced converter', *IEEE Transactions on Power Delivery*, vol. 20, no. 1, January 2005

the leakage inductance L_2 of the tertiary winding of the existing transformer, and the VSC as illustrated in Figure 12.15.

The basic equivalent circuit close to the VSC system has a series resonant circuit and a parallel circuit. The equivalent impedance Z_{eq} of the circuit is given by (12.3).

$$Z_{eq} = jX_{eq} = j\left\{ \frac{\omega L_2 + \omega L_1(1 - \omega^2 L_2 C)}{1 - \omega^2 L_2 C} \right\} \tag{12.3}$$

The condition of series resonance is that the equivalent reactance X_{eq} of the circuit is zero.

$$\omega L_2 = \omega L_1 \left(1 - \omega^2 L_2 C\right) = 0 \tag{12.4}$$

The series resonance frequency f_{ser} is as follows:

$$f_{ser} = \frac{1}{2\pi\sqrt{\dfrac{L_1 L_2 C}{L_1 + L_2}}} = 6{,}962 \text{ Hz} = 139.3\text{th} \tag{12.5}$$

Furthermore, the condition of parallel resonance is that the equivalent reactance X_{eq} of the circuit is infinite.

$$1 - \omega^2 L_2 C = 0 \tag{12.6}$$

The parallel resonance frequency f_{par} is as follows:

$$f_{par} = \frac{1}{2\pi\sqrt{L_2 C}} = 6{,}681 \text{ Hz} = 133.6\text{th} \tag{12.7}$$

where the leakage impedance of the tertiary winding of the existing transformer is $\%Z_t = 32.8\%$ on a 1,000 MVA base.

When the sample circuit of Figure 12.10 is modeled in detail and performed by EMTP, the series resonance frequency f_{ser} is 135.8th and the parallel resonance frequency f_{par} is 130.1th. Each result of the EMTP analysis corresponds to each analytical result of (12.5) and (12.7) approximately. Therefore, it is verified that the fundamental principle of the series resonance and the parallel resonance occurs by means of the circuit elements of the basic equivalent circuit of Figure 12.15, namely the leakage inductance L_1 of the converter transformer, the capacitance C of the cable, and the leakage inductance L_2 of the tertiary winding of the existing transformer.

The series resonance frequency f_{ser} calculated by the detailed circuit (Figure 12.10) is 135.8th. At first, a series resonance occurs when the voltage generated from the VSC includes around 135th order harmonic voltage. The series resonance magnifies a harmonic current I_{ln} generated from the VSC on the secondary side of the converter transformer. The harmonic current I_{ln} on the secondary side of the converter transformer (L_1) is equal to a harmonic current I_{hn} on the primary side of the converter transformer (L_1) as illustrated in Figure 12.16. Then, the 131st order parallel resonance exists in the f–Z characteristic viewed from the primary side of the converter transformer. An impedance Z_{hn} around the 135th is close to the 131st. Since the voltage on the primary side of the converter transformer is determined by $Z_{hn} \times I_{hn}$ and with both Z_{hn} and I_{hn} around the 135th harmonic magnified, then the 135th order harmonic voltage is magnified.

Therefore, it is clarified that the principle of the high-order harmonic resonance phenomenon is as follows:

1. When a very small harmonic voltage is generated from a VSC on the secondary side of the multiple converter transformer, a series resonance that exists in the frequency close to the small harmonic voltage magnifies a harmonic current.
2. The magnified current on secondary side of the converter transformer flows to the primary side through its leakage reactance.

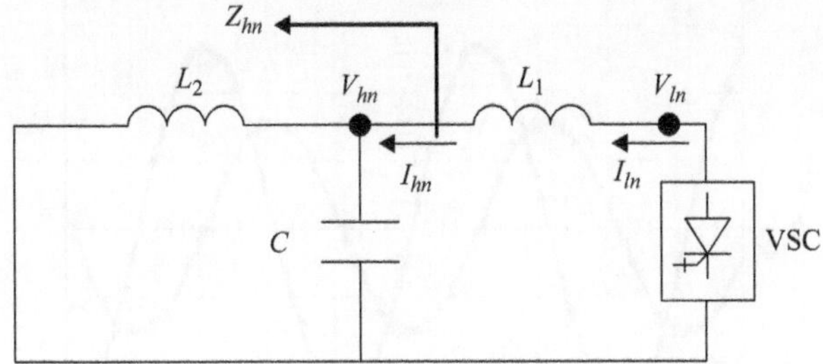

Figure 12.16   Series resonance and parallel resonance of high-order harmonic resonance

© IEEE 2005. Reprinted with permission from Temma, K., Ishiguro, F., Toki, N., Iyoda, I., Paserba, J.J., 'Clarification and measurements of high frequency harmonic resonance by a voltage sourced converter', *IEEE Transactions on Power Delivery*, vol. 20, no. 1, January 2005

3. When a parallel resonance frequency is close to the series resonance frequency, a large harmonic voltage could occur by the magnified current and the resonance impedance of these resonant circuits.

12.4.3   Field test

12.4.3.1   Field test condition

In order to verify the high-order harmonic resonance phenomenon, a field test for the phenomenon is performed. A 53 MVA VSC system with a 4-stage-multiple converter transformer is connected to a tertiary side of an existing transformer through a cable 500 meters long, in common with the sample circuit of Figure 12.10.

12.4.3.2   Field test results

The phase voltage V_{m1} and current I_{m1} are measured on the primary side of the multiple converter transformer when the reactive power outputs are +37.5 Mvar and 0 Mvar respectively in a STATCOM operation mode for the condition. The measurements of the field test are shown in Figure 12.17. High frequency voltages around 130th are found on the primary side of the converter transformer as shown in the field test results of Figure 12.17.

These field test results show that a very small harmonic voltage is magnified by a series and parallel resonant circuits as clarified in the simulation and analytical results when a grid has the series and parallel resonant circuits and both resonant circuits are close in frequency. Thus it is confirmed by analysis, simulation, and measurements that a high-order harmonic resonance phenomenon can occur and is a practical matter to be dealt with.

1. Voltage and current at VSC output $Q = +37.5$ Mvar
2. Voltage and current at VSC output $Q = 0$ Mvar
3. Enlarged view of voltage and current at VSC output $Q = 0$ Mvar

axis: $V = 20$ kV/div, $I = 300$ A/div; horizontal axis: Time $= 5$ ms/div (Figure 12.17(a) and (b)), Time $= 1$ ms/div (Figure 12.17(c)).

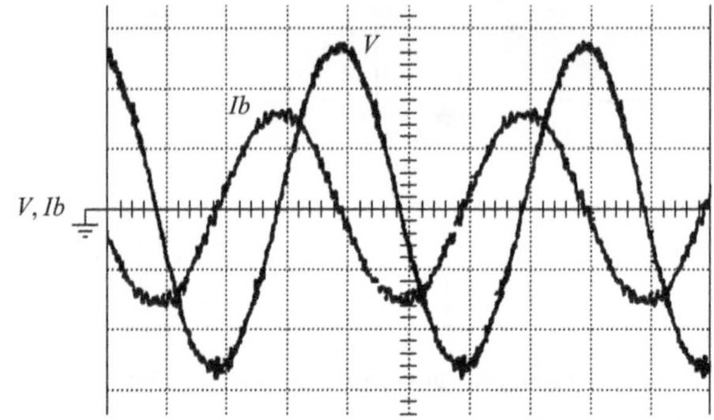

Vertical axis: V = 20 kV/div, I = 300 A/div   Horizontal axis: Time = 5 ms/div

(a)

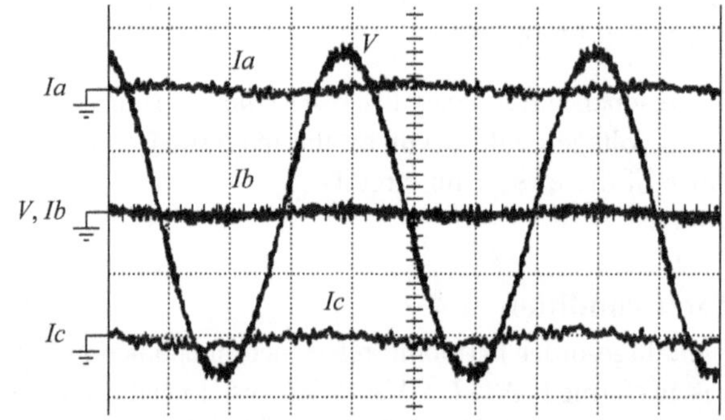

Vertical axis: V = 20 kV/div, I = 300 A/div   Horizontal axis: Time = 5 ms/div

(b)

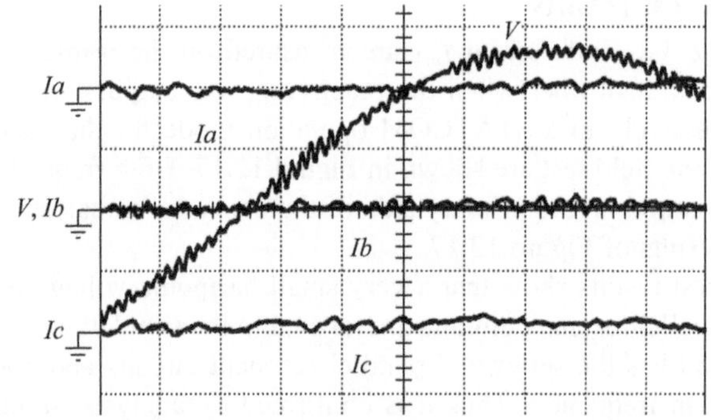

Vertical axis: V = 20 kV/div, I = 300 A/div   Horizontal axis: Time = 1 ms/div

(c)

Figure 12.17   Measurement results of field test of 53 MVA voltage-sourced converter

© IEEE 2005. Reprinted with permission from Temma, K., Ishiguro, F., Toki, N., Iyoda, I., Paserba, J.J., 'Clarification and measurements of high frequency harmonic resonance by a voltage sourced converter', *IEEE Transactions on Power Delivery*, vol. 20, no. 1, January 2005

12.4.4   Considerations and countermeasures

High-order harmonic resonance phenomenon depends on a local condition at the grid connection point of the VSC system. The local condition is caused by the series and parallel resonant circuit of the reactance of converter transformer, the capacitances of cables, and the reactance of existing transformer. It has been considered in the past that high-order harmonics are reduced immediately and no critical problem will be caused. However, the advancements of technology has decreased the impacts of the factors that reduce harmonics, such as losses of a cable and a transformer, in this case with VSC and PWM, therefore the problems have begun to actualize. Such phenomenon has been reported by some VSC-based systems recently. The influence on high-order harmonic resonance by the cable and countermeasures for them is described in this section.

12.4.4.1   Influence on phenomenon by length of cable

A capacitance of a cable becomes larger as the cable length is longer. As the capacitance is larger, the orders of series resonance and parallel resonance move lower. In general, amplitude of a harmonic voltage tends to increase when the harmonic order becomes low. Therefore, it is assumed that the harmonics increase as a resonance point transitions to a low frequency. Simulations of the total harmonic distortion (THD) are performed to show the impact of cable length. The THDs are the phase voltages on the primary side of the multiple converter transformer illustrated in Figure 12.10. The parameter of these simulations is the length of the cable. The results of the simulations in Table 12.10 indicate that the THD of the voltage increases with the larger capacitance of the longer cable. It is verified that the magnitude of the high-order harmonic resonance tends to increase as the length of the cable becomes longer. It should be noted that the same phenomenon would occur for anything else that increases the capacitance between the power system and the VSC such as cables in parallel.

When there is no cable between the existing transformer and the converter transformer, there is no significant capacitance. It is assumed that the high-order harmonic resonance phenomenon will not occur because of the absence of the resonant circuit. Figure 12.18 shows the simulation result with no cable. It is clarified that a capacitance of a cable is one of the causes of high-order harmonic resonance as confirmed in Figure 12.18. Thus, the resonance can be prevented from

Table 12.10   Relation between cable length and THD of
output voltage of VSC

Length (m)	0	200	390	500	700	900
THD^a (%)	1.54	1.84	5.19	8.50	11.60	20.16

[a]THD of harmonic voltages while connecting to grid.
© IEEE 2005. Reprinted with permission from Temma, K., Ishiguro, F., Toki, N., Iyoda, I., Paserba, J.J., 'Clarification and measurements of high frequency harmonic resonance by a voltage sourced converter', *IEEE Transactions on Power Delivery*, vol. 20, no. 1, January 2005

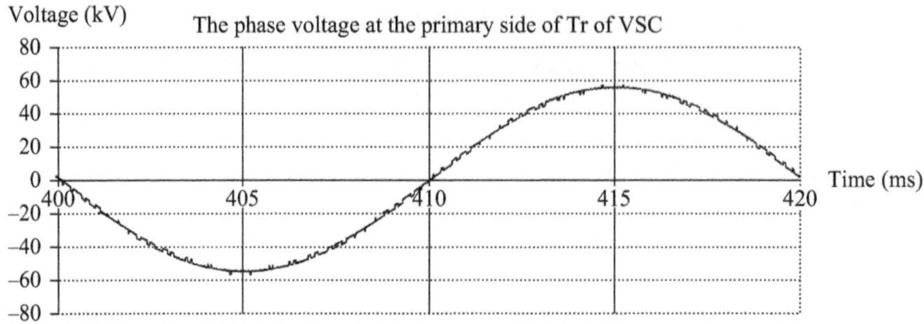

Figure 12.18   *EMTP results with no cable*

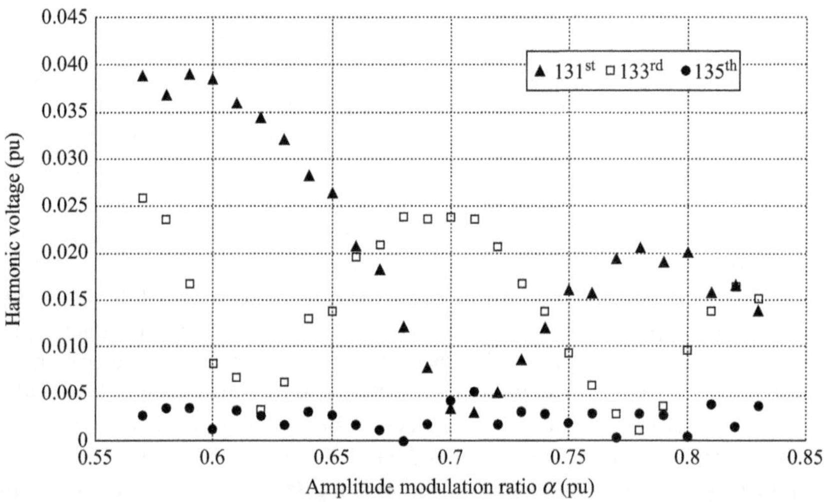

Figure 12.19   *Harmonics voltages in the VSC operating range*

© IEEE 2005. Reprinted with permission from Temma, K., Ishiguro, F., Toki, N., Iyoda, I., Paserba, J.J., 'Clarification and measurements of high frequency harmonic resonance by a voltage sourced converter', *IEEE Transactions on Power Delivery*, vol. 20, no. 1, January 2005

occurring when there is no cable, or minimized by significantly reducing the capacitance by the use of, for example, an aerial line.

12.4.4.2   Operating range of VSC and harmonics

The 130th, 133rd, and 135th harmonic voltages on the primary side of the multiple converter transformer are investigated. The parameter of the simulation is the amplitude modulation ratio α of the VSC. The condition of the operating range of the amplitude modulation ratio α is from 0.59 to 0.83. The circuit breaker on the primary side of the converter transformer is opened in order to investigate the net generated harmonic voltages as a function of the amplitude modulation ratio α. Figure 12.19 shows the result of the simulations. The exact domination of various

*Table 12.11   Simulation results for harmonic voltage reduction
by a high-pass filter*

Capacity of high-pass Filter (MVA)	0	1.0	1.5	2.0
THD of harmonic voltage (%)	8.50	3.12	1.94	1.61

© IEEE 2005. Reprinted with permission from Temma, K., Ishiguro, F., Toki, N.,
Iyoda, I., Paserba, J.J., 'Clarification and measurements of high frequency
harmonic resonance by a voltage sourced converter', *IEEE Transactions on
Power Delivery*, vol. 20, no. 1, January 2005

harmonic voltage orders is different through the examined range of α. However, the
dominant harmonic voltages are all around the 130th order for the entire range of α.
Therefore, this analysis shows that modifying the VSC control is not practical and
effective means to prevent this high-order harmonic resonance phenomenon.

12.4.4.3   Countermeasure for high-order harmonic resonance

It is difficult to estimate the amount of high-order harmonics in the pre-manufacturing/
design stages exactly. When small deviations of a leakage inductance of a transformer
occur or the exact length of cable may not be precisely defined in advance, the resonant and anti-resonant point may move and the amount of the high-order harmonics
may change because of these reasons. Furthermore, it would be overly conservative to
consider the maximum conditions of the high-order harmonics and to design and
manufacture a multiple converter transformer, a cable, cable terminations, a power
transformer, protective relays, and the VSC to meet the potentially higher levels of
high-order harmonic and the respective impacts on insulation and control systems.

One of the countermeasures for the high-order harmonic resonance is to install a
harmonic filter on the primary side of the converter transformer. The simulation of
the THD of the harmonic voltages with the harmonic filter is performed. The impact
on the constraint for high-order harmonic resonance phenomenon is verified and the
capacity of the harmonic filter can be estimated. The parameter of the simulation is
the capacity of the harmonic filter. The results for THD for the capacity of each
harmonic filter are shown in Table 12.11. The criterion of the THD of the harmonic
voltages is within 2%. A 1.5 to 2.0 MVA harmonic filter should be installed
approximately by simulation results.

Thus, it is evaluated that high-order harmonic resonance can be restricted by a
harmonic filter with a relatively small capacity compared to the capacity of the
VSC. The example filter size in this study is about 4% of the rating of the VSC.

The best method to avoid the resonance phenomenon in the field is to adequately perform a power system analysis in advance and to consider the accuracy of
power system parameters used in the analysis.

References

[1]   Hingorani N. G., 'High power electronics and flexible AC transmission
      system', *IEEE Power Engineering Review*, vol. 8, no. 7. pp. 3–4, July 1988.

[2]   Mori S., 'Initiatives perspectives by the power industry towards a low carbon emission society', Keynote address of Paris Session CIGRE 2010.

[3]   Yokoyama A., 'Optimal control demonstration project for the next-generation transmission and distribution system', *Electrical Review (Denki Hyoron)*, vol. 95, no. 10, pp. 26–29, 2010 (in Japanese).

[4]   Dennis W., 'Power electronics for wind energy applications', presentation slide of IEEE northern Canada section, Edmonton, September 17, 2012.

[5]   Hingorani N. G., Gyugyi L., *Understanding FACTS: Concepts and Technology of Flexible ac Transmission Systems*, Wiley-IEEE Press, NJ, United States, 1999.

[6]   Temma K., Kitayama M., 'STATCOM technology for smart grid and the next-generation transmission and distribution system', *Smart Grid* (technical journal), April, pp. 8–12, 2012 (in Japanese).

[7]   Akedani T., Hayashi J., Temma K., Morishima N., '450MVA STATCOM installation plan for stability improvement', *Proceeding of 2010 CIGRE*, Paper B4-207, Paris Session 2010.

[8]   Matsuda T., Shimonosono T., Harada H., Temma K., Morishima N., Shimomura T., 'Factory test and commissioning test of 450 MVA-STATCOM project', *Proceeding of 2014 CIGRE*, Paper B4-204, Paris Session 2014.

[9]   Fujii T., Temma K., Morishima N., Akedani T., Shimonosono T., Harada H., '450MVA GCT-STATCOM for stability improvement and over-voltage suppression', *International Power Electronics Conference*, IEEJ, IPEC-Sapporo, Japan, pp. 1766–1772, 2010.

[10]   Yagi M., Takano T., Sato T., Iyoda I., Temma K., 'Role and new technologies of STATCOM for flexible and low cost power system planning', *Proceedings of 2002 CIGRE*, Paper SC14-107, Paris Session 2002.

[11]   Yonezawa H., Imura H., Shinki Y., Temma K., Funahashi S., Teramoto H., 'Improvement of voltage stability with STATCOM', *Proceedings of the Technical Meeting on Power Engineering of IEEJ*, PE-04-1, pp. 1–6, 2004 (in Japanese).

[12]   Sato T., Matsushita Y., Temma K., Morishima N., Iyoda I., 'Prototype test of STATCOM and BTB based on voltage source converter using GCT thyristor', *Transmission and Distribution Conference and Exhibition 2002 Asia Pacific*. IEEE/PES, pp. 2037–2042, 2002.

[13]   Anjiki M., Nagatomo Y., Taketa K., Temma K., Masaki K., Fujii T., Morishima N., 'Development of STATCOM for transient stability improvement', *Proceedings of the Technical Meeting on Power Engineering of IEEJ*, PE-11-187, pp. 27–30, 2011 (in Japanese).

[14]   Imanishi T., Nagatomo Y., Iwasaki S., Masaki K., Fujii T., Ieda J., '130 MVA-STATCOM for transient stability improvement', *International Power Electronics Conference*, IEEJ, IPEC-Hiroshima, Japan, pp. 2663–2667, 2014.

[15]   Mori S., Matsuno K., Hasegawa T., Ohnishi S., Takeda M., Seto M., Murakami S., Ishiguro F., 'Development of a large static VAr generator using self-commutated inverters for improving power system stability', *IEEE Transactions on Power Systems*, vol. 8, pp. 371–377, 1993.

[16] Reed G., Paserba J., Croasdaile T., Takeda M., Hamasaki Y., Aritsuka T., Morishima N., … Haas D., 'The VELCO STATCOM-based transmission system project', *2001 IEEE PES WM*, vol. 3, pp. 1109–1114, 2001.

[17] Kundur P., Paserba J., Ajjarapu V., Andersson G., Bose A., Canizares C., Hatziargyriou N.… Vittal V., 'Definition and classification of power system stability', IEEE/CIGRE joint task force on stability terms and definitions. *IEEE Transactions on Power Systems*, vol. 19, no. 3, pp. 1387–1401, 2004.

[18] Temma K., Ishiguro F., Toki N., Iyoda I., Paserba J. J., 'Clarification and measurements of high frequency harmonic resonance by a voltage sourced converter', *IEEE Transactions on Power Delivery*, vol. 20, pp. 450–457, 2005.

Chapter 13

Application of SVC to cable systems

*Y. Tamura**

13.1 AC cable interconnection to an island

Interconnection to the island from the mainland has advantages as follows. The electric power on remote islands, in general, has been provided by power generation on the islands, which requires maintenance and fuel transport for the generators. If the mainland and the remote island are connected by submarine cables and power is transmitted from the mainland, there are economic advantages since the costs of maintenance and fuel for generators on the island can be reduced.

However, in long-distance AC cable power transmission, voltage problems such as variations on the load side caused by the capacitance of the cable and harmonic resonance phenomena must be resolved [1–8]. Among these problems, voltage variation on the load side caused by switching on and off the long-distance AC cable cannot be ignored.

The Kyushu Goto Island chain was connected to the mainland and operation started through 66 kV AC cables in 2005 as illustrated in Figure 13.1. The short-circuit capacitance of the Goto Island system is several hundred megavolt amperes, while the 66 kV AC cable linked to the mainland has a capacitance of approximately 20 Mvar per one circuit. Therefore, voltage variations occur on the island as a result of switching one circuit on or off. In particular, in case if fault occurs, the existing voltage regulation equipment such as transformer taps or phase-modifying equipment control cannot suppress the voltage variation immediately because of slow response. Thus, an Static Var Compensator (SVC) was installed in the Okuura switching station on Fukue Island in the Goto Islands as illustrated in Figure 13.1 [9, 10].

13.2 Typical example of voltage variations in an island

The causes of voltage variations include transmission line faults on an island, tap control in a transformer linked to the mainland, and phase-modifying equipment control on an island. The greatest voltage variations occur when switching on or off one circuit of the AC cables linked to the mainland. Figure 13.2 shows an example of

*Senior Engineer, Toshiba Co., Japan

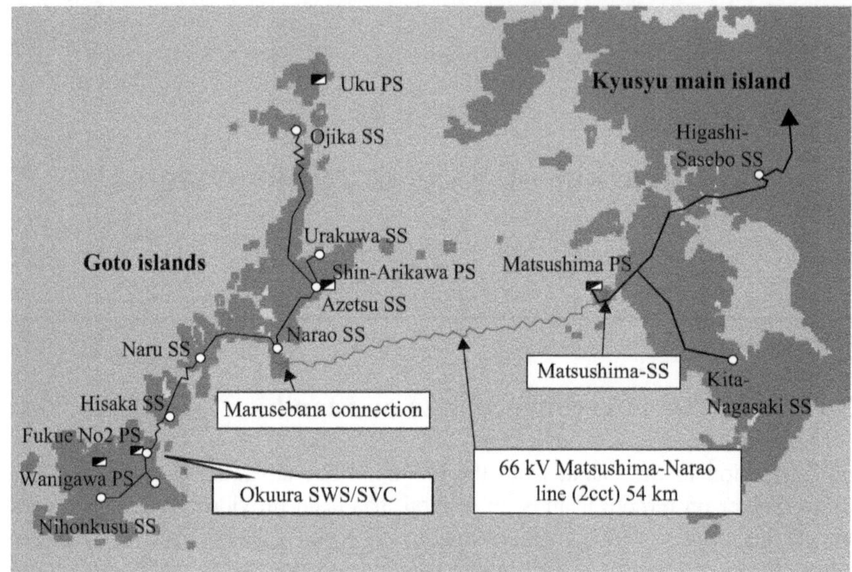

Figure 13.1   Interconnection from the mainland to the island

© 2013 Wiley Periodicals, Inc. Reprinted with permission from Yuji Tamura, Shinji Takasaki, Yasuyuki Miyazaki, Hideo Takeda, Shoichi Irokawa, Kikuo Takagi, Naoto Nagaoka, 'Voltage control of static var compensator for a remote system interconnected by long-distance AC cables', *Electrical Engineering in Japan*, pp. 19–30, November 2013

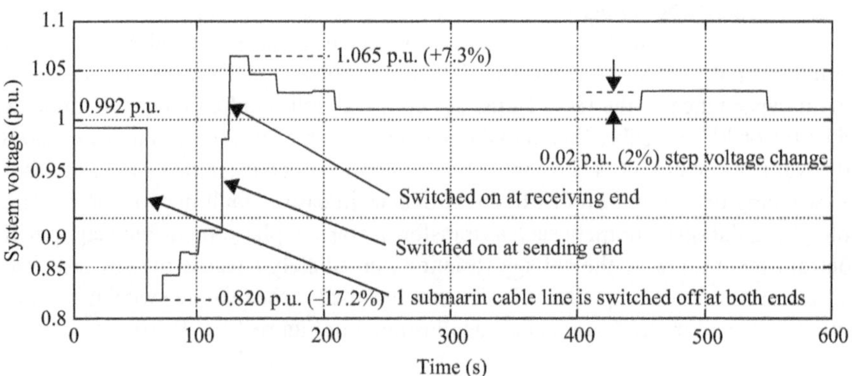

Figure 13.2   Typical voltage variation caused by long distance AC cable

voltage variations on an island. This waveform is a calculated example for a voltage of 66 kV at the Okuura switching station without the SVC, using an analysis model to be described below. When a fault is detected in an overhead transmission line segment among two circuits of the 66 kV interconnection to the mainland, the circuit which includes faulty overhead transmission line is opened at both ends of the interconnection line including the cable segment. Then the circuit is reconnected after 60 seconds. In this example, the voltage after the cable is disconnected drops by 17.2%, and after restart it rises to 7.3% above the value compared to the voltage before the fault. There are also voltage variations during normal operation due to load fluctuations on the islands, and Figure 13.2 shows a voltage variation of 2% as an example.

13.3   The required control function for the SVC

SVC can suppress the voltage variations in the network system; however, the optimized voltage control method of the SVC should be developed for the specific scenarios to be activated. The scenarios to which the system voltage control (automatic voltage regulator or AVR) of the SVC should be functionalized properly are summarized as the following three points.

1.   When one circuit of the 66 kV AC cable linked to the mainland is opened due to a fault, a voltage drop due to a loss of approximately 20 Mvar of capacitance in the cable must be suppressed.
2.   During a fault in the 66 kV transmission line connecting to the mainland, the voltage rises due to tap control in the transformer linking to the mainland side, from the time at which the AC cable circuit is opened to the time when it is closed again 60 s later, and an overvoltage then occurs immediately after restart. This must be suppressed.
3.   The AVR of SVC should not be excessively operated to the normal voltage variations due to load variations on the islands, the transformer taps connected to the mainland, and phase-modifying equipment operation on the islands.

The SVC capacity is set to the capacity that can be continuously regulated from a phase lead (capacitive) of 20 Mvar to a phase lag (inductive) of 10 Mvar so as to be able to adequately compensate the cable capacitance of 20 Mvar and resolve issues (1) and (2) above.

The details of the voltage control method for an SVC specific to AC power transmission to remote islands that resolves the above issues are given below.

13.4   V-I characteristics of the SVC

Figure 13.3 shows V-I characteristics for the AVR function of the SVC used. The black broken line represents the SVC V-I characteristic, and its intersection with the V-I characteristic of the system shown by a dotted line represents the SVC operation point [11–14]. The use of linear characteristics [9, 11, 13, 14] is common for typical SVC V-I characteristics, but here a broken-line characteristic with a dead band is used.

The slope of the V-I characteristic of the system varies depending on the system conditions and the conditions of the existing voltage control equipment; if the system voltage drops, the V-I characteristic of the system trends downward and the SVC operating point shifts toward the phase-lead side (capacitive output side). If the system voltage rises, then the operating point shifts toward the phase-lag side (inductive side). The SVC continues output equivalent to the steady-state current reference value I_{ref} of the SVC until the boundary of the $+/-D1$ dead band is exceeded [14]. However, if there are substantial changes in the system

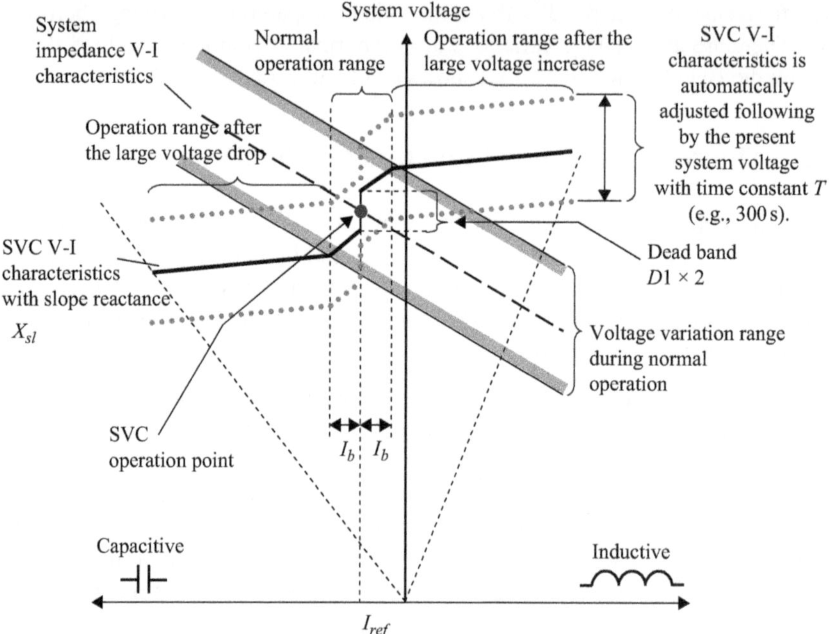

Figure 13.3   SVC output characteristics

voltage and the dead band is exceeded, then the system voltage variations are suppressed by the SVC.

The SVC response time is fast, a few tens of milliseconds, but on the other hand, the existing voltage control equipment involves slow control on the order of seconds, in which a response is generated by slowly integrating the variation from the control reference voltage. If the SVC continues output, then the existing voltage control equipment does not operate. Thus, after the start of SVC voltage control, the SVC voltage reference value must track the system, and exert control to steadily restore the SVC operation point to the steady-state current reference value I_{ref}. Thus, immediately after a large voltage variation occurs, the SVC suppresses voltage variations, but then, as time passes, the SVC output I_{svc} is restored to the steady-state value I_{ref}. For this reason, when the existing voltage control equipment operates, the system voltage variation is kept within a set range, and coordinated control between the SVC and the existing voltage control equipment can be performed automatically. Moreover, the SVC standby capacity can be maintained at a higher level by being prepared for the next voltage variation.

When a system voltage variation that slightly exceeds the dead band $+/-$DI, when the SVC output current is in the range of $I_{ref} +/- I_b$ as indicated in Figure 13.3, the inclination of the slope reactance is set to be large. For a relatively small voltage variation that exceeds the dead band $+/-$DI, the sensitivity is low, but the SVC can respond. Such broken-line characteristics allow mild voltage control for a small disturbance and strong voltage control for a large disturbance, with small disturbances and large disturbances regulated separately.

13.5   Automatic Voltage Regulator (AVR) of the SVC

The SVC AVR function that realizes the V-I characteristic in Figure 13.3 is explained in Figure 13.4.

The AVR function can be broadly subdivided into four parts.

(1) Function for calculating the voltage control quantity ΔV_{1b}

The details of the signal processing are explained using Figure 13.5(a). The relatively slow voltage variation by first-order lag with a time constant T of several hundred seconds at the system voltage V is extracted, and the result is used as the target voltage V_f. Subtracting V_f from V, fast voltage variation ΔV_{1a} is calculated,

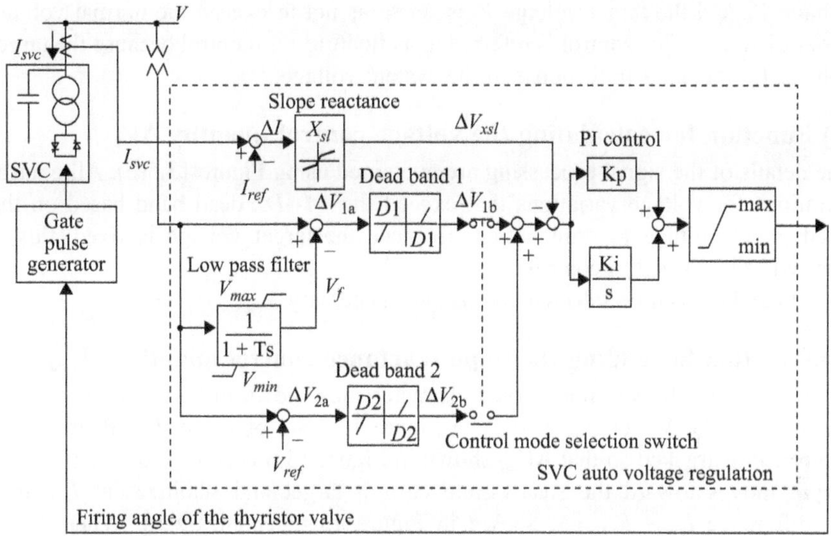

Figure 13.4   *Block diagram of the SVC voltage control*

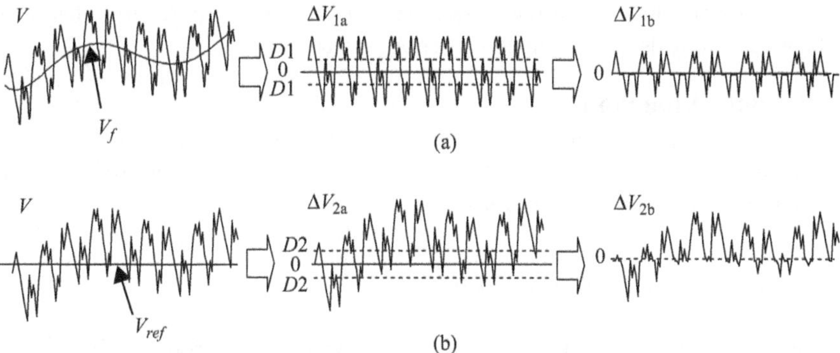

Figure 13.5   *Behavior of the variables inside the block diagram: (a) Floating voltage reference; (b) Fixed voltage reference*

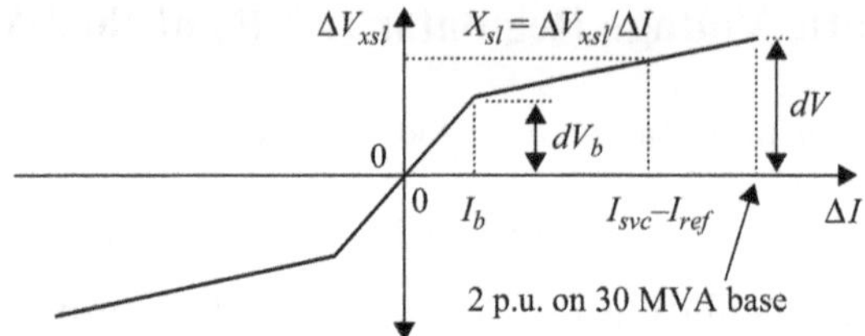

Figure 13.6   Slope reactance of the SVC voltage control

and the control quantity ΔV_{1b} is output in the case of voltage variations for which the $+/-D1$ dead band is exceeded. The limiters V_{max} and V_{min} are set for the target voltage V_f, and the target voltage V_f is set so as not to exceed the normal voltage variation range. This control is referred to as floating V_{ref} control because the target voltage V_f fluctuates in response to the system voltage.

(2) Function for calculating the voltage control quantity ΔV_{2b}

The details of the signal processing are explained using Figure 13.5(b). All control quantities for voltage variations that exceed the $+/-D2$ dead band based on the fixed target voltage V_{ref} are output. Because the target voltage is fixed, this is referred to as fixed V_{ref} control.

(1) and (2) can be selected and used as necessary.

(3) Function for adding the slope reactance control quantity ΔV_{xsl}

By applying the broken-line curved characteristics shown in Figure 13.6 to X_{sl} in Figure 13.4, the V-I characteristics in Figure 13.3 are obtained. For floating V_{ref} control, V_f is tracked so that ΔV_{xsl} shown in Figure 13.6 is zero, and thus the SVC output moves toward the steady-state current target and stabilizes at I_{ref} until the difference $I_{svc} - I_{ref} = \Delta I$ shown in Figure 13.4 becomes zero. When X_{sl} in Figure 13.4 is zero, that is, when $dV = 0$ or $dV_b = 0$ in Figure 13.6, the control quantity ΔV_{xsl} is zero regardless of ΔI. Consequently, the SVC output does not return to I_{ref} even under floating V_{ref} control. That is, control to restore the SVC output to the steady-state current reference level I_{ref} after suppression of system voltage variations does not function when X_{sl} is zero.

(4) PI control function

The deviation whose input is the control quantity described in items (1)–(3) is input to PI (proportional integral) control, and the control angle of the thyristor valve is regulated so that this is equal to zero.

13.6   Transient analysis model

Simulations using an effective value analysis program are performed. Figure 13.7 shows an overview of the system configuration that provides input for the network model represented in the simulations. The simulation includes voltage control models

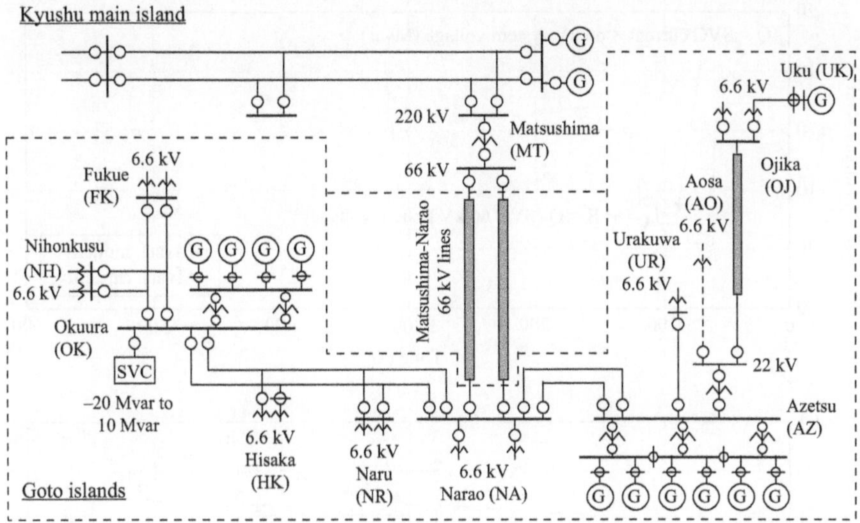

Figure 13.7   AC network model

of the existing phase-modifying equipment on the islands, and tap control models for the transformer connected at the mainland as well as for the transformer on the islands, although they are not indicated in Figure 13.7. The voltage variations are evaluated at the primary electric power stations on the islands using the resulting analysis model.

The parameters used to evaluate the voltage variation suppression effect due to the SVC are explained using Figure 13.8. Figure 13.8 represents the case in which the SVC is connected in the simulation of cable disconnection and reconnection and the occurrence of small voltage variation of 2% as shown in Figure 13.2. The floating V_{ref} control of Case 3 in Figure 13.9 and Table 13.1 (explained below in Section 13.7) is used for the V-I characteristics of the SVC. The evaluation of the effect of connection of the SVC when one circuit of the 66 kV AC cable connected to the mainland is disconnected or reconnected includes the voltage V_S before disconnection in Figure 13.8, the minimum value V_{smin} of the system voltage when the circuit is open, the maximum value V_{smax} of the system voltage after reconnection, and the time T_{tap} until the tap response of the transformer linked to the mainland after reconnection. The response to the small voltage variation of about 2% is evaluated by the values of V_{d1} and V_{d2}. The 2% variation is represented by the load connection and disconnection in the network model. V_{d1} is voltage increase variation when the network system voltage is increased. V_{d2} is voltage decrease variation when the system voltage is decreased after 100 seconds. Both V_{d1} and V_{d2} are comparisons to the voltage level just before 2% voltage step increase.

13.7   Control parameter settings survey

Table 13.1 gives details of the set values of the V-I characteristics of the SVC compared in Figure 13.9. Using the characteristics in Cases 1 through 4, floating

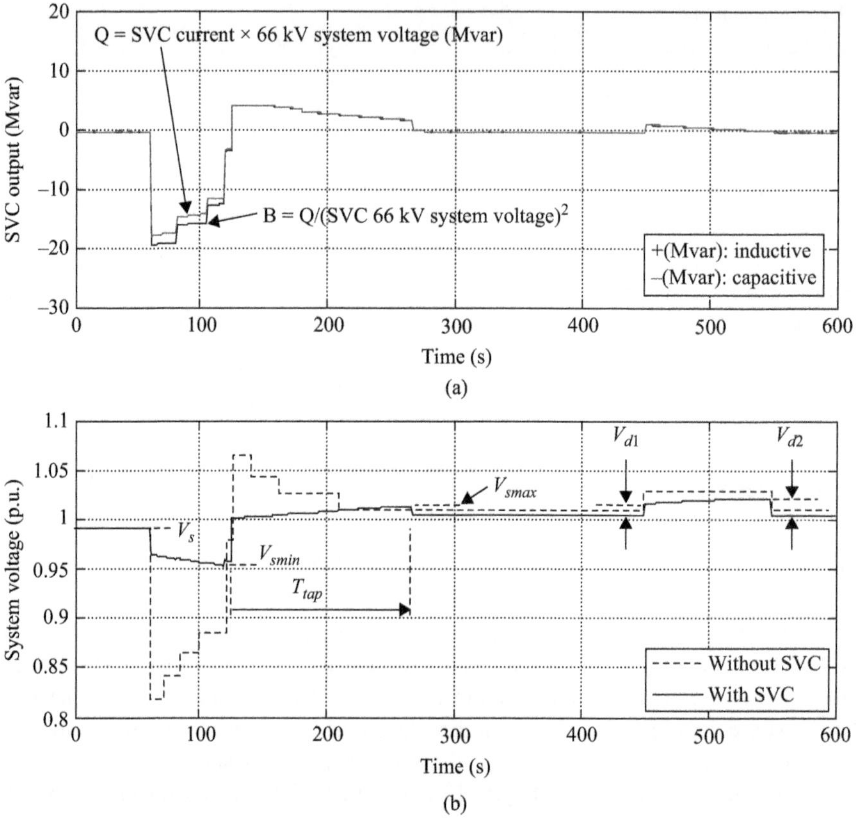

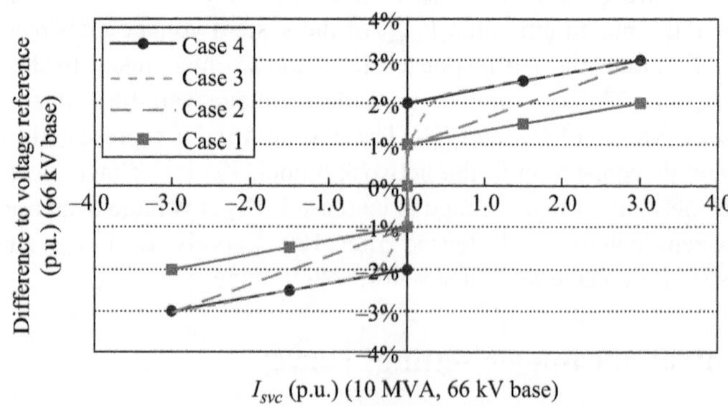

Figure 13.8   Typical voltage variation with an SVC: (a) SVC output; (b) Voltage variation with SVC

Figure 13.9   Parameters for the slope reactance

Table 13.1 Parameters for the slope reactance

(a) Slope reactance parameters

Parameters	Case 1	Case 2	Case 3	Case 4
$D1, D2$	$+/-1\%$	$+/-1\%$	$+/-1\%$	$+/-2\%$
dV	1%	2%	2%	1%
dV_b	Not used	Not used	1%	Not used
I_b	Not used	Not used	0.4 p.u.	Not used

(b) Voltage control parameters (common to Cases 1 to 4)

Parameters	Setting	Remarks
T	120 s	Time constant for floating voltage reference control
V_{max}	1.05 p.u.	Limit for floating voltage control
V_{min}	0.95 p.u.	Limit for floating voltage control
I_{ref}	0 p.u.	Standby operation point of SVC

© 2013 Wiley Periodicals, Inc. Reprinted with permission from Yuji Tamura, Shinji Takasaki, Yasuyuki Miyazaki, Hideo Takeda, Shoichi Irokawa, Kikuo Takagi, Naoto Nagaoka, 'Voltage control of static var compensator for a remote system interconnected by long-distance AC cables', *Electrical Engineering in Japan*, pp. 19–30, November 2013.

V_{ref} control and fixed V_{ref} control are compared, including the discrepancy between the dead band and the X_{sl} slope and the broken-line case. In floating V_{ref} control, the first-order lag filter time constant in Figure 13.4, used to calculate the target voltage V_f by extracting the long-period oscillation of the system voltage, is set to 120 s, twice the duration 60 s between the time at which one circuit of the 66 kV AC cable linked to the mainland is opened and the time at which it is reconnected, so as not to respond unnecessarily to a voltage drop during that period. The upper and lower bound voltages V_{max} and V_{min} of the first-order lag control function in Figure 13.4 are set 1.05 p.u. and 0.95 p.u. so that V_f is within the normal range of voltage variation. The current reference value I_{ref} is set to 0, and the SVC output I_{svc} immediately before the opening of circuit in the cable is set to 0. The SVC output after the SVC response is set to return to I_{ref}, i.e., to zero. The initial values of V_{ref} and of V_f immediately before opening one circuit in the cable are adjusted to the system voltage immediately before the circuit is opened in the simulation.

13.8   Comparison of the simulation results

Figures 13.10–13.12 show the results of the simulation. With respect to the voltage variation suppression effect when one circuit in the cable is opened, comparing Figure 13.10(a) for floating V_{ref} control and Figure 13.10(b) for fixed V_{ref} control, the greater voltage variation suppression effect is found for fixed V_{ref} control, with its smaller difference between V_{smin} and V_{smax}. However, in fixed V_{ref} control, if the dead band is small, as in Cases 1 and 2 in Figure 13.11(b), then all voltage variations are suppressed using only the SVC, and transformer tap control may be

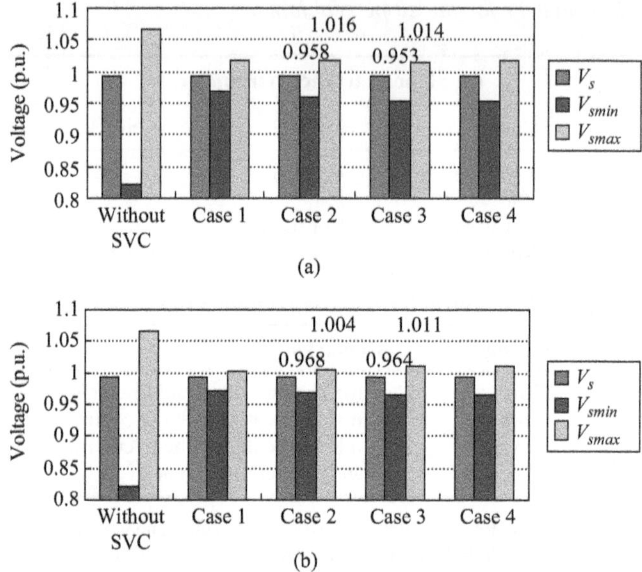

Figure 13.10   *Voltage variations in large event (simulation results): (a) Floating voltage reference control; (b) Fixed voltage reference control*

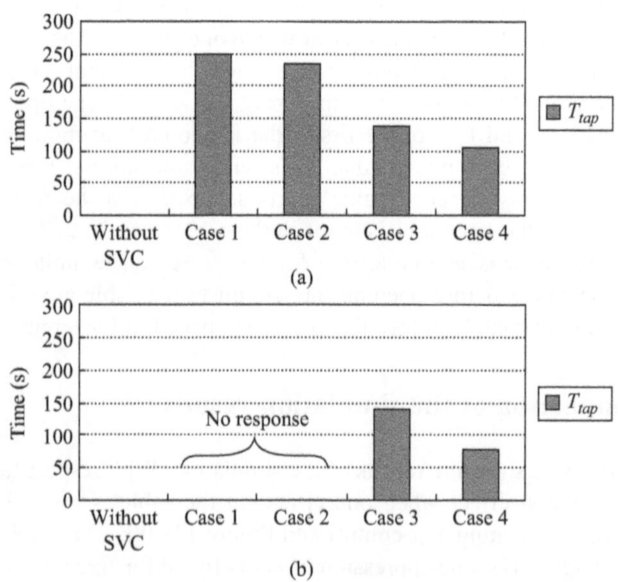

Figure 13.11   *Time to tap operation (simulation results): (a) Floating voltage reference control; (b) Fixed voltage reference control*

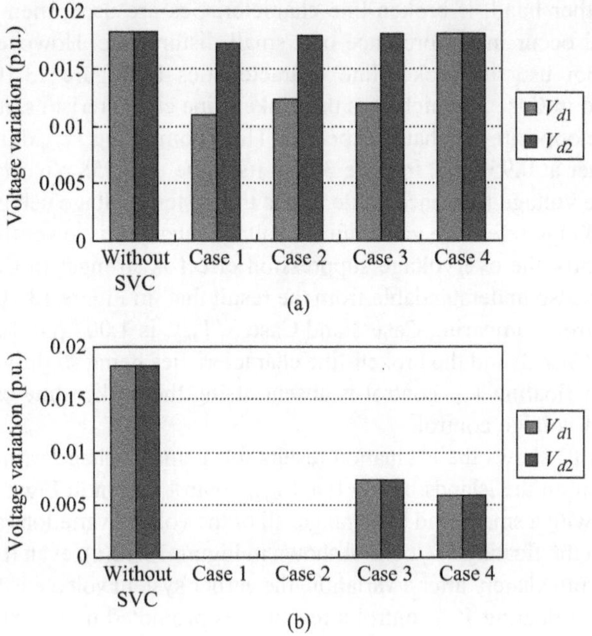

Figure 13.12   Voltage variations in small event (simulation results): (a) Floating voltage reference control; (b) Fixed voltage reference control

inhibited. For floating V_{ref} control, the SVC output automatically returns to I_{ref} after voltage variation suppression, and thus transformer tap control is not inhibited.

For floating V_{ref} control, the voltage variation suppression effect in Figure 13.10(a) shows no significant difference in Cases 1–4. However, there is a large difference in the transformer tap response time in Figure 13.11(a). When comparing Cases 1 and 2, the transformer tap response is faster in Case 2, with a higher set value for dV in the slope reactance X_{sl}. This is because at higher values of dV, i.e., of the slope reactance X_{sl}, the integrated voltage value in transformer tap control is larger due to the increase in the control quantity ΔV_{xsl} in Figure 13.4. When comparing Cases 2 and 3, the transformer tap response time is faster in Case 3, with the slope reactance as a broken line. This is because the ΔV_{xsl} control quantity is greater than in Case 2 in the broken-line region. Comparing Cases 3 and 4, the transformer tap response time is faster in Case 4 with its larger dead band $D1$. In Case 4, the control quantity ΔV_{xsl} falls. However, the system voltage when the SVC output is restored to the normal value for I_{ref} stabilizes at a higher level, representing the increase in the dead band range, which is the reason that transformer tap control is enhanced.

Thus the response of the existing voltage control equipment becomes greater as the SVC dead band increases. Furthermore, if the dead band ranges are the same, then it is greater for a larger slope reactance X_{sl}, and if broken-line characteristics are used in the slope reactance, then it is greater further.

On the other hand, if broken-line characteristics are used, then mild voltage control should occur in the presence of a small disturbance. However, in Case 2, which does not use the broken-line characteristics in Figure 13.10(a), V_{max} is 1.016 p.u., and in Case 3, which uses the broken-line characteristics, it is smaller at 1.014 p.u., the opposite of what is expected. Thus, comparing V_{min} during a voltage drop, it is lower at 0.953 p.u. in Case 3 than its value of 0.958 p.u. in Case 2. That is, because the voltage reference value tracks the system voltage using floating V_{ref} control, the voltage reference value during voltage recovery is lower than in Case 3, and consequently the overvoltage suppression effect is stronger in Case 3 than in Case 2. This is also understandable from the result that, in Figure 13.10(b), utilizing fixed V_{ref} control, comparing Case 2 and Case 3, V_{max} is 1.004 p.u. for Case 2 and 1.011 p.u. for Case 3, and the broken-line characteristics being slightly higher. Thus if the effect of floating V_{ref} control is absent, using the broken-line characteristics results in mild voltage control.

Figure 13.12 shows the evaluation results for a small voltage variation due to a load fluctuation on the islands. In the fixed V_{ref} control shown in Figure 13.12(b), in Cases 1 and 2 with a small dead band range, all of the voltage variations are controlled by the SVC. In the floating V_{ref} control shown in Figure 13.12(a), even if some control is performed immediately after a variation, the earlier system voltage is then restored.

Thus under floating V_{ref} control a response is promoted in the existing voltage control equipment, which is an improvement in the sense of avoiding an excessive response to a small voltage variation.

13.9   The applied control parameters

Table 13.2 lists the control parameters used in the SVC in a real system. These are set values selected through parameter studies based on effective value analysis. The setting of the AVR control parameters is satisfied for remote island power transmission. A survey of the settings is given below.

Table 13.2   Applied parameters for SVC voltage control

Parameters	Setting	Remarks
$D1$	+/−0.8%	Dead band for floating voltage control
T	300 s	Time constant for floating voltage control
V_{max}	1.05 p.u.	Limit for floating voltage reference control
V_{min}	0.95 p.u.	Limit for floating voltage reference control
X_{sl}	$dV = 2\%$	Slope reactance on $I_{svc} = 2$ [p.u.], 30 MVA base
	$dV_b = 0$	Not used
	$I_b = 0$ p.u.	Not used

© 2013 Wiley Periodicals, Inc. Reprinted with permission from Yuji Tamura, Shinji Takasaki, Yasuyuki Miyazaki, Hideo Takeda, Shoichi Irokawa, Kikuo Takagi, Naoto Nagaoka, 'Voltage control of static var compensator for a remote system interconnected by long-distance AC cables', *Electrical Engineering in Japan*, pp. 19–30, November 2013.

(1) Voltage Control Method (fixed or floating)

Floating V_{ref} control was confirmed to be effective. The voltage on the islands, at the receiving side of the long-distance transmission system, varies significantly with to the time slot and season (load). Fixed V_{ref} control is effective in suppressing small voltage variations, but its operation requires that the reference voltage pattern be determined in advance, and creating this pattern is extremely difficult.

(2) V_f Time Constant T

If V_f responds too quickly when the AC cable circuit is disconnected due to a transmission fault, the reference of the voltage V_f also drops when control of the voltage drop due to disconnection is performed. The voltage reference at reconnection is set to the voltage before the fault to the extent possible, and thus a sufficiently long time to reconnection of 300 s, greater than the time to close the circuit (60 s) after a fault, is set.

Compared to the value of 120 s used in the effective value analysis in Section 13.3, the speed of the response of the existing voltage control equipment after reconnecting the AC cable, i.e., the time until the SVC voltage is restored to the standby operation point after reconnection, is longer. However, since priority is accorded to maintaining the SVC target voltage unchanged from disconnection until reconnection 60 s later, a longer time of 300 s is set.

(3) Dead Time Range $D1$

If the SVC is used only for large voltage variations during a transmission fault or when the cable circuit is open, its usage frequency is low. Thus the dead band values are set low and can be used even for voltage variations during normal operation, as when switching phase-modifying equipment on or off, or changing the taps in the transformer connected to the mainland. The dead band value of $D1 = +/-0.8\%$ is set as the range in which the SVC does not operate in response to voltage variations due to load fluctuations during normal operation.

(4) Slope Reactance

Voltage variations that occur when switching phase-modifying equipment on or off, or during tap control of the transformer connected to the mainland, exceed the dead band of $+/-0.8\%$. Thus if the broken-line characteristics are used for these voltage variations, the response will not be excessive and mild voltage control can be applied. Further, the time until the SVC output returns to the normal value I_{ref} can be reduced. However, in evaluating the actual effect, assuring good voltage variation suppression even at an operating point that exceeds $+/-0.8\%$ was given priority, and the broken-line characteristic was not adopted.

13.10 Verification by the transient analysis

Figure 13.13 shows the results of effective value analysis of the voltage variations on the islands when one circuit in the 66 kV cable connected to the mainland is

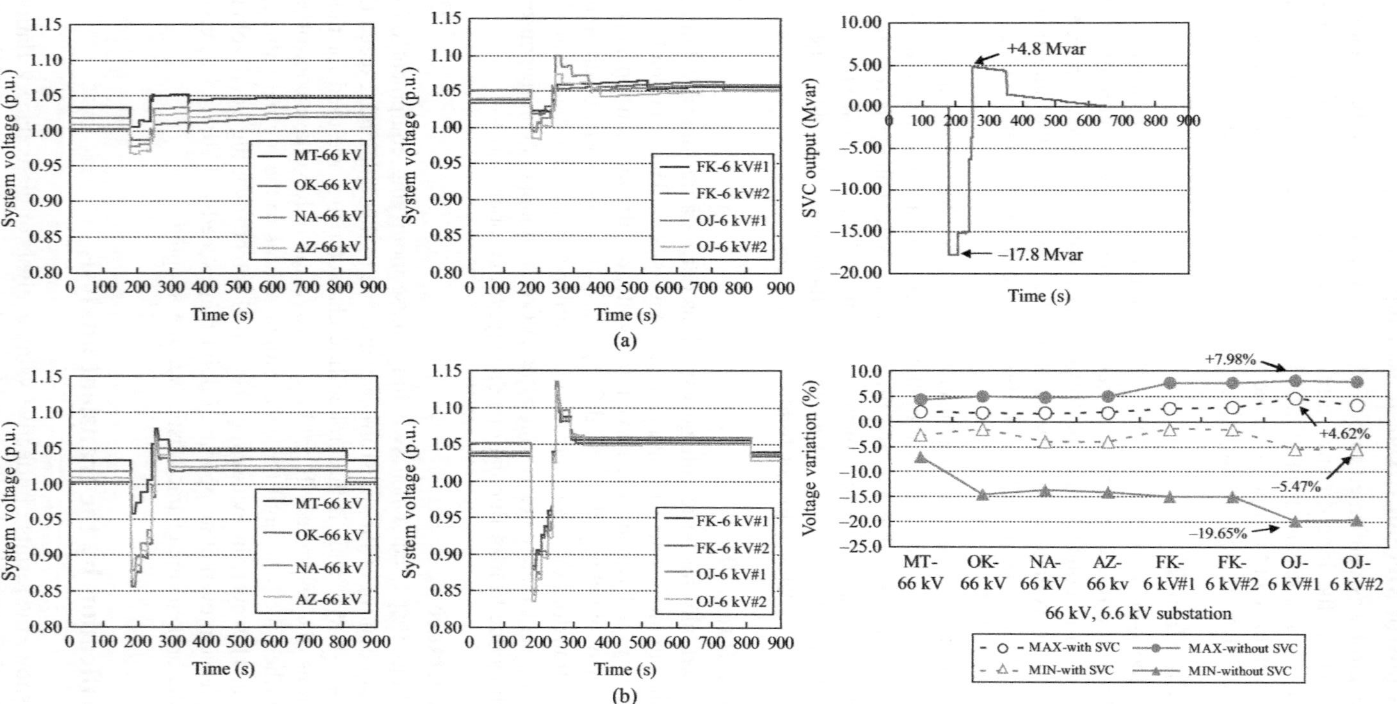

Figure 13.13 *Voltage variation in off/on of a long-distance submarine cable (simulation result): (a) SVC output and voltage variation in Goto islands with SVC; (b) Voltage variation in Goto islands without SVC; (c) Summary of voltage control results*

opened and then reconnected 60 s later, with the SVC AVR function and settings described above taken into consideration. The system model uses heavy load conditions, and the results are for the conditions with the greatest voltage variation. When the SVC is connected in Figure 13.13(a), the voltage variations for the 66 kV, 6 kV systems on the islands are greatly reduced from those when the SVC is not present in Figure 13.13(b). A voltage drop of -19.65% and a voltage rise of $+7.98\%$ without the SVC are reduced to a voltage drop of -5.47% and a voltage rise of $+4.62\%$ when the SVC is used as illustrated in Figure 13.13(c).

The SVC output range extends from a phase lead of 17.8 Mvar to a phase lag of 4.8 Mvar, as shown in Figure 13.13(a). This is within the range of 20 Mvar for the phase lead (capacitive) and 10 Mvar for the phase lag (inductive) of the assumed rated swing range. After SVC mitigates large voltage variations, the SVC output (Mvar) returns to zero (i.e. I_{ref}) gradually, by enhancing the operation of the existing voltage control equipment in the network, as observed after around 250 seconds in Figure 13.13(a).

Figure 13.13(c) shows the voltage variation range for each electric power station. The abbreviations for the stations are based on Figure 13.7. For the SVC connection site, placing the receiving end of the 66 kV cable connected to the mainland close to the targets of reactive power compensation is effective, but due to site conditions and constraints on installation space, the Okuura system switching station (abbreviated as "OK" in the figure) is used.

However, a voltage variation suppression effect similar to that at the receiving end of the cable linked to the mainland is obtained throughout the islands, indicating that the location is suitable for the SVC connection point.

13.11   Verification at the commissioning test

As part of the SVC system commissioning test, a test was performed in which one circuit of the 66 kV AC cable to the mainland was disconnected and reconnected.

Figure 13.14(a) shows the results of measurements during system commissioning test, and Figure 13.14(b) shows the results of transient analysis. The waveform characteristics show good agreement in the two cases. Based on a comparison with Figure 13.14(c), the results of the transient analysis without the SVC confirm the voltage variation suppression effectiveness of the SVC. Figure 13.15 summarizes the voltage variation ranges at each electric power station using the voltage before the test as a reference. As shown in Figure 13.15(a), the measurements and the analysis agree, confirming the validity of the analysis model.

Furthermore, a comparison of the results of analyses with and without the SVC, as shown in Figure 13.15(b), shows that the voltage variation suppression effect was obtained at all of the electric power stations. Because a light load condition was used when performing testing of the actual system, the voltage variation range is small compared to the results shown in Figure 13.13, under the most severe conditions described above. However, the voltage drop of -7.88% and voltage rise

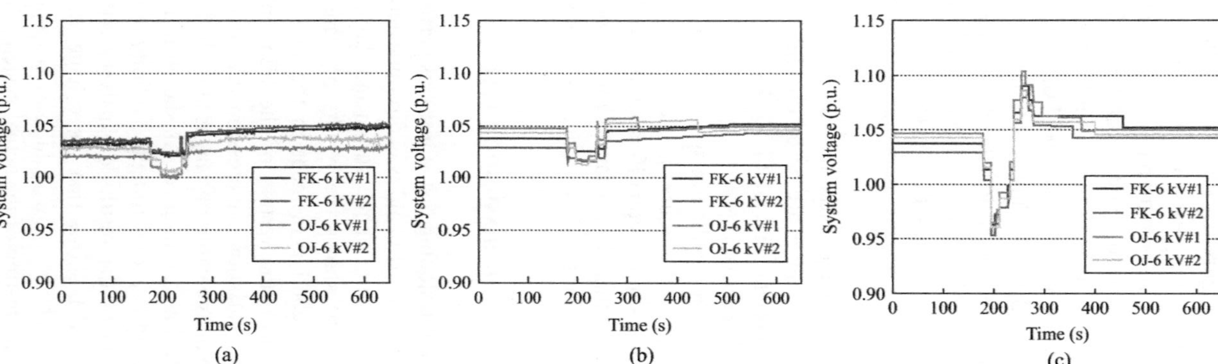

Figure 13.14   Comparison of simulation results and measurement results (voltage charts): (a) Measurement result with SVC; (b) Simulation result with SVC; (c) Simulation result without SVC

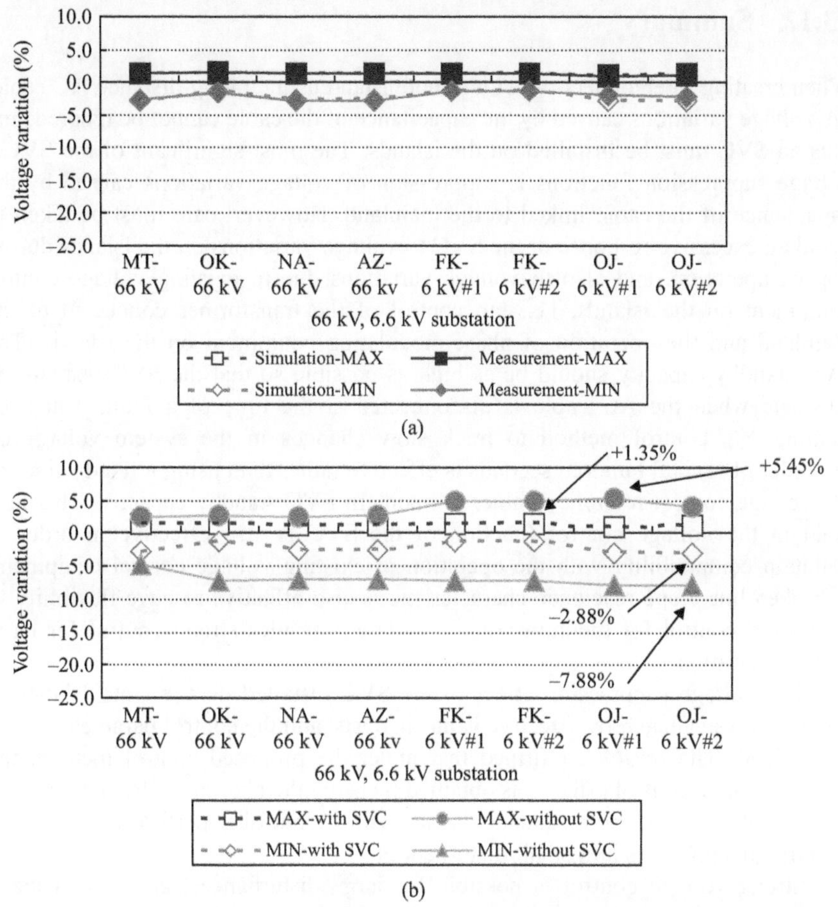

Figure 13.15   *Comparison of simulation results and measurement results (Max./Min. variations): (a) Comparison between simulation result and measurement result; (b) Simulation results with SVC and without SVC*

© 2013 Wiley Periodicals, Inc. Reprinted with permission from Yuji Tamura, Shinji Takasaki, Yasuyuki Miyazaki, Hideo Takeda, Shoichi Irokawa, Kikuo Takagi, Naoto Nagaoka, 'Voltage control of static var compensator for a remote system interconnected by long-distance AC cables', *Electrical Engineering in Japan*, pp. 19–30, November 2013

of +5.45% without the SVC were reduced to a voltage drop of −2.88% and a voltage rise of +1.35% with the SVC.

The test of switching disconnection and reconnection of one circuit in the AC cable connected to the mainland in the absence of the SVC was not performed because the voltage variation was substantial.

13.12   Summary

When creating interconnection with remote island using a long-distance AC cable, the voltage variations caused by the capacitance of the cable cannot be ignored, and thus an SVC must be installed on the islands. The most significant of the SVC's voltage suppression functions is suppression of voltage variations caused by the capacitance of the cable linked to the mainland. However, care must be taken to avoid an excessive response of the SVC to voltage variations on the islands during normal operation, and also to voltage variations due to existing voltage control equipment on the islands, i.e., tap control of the transformer connected to the mainland and the operation of phase-modulating equipment on the islands. The SVC standby capacity should be as high as possible so that the SVC response is adequate when the AC cable is disconnected at the time of a fault. Thus, the floating V_{ref} control method to track slow changes in the system voltage on the order of several hundred seconds is effective rather than using a fixed value for the voltage control reference value, to maintain SVC standby capacity. The dead band in the voltage control function of the SVC is also effective in order to maintain compatibility with the operation of existing voltage control equipment. A broken-line slope reactance characteristic is also effective to have flexibility to have mild control for medium voltage variation besides strong control for large voltage variation.

The voltage suppression effect of the SVC installed on the Goto Islands in Kyushu was evaluated by effective value analysis, and the control parameters were determined. The results confirmed that under the proposed control method, the desired voltage control effect was obtained by using the chosen control parameters. In the broken-line slope reactance characteristics, control is performed separately for large disturbances and small disturbances.

Strong voltage control is possible for large disturbances, and mild voltage control for small disturbances. But giving priority to good voltage suppression even for a small disturbance was judged unnecessary for the SVC on the Goto Islands. Effective value analysis confirmed that for a small disturbance, maintaining a relatively mild voltage suppression effect while fostering a response by the existing voltage control equipment was effective.

Under the most stringent load conditions, with disconnection of one circuit in the AC cable connected to the mainland, and reconnection after 60 s, a voltage drop of -19.65% and a voltage rise of $+7.98\%$ without the SVC were reduced to a voltage drop of -5.47% and a voltage rise of $+4.62\%$ with the SVC. The desired effects were verified by the effective value analysis with the control values that were ultimately chosen.

The results of actual system commissioning test in 2007 have verified the SVC voltage suppression function. The measurements and results of prior simulations agreed well when one circuit in the cable connected to the mainland was disconnected and then reconnected. Thus the validity of the effective value analysis model was verified.

The SVC with the voltage suppression function described in this chapter started operation in 2007, and has contributed to improved power quality on the Goto Islands.

References

[1] Anan F., Ikumi S., Shimada S., Yamamoto K., Hosokawa O., Nishiwaki S. *et al.* "Systemization of Various Special Phenomena in Long Cable Transmission System," 2006 National Convention Record Institute of Electrical Engineers of Japan, No. 7-127 (2006-3) (in Japanese).

[2] Iiyama K., Eguchi T., Shimada S., Hikami T., Hosokawa O., Yamamoto K. *et al.* "A Current Zero-Miss Phenomena and It's Countermeasure in a Cable Transmission System," 2006 National Convention Record, Institute of Electrical Engineers of Japan, No. 7-126 (2006-3) (in Japanese).

[3] Tamura Y. "Static Var Compensator in a System Interconnected with Long Distance AC cables," Ph.D. Thesis, Doshisha University, March, 2012.

[4] Tamura Y., Takasaki S., Miyazaki Y., Takeda H., Irokawa S., Takagi K. *et al.* "A Voltage Control Method of Static Var Compensator for the Remote System Interconnected by Long Distance AC Cables," *IEEJ Trans. PE*, Vol. 131, No. 11, pp. 896–904 (2011) (in Japanese).

[5] Tamura Y., Takasaki S., Miyazaki Y., Takeda H., Irokawa S., Takagi K. *et al.* "Voltage Control of Static Var Compensator for a Remote System Interconnected by Long Distance AC Cables," *EEJ*, Vol. 186, No. 3, 2014.

[6] Tamura Y., Miyazaki Y., Takeda H. "Static Var Compensator for Long-Distance AC Cable Transmission System," Toshiba Review, Vol. 63, No. 8 (2008) (in Japanese).

[7] Anan F., Ikumi S., Shimada S., Nishikawa S., Noro Y., Yokota T. *et al.* "Countermeasures for Substation Equipment Against Special Phenomena in Japan's longest (54 km) 66 kV AC Cable Transmission System," Power Engineering Society General Meeting, 2004, IEEE, pp. 490–495, v61. 1 (2004).

[8] Iiyama K., Eguchi T., Shimada S., Hikami T., Noro Y., Yamamoto K. *et al.* "Countermeasures for Harmonic Resonance in Japan's Longest AC cable Transmission System," 2006 National Convention Record Institute of Electrical Engineers of Japan, No. 7-128 (2006-3) (in Japanese).

[9] "The Investigation Committee on Control Technologies of Power Electronic Equipment for Power System Applications," IEEJ Technical Report, No. 1084 (2007) (in Japanese).

[10] Tamura Y., Kanatake S., Shibayama T., Noro Y., Miyazaki Y., Takeda H. "SVC System Application in Remote Power System Connected via Long Distance AC Cable Transmission," 2008 National Convention Record, Institute of Electrical Engineers of Japan, No. 6-236 (2008-3) (in Japanese).

[11] CIGRE WG 38-01, Task Force No. 2 on SVC: "Static Var Compensators," CIGRE Technical Brochure 25 (1986).

[12] CIGRE WG 38-05, Task Force No. 4 on SVC: "Analysis and Optimization of SVC Use in Transmission System," CIGRE Technical Brochure 77 (1993).

[13] Electric Technology Research Association, Subcommittee on Electric Power System Operation Technology: Electric Power Systems Stabilization Operation Technology, Vol. 47, No. 1 (1991).

[14] Irokawa S., Anderson L., Pritchard D., Buckley N. "A Coordination Control between SVC and Shunt Capacitor for Windfarm," CIGRE 2008, B4-307 (2008-8).

Chapter 14
Transients on grounding systems
*S. Visacro**

14.1 Introduction: power system transients and grounding

Grounding plays a relevant role in the operation of electrical power systems. Notably, when electrical systems are subjected to transient phenomena resulting from internal or external events, the response of these systems might be strongly influenced by the behavior of their ground terminations [1].

In order to assess how the ground terminations influence this response, one must first understand the transient response of the grounding electrodes, which is quite different from that observed when electrodes are subjected to slow phenomena, such as short circuits. This chapter addresses this issue throughout five sections, by means of a conceptual approach illustrated with the analysis of a specific application.

After this introduction, section 14.2 explains the basic concepts of grounding, based on a simplified general approach. Section 14.3 addresses the transient behavior of grounding electrodes. The parameters that describe the electrode response when subjected to transient currents are introduced. Based on experimental evidences and on theoretical considerations this behavior is explored and the factors that influence it are discussed.

Results of numerical simulation of the transient response of electrodes are presented in section 14.4 and this response is discussed based on sensitivity analyses developed from such results.

Finally, section 14.5 presents a case example, which illustrates the influence of grounding electrodes on the lightning performance of transmission lines.

Simplified approaches are used to develop the concepts presented in this chapter, based on the representation of grounding systems by means of distributed circuits. However, the results used to provide support for the discussions about the transient response of grounding electrodes and corresponding impacts arise from experimental tests and from numerical simulations, all using the Hybrid Electromagnetic Model (HEM), as discussed in section 14.4.

---

*Professor, Federal University of Minas Gerais, Brazil

14.2   Basic considerations on grounding systems

Any electric unbalance of grounded electrical systems results in an impression of currents on their ground terminations, composed by their grounding systems.

As discussed in [2, 3], a grounding system consists of three components: (i) the metallic conductors that drive the current to the electrodes, (ii) the metallic electrodes buried in the soil, and (iii) the earth surrounding the electrodes. The last one is the most relevant component. Figure 14.1 illustrates such components.

The arrangement and dimension of electrodes depend very much on the application. For instance, this arrangement can vary from a simple buried vertical rod, connected to the transformer of a distribution line, to horizontal electrodes buried a fraction of a meter deep in the soil composing the counterpoise wires of transmission-line tower footings, or the typical grounding grids of power system substations. In fact, the electrode consists of any buried metallic body responsible for dispersing current to the soil. The area covered by such systems may vary widely, from a few to thousands of square meters, depending on the application and on the soil resistivity value.

The specific characteristic of the soil surrounding the electrodes has great influence on their response and, in this respect, the macroscopic electromagnetic parameters of the soil, mainly its resistivity, can vary substantially, depending on the local properties of this medium.

The function of a grounding system also depends very much on the application, but, in most cases, this function is geared toward ensuring safety or an expected performance of the electrical system when subjected to electrical unbalances, short circuits, or transients [3]. Specifically in transient-related conditions, this function consists typically of providing a low-impedance path for the flow of currents resulting from corresponding electrical stresses toward the soil, in addition to ensuring a smooth distribution of the electric potentials on the ground surface in the region where the electrodes are buried, due to safety concerns [4].

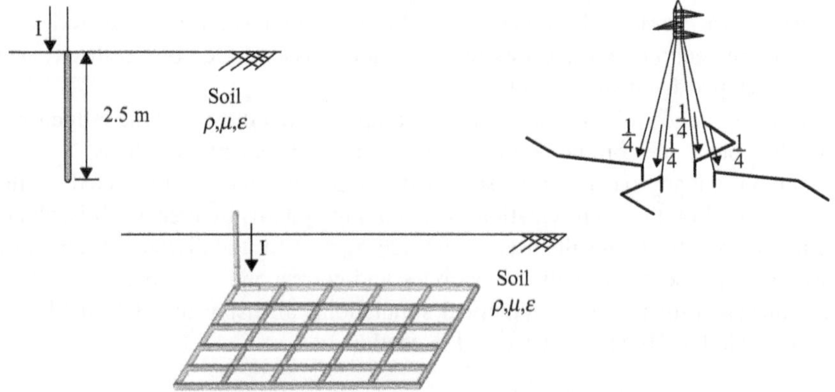

Figure 14.1   Illustration of grounding system components

The simplest way to express the response of grounding systems consists of the *grounding resistance* R_T. To establish the concept of this parameter, a reference to the so-called "remote earth" is required. It corresponds to a place in the ground, very distant from the electrodes, where the electric potential is null. When an electric current is impressed on electrodes, it flows toward the remote earth. The grounding resistance is defined as the ratio between the potential rise of the electrodes in relation to the remote earth V_T and the impressed current I_T (14.1) [5]. Frequently, this V_T is referred to as grounding potential rise *GPR*.

$$R_T = \frac{V_T}{I_T} \tag{14.1}$$

This definition fits only for slow varying currents, for which the reactive and propagation effects are negligible, as discussed in section 14.3.

Several factors influence the grounding-resistance value, though they can all be summarized in two main parameters: soil resistivity ρ and a geometrical factor K, as in the following equation.

$$R_T = K \cdot \rho \tag{14.2}$$

Equation (14.2) shows that the grounding resistance is proportional to the electrical resistivity of the soil. In soils, which are stratified in layers, this parameter becomes an apparent resistivity. Factor K is determined from the geometry and dimension of electrodes. Figure 14.2 illustrates this factor for some specific electrodes [3, 5].

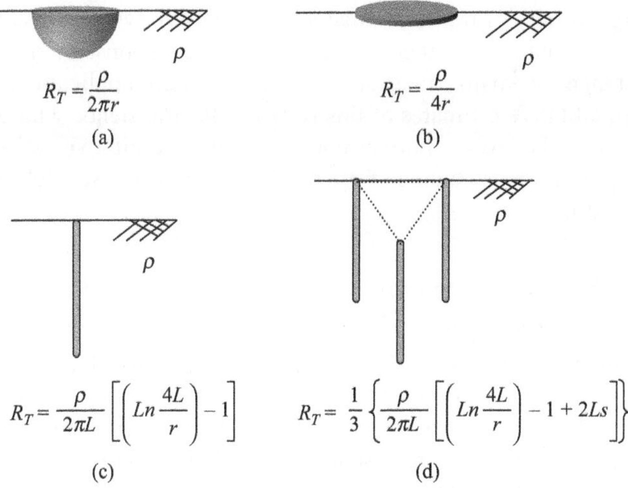

Figure 14.2   Analytical expressions for the grounding resistance of some electrode arrangements denoting the geometrical factor K: (a) metallic hemisphere of radius r, (b) vertical rod of length L and radius r, (c) metallic disc of radius r laying on soil surface, (d) three connected vertical rods buried at the vertices of a triangle of sides s

Soil resistivity depends on a large number of factors, though the most influential are soil type and the content of water and mineral salts. In general, the increase of water content leads soil resistivity to decrease. This decrease is very pronounced when the soil is primarily very dry (percentage of water content in weight w lower than 2%) and tends to saturate after w is higher than 18%. In natural conditions, at the usual depth in which electrodes are buried (below 0.4 m), rarely do soils present water content lower than 4% and higher than 15%. Soil resistivity also decreases as salt concentration increases. The resistivity varies a lot with the soil type and with its relative porosity per volume, which is closely related to the soil's capability to retain water and salts.

In most cases, resistivity value ranges from 50 to 2,000 Ωm in natural soil conditions, though it is not unusual to find soils with resistivity values as low as 10 Ωm (for instance, very wet and organically rich soils close to swamps) or around 4 kΩm, for instance, very dry rocky soils. In extreme cases, resistivity values ranging from 6 to 20 kΩm can be found, notably for sandy salt-deprived soils.

14.3   The response of grounding electrodes subjected to transients currents

14.3.1   Introduction

The study of the transient response of electrodes can be developed following both time-domain and frequency-domain approaches. Due to didactic concerns, this section begins with a frequency-domain approach.

The large variety of transient phenomena in power systems makes for a very wide frequency-component range. Understanding the response of electrodes in the frequency range of major interest for power-system applications allows for developing qualitative estimates of this response for transients, whose frequency-component range is known. Furthermore, accurate quantitative estimates of this transient response can be obtained from the frequency response of electrodes using Fourier transform.

14.3.2   Behavior of grounding electrodes subjected to harmonic currents

Any termination to earth presents resistive, capacitive, and inductive effects. When subjected to harmonic currents above a few kilohertz, two main assumptions ordinarily valid when electrodes are subjected to slow varying currents are no longer valid: the possibility to represent the grounding system by means of a simple resistance R_T and the assumption of equipotentiality along the electrodes. In this frequency range, reactive effects become relevant and the electric potential can vary pronouncedly along the electrodes.

It is possible to denote this aspect by means of a simplified approach based on the representation of the grounding system by an equivalent circuit, as illustrated in Figure 14.3 for a vertical rod buried in the soil. The rod can be divided into a set of

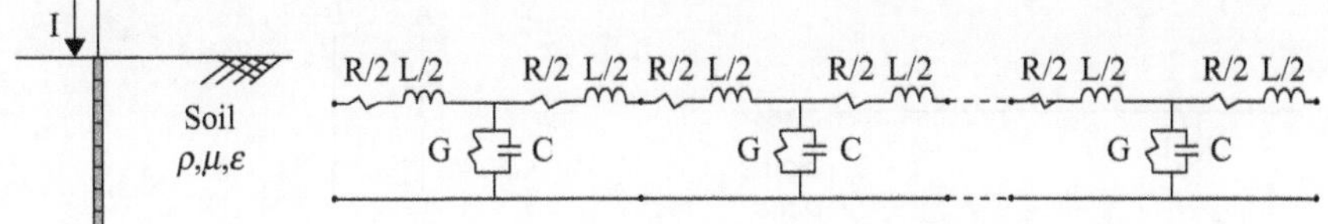

Figure 14.3   Representation of a grounding system (rod) by an equivalent electric circuit

connected short-length elements. Each element, along with the surrounding soil, is represented by an element of the circuit indicated to the right of Figure 14.3.

In this circuit, parameters R and L correspond to the resistance and inductance of the longitudinal impedance of the element. They are associated respectively with the internal losses in the metallic electrode and with the energy stored in the magnetic field yield by the flow of current along it. Parameters G and C correspond to the conductance and capacitance of the transversal admittance of the element in relation to the remote earth, which are associated, respectively, with the losses and electric energy stored in the soil, while the current is spread from the element to the remote earth [4].

The upper bar in the circuit represents the electrode where the longitudinal impedance of the elements is distributed. The inferior bar corresponds to the remote earth where the electric potential has a null value. The leakage current flowing into the soil causes a potential rise of both the electrode and the surrounding earth in relation to the remote earth. The flow of the longitudinal current generates a voltage drop along the electrode. The voltage source impresses a current from the electrode (point A) to the remote earth represented by point B. The magnitude of this voltage equals the potential rise at the current injection point yielded by the total current dispersed to the soil toward the remote earth [3].

The circuit of Figure 14.4 shows an improved representation of the grounding system on the left, since it complements the previous representation including the electromagnetic coupling effect among the elements the system was divided into [3].

The response of the grounding system to an impressed current can be determined in terms of the grounding potential rise calculated as the solution of the complex circuit consisting of a series of elementary circuits similar to that of Figure 14.1, connected according to the topology of electrodes, and taking into account mutual effects among all the grounding elements. This response can be also expressed by the harmonic impedance $Z(\omega)$ seen from the point where current is impressed on the electrodes. This impedance is given by the ratio between the potential developed at the electrode V_T in relation to a remote earth and the impressed current I_T. $Z(\omega)$ is also known as "complex grounding impedance" [4].

$$Z(\omega) = \frac{V_T(\omega)}{I_T(\omega)} \tag{14.3}$$

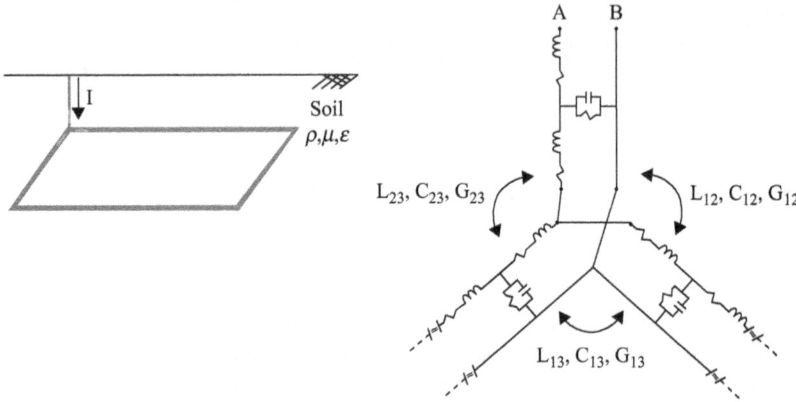

Figure 14.4   Representation of a grounding system by an equivalent circuit, including mutual effects among the elements of the system

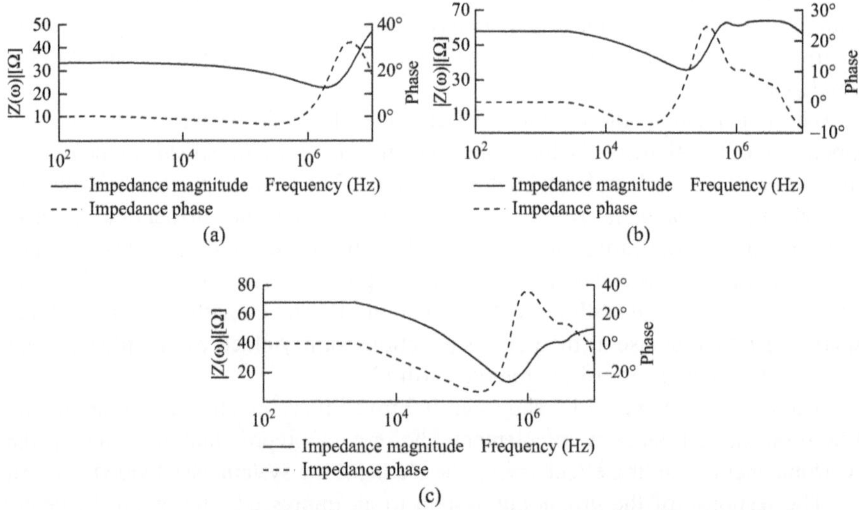

Figure 14.5   Harmonic grounding impedance within the frequency range of interest in power-system applications: (a) 3-m-long vertical rod buried in a 100-Ωm soil; (b) 30-m-long horizontal electrode buried 0.5 m deep in a 1,000-Ωm soil; (c) 400-m^2 square grid buried 0.5 m deep in a 3-kΩm soil, comprising 25 square meshes of 16 m^2. Electrode radius: 0.5 cm. Variation of the electric resistivity and permittivity of soil given by the Visacro-Alipio expressions [6]

Note that, as a solution to a circuit comprising reactive elements, the impedance varies with frequency. Figure 14.5 illustrates how the impedance of different arrangements of electrodes buried in different soils behaves as frequency increases.

Note that, in all cases, at the low-frequency range the harmonic impedance is basically constant and equivalent to a resistance (phase angle approximately null). In this range, reactive effects are practically negligible: the inductance yields no voltage drop along the electrode and, basically, no capacitive currents flow into the soil. In addition, the internal resistance yields no voltage drop in this range, as skin effect is negligible and for the typical conductivity of electrodes, the resistance per unit length is too low. As a result, the corresponding reactances and resistances are dropped from the equivalent circuit, which becomes simply a set of conductances, including mutual conductive effects among them [7].

This denotes that, indeed, the grounding resistance R_T consists simply of a particular case of the harmonic impedance $Z(\omega)$ and this induces us to call it low-frequency grounding resistance [4]. Furthermore, at the low-frequency range, all the electrodes have the same electric potential.

While frequency rises, above a few kilohertz, the capacitive currents in the soil become relevant (see the negative impedance phase, whose magnitude rises with increasing frequency) and this results in decreasing harmonic impedance. Since inductive effects are not important in this range, the effect of the parallel G-C circuits prevails, and capacitive currents flowing in the soil contribute to reduce the impedance against that observed at the low-frequency range, within which only conductive currents flow into the soil for the same amplitude of applied voltage.

As frequency increases further, inductive effects become relevant. First, their interaction with capacitive effects results in additional decrease of impedance. Note this decrease while the negative phase begins to diminish, as a result of the inductive effects. At the limit of this process, around the frequency where the phase crosses the null line, the harmonic impedance reaches the lowest value.

Then, the inductive effect begins to prevail (positive impedance phase) and the impedance increases continuously to a few megahertz. All through this range, the voltage drop along the electrodes becomes significant and the equipotential assumption is no longer valid for electrodes.

Within determined limits, the equivalent circuit can also be used to consider the propagation effects in the soil. In fact, a sequence of distributed circuit elements such as that corresponding to a long electrode in Figure 14.2 can be used to represent the electrode as a uniform transmission line embedded in the soil, if it is assumed that the circuit parameters R, L, G, and C correspond to per unit length parameters [4].

The current and voltage propagating along the electrodes are indicated in (14.4) and (14.5) [8].

$$I(\omega) = I_0 e^{-\alpha z} \cos(\omega t - \beta \cdot z) \tag{14.4}$$

$$V(\omega) = Z \cdot I_0 e^{-\alpha z} \cos(\omega t - \beta \cdot z - \theta_z) \tag{14.5}$$

The propagation constant γ and associated attenuation and phase constants, respectively, α and β, can be determined from the square root of the product of the longitudinal impedance by the transversal admittance per unit length (14.5).

$$\gamma = \alpha + j\beta = \sqrt{(R + j\omega L)(G + j\omega C)} \tag{14.6}$$

The magnitude Z and the phase θ_Z of the intrinsic impedance of the line are determined from the square root of the ratio between the longitudinal impedance and the transversal admittance per unit length.

Note that, if internal losses in the electrode (associated with R) and the losses in the soil (associated with G) are negligible, no attenuation would be observed in the current and voltage waves propagating along the electrode (null value for α).

At the high-frequency range, in which propagation effects are relevant, in representative conditions of soil R frequently has a negligible value (compared with that of the inductive reactance ωL), and the conductance G has always very significant values compared with that of the capacitive susceptance ωC. Thus, from (14.6) it becomes evident that, in these conditions, G is effectively the parameter responsible for attenuation along the electrode. Since G is basically proportional to the apparent soil conductivity σ, it is clear that attenuation increases with increasing conductivity or decreasing soil resistivity.

The most important propagation effect in grounding electrodes consists of the strong attenuation the current and voltage waves are subjected to along the electrode, which becomes more intense with decreasing soil resistivity, as explained above.

To account for this effect, it is common to refer to the space constant of the electrode C_S, which corresponds to the length of electrode required for the waves to attenuate to a certain amplitude: 0.368 of the wave amplitude at the point the current is impressed on the electrode. This value derives from the term "$e^{-\alpha z}$" multiplying the current and voltage waves, when it becomes "e^{-1}". If (14.4)–(14.6) are referred to, this would correspond to condition (14.7).

$$I_0\, e^{-\alpha z} = I_0\, e^{-1} = I_0\, e^{-\alpha \cdot \frac{1}{\alpha}} \qquad \therefore \quad C_S = z = \frac{1}{\alpha} \tag{14.7}$$

Note that the attenuation constant is the inverse of the space constant. The shorter is this constant, the higher is the wave attenuation.

Figure 14.6 illustrates how the attenuation of the current and voltage signals propagating along an electrode buried in soils of different resistivity values behaves as a function of frequency, considering different values of soil resistivity.

Note the significant decrease of C_S with decreasing soil resistivity and increasing frequency. This means that attenuation significantly increases not only with the decrease of soil resistivity but also with the frequency increase.

One frequency-related effect that affects strongly the transient response of grounding electrodes is the frequency dependence of soil resistivity and permittivity, as discussed in the next subsection.

14.3.3   The frequency dependence of soil resistivity and permittivity

Though the magnetic permeability of soil has a constant value similar to the permeability of air, experimental results have shown that the soil resistivity ρ and permittivity ϵ are strongly frequency dependent [9–15].

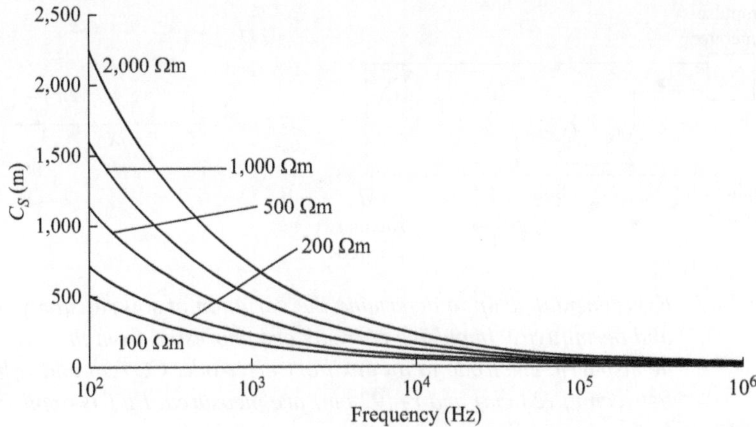

Figure 14.6   Variation of the space constant of electrodes buried in different soils as a function of frequency and resistivity (relative permittivity ε_r equal to 20)

Until recently, this effect had been disregarded due to the lack of reliable general formulations expressing this frequency dependence. In a conservative approach, soil resistivity had been assumed to be the value measured by using conventional measuring instruments, which employ low-frequency signals, and the relative permittivity of soil had been assumed to vary from 4 to 81, according to the soil humidity [4].

Recently, Visacro et al. [6, 14] proved the relevance of this effect in the response of grounding electrodes and developed a new experimental methodology to determine this frequency dependence in the range between 100 Hz and 4 MHz, based on measurements performed in field conditions [6], as illustrated in Figure 14.7.

Typical results obtained from the application of this methodology are shown in the curves of Figure 14.8, where both parameters, soil resistivity and permittivity, show strong variation within the frequency interval between 100 Hz and 4 MHz, both decreasing with increasing frequency.

Similar results obtained for a large number of tested soils show that the resistivity decrease is very sensitive to the value of the low-frequency resistivity and is more pronounced with increasing low-frequency resistivity. Soil permittivity is less sensitive to the variation of this resistivity.

Comparison of experimental and simulated results proved the consistency of this methodology. The measured potential rises of different arrangements of electrodes buried in several soils developed in response to the impression of impulsive currents were compared with those simulated using the frequency dependent resistivity and permittivity values provided by the application of the methodology in the same soils, as considered in Figure 14.9 [12, 13, 16].

Note that the simulated and experimental GPRs practically match, while the simulated GPR under the assumption of constant values for soil parameters leads to

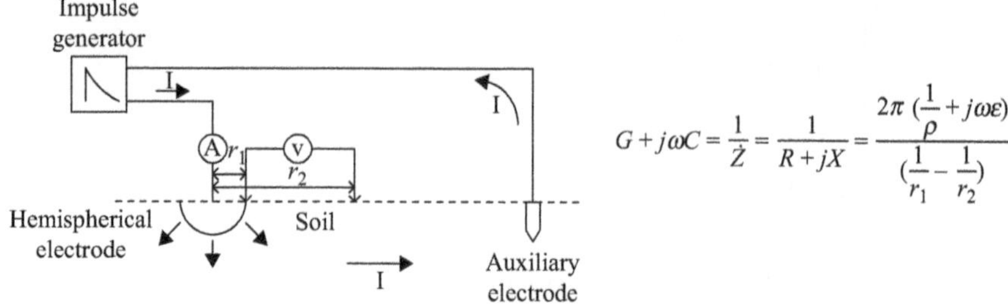

$$G + j\omega C = \frac{1}{\dot{Z}} = \frac{1}{R + jX} = \frac{2\pi\,(\frac{1}{\rho} + j\omega\varepsilon)}{(\frac{1}{r_1} - \frac{1}{r_2})}$$

Figure 14.7   Experimental setup to determine the variation of soil resistivity and permittivity. Impulsive currents are impressed from the hemispheric electrode to an auxiliary electrode. Current and voltage between r_1 (23 cm) and r_2 (92 cm) are measured. FFT is applied to both waves to determine the components of voltage and current for each frequency in the range from 100 Hz to 4 MHz and the corresponding complex harmonic impedances Z are calculated. Considering the hemispherical symmetry of the problem, the expression above can be used to determine the admittance at each frequency and, then, the corresponding resistivity and permittivity [14]

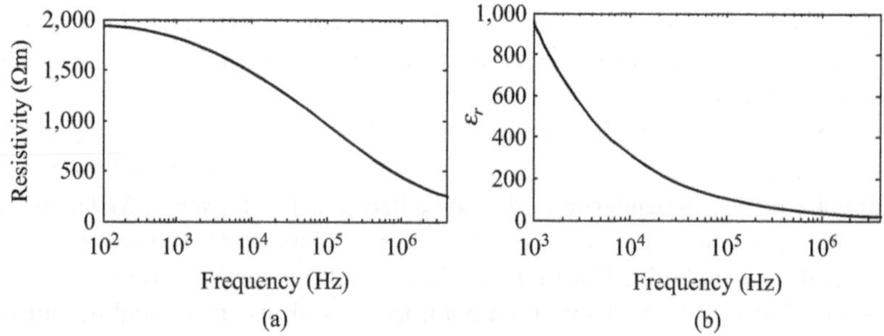

Figure 14.8   Measured variation of soil parameters in the frequency range of 100 Hz to 4 MHz for soil of about 2,000-Ωm low-frequency resistivity: (a) resistivity; (b) relative permittivity

very significant errors, with the amplitude of the grounding potential rise much higher than the measured one.

After the methodology consistency was demonstrated, it was applied to a large number of different soils in natural conditions. In general, the impact of this frequency dependence is decreasing the grounding impedance in the high-frequency range against that obtained under the assumption of constant soil parameters. This decrease becomes more significant with increasing low-frequency soil resistivity.

Based on the experimental curves determined for soil resistivity and permittivity variation in frequency spectrum, the Visacro-Alipio expressions (14.8) and

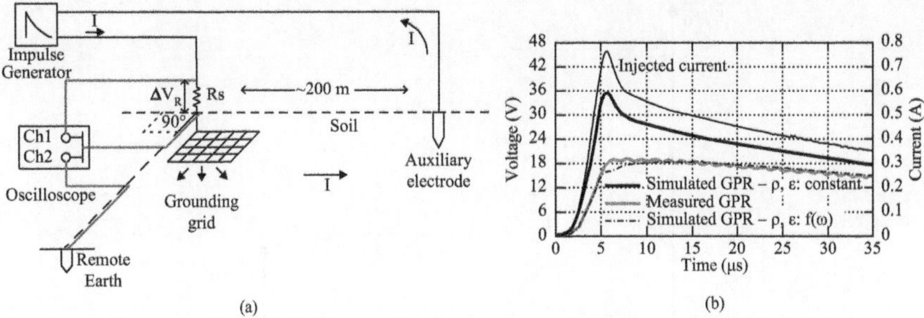

Figure 14.9   *Experimental setup (a) along with simulated and measured grounding potential rise (b) of a 20 × 16-m grid with 20 4 × 4-m meshes buried 0.5 m deep in a 2,000-Ωm low- frequency-resistivity soil (electrode radius = 0.5 cm) due to the impression of an impulsive current with 4 μs front time at the grid corner*

(14.9) were derived following a conservative approach to predict the frequency dependence of both parameters, as described in [6].

$$\rho = \rho_0 \left\{ 1 + \left[1.2 \cdot 10^{-6} \cdot \rho_0^{0.73} \right] \cdot \left[(f - 100)^{0.65} \right] \right\}^{-1} \tag{14.8}$$

$$\varepsilon_r = 7.6 \cdot 10^3 \, f^{-0.4} + 1.3 \tag{14.9}$$

ρ and ε_r are the soil resistivity and relative permittivity at frequency f in Hz. ρ_0 is the soil resistivity at 100 Hz. Expressions (14.8) and (14.9) are applicable to the 100-Hz to 4-MHz range. The soil resistivity is basically constant below 100 Hz.

Even in a very simplified way, it is worth commenting on the backgrounds of the frequency dependence of soil parameters. This property comes out from the behavior of the complex permittivity of soils (14.10), as discussed in details in [6, 15]:

$$\varepsilon = \varepsilon' - j\varepsilon'' \tag{14.10}$$

In a macroscopic interpretation, the conductivity of a material is associated with heating losses caused by the impressed electric field and permittivity is associated with the polarizability of the material, responsible for the accumulation of energy in the electric field impressed to it. While conductivity is a real number, permittivity is a complex number, as indicated by expression (14.10). The real component of permittivity ε' measures the pure polarizability of the material, while the imaginary component ε'' measures the losses occurred during the polarization processes [6, 12, 15].

Figure 14.10 displays the qualitative relative variation of the permittivity components of soil permittivity with increasing frequency. As explained in [15], this variation is different from the decrease by steps of ε' observed in homogeneous materials, corresponding to the exclusion of the contribution of polarization processes of larger inertia while the frequency increases. In soils, a continuous and

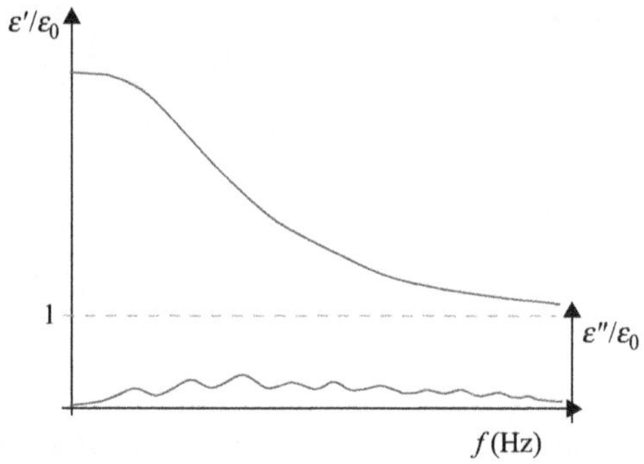

Figure 14.10   *Expected variation of the permittivity components in*
real soils

smooth decrease of ε' is observed, due to an almost continuous distribution of relaxation frequencies. Additionally, the discrete peaks of ε'' observed in homogeneous materials during the transition of ε' by steps are replaced by a slow varying value of ε''. Such peaks are associated to the maximum losses during the exclusion of each polarization process while frequency increases.

Considering this complex permittivity in the Maxwell equation expressed in frequency domain yields:

$$\vec{\nabla} \times \vec{H} = \sigma_0 \vec{E} + j\omega\varepsilon\vec{E} \tag{14.11}$$

$$\vec{\nabla} \times \vec{H} = \sigma_0 \vec{E} + j\omega(\varepsilon' - j\omega\varepsilon'')\vec{E} = (\sigma_0 + \omega\varepsilon'')\vec{E} + j\omega\varepsilon'\vec{E} \tag{14.12}$$

$$\vec{\nabla} \times \vec{H} = \sigma\vec{E} + j\omega\varepsilon'\vec{E}, \quad \sigma = \sigma_0 + \omega\varepsilon'' \tag{14.13}$$

Equation (14.13) expresses clearly that the effective conductivity σ comprises a constant component, corresponding to the low-frequency conductivity σ_0 (associated to the ability to transport electric charges), and a frequency-dependent component $\omega\varepsilon''$ (associated to losses in the electric polarization processes). This component is responsible for increasing σ (or decreasing soil resistivity ρ) with increasing frequency [6, 12, 15].

14.3.4   *Behavior of grounding electrodes subjected to impulsive currents*

In addition to providing the means to analyze the transient response of electrodes in a qualitative perspective, in certain transient analyses it can be reasonable to represent the grounding behavior in a simplified way by the magnitude of the grounding impedance at a specific frequency within the range of the most representative frequencies of the transient under analysis. Nevertheless, determining the accurate transient response of the electrodes by frequency-domain approaches requires considering this response in a wide range of frequency components to allow calculating the time-domain response using Fourier transform.

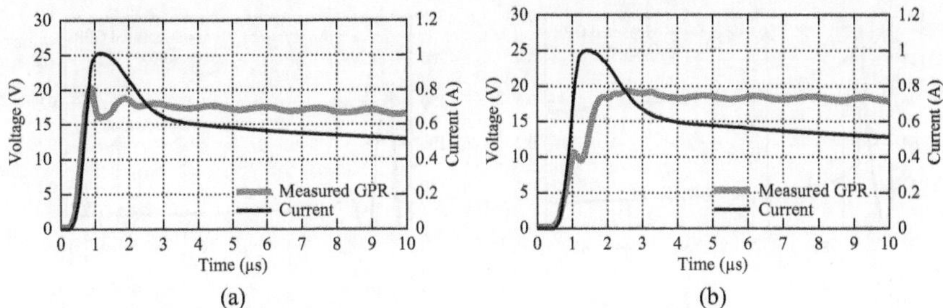

Figure 14.11   Measured GPRs of a 20 × 16-m grid, comprising 20 meshes of 4 × 4 m, buried in a 2,000-Ωm low-frequency resistivity soil and subjected to impulsive 0.7-µs front-time currents: current impressed at the grid corner (a) and at the grid center (b)

Due to the large variety of transient occurrences in power systems, frequency-domain approaches associated with the application of Fourier transform consist of a powerful tool to determine the transient response of electrodes. This variety makes it very difficult to address the time response of grounding electrodes in general approaches. Nevertheless, there is a class of relevant transient phenomena, consisting of impulsive currents and, notably, those related to lightning events, which it worths to address.

As mentioned, in general the primary response of electrodes subjected to impulsive currents is expressed by the grounding potential rise GPR. The experimental results of Figure 14.11 illustrate this response for a grounding grid subjected to the impression of impulsive currents with the same front time at the grid corner and center [16–18].

Note the difference on the GPR curve according to the position of current impression on the grid.

In engineering applications, this response is commonly expressed in a simplified way by means of the so-called *impulse impedance* Z_P, defined as the ratio between the peak values of the grounding potential rise V_P and impressed current I_P (14.14).

$$Z_P = V_P/I_P \tag{14.14}$$

Figure 14.12 illustrates the concept of this impedance, referring to experimental GPR curves obtained for a horizontal electrode subjected to different impulsive current waves [19].

Note that the waveforms of current and voltage are relatively similar in each case, though their peaks are not simultaneous due to reactive and propagation effects in soil, notably for the wave with the shortest front time. Furthermore, quite different values of Z_P are found for the same grounding electrode buried in the same soil, due to the different impressed current waves. This is explained by the difference in the frequency content of the current waves in the high-frequency range, derived from the different front times. The wave with shortest front time,

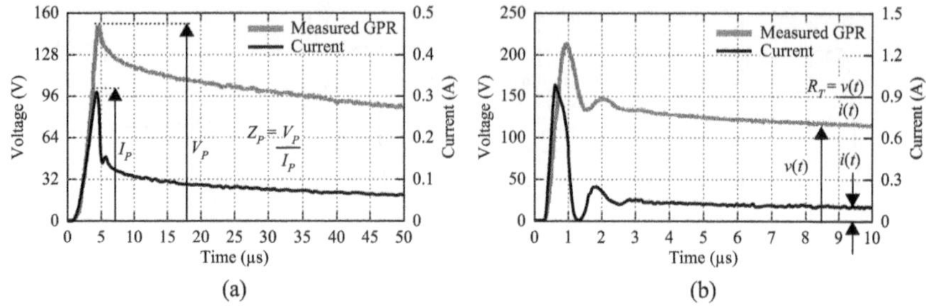

Figure 14.12   *Measured GPRs of a 6-m-long horizontal electrode buried 0.5 m
deep in a 6.9-kΩm low-resistivity soil. Front time of current
waves and calculated Z_P: (a) 4 μs and 477 Ω and (b) 0.4 μs and
212 Ω. $R_{LF} = 1,536$ Ω*

whose superior frequency components have higher values, shows significantly
lower impulse impedance (212 Ω), about the half-value of that of the long front-
time wave (477 Ω). This is attributed to both the effect of significant capacitive
currents flowing in this high-resistivity soil and the pronounced decrease of soil
resistivity due to the frequency dependence effect. Nevertheless, in both cases the
same value of 1,536 Ω is obtained for the low-frequency resistance R_T, as the ratio of
instantaneous values of voltage $v(t)$ and current $i(t)$ at the wave tails $[R_T = v(t)/i(t)]$.
It is worth commenting that, though the impulse impedance is much lower than R_T in
both cases, this result is specific for the condition of very high resistivity of soil and
short electrode. In low-resistivity soils, long electrodes commonly show the opposite
result ($Z_P \gg R_T$), as discussed in [17, 20].

In spite of the nonsimultaneous peaks of current and voltage, the impulse
impedance is very useful when expedite estimates of GPR are required for some
defined transient occurrences, such as in lightning protection applications. In this
case, it is of interest to learn the grounding potential rise specifically for the
impression of representative current waves of first and subsequent return-strokes on
electrodes. Considering that Z_P shows specific values for the first-stroke and for the
subsequent-stroke currents, their GPRs can be promptly obtained from the product
of their assumed peak-current by the corresponding value of Z_P [2, 4].

One fundamental aspect in the analysis of the response of grounding electrode
subjected to impulsive currents is the attenuation the current and voltage waves
propagating along the buried electrodes are subjected to. Because of attenuation,
the current that is dispersed to the ground along the electrode presents a nonuniform
distribution, meaning that the linear current density (A/m) decreases along the
electrode. While the current and voltage waves propagates, the loss of energy
mainly associated with the electric field in the soil yields attenuation of their
amplitude and also reduction of their front steepness, as illustrated in Figure 14.13.
This last effect is derived from the more intense attenuation the higher frequency
components associated to the wave front, as discussed in section 14.3.2 [4].

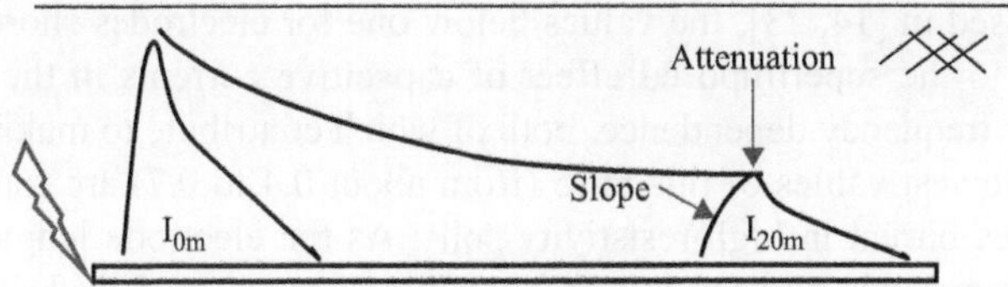

Figure 14.13   Attenuation of the impulsive current wave propagating along the electrodes attributed to losses in the soil

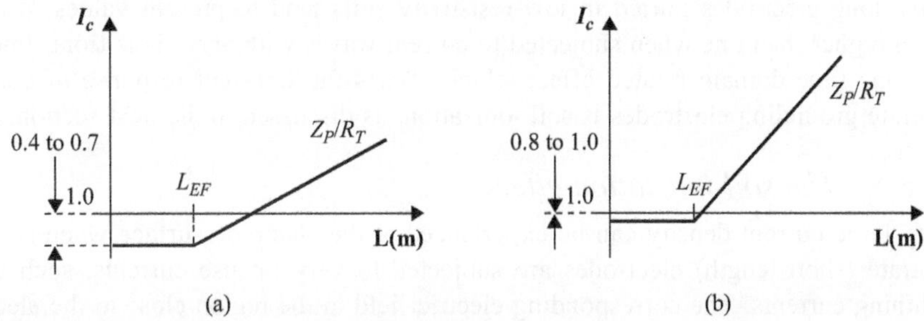

Figure 14.14   Typical profile of the impulse coefficient I_C as a function of electrode length in a high (a) and a low (b) soil resistivity, considering typical lightning current waves

The parameter used to quantify the attenuation in time-domain analysis is the *Effective Length of Electrode L_{EF}*, the counterpart of the space constant in the frequency-domain analysis. As discussed in [4, 21, 22], L_{EF} corresponds to a limiting length. Increasing the electrode length further does not yield reduction of the grounding impedance value, considering that for this length high-frequency components of current associated with the wave front are strongly attenuated and present negligible amplitude.

As commented in section 14.4.3, attenuation increases with soil conductivity and frequency rise, considering that both parameters are responsible for increasing ground losses. Thus, the value of L_{EF} depends on the front time of the impulsive current, as this time defines the superior frequency content of the wave. The decrease of both, front time and soil resistivity, leads to shorter values of L_{EF} [4].

With respect to grids, it is usual to refer to an effective area as an extension of the concept of the effective length of electrodes. This specific issue is discussed in [16].

A very useful parameter in lightning protection application is the *Impulse Coefficient I_C*, given by the ratio of impulse impedance and the low-frequency resistance (14.15).

$$I_C = Z_P/R_T \tag{14.15}$$

Figure 14.14 depicts the typical profile of I_C as a function of the electrode length and denotes that this coefficient varies from a value below one for short electrodes to values higher than it for electrodes longer than L_{EF} [14, 18].

As discussed in [14, 23], the values below one for electrodes shorter than L_{EF} are attributed to the superimposed effect of capacitive currents in the soil and of soil-resistivity frequency dependence, both of which contribute to making Z_P lower than R_T. The lowest values of this ratio (from about 0.4 to 0.7) are found for very short electrodes buried in high-resistivity soils. As the electrode length increases, the impulse impedance and the low-frequency resistance tend to decrease. However, for electrodes longer than L_{EF}, while the low-frequency resistance continuously decreases by increasing electrode length, the value of Z_P remains constant, leading to increasing values of I_C above one, as discussed in [23–26]. Thus, long electrodes buried in low-resistivity soils tend to present values of I_C much higher than one when subjected to current waves with very short front time.

One time-domain related effect which affects the transient response of concentrate grounding electrodes is soil ionization, as discussed in the next section.

14.3.5   The soil ionization effect

Very large current density can be experienced at the electrode surface when concentrate (short-length) electrodes are subjected to very intense currents, such as lightning currents. The corresponding electric field in the region close to the electrode surface can exceed a critical limit, above which an ionization process takes place in the soil leading to electrical discharges in this medium. The so-called critical electric field E_0 ranges from 0.2 to 1.7 MV/m, depending on soil resistivity and humidity [27, 28].

The overall effect of soil ionization is an equivalent increase of the electrode dissipation area and a consequent reduction of the impedance Z_P [4, 29, 30], as illustrated in Figure 14.15.

The low-impulsive voltage (50 kV) applied to a short electrode buried in a low-resistivity soil yielded a current with a peak value about 2 A, corresponding to an impulse impedance of about 25 Ω. As the voltage was increased to 5 kV (100-fold times), ionization process took place and current was increased to about 410 A (almost 200-fold times), leading the impedance to decrease to about the half-value.

Two main complexities are involved when this nonlinear effect is taken into account: defining the critical electric-field value, corresponding to the onset of the ionization process, and considering the nonlinear distribution of the effect along the

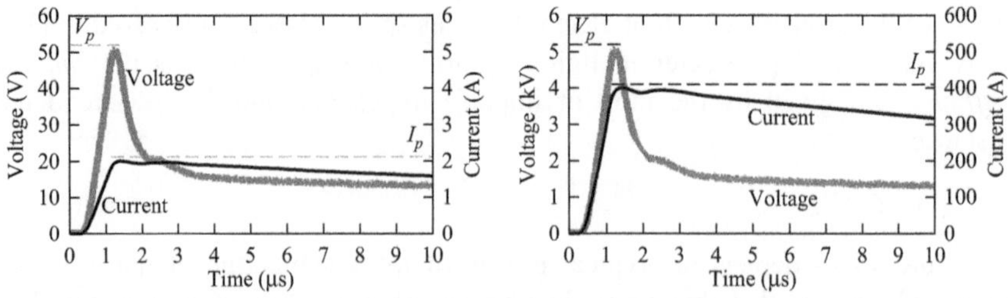

Figure 14.15   The effect of soil ionization

electrode due to propagation effects (attenuation of dispersed current along the electrode).

Nevertheless, the dynamics of soil ionization have already been modeled by different approaches [31–34] and certain references recommend a value of 400 kV/m for E_0 [34], though values varying from 200 kV/m to more than 1,000 kV/m have been reported in technical literature [27, 28, 35].

According to certain approaches, when the electric-field configuration is not substantially disturbed near the electrode by ionization process, assuming a given value for E_0, it is possible to consider this effect by means of an equivalent increase of the electrode transversal section, allowing this increase to vary along the electrode length [32, 33]. This allows estimating the impact of the effect in the electrode response. However, for very concentrate electrodes, in which the effect is more relevant, the ionized channels may entirely alter the field configuration, making it difficult to estimate this impact.

It is reasonable to ponder that the ionization effect is important for concentrate electrodes, typically used in low-resistivity soils, from which low impedance values are obtained easily. In high-resistivity soils, long electrodes are usually required to ensure low grounding impedance values. Thus, their current density tends to be low, reducing the probability of soil ionization occurrence. In applications involving long electrodes, such as the common counterpoises of transmission lines, neglecting the favorable effect of soil ionization is considered prudent [4, 36].

14.4 Numerical simulation of the transient response of grounding electrodes

14.4.1 Preliminary considerations

This section presents a set of numerical results intended to allow establishing quantitative references in relation to the concepts described in previous sections. The evaluations focused on typical electrode arrangements of high-voltage transmission lines in order to articulate the results of this section with the developments of the next section about the impact of the transient behavior of grounding electrodes on the lightning response of transmission lines.

All computational simulations were carried out using the Hybrid Electromagnetic Model (HEM). This rigorous full-wave frequency-domain electromagnetic model uses fast Fourier transform to convert the results to time domain. It is called hybrid because, in spite of being formulated by means of electric field integral equations, it provides results in terms of circuital quantities (voltages and currents) that are more useful in engineering applications [37].

In the application of this model, the simulated system is partitioned into n elements of different length, usually about 400 to 1,000 elements, depending on the case. The model uses full $n \times n$ matrices expressing the electromagnetic coupling among such elements to yield results. Preprocessing is used to determine the number of frequencies and of elements required to ensure quality to the simulated results. In regular applications, about 400 to 500 frequencies are processed in each simulated condition (usual upper limit of 5 MHz for lightning-related problems and a maximum frequency step of 12.5 kHz). Fast Fourier transform is used to obtain results in time domain. It takes a few hours to process a complex simulated condition in regular personal computers. Details on the model formulation are found in [37].

Specific applications of the model related to the lightning response of electrodes and lightning response of lines hit by direct strikes can be found in [13–16] and in [38–40], respectively. In particular, the use of the HEM model to calculate grounding potential rise and lightning overvoltage experienced across insulator strings of transmission lines was validated by comparing simulated and measured results, respectively in [12, 16] and [18].

The representative current waves of first and subsequent return-strokes of Figure 14.16 were used in simulations. As discussed in [41–44], waveforms of first-return-stroke currents include an initial concavity followed by an abrupt rise around the half-peak value and several subsidiary peaks, with the second peak usually turning out to be the highest one. On the other hand, the waveform of most subsequent stroke currents presents a single peak and a relatively smooth shape. The waves of Figure 14.16 [45] reproduces such patterns and also the median parameters of currents measured at Morro do Cachimbo Station (MCS), in Brazil, comprising about 47 flashes and 138 subsequent stroke records [46, 47].

Very similar waveforms and time parameters were found after processing the records of Berger's measurements [48, 49], though their median peak-current values were significantly lower, about 50% and 30% higher, respectively, for currents of first and subsequent return-strokes. Table 14.1 indicates the parameters of such currents.

In all simulations, frequency dependence of soil parameters was taken into account by means of the Visacro-Alipio expressions (14.8) and (14.9). Soil ionization was disregarded, since in practical conditions of transmission-line tower-footings the

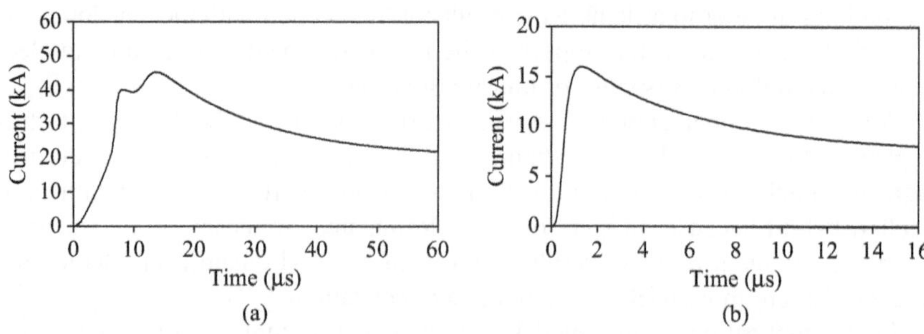

Figure 14.16   Representation of typical waveforms of (a) first- and (b) subsequent-return-stroke currents with median parameters of currents measured at MCS. The waveforms were reproduced using Heidler functions, according to the procedures described in [41] and [45] for first- and subsequent-strokes

Table 14.1   Median parameters of lightning currents measured at MCS station [46]

Event	I_{P1} (kA)	I_{P2} (kA)	T_{10} (µs)	T_{30} (µs)	T_{50} (µs)	dI/dt_{MAX} (kA/µs)
FIRST	40.3	45.3	5.6	2.9	53.5	19.4
SUB	16.3	–	0.7	0.4	16.4	29.9

I_{p1} and I_{p2} refer to the first and second return-stroke current peak value, T_{10} and T_{30} correspond, respectively, to the time between 0.1 I_{p1} and 0.9 I_{p1} and between 0.3 I_{p1} and 0.9 I_{p1}, T_{50} is the time interval required for the current to decay to 0.5 I_{p2}, and dI/dt_{MAX} is the maximum time derivative.

use of long electrodes results in low linear density of current dispersed along the electrodes, decreasing the probability of occurrence of this effect.

The model was applied to determine the grounding potential rise experienced at the electrodes due to the impression of the lightning current at the top of an isolated tower (all the current flows to the tower-footing electrodes), as denoted in Figure 14.17, which also represents the typical arrangement of transmission line electrodes consisting of counterpoise wires.

14.4.2   General results of the response of grounding electrodes

GPR curves simulated for a 1,000-Ωm low-frequency soil resistivity and corresponding impulse impedances of first and subsequent strokes are presented in Figure 14.18 to illustrate how the variation of the electrode length affects both parameters.

Note that increasing the electrode length decreases the GPR peak, though this trend is diminished as the length approaches L_{EF}. This qualitative behavior is also observed for soils of lower and higher resistivity, though in these cases the saturation of the GPR decrease would occur, respectively, for shorter or longer electrode length, according to the value of L_{EF} for that resistivity. Also, it is clear that, though the decrease of GPR peak is saturated with increasing electrode length, this increase still reduces the GPR values at the wave tail, whose amplitude is governed by the low-frequency resistance.

In addition to provide means to estimate the impulse impedance for specific conditions of soil resistivity and electrode length, Z_P curves similar to those of Figure 14.18(c) and 14.18(d) allow determining the effective length of the electrodes. L_{EF} can be identified by inspection as the length in which the flat region, corresponding to the saturation of the impedance decrease with increasing electrode length, begins. As discussed in [25], the effective lengths L_{EF} of horizontal electrodes for currents of first and subsequent return-strokes are respectively about 13, 23, 35, 48, 80 m and 8, 12, 18, 23, 36 m for soils of low-frequency resistivity of 100, 300, 600, 1,000 and 2,000 Ωm, considering frequency-dependent soil parameters.

Knowledge of the low-frequency grounding resistance of the tower-footing electrodes is very useful, as measuring tower-footing impedance in field conditions is a very difficult task for engineering personnel. Usually grounding resistance is

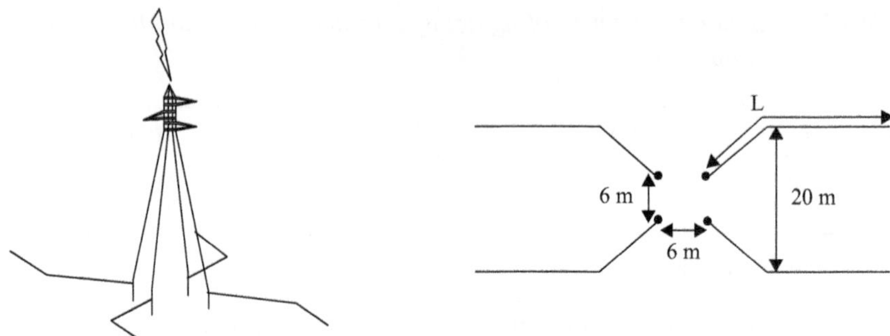

Figure 14.17   Representation of the simulated condition to determine the GPR and of the grounding electrode arrangements. The 0.5-cm radius electrodes are buried 0.6 m deep in the soil. Their lengths L vary according to soil resistivity. The buried metallic parts of the tower were represented by 3-m-long cylindrical vertical rods with 10-cm diameter

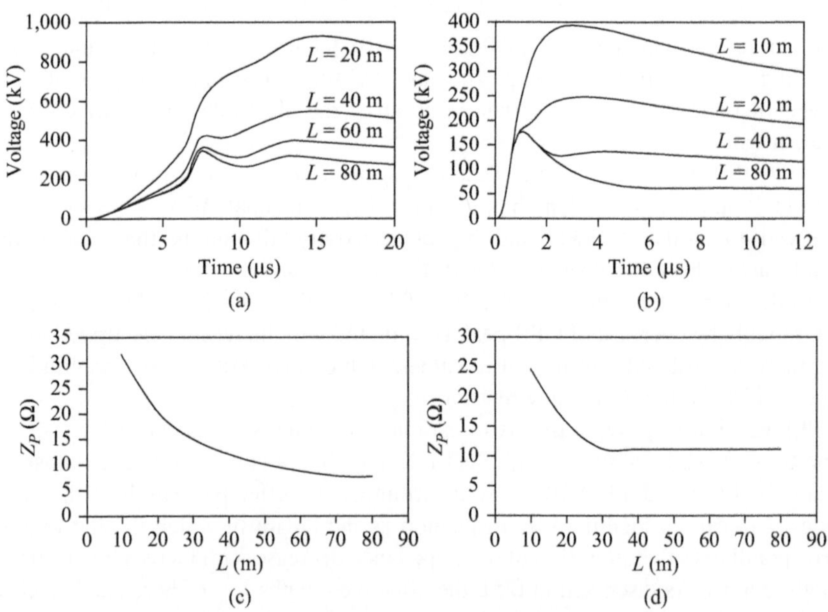

Figure 14.18   Grounding potential rise of electrodes as a function of electrode length for a 1,000-Ωm soil, developed in response to the impression of currents of (a) first and (b) subsequent return-stroke currents and corresponding impulse impedances (c) and (d)

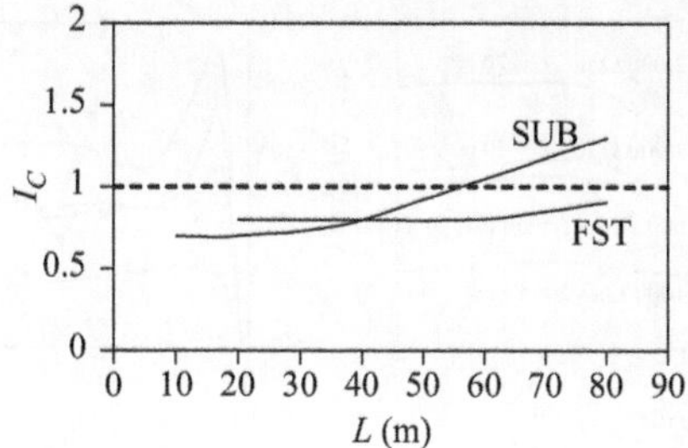

First strokes

L (m)	Z_P (Ω)	R_T (Ω)	$I_C = Z_P/R_T$
20	20.5	24.3	0.8
40	12.1	14.9	0.8
60	8.8	11	0.8
80	7.6	8.8	0.9

Subsequent strokes

L (m)	Z_P (Ω)	R_T (Ω)	$I_C = Z_P/R_T$
10	24.8	37.1	0.7
20	15.6	24.3	0.7
40	11.2	14.9	0.8
80	11.1	8.8	1.3

Figure 14.19   Calculated impulse coefficient for first (FST) and subsequent (SUB) strokes as a function of electrode length for the conditions of Figure 14.18 (low-frequency soil resistivity of 1,000 Ωm)

Table 14.2   Length of electrodes used for each soil resistivity

ρ_0 (Ωm)	100	600	1,000	2,000
L (m)	5	30	50	70

measured instead and the impedance is estimated according to the electrode configuration and soil resistivity. In this respect, it is very interesting to learn the impulse coefficient I_C of such configurations to allow prompt calculation of Z_P from the measured low-frequency resistance. Figure 14.19 depicts this coefficient specifically for the conditions of Figure 14.18.

In the design of transmission lines, the requirements to achieve reasonable lightning performance lead to adopting different lengths of electrodes, as explained in the next section. Longer electrodes are used in soils of higher resistivity to ensure that the grounding impedance will not to exceed predetermined limits. Table 14.2 indicates typical lengths of counterpoise wires used in real 138 kV lines for a set of soil-resistivity values.

14.4.3   Grounding potential rise of electrodes subject to lightning currents

The simulated GPRs developed in response to the impression of representative currents waves of first and subsequent strokes to the top of the 138 kV line, whose counterpoise wires' lengths are indicated in Table 14.1, are shown in Figure 14.20.

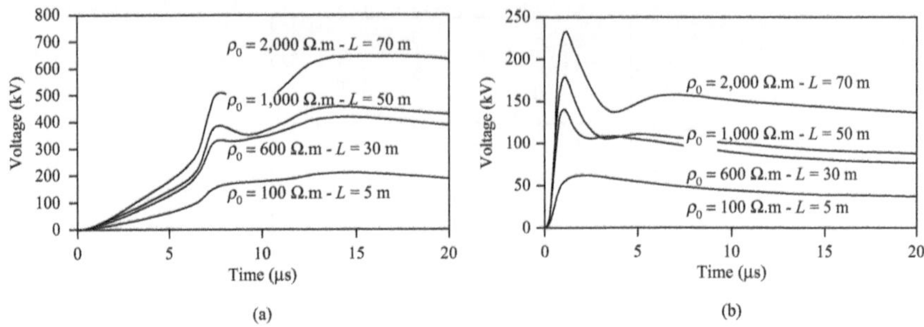

Figure 14.20   Simulated GPR of the 138-kV-line tower footing considering different values of soil resistivity ρ_0: (a) first stroke and (b) subsequent stroke

Table 14.3   Calculated impulse impedance, resistance, and impulse coefficient of electrodes (curves of Figure 14.20)

ρ_0 (Ω m)	L (m)	R_T (Ω)	First stroke		Subsequentstroke	
			Z_P (Ω)	I_C	Z_P (Ω)	I_C
100	5	5.1	4.6	0.9	3.7	0.73
600	30	11.0	9.3	0.85	9.1	0.83
1,000	50	12.5	10.1	0.81	11.6	0.93
2,000	70	19.5	14.7	0.75	15.0	0.78

14.4.4   Impulse impedance and impulse coefficient for first and subsequent return-stroke currents

From the GPR curves in Figure 14.20 and the corresponding impressed currents, the impulse impedance was calculated, along with the low-frequency resistance and the impulse coefficient (Table 14.3).

Note that the grounding-impedance increase due to the higher soil-resistivity values was partially offset by increasing the electrode length. While soil resistivity increased 20-fold times, resistance increased by only about 400% and grounding impedance by about 300% and 400%, respectively, for first and subsequent stroke currents. Though increasing further the electrode length would lead to additional decrease of the grounding resistance, the impedance would show small variation, since the lengths of electrodes are relatively close to the effective length. As expected, in all cases the impulse coefficient is lower than one, as L is always shorter than L_{EF}.

14.5 Case example: analysis of the influence of grounding electrodes on the lightning response of transmission lines

Lightning is a frequent cause of transmission line outages. Though shielding failures leading to lightning strikes to the phase conductors can contribute to a number of faults, the backflashover largely prevails as the main cause of line outages, as discussed in [39, 50, 51]. This is the reason this mechanism, mainly governed by the value of tower-footing grounding impedance, is the main focus of techniques intended to improve the lightning performance of transmission lines.

Backflashover can occur due to direct lightning strikes to the towers or to shield wires at the towers' vicinities, causing high overvoltages and, eventually, electrical discharges across insulators. The occurrence of such discharges depends on a balance between the insulator withstand and the amplitude of the lightning overvoltage the insulators experience in response to the injection of lightning currents into the soil through tower-footing electrodes. Thus, decreasing tower-footing grounding impedance is a very effective means to reduce this overvoltage and, in turn, to decrease backflashover occurrences. Figure 14.21 illustrates the effectiveness of this practice to reduce lightning overvoltage, based on the simulation of a lightning strike to the top of an isolated tower of a 138 kV line, using the HEM model and the representative current of a first return-stroke of Figure 14.16(a).

Note that the increase of the tower-footing impedance (and, accordingly, the grounding resistance) from 20.5 to 7.6 Ω (resulting from increasing L from 20 to 80 m) yielded a reduction of only about 15% of the overvoltage experienced across the upper insulator string. However, the reduction was much more significant at the wave tail.

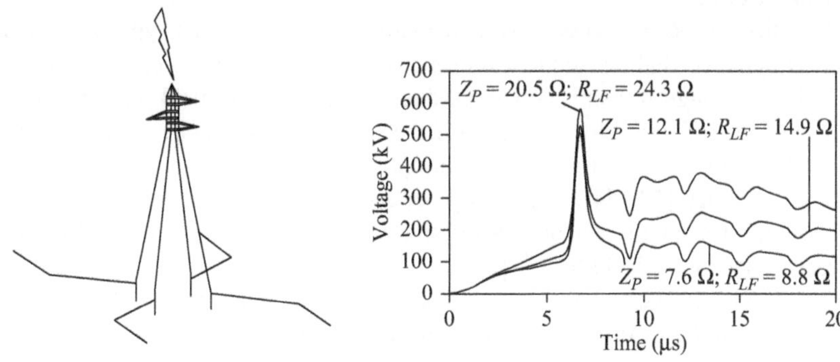

Figure 14.21   Simulated lightning overvoltage across the upper insulator string of a real 138-kV line due to a strike to the top of a 30-m-high isolated tower. Impressed first-stroke current of Figure 14.16(a). L = 20, 40, and 80 m. Soil resistivity: 1,000 Ωm

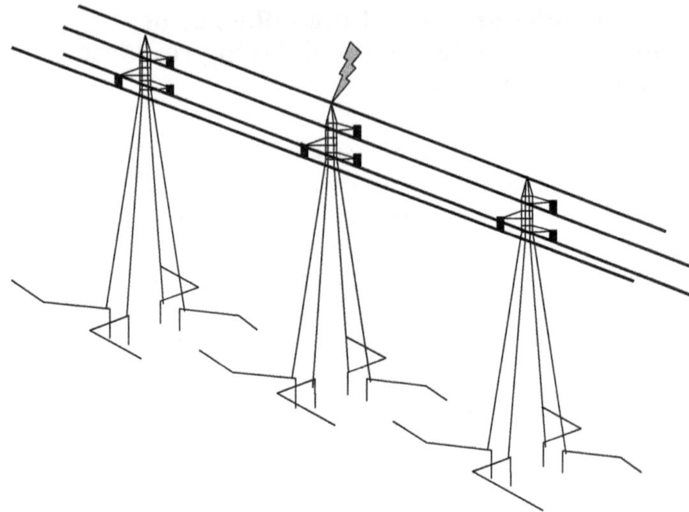

Figure 14.22   Representation of the simulated event: direct lightning strike to the top of a tower flanked by adjacent towers with line conductors impedance-matched beyond them

In this section, the overvoltages developed across insulators of the 138 kV transmission line represented in Figure 14.22 were simulated, considering the resistivity values and respective counterpoise wire length of Table 14.1. In this configuration, two adjacent towers with the same geometry and dimension as those of the stricken one were included and all the conductors were impedance-matched 30 m beyond the two 400-m spans. Though the two adjacent towers have strong influence on the amplitude of lightning overvoltage, additional towers would have only a slight effect on it.

Only first stroke currents were considered, as usual in this type of analysis, since it is believed that subsequent return-strokes are not able to cause backflashover in high voltage transmission lines due to their median peak current around one-third of that of first return-strokes [52]. Some recent works have demonstrated that for 69- and 138-kV lines this assumption is questionable in certain usual conditions of lines [53, 54]. Frequency-dependent soil parameters were used, as they have significant impact on the lightning response of lines installed on soils with resistivity above 300 Ωm [55].

The simulated overvoltages across the upper insulator strings of the line due to the representative first stroke current of Figure 14.16(a) are depicted in Figure 14.23.

Note that, with the exception of the case of very low tower-footing grounding impedance about 5 Ω (resulting from a very low-soil resistivity—100 Ωm), the amplitude of the lightning overvoltages is relatively similar, as a result of the longer electrodes defined for soils with higher resistivity. Table 14.4 compares the peak values of the developed overvoltages. On the other hand, though the procedure of

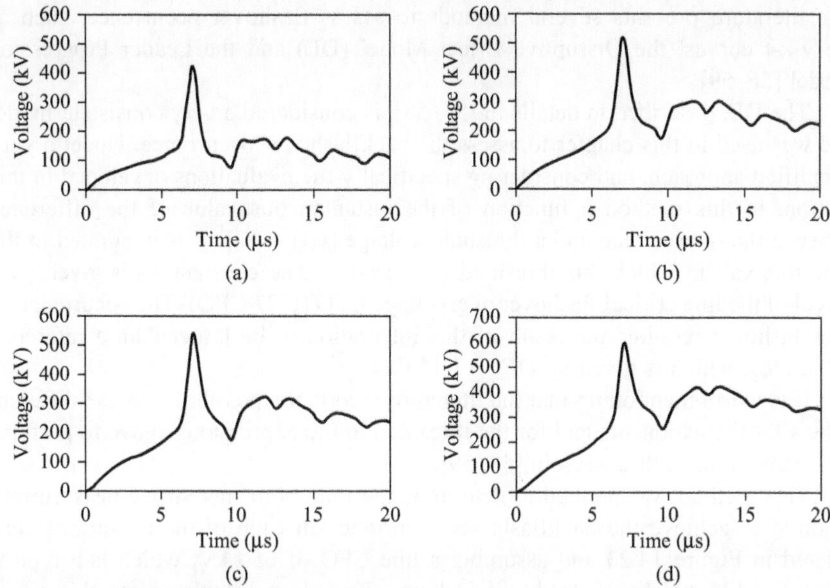

Figure 14.23   *Voltage developed across upper insulator string of 30-m-high tower of the 138-kV line. (First stroke—MCS). Frequency-dependent soil parameters of resistivity ρ_0 of 100 (a), 600 (b), 1,000 (c), and 2,000 Ωm (d). Two adjacent towers included (400-m span length)*

Table 14.4   *Peak of the overvoltage V_P developed across the upper insulator string of the 138-kV line for the conditions of Figure 14.22*

ρ_0 (Ωm)	L (m)	V_P (kV)
100	5	424
600	30	521
1,000	50	542
2,000	70	597

increasing the electrode length is effective to reduce the impulse impedance, the amplitude of the overvoltage in the wave tail is still even higher for soils with higher resistivity values. This occurs because in these conditions, the low-frequency tower-footing resistance is still even larger than those of lower resistivity soils, as shown in Table 14.3. This resistance greatly influences the amplitude of the overvoltage at the wave tail.

The occurrence of flashover across insulators does not depend only on the overvoltage peak value but also on the whole overvoltage wave amplitude.

The literature presents several methods to assess flashover occurrence, such as the v-x-t curves, the Disruptive Effect Model (DE) and the Leader Progression Model [56–59].

The DE, described in details in [57, 58], is considered a very consistent model and was used in this chapter to assess the backflashover occurrence. Under a very simplified approach, and considering specifically the evaluations developed in this section, in this method a function of the instantaneous value of the difference between the overvoltage and a threshold voltage $[v(t) - V_0]^{1.36}$ is integrated in the time interval, in which this threshold is exceeded. The constant V_0 is given by a parcel of the line critical flashover overvoltage CFO (0.77·CFO). The occurrence of backflashover requires the result of this integration to be larger than a reference value DE_b, which is given by $DE_b = 1.15 \cdot \text{CFO}^{1.36}$.

It is worth mentioning that the literature reports the possibility to use different values for the exponents and for the threshold in the expressions above to perform this evaluation, as discussed in [58, 59].

This method was applied to determine the minimum first-stroke peak current required to achieve the backflashover occurrence in each of the conditions considered in Figure 14.23 and assuming a line CFO of 650 kV, which is a typical value for 138 kV lines. Table 14.5 shows the values calculated for this peak-current, known as critical current I_{Crit}.

Note that all first-stroke currents of lightning strikes to the tower, whose peak current were higher than I_{Crit} would lead the line to a backflashover [60].

In spite of the longer electrode length with increasing soil resistivity, the critical currents in higher resistivity soil show lower value, meaning that the probability of backflashover has increased. The last column of Table 14.4 shows the percentage of first-stroke lightning currents exceeding the critical peak current according to the cumulative distribution of first-stroke peak-currents determined from the currents measured at Morro do Cachimbo Station [61].

Note that, according to Table 14.5, the lightning performance of most towers is very poor: about 23% and 43% of the strikes to the tower would lead to backflashover and probably to line outage for the 1,000 and 2,000 Ωm soils, respectively.

A rehearsal of the lightning performance of the line was developed, in terms of its expected outage rates under certain conditions. The line was supposed to be crossing a region with relative intense lightning activity, leading to an expected number of 20 flashes striking the line per 100 km per year. Assuming, as

Table 14.5   *Critical first return-stroke peak-currents for the*
138-kV line CFO: 650 kV

ρ_0 (Ωm)	L (m)	I_{Crit} (kA)	$I_P > I_{Crit}$ (%)
100	5	101	4
600	30	67	19
1,000	50	63	23
2,000	70	49	43

*Table 14.6   Expected outage rate of the 138-kV line (assumed: 20 strikes/
100 km/year)*

Hypotheses for the distribution of ρ (%) along the line				Outages/100 km/year
100 Ωm $L = 5$ m	600 Ωm $L = 30$ m	1,000 Ωm $L = 50$ m	2,000 Ωm $L = 70$ m	
100	0	0	0	0.44
0	100	0	0	2.2
0	0	100	0	2.7
0	0	0	100	5.1
20	40	30	0	1.8
15	35	35	15	2.6
0	40	30	20	2.7

recommended in [54], that only 60% of the flashes would strike to the towers or to the shield wire in their vicinities, with potential to cause backflashovers, the expected outage rates of the line were calculated considering the distributions of soil resistivity along the line shown in Table 14.6.

With the exception of the low-resistivity soil (100 Ωm), in all cases of uniform distribution of soil resistivity along the line the estimated outage rates are relatively high, exceeding a rate of 2 outages/100 km/year. This rate is considered an acceptable limit for 138 kV lines in certain countries.

This low lightning performance of the towers installed on higher resistivity soils results from the high values of the grounding impedances of their tower-footing electrodes. Moreover, one has to recognize also the very severe conditions determined by the peak current distribution of MCS used in the calculations, as the median peak current (45 kA) is about 50% higher than that of Berger's distribution. Therefore, only for one of three assumptions of nonuniform distributions of soil resistivity along the line, the outage rate was lower than this threshold.

Anyway, this picture recommends working to improve the lightning performance of this line and this improvement could be achieved by reducing the tower-footing grounding impedances for the high-resistivity soils.

In order to demonstrate the impact of this impedance reduction, the lengths of the electrodes of towers installed on higher resistivity soils were increased, as indicated in Table 14.7. The results show that the increase of electrode lengths led to lower first-stroke impulse impedance and larger critical currents. The same steps described previously in the rehearsal developed to obtain the data of Table 14.6 were repeated. This resulted in lower lightning overvoltage amplitudes, decreasing the backflashover outage rates of the line, as shown in Table 14.8.

Note that the calculated number of outage rates of the line decreased very significantly, achieving acceptable levels for the three usual nonuniform distributions of soil resistivity along the lines, though this rate is still high specifically for the 2,000-Ωm soil resistivity.

Table 14.7   *Critical first return-stroke peak-currents for the 138-kV line CFO: 650 kV*

ρ_0 (Ωm)	L (m)	V_P (1st stroke)(kV)	I_{Crit} (kA)	$I_P > I_{Crit}$ (%)*
100	5	424	101	4
600	40	498 [521]	77 [67]	11.6 [19]
1,000	70	522 [542]	73 [63]	14.1 [23]
2,000	100	575 [597]	60 [49]	26.1 [43]

*The value before increasing the electrode length is shown in brackets.

Table 14.8   *Expected outage rate of the 138-kV line with longer electrodes (assumed 20 strikes/100 km/year)*

Hypotheses for the distribution of ρ (%) along the line				Outages/100 km/year*
100 Ωm $L = 5$ m	600 Ωm $L = 40$ m	1,000 Ωm $L = 70$ m	2,000 Ωm $L = 100$ m	
100	0	0	0	0.44
0	100	0	0	1.4 [2.2]
0	0	100	0	1.7 [2.7]
0	0	0	100	3.1 [5.1]
20	40	30	0	1.2 [1.8]
15	35	35	15	1.6 [2.6]
0	40	30	20	1.7 [2.7]

*The value in brackets corresponds to the outage rate before increasing the electrode length.

The significant decrease of the outage rates derived from the increase in the electrode length only (and, therefore, from the decrease in tower-footing impedance) demonstrates that this technique can be very effective to improve the lightning performance of transmission lines.

References

[1]   Meliopoulos A. P. *Power system grounding and transients*. New York: Marsel Dekker Inc.; 1988.

[2]   Visacro S. Low-frequency grounding resistance and lightning protection. Chapter 9: 475–502. *Lightning protection*. Cooray V. (Ed.), London: IET; 2009.

[3]   Visacro S. *Grounding and earthing: Basic concepts, measurements and instrumentation, grounding strategies*. (in Portuguese). São Paulo: ArtLiber Edit.; 2002.

[4] Visacro S. 'A comprehensive approach to the grounding response to lightning currents'. *IEEE Trans. Power Delivery*. Jan 2007; **22**: 381–386.

[5] Rudenberg R. 'Fundamental considerations on grounding currents'. *Electrical Engineering*. Vol. 64, pp. 1–13, Jan 1945; **6** (1).

[6] Visacro S., Alipio R. 'Frequency dependence of soil parameters: experimental results, predicting formula and influence on the lightning response of grounding electrodes'. *IEEE Trans. Power Delivery*. Apr 2012; **27** (2): 927–935.

[7] Heppe R. J. 'Computation of potential at surface face above an energized grid or other electrode, allowing for non-uniform current distributions'. *IEEE Trans. Power Delivery*. 1998; **13**: 762–767.

[8] Ramo S., Whinnery J. R., Duzer T. V. *Fields and waves in communication electronics*, New York: J. Wiley and Sons; 1965.

[9] Smith-Rose R. L. 'The electrical properties of soils for alternating currents at radiofrequencies'. *Proc. Royal Society*. 1933; **140** (841A): 359–377.

[10] Scott J. H. *Electrical and magnetic properties of rock and soil*. U.S. Geol. Survey, Dept. of the Interior, Washington, 1966.

[11] Longmire C. L., Smith K. S. *A universal impedance for soils*. Topical Report for Period July 1–September 30, Defense Nuclear Agency, Santa Barbara, California, 1975.

[12] Visacro S., Portela C. M. 'Soil permittivity and conductivity behavior on frequency range of transient phenomena in electric power systems'. *Int. Symp. High Voltage Engineering* (ISH 1987), paper 93.06.

[13] Portela C. M. 'Measurement and modeling of soil electromagnetic behavior'. *IEEE Int. Symp. Electromagnetic Compatibility* (IEEE EMC 1999): 1004–1009.

[14] Visacro S., Alipio R., Murta Vale M. H., Pereira C. 'The response of grounding electrodes to lightning currents: the effect of frequency-dependent soil resistivity and permittivity'. *IEEE Trans. Electromagnetic Compatibility*. May 2011; **53** (2): 401–406.

[15] Alipio R., Visacro S. 'Modeling the frequency dependence of electrical parameters of soil'. *IEEE Trans. Electromagnetic Compatibility*. 2014; doi:10.1109/TEMC.2014.2313977.

[16] Visacro S., Alipio R., Pereira C., Guimarães M., Schroeder M. A. 'Lightning response of grounding grids: Simulated and experimental results'. *IEEE Trans. Electromagnetic Compatibility*. 2014; doi 10.1109/TEMC.2014.2362091.

[17] Visacro S., Rosado G. 'Response of grounding electrodes to impulsive currents: an experimental evaluation'. *IEEE Trans. Electromagnetic Compatibility*. Feb 2009; **51**: 161–164.

[18] Visacro S., Guimarães M., Araujo L. 'Experimental impulse response of grounding grids'. *Electric Power Systems Research*. 2013; **94**: 92–98. doi.org/10.1016/j.epsr.2012.04.011.

[19] Rodrigues B. D., Visacro S. 'Portable grounding impedance meter based on DSP'. *IEEE Trans. Instrumentation & Measurement*. 2014; 10.1109/ TIM.2014.2303532.

[20] Visacro S., Pinto W. L. F. 'Is grounding impedance really larger than low frequency resistance?' Conference Grounding and Earthing (GROUND'2006): 185–188.

[21] Gupta B. R., Thapar B. 'Impulsive impedance of grounding grids'. *IEEE Trans. Power App. Syst.* Nov/Dec 1980; **PAS-99** (6): 2357–2362.

[22] Bewley L. V. 'Theory and test of the counterpoise'. *Electrical Engineering.* Aug 1934; 1163–1172.

[23] Alípio R., Visacro S. 'Frequency dependence of soil parameters: Effect on the lightning response of grounding electrodes'. *IEEE Trans. Electromagnetic Compatibility.* 2013; **55**: 132–139.

[24] Grcev L. 'Impulse efficiency of ground electrodes'. *IEEE Trans. Power Delivery.* Jan 2009; **24** (1): 441–451.

[25] Alípio R., Visacro S. 'Impulse efficiency of grounding electrodes: Effect of frequency-dependent soil parameters'. *IEEE Trans. Power Delivery.* 2014; **29**: 716–723.

[26] Grcev L. 'Lightning surge efficiency of grounding grids'. *IEEE Trans. Power Delivery.* Jul 2011; **26** (3): 223–237.

[27] Oettle E. E. 'A new general estimation curve for predicting the impulse impedance of concentrated earth electrodes'. *IEEE Trans. Power Delivery.* 1988; **3** (4): 2020–2029.

[28] Mousa A. M. 'The soil ionization gradient associated with discharge of high currents into concentrated electrodes'. *IEEE Trans. Power Delivery.* Jul 1994; **9** (3): 1669–1677.

[29] Sekioka S., Sonoda T., Ametani A. 'Experimental study of current-dependent grounding resistance of rod electrode'. *IEEE Trans. Power Delivery.* Apr 2005; **20** (2): 1579–1576.

[30] Geri A. 'Behavior of grounding systems excited by high impulse currents: The model and its validation'. *IEEE Trans. Power Delivery.* Jul 1999; **14** (3): 1008–1017.

[31] Liew A. C., Darveniza M. 'Dynamic model of impulse characteristic of concentrated earths'. *IEE Proc.* Feb 1974; **121** (2): 123–135.

[32] Visacro S., Soares Jr. A. 'Sensitivity analysis for the effect of lightning current intensity on the behavior of earthing systems'. *Int. Conference Lightning Protection (ICLP* 1994), paper R3a-01\(1-5).

[33] Almeida M. E., Correia de Barros M. T. 'Accurate modelling of rod driven tower footing'. *IEEE Trans. Power Delivery.* Jul 1996; **11** (3): 1606–1609.

[34] CIGRE C4.33.01 'Guide to procedures for estimating the lightning performance of transmission lines'. *CIGRE Technical Brochure 63*, Oct 1991.

[35] Lima J. L. C., Visacro S. 'Experimental developments on soil ionization: New findings'. *Int. Conference on Grounding and Earthing* (GROUND 2008): 174–179.

[36] Visacro S., Hermoso B., Almeida M. T., Torres H., Loboda M., Sekioka S., Geri A., Chisholm W. 'The response of grounding electrodes to lightning currents – summary report CIGRE WG C4.406'. *Electra.* 2009; **246**: 18–21.

[37] Visacro S., Soares Jr. A. 'HEM: A model for simulation of lightning-related engineering problems'. *IEEE Trans. Power Delivery*. Apr 2005; **20** (2): 1026–1208.

[38] Soares Jr. A., Schroeder M. A. O., Visacro S. 'Transient voltages in transmission lines caused by direct lightning strikes'. *IEEE Trans. Power Delivery*. Apr 2005; **20** (2): 1447–1452.

[39] Visacro S., Silveira F. H., De Conti A. 'The use of underbuilt wires to improve the lightning performance of transmission lines'. *IEEE Trans. Power Delivery*. Jan 2012; **27** (1): 205–213.

[40] Silveira F. H., De Conti A., Visacro S. 'Lightning overvoltage due to first strokes considering a realistic current representation'. *IEEE Trans. Electromagnetic Compatibility*. Nov 2010; **52** (4): 929–935.

[41] Visacro S. 'A representative curve for lightning current waveshape of first negative stroke'. *Geophysical Research Letters*. Apr 2004; **31**: L07112.

[42] Visacro S. *Lightning: An engineering approach*. (in Portuguese). São Paulo: ArtLiber Edit.; 2005.

[43] Visacro S., Vale M. H., Correa G., Teixeira A. 'Early phase of lightning currents measured in a short tower associated with direct and nearby lightning strikes'. *Journal on Geophysical Research*. 2010; **115** (D16104): 1–11. doi:10.1029/2010JD014097.

[44] Rakov V., Uman M. A. *Lightning physics and effects*. Cambridge: Cambridge University Press; 2003.

[45] A. R. Conti, S. Visacro, 'Analytical representation of single- and double-peaked lightning current waveforms'. *IEEE Trans. Electromagnetic Compatibility*. 2007; **49**: 448–451.

[46] Visacro S., Schroeder, M. A. O., Soares A. J., Cherchiglia L. C. L., Sousa V. J. 'Statistical analysis of lightning current parameters: Measurements at Morro do Cachimbo Station'. *Journal on Geophysical Research*. 2004; **109** (D01105): 1–11. doi:10.1029/2003JD003662.

[47] Visacro S., Mesquita C. R., De Conti A. R., Silveira F. H. 'Updated statistics of lightning currents measured at Morro do Cachimbo Station'. *Atmospheric Research*. 2012; **117**: 55–63.

[48] Berger K., Anderson R. B., Kröninger H. 'Parameters of lightning flashes'. *Electra*. 1975; **41**: 23–37.

[49] Anderson R. B., Eriksson A. J. 'Lightning parameters for engineering application'. *Electra*. 1980; **69**: 65–102.

[50] Visacro S. 'Direct strokes to transmission lines: Considerations on the mechanism of overvoltage formation and their influence on the lightning performance of lines'. *Journal of Lightning Research*. 2007; **1**: 60–68.

[51] Brown G. W. 'Lightning performance II – updating backflash calculations'. *IEEE Trans. Power App. Syst.* Jan/Feb 1978; **PAS-97** (1): 39–52.

[52] IEEE PES Transmission and Distribution Com. 'IEEE guide for improving the lightning performance of transmission lines'. *IEEE Standard 1243*, Dec 1997.

[53] Silveira F. H., Visacro S., De Conti A. R. Lightning performance of 138-kV transmission lines: The relevance of subsequent strokes'. *IEEE Trans. Electromagnetic Compatibility*. 2013; **55**: 1195–1200.

[54] Silveira F. H., Visacro S., De Conti A., Mesquita C. R. 'Backflashovers of transmission lines due to subsequent lightning strokes'. *IEEE Trans. Electromagnetic Compatibility*. Apr 2012; **54** (2): 316–322. doi:10.1109/TEMC.2011.2181851.

[55] Visacro S., Silveira F. H., Xavier S., Ferreira H. B. 'Frequency dependence of soil parameters: The influence on the lightning performance of transmission lines'. *Int. Conference on Lightning Protection* (ICLP 2012).

[56] Hagenguth J. H. 'Volt-time areas of impulse spark-over'. *AIEE Trans*. 1941; **60**: 803–810.

[57] Hileman A. H. *Insulation coordination for power systems*. Boca Raton, FL: CRC Press; 1999.

[58] Darveniza M., Vlastos A. E. 'The generalized integration method for predicting impulse volt-time characteristics for non-standard wave shapes— A theoretical basis'. *IEEE Trans. Electrical Insulation*. Jun 1988; **23** (3): 373–381.

[59] Chisholm W. A. 'New challenges in lightning impulse flashover modeling of air gaps and insulators'. *IEEE Electrical Insulation Magazine*. 2010; **26** (2): 14–25.

[60] IEEE Working Group on Lightning Performance of Transmission Lines. 'A simplified method for estimating the lightning performance of transmission lines'. *IEEE Trans. Power App. Syst.* Apr 1985; **104** (4): 919–932.

[61] Visacro S., Mesquita C. R., Dias R., Silveira F. H., De Conti A. 'A class of hazardous subsequent lightning strokes in terms of insulation stress'. *IEEE Trans. Electromagnetic Compatibility*. Oct 2012; **54** (5): 1028–1033. doi:10.1109/TEMC.2012.2187339.